The Entropy Production—as Cornerstone in Applied Nonequilibrium Thermodynamics—Dedicated to Professor Signe Kjelstrup on the Occasion of Her 75th Birthday

The Entropy Production—as Cornerstone in Applied Nonequilibrium Thermodynamics—Dedicated to Professor Signe Kjelstrup on the Occasion of Her 75th Birthday

Guest Editors

Dick Bedeaux
Fernando Bresme
Alex Hansen

MDPI

Basel • Beijing • Wuhan • Barcelona • Belgrade • Novi Sad • Cluj • Manchester

Guest Editors
Dick Bedeaux
PoreLab
Department of Chemistry
Norwegian University of Science and Technology
Trondheim
Norway

Fernando Bresme
Department of Chemistry
Imperial College London
London
UK

Alex Hansen
PoreLab
Department of Physics
Norwegian University of Science and Technology
Trondheim
Norway

Editorial Office
MDPI AG
Grosspeteranlage 5
4052 Basel, Switzerland

This is a reprint of the Special Issue, published open access by the journal *Entropy* (ISSN 1099-4300), freely accessible at: https://www.mdpi.com/journal/entropy/special_issues/815Y1IZPQ5.

For citation purposes, cite each article independently as indicated on the article page online and as indicated below:

Lastname, A.A.; Lastname, B.B. Article Title. *Journal Name* **Year**, *Volume Number*, Page Range.

ISBN 978-3-7258-5395-3 (Hbk)
ISBN 978-3-7258-5396-0 (PDF)
https://doi.org/10.3390/books978-3-7258-5396-0

Contents

Preface

The entropy production of nonequilibrium transport processes is at the heart of all of Signe Kjelstrup's contributions to science, from fuel cells and batteries to shock waves. It is always used to define the independent system variables and thus the coupled flux–force relations in the system of interest, whether the system is a homogeneous electrolyte, a heterogeneous electrochemical interface, a catalyst, or a porous medium. To her, typical questions concern the following: how many molecules are needed to write thermodynamic equations for a volume element? Local equilibrium conditions, do they apply to evaporation at interfaces? The entropy balance, according to Kjelstrup, is an underused balance equation in the modeling of applications relating to chemical reactors or batteries. Is that so? Should entropy production, as computed precisely from the flux–force relations, be linked to exergy analysis, say on oil platforms or for the liquefaction of air. Entropy production minimization is essential for work on the energy efficiency of process plants, and maybe also in understanding designs in nature, like in the seal nose.

With this Special Issue, we hope to encourage a large efforts in this important field. The purpose of the Special Issue is to demonstrate the strength and importance of the theory of nonequilibrium thermodynamics, with its fundamental basis in entropy production, when used to improve the understanding of processes essential to mankind, industry or nature at large. The topics cover fundamental aspects and applications, of all sorts, for instance the applications Kjelstrup worked on, fuel cells, batteries, thermoelectric devices, chemical reactors, distillation columns, reindeer or seal noses.

Dick Bedeaux, Fernando Bresme, and Alex Hansen
Guest Editors

entropy

MDPI

Review

Complexity, Uncertainty, and Entropy: Applications to Food Sensory Perception and Other Complex Phenomena

Luka Sturtewagen [1], Harald van Mil [2] and Erik van der Linden [1,*]

[1] Laboratory of Physics and Physical Chemistry of Foods, Wageningen University and Research Center, Bornse Weilanden 9 (Building 118), 6708 WG Wageningen, The Netherlands; luka.sturtewagen@gmail.com
[2] Mathematical Institute, Leiden University, 2333 ZA Leiden, The Netherlands; h.g.j.van.mil@math.leidenuniv.nl
* Correspondence: erik.vanderlinden@wur.nl

Abstract: Complexity has been studied in various areas of science, such as ecology and sensory science. One important aspect is the quantification of the complexity of a system. There exist a multitude of different approaches. One approach relates complexity to information, with its measure introduced by Shannon. This is equal to the negative value of entropy, or uncertainty. In this review, we discuss the different approaches that are used to quantify and measure complexity in the realm of food sensory perception. We address how the food sensory field could benefit from an approach that is based on an information-theoretical footing, to allow for one quantitative measure, thus improving comparison and reproducibility among different laboratories. Reversely, the review is intended to inspire the physics community to explore complex phenomena, such as sensory perception, in terms of a general information-theoretical measure, such as entropy and other fields.

Keywords: entropy; complexity; sensory; information; diversity; ecology

1. Introduction

The concept of complexity may enable the crossing of boundaries between traditional scientific disciplines by identifying according measures to allow for the analysis and interpretation of high-dimensional phenomena. The field of food sensory perception is such a field that crosses different disciplines, from chemistry and physics to psychology, and, indeed, the concept of complexity is an important concept in this field.

In defining the concept of complexity in relation to food sensory perception, Köster et al. [1] stated that the concept of complexity is either defined a priori on the basis of information-theoretical considerations, or measured, with the help of rating scales or paired comparison, by a panel.

Commenting at this point on the use of the term "information-theoretical", it should be noted that if something is more complex, the amount of information used to describe it will be larger. In this sense, one may equate complexity with information. The quantification of the term "information" is based on the fact that when a certain amount of information on a system has become observable, it is this information that is not hidden anymore. Before such an observation became possible, this information was hidden within the system. Because that amount of information was hidden, it added to the uncertainty of describing the system. In other words, the uncertainty that is related to describing a system equals the negative value of the information available in the system. Now, the uncertainty in describing a system is determined by the number of possibilities that a system has internally.

In fact, Shannon argued that an expression for this uncertainty equals the expression for the entropy of a system, analogous to the entropy as a function of the number of possible states that a system can attain in physical descriptions of a system. It is noted that the thermodynamic concept of entropy is well defined for physical systems, and relates to the number of possible states for such a system. For "a person consuming food", the principal number of possible states, and a thermodynamic interpretation of the entropy and information of "a person consuming food", are not available. Shannon strictly applied the information measure to the problem of efficient and errorless communication between a sender and a receiver, and explicitly expressed his concerns towards the possibility of erroneously extrapolating his arguments to situations outside of the communication area. Nonetheless, discarding this warning, many people have utilised a direct extrapolation of the Shannon information measure in other areas, including the area of food sensory science, and we will use this extrapolation as well. Some more background information on information and entropy can be found in the introduction by Sturtewagen et al. [2]. Using the concept of the number of possible states of a system, and the above-explained Shannon information measure, one can use the observed number of bits in an observed pattern as a measure for the information. Such an approach, and its relation to complexity, has been conducted in many areas of science, in different model systems and interpretation schemes. The popularity of using the Shannon information measure is partially due to its mathematical properties, which also give this measure a meaning in a practical sense. An example of how this information-theoretical measure was recently applied to sensory time–intensity evaluations during chewing of a food, in relation to the moment of swallowing, can be found in Sturtewagen et al. [2]. It was found that using this measure, the information obtained from sensory perception during chewing until the moment of swallowing showed one master curve for all participants in the study, when plotted versus individually normalised chewing time.

Apart from the information-theoretical considerations in the above paragraph, it is important to realise that the term "information" is related to the context. Namely, the information coming to a perceiving individual partially comes from the information available in a given context, e.g., the surroundings of the individual. This contextual information is processed in the individual together with information coming directly from the food stimuli. The processing of this food-related information with the contextual information may lead to interference, and the information, after being processed, contains a contextual part. If complexity is seen as a perceived signal or perceived indicator of important food sensory markers, it would be expedient to define this context in identifying the complexity of the perception.

Apart from using the above quantification of complexity in an information-theoretic way, other routes are used. The most often-used alternative is one according to a much-cited paper by Berlyne [3], in which something becomes more complex when there are (1) more distinguishable elements, (2) when there is more dissimilarity between the elements, and (3) when the elements cannot be grouped into a single class or unit. The study by Berlyne was performed on visual stimuli, which are just one of the types of stimuli that are relevant in food sensory science. The multifaceted description is hypothesised to link the stimulus' complexity to its arousal potential and hedonic value [3–5] by following an "inverse U-shaped-curve", which is often being used throughout food sensory science.

We note that (1) suggests a linear relation between complexity/information and the number of elements, N, while the expression of information based on Shannon's arguments contains a logarithmic term instead of a linear term. Apart from these two routes of quantifying complexity, there exist multiple other, albeit similar, connotations and measures for complexity. This makes interpretation and interlaboratory comparisons of the measure ambiguous. When the complexity of a stimulus, such as sensory perception of a (model) food product or beverage, is studied, there are different views on what a complex stimulus entails [6]. For example, in a study on the perceived complexity of wine, Parr et al. [7] concluded that wine experts and wine novices had different interpretations of the term "perceived complexity". To make matters worse, perceived complexity and sensory complexity are often used interchangeably [8]. In food sensory science, complexity has been studied mostly with regard to flavour and aroma perception. The complexity of flavours has been studied both in foods, such as yoghurts [9], chocolate [10], and various foods with added spices [11], and in drinks, such as orange juices [12], beers [13] and wine [14]. Here, the scale of complexity is related to the stimulus. Much of the perceived complexity or the sensation of complexity in flavour and aroma research is based on research in fragrance [15,16] and scent [17]. Different measures of complexity have been related to key sensory variables, such as sensory-specific satiety [18], the expected satiating capacity of food products [6], arousal potential, liking [12,13,19–25], or the food-neophobic scale [11]. Many studies show a linear relationship between perceived complexity and the number of stimuli, number of compounds [9,11,18,23,24], difficulty of describing or identifying the stimulus [9,11,23], or inverse of simplicity [10,15,20–22]. Such linear relations might point to an additive property of the complexity measure used. However, some authors do not find, or expect, a linear relationship when interactions between components are involved [8,26].

The different approaches in the literature to quantifying complexity and its use for sensory perception, in relation to the measure of complexity as is known from the physics field, will be the main focus of this review paper. We will refrain from giving experimental details for the examples addressed or calculating other possible measures of complexity for specific experiments. We also refrain from discussing relations between instrumental and sensory measures. This topic is beyond the scope of the current review. Usually, instrumental measures also only represent a pale reflection of sensory reality. This is why sensory panels are a very important tool in assessing the quality of food and new developments thereof, and why we have focused on sensory measurements alone.

Taking Berlyne's view as a starting point, we first review some of the food sensory literature and identify the type of observation and context on which a measure of complexity is based. Then, we explore analogies with similar measures that are utilised in the field of ecology, like species richness and diversity. New results developed in the ecological and socio-economic contexts that deal with similar complications as in the sensory area are mapped to the context of food sensory science, in order to relate the different classes of complexity observed. After this, we treat, in more depth, the specific information transfer between food and individuals in relation to complexity. In the subsequent section, we treat sensory experiments performed to extract complexity and information by means of identifying perceived dissimilarities. In fact, this resembles measuring the uncertainty and entropy more directly, i.e., the negative of information. We end with a discussion of future directions.

2. Complexity in Food Sensory Science

The concept and measures of complexity have relevance at different levels in food sensory science, as discussed in [27]. In order to structure the discussion below, we need a generic measure for complexity that is rich enough, with links to empirical science and statistics, and can at least partially accommodate some of the critiques of existing measures. As mentioned in the introduction, one such a candidate is information entropy, by means of assuming the validity of generalising the information entropy in Shannon's theory of communication [28]. The most common interpretation of entropy is uncertainty, and as such, the term has been introduced as a measure for surprise in food sensory science, or as a measure for biodiversity in ecology. While applications of information entropy within complexity science are manifold, its application within the behavioural sciences is not without criticism. Luce [29] challenges the usefulness of information theory in psychology due to its inability to allow for any structural relations between stimuli. Luce refers frequently to papers by Laming [30,31], and reiterates Laming's criticism that string size (the maximum number of possibilities for a parameter), which forms an important element within the context of information theory to deal with redundancies, is seldom considered in psychological experiments. Laming, however, concludes in his paper [30] that although, in most cases, information theory is not very useful, a notable exception might be found in the investigation of complex systems when a detailed description of the underlying mechanism is not available. In this paper, we will use Shannon information entropy, and the properties of complexity, as discussed by Berlyne, as concepts to structure our discussion of the use of a measure of complexity and its interpretations in food sensory science. This approach allows for an investigation of the relations between the different measures of complexity that are used in the food sensory literature.

2.1. Complexities of Different Sorts: Stimulus, Sensory, Perceived, and Descriptive Complexity

To allow for a more structured classification of the complexities used, we will first discuss four basic contexts in which the concept or a measure of complexity is used in food sensory science, i.e., stimulus, sensory, perceived, and descriptive complexity. We note that representations are numerical (both discrete (counts) and continuous (slider, Visual Analogue Scale (VAS)), or categorical (x-level Likert scale).

We start with the stimulus complexity (numeric) of a food/sample, which, in most cases, is related to the number of ingredients, like flavour, aroma, textures or visual cues, but in some cases, is related to the heterogeneity of the food. As such, stimulus complexity might correlate food properties with behavioural properties like acceptance.

Sensory complexity (numeric), like stimulus complexity, is related to the number of ingredients or attributes that can be distinguished by the study subjects or panel. One might view this as a subjective representation of the stimulus complexity, and therefore a property of the test subject.

Perceived complexity (numeric or categorical) is measured on a numerical scale using the Visual Analogue Scale (VAS), where the subject needs to score the perceived level of complexity between the anchors "not at all" and "very much"; or in a categorical version, like an x-level Likert scale, between the anchors "not at all complex" and "extremely complex". This directly relates a property of the test subject to a measure of complexity.

As a fourth context, we have descriptive complexity, which measures the difficulty of describing a sensation (numerical and categorical). Here, the length of the description of the sensation can be a numerical measure for the complexity of the sample, or an x-level Likert scale, with anchors between easy and very difficult. This is a subjective measure, and the property of the subject serves as the observed variable for the latent variable complexity.

Some authors combine the above different classes of complexity into one measure. An example of this is in a combination of sensory complexity, compound perceived complexity, and descriptive complexity [9].

2.2. Food Sensory Studies and Different Sorts of Complexity

Olabi et al. [11] compared two versions of the same foods, in neophilic and neophobic subjects, with one sample having a more complex flavour. The complexity of the flavour was controlled by adding more flavours, in which case we can talk about stimulus complexity. Subjects scored the samples based on perceived complexity, acceptability, familiarity, and expectation. The perceived complexity was measured using the 2-AFC (Alternative Force Choice) test. The 2-AFC test is a two-level sensory test used to select the most complex sample, followed by a three-level test to test how sure the subject is regarding the selection. Six out of eight flavoured samples were considered to be more complex, and the acceptability of the more complex samples was higher for the neophilic group.

Weijzen et al. [18] studied the effects of intensity and complexity on sensory-specific satiety and food acceptance of various chicken soups and snacks, such as chocolate bars with incorporated nuts, with different flavours and textures. The set of samples was selected based on the samples' perceived complexity and perceived intensity on a nine-level Likert scale. Participants assessed the complexity of the products by asking the panellists directly how complex the product was (perceived complexity, Likert scale), how many different ingredients it contained (sensory complexity, Likert scale), and how difficult it was to describe it (descriptive complexity). From their results, the authors concluded that perceived complexity was strongly correlated with sensory complexity and that descriptive complexity was inversely related to familiarity. They concluded that intensity increases, and perceived complexity decreases, sensory-specific satiety.

Ruijschop et al. [9] investigated the influence of complexity on satiation and food intake. Their samples consisted of two types of strawberry-aromatized sweetened yoghurts: one with a single-component strawberry aroma, and one with a multi-component strawberry aroma (stimulus complexity). They assessed the complexity by asking the panellists about the difficulty of describing the taste (descriptive complexity, VAS), the perceived number of ingredients (sensory complexity, VAS), and the perceived level of complexity (perceived complexity, VAS). Their study indicated that the more complex aroma had higher perceived complexity scores, but had similar intensity to the single-component aroma. Pleasantness, compared with the single-component strawberry aroma, showed enhanced satiation.

Paulsen et al. [32] indicated that for the composite properties of meals, harmony and complexity can be used to describe a meal or a dish, such as fish combined with a sauce, but it is difficult to relate specific sensory attributes to such descriptions. They defined complexity as the sensation of many tastes and flavours which is conceptually close to sensory complexity, but is a measure like perceived complexity (intensity of complexity). This study indicated that bitter, acid, and salty sauces significantly affected perceived complexity.

Lévy et al. [12] studied the perceived complexity of orange juices to which they added different flavour extracts. Perceived complexity was considered as a measure of a combination of scores on a nine-point scale for three attribute dimensions, using the anchors of simple–elaborateness, straight–confusion, and homogeneity–scattered. The optimal level of complexity was found to be related to the most-liked product. The differences between the scores of the complexity attributes for a product, and the scores of those attributes for the most-liked product, was defined as the perceived complexity. The difference divided by the complexity of the most-liked product led to a scaled relative perceived complexity. Then, the sums of the scores were used as a measure of the perceived complexity, grouped around 0, representing the optimal complexity. The first two principal components in a principal component analysis indicated an almost linear relation with the simple–elaborateness and homogeneity–scattered attributes. The straight–confusion attribute correlated more with the second principal component only. From this, we can conclude that their use of the term "perceived complexity" belongs to the perceived complexity class. The study concluded that after exposing the subjects to samples with higher complexity, their optimal complexity perception was shifted upwards, without effects on low complexity.

Soerensen et al. [10] wanted to predict long-term consumer liking of chocolate. They defined perceived complexity on a 15 cm line scale with nine verbal anchors, ranging from extremely simple to extremely complex. Although sensory complexity was discussed by them, it was not measured, but described as the number of perceived flavours, and the number of perceived flavours was not recorded. They pointed out that this would have been a hard task, as there have been more than 800 flavour compounds found in chocolate. At the same time, to complicate matters even more, these compounds are not released at the same rate and the same time. Their release is dependent on eating behaviour and oral processing. This causes flavour perception to be a dynamic phenomenon. The problem becomes even more difficult because of interpersonal differences in perception and differences in previous experiences between panellists. Jellinek and Koster [15] discussed this problem in terms of the many receptors involved at different times during perception, and determined the problem to lie in the area of physiological complexity.

There has been a lot of confusion between the complexity of a stimulus and the complexity of a sensation [8]. The number of different sensations is not always consistent with the number of stimulus components. The perceived complexity of a sensation does not always correlate very well with the number of compounds [15] or stimulus complexity. A single compound may be perceived as more complex than a multi-compound mixture. Therefore, the number of ingredients alone is not enough of an indication of what the perceived complexity of a product will turn out to be.

The complexity of texture perception was studied by Larsen et al. [33] and Marcano et al. [6]. Larsen et al. [34] defined "textural complexity", or sensory complexity according to our classification, to be proportional to the number of texture attributes sensed during mastication of a food product, from the first bite through to the point of swallowing. The number of texture attributes is established by asking the panel to name all the attributes they perceive while consuming the sample. For the sensory assessment, they used qualitative descriptive analysis (QDA), texture profiling (TP), and temporal dominance of sensations (TDS). For their study, they used model foods with layers of different mechanical properties created by incorporating various gelled discs and seeds. In later studies, they analysed the food bolus at different stages of the chewing process [35], conducted satiation experiments [33], and assessed mechanical properties using a texture analyser [36]. They saw a "clear upward trend" in the number of unique descriptors for the different samples during the QDA. From this, they concluded that their samples had different perceived texture complexities. They concluded from the TDS that for more texturally complex samples, there were more texture attribute changes, and more texture attributes were perceived as dominant.

Marcano et al. [6] studied the texture perception of different cheese pies with visible particles added to the recipe. They hypothesised that these particles would increase the perceived complexity of the texture and the expected satiety. During their experiments, consumers were asked to rate the perceived complexity on a nine-point scale, from not complex at all to very complex. Moreover, panellists were asked to indicate which characteristics they included in their assessment of the perceived complexity of the pies. Using a flash profiling set-up, panellists were asked to make a list of attributes they used to describe the differences between the different cheese pie samples. Then, the panellists rated the samples based on a 0–10 score of their selected attributes. A high complexity coincided with a lack of homogeneity, as indicated by the use of the terms "heterogeneity" and "texture". The authors concluded that a higher perceived complexity coincided with a higher satiation expectation. The most frequently mentioned terms for the complexity were related to the particle characteristics and to the texture (it is not clear what is meant by "texture", the generic term used).

Reverdy et al. [24] studied the effects of the intensity and complexity of different food products on children, and investigated how sensory education had an effect on their food preferences. They varied the ingredients of the products and asked the children to indicate which sample had more tastes, and which sample was more surprising, to assess the complexity of the different products.

The complexity of visual perception was studied by Mielby et al. [21,22]. They assessed the preference of adolescents for pictures of fruits and vegetables mixed in different ratios and cut in different patterns. They studied how the perceived complexity of the different mixes influenced preference for the mixes. The definition of complexity they used was proposed by the panel, and was defined as "the more energy you spend to overview the picture of the mix, the more complex it is". This definition shows some similarities to Kolmogorov complexity, or minimal description length. The Kolmogorov complexity of a string is defined as the minimal length of a programme on a universal Turing machine to obtain a string x [37].

The complexity of taste perception was studied in [27]. The authors looked into how the complexity of food products made it more difficult for sensory panels to assess the products. The products they investigated were mixtures of different tastes (sucrose, sodium chloride, citric acid, and caffeine) in water, iced tea, and tomato soup. They indicated that the definition of complexity involves more than just counting the number of constituents and taste sensations. The sensory sensation can change significantly when two or more components of qualitatively different stimuli are mixed. They can enhance, suppress, or mask one another, and react chemically, physiochemically, or biophysically. Thus, a linear relationship between the number of constituents and the perceived complexity of the product does not exist. Bitnes et al. also noted that by increasing the complexity of a product by adding a new stimulus dimension, such as texture, the product became more difficult to evaluate.

Giacalone et al. [13] based their definition of complexity on the one proposed by Berlyne [3], and defined complexity to be related to the number of discernible elements within a stimulus, and to the degree to which these elements coexist or conflict. They noted that complexity is closely related to perceived ambiguity of a stimulus, and to the cognitive effort necessary for its interpretation. Therefore, perceived complexity is not as straightforward as counting the number of compounds. We tend to perceive complex odours as unique stimuli, rather than as a muddle of components. Thus, the relationship between the chemical complexity of a mixture leading to a given flavour, as defined by Jellinek and Köster [15], and the perceived complexity of a flavour, is rapidly lost with an in increase in the number of flavour components.

Herrmann et al. [17] researched how scents influence cognition and spending on groceries. They developed scents varying in terms of complexity. They based this on the idea that more complex olfactory stimuli contain more information to be processed, thereby decreasing the ease of processing the information before deciding to buy, compared to simple scents. They exposed people to scents with different complexities and asked them to perform cognitive tasks, such as solving anagrams, during exposure to the scent. The amount of anagrams solved decreased with increasing complexity of the scent, indicating that more complex ambient scents decreased processing fluency. They found that simpler scents had more of an impact on consumers' spending in retail situations.

3. Complexity in Sensory Science, Ecology, and Socio-Economics

The science of complexity crosses discipline boundaries, resulting in the reincarnation of measures of complexity in different guises within different contexts, which call for specific representations. For instance, Shannon information entropy has been proposed as a measure for biodiversity, in which context it is widely used and debated [38], along similar lines as in the behavioural sciences.

We note an interesting overlap between the discourse on the use of Shannon entropy in sensory science and biodiversity. In biodiversity, diversity is equated to uncertainty. The greater the uncertainty in predicting the outcome of a random sample from an ecosystem, the greater the diversity. Different measures of diversity have been proposed that bear different names, but most of them can be linked to each other using Shannon entropy. One of these measures is richness, which can just measure the number of species, or its equivalent number of selected attributes, or the number of ingredients in a recipe. A critique related to the measure of richness is that it does not take the distribution of species into account. This led to the introduction of measures such as Shannon entropy, linking diversity to the uncertainty of interpretation of entropy. The next level of critique is that two systems could have the same level of entropy, and thus complexity, but that these two systems could consist of a population with common species only, or a population consisting of common and rare species together. This critique bears similarities with the critique of Luce and Laming, in that all sensations are measured with equal weight, and therefore do not allow the incorporation of partial new experiences into the measure of complexity. These incorporations are nonetheless important for perception, as put forward by Berlyne in his contributions 2 and 3 to complexity. Of course, these incorporations should also be considered from the intuitive biologist's perspective on diversity.

In order to explain complexities other than sensory and ecological ones, for example, regarding traffic jams, fractures, and the like, one can resort to ways of expanding on the use of the term "information". As such, extending the concept of entropy, so-called Tsallis entropy was introduced [39]. Depending on an order parameter alpha, rare species will contribute more or less to the diversity measure compared to a so-called uniform case of alpha = 1, which resembles the special Shannon entropy case. The introduction of Tsallis entropy has been argued to allow for the introduction of more structural information of the system into the analysis, facilitating explanations for the existence of power laws in, for example, socio-economic systems or biological systems. However, a major criticism of Tsallis entropy is that it does not represent an extensive measure, which contrasts with the extensive character of the state function of entropy [40], contrasting with the according mathematical convenience in the expression for the entropy, based on information-theoretical reasoning. This criticism is further substantiated by the fact that one can also explain power laws without an extended expression for entropy, by including structural information, but by means of a structural fractality within biological [41] and socio-economic systems [42]. This fractal structure facilitates the transport of matter and information throughout the system with minimised effort. In these analyses, entropy remains an extensive function, and there is no need to introduce a fitting parameter within the entropy itself that makes the entropy non-extensive. In other words, the structure within the system should be taken into consideration in the analysis of the system, in a manner that is self-consistent with the main underlying basis of extensivity of entropy. It is noted that the elegancy of this fractality approach lies in the fact that the fractality of such structures guarantees an entropy production that is smooth over a vast range of length scales, without any sinks or holes (discontinuities in the entropy production), leading to efficiencies in, for example, catalysis reactions in fractal geometries [43], or the maintenance of a constant temperature in the nose of a reindeer [44].

At this point, we like to refer to the third complexity property of Berlyne. Elements that can be grouped into one unit, with the equivalent of a population of organisms grouped into one species, are considered to have lower diversity than elements that cannot be grouped into one unit, with the equivalent population of a different species. For example, a population of 10 dogs is considered to have less diversity than a population of three dogs, four finches, and four ferns. This means that one needs to take into account the relations between species, as well as their specific ecosystems. Similarly, one needs to take into account attributes in sensory science for describing more complex interactions. For both, there exist different measures, both theoretical and empirical, that can add weight to the probabilities. In a recent contribution [45], it was shown that adding a similarity matrix, the elements add weight to the probability of observing a particular species in an ecosystem. Moreover, the authors' proposed measure for diversity has the property of having a linear relation with the number of species present or the richness, but is additionally corrected for species distribution, rareness, and similarity. This linearity is important for the interpretation of the according complexity measure and value comparison. It is noted that such linearity is at odds with the use of the entropy/information expression as introduced by Shannon.

4. Information Transfer During a Food Sensory Experiment

In food sensory science, experiments often contain samples that need to be tested by a panellist, who can either be a trained or untrained panellist. A typical experiment has at least two components, the sample and the panellist.

Regarding the sample part, we need to assume that any probability distribution of sample attributes remains the same during the experiment. This implies that the samples should be physically and chemically stable during the experiment. To produce a stable sample, and ensure that all samples are used throughout the experiment, a robust and well-described process is required. This information should be encoded in a recipe. We could say that the sample is a manifestation of the recipe which, in a poetic sense, contains the message of its creator. Elements from the process have found their way into complexity measures in food sensory science. Most frequently, this has been along the dimension of number of ingredients. In a typical recipe, the number of food ingredients is encoded by the mass fraction or volume fraction. The use of merely the number of ingredients is not without criticism, as already pointed out in Section 2.

Regarding the perception part of the experiment, similar considerations apply as for the sample part. Panellists are asked to recognise the number of ingredients or sensations. The frequency of occurrence of perception of ingredients versus the total number of ingredients could be a measure of complexity, as described with information theory; however, like for the sample part above, this is still sensitive to criticisms, e.g., as espoused by Luce [29] and Laming [30], and discussed in the introduction of Section 2. The mapping of the sample and of the recipe to the sensory perception output produced by the panellist is an acceptable analysis [30]. As a second problem, there is the required incorporation of the context, i.e., the information coming from the environment and circumstances outside of the panellist, and beyond the stimuli directly initiated by the food.

We conclude that we have probabilities of different sorts for the input (sample) and output (perception) of the experiment. These inputs and outputs allow for the use of the concept of mutual information. Importantly, for the perception part of the experiment, different panellists may use different alphabets to encode their sensory experiences, for example, to distinguish the connoisseur from the ignoramus. A connoisseur has a more sophisticated encoding apparatus, using a better-trained sensory network, which allows for better discrimination of recipes than is possible for the ignoramus. The latter usually encodes with fewer attributes, which are, moreover, less capable of distinguishing between recipes. The use of a larger vocabulary (of the connoisseur) means that a long string of information made by the connoisseur contains potentially more information than an equal length of string made by the ignoramus. Whether or not they are connoisseurs, overall, the training of panellists will influence the magnitude of any measure of complexity. The higher density of information of strings produced by connoisseurs likely makes the processing of the information slower. Thus, the time required for the processing of information, and the according energy use, is likely to be explicitly coupled with complexity [46].

Normalising alphabet size allows us to define redundancy in the amount of information, which can be compressed due to the underlying structure of the information. In this way, we can compare what the structure is like on the input side and on the output side.

The goal of the experiments is to relate a particular set of sample properties to perceived properties. Examples of perceived properties are arousal, satiety liking, and the sensation of a complex experience at the perceptual level. These form a mapping from one kind of property to another kind of property. Sometimes, these different kinds of properties cannot be related to each other, and can be considered independent. The concept of complexity is relevant to both sample and receiver properties. An example is the process of sample production and the actual sample properties, as in the sequence below:

recipe process -> sample properties -> *sensory process* -> *experience*

where the regular letter type refers to the sample part, and the italic letter type refers to the perceptual part. In most cases of sensory experiments, there are relations between the normal and italic types, but also among the normal type and among the italic type. Because not all kinds of properties can be related, one needs to select a set of properties that are deemed of interest, and form a hypothesis. If complexity is a relevant property, then a measure for it needs be constructed.

In this paper, we investigate the use of the concept of complexity and the property of complexity, in the context of sensory science. Information, or entropy, is often used as a measure for complexity in the sensory literature. In this way, we hope to provide some structure for the ongoing discourse of complexity in sensory data. However, complexity and information are not necessarily the same, as will become more clear below.

According to Berlyne [3,5], each individual strives for an optimal level of arousal (inverted U-shaped curve). The arousal potential of a stimulus is dependent on the degree of novelty, complexity, and intensity of the stimulus. The arousal potential indicates in which direction the hedonic value or liking of a product might change with repeated exposure. Products with a low arousal potential tend to show a decrease in liking, while products with a high arousal potential tend to show an increase in liking.

Sulmont-Rossé et al. [25] hypothesise that there may be two opposing processes taking place during repeated exposure to the same food. On the one hand, there is a decrease in uncertainty regarding the safety and identity of the food, while on the other hand, there is an increased feeling of boredom. A novel stimulus or product is more commonly associated with uncertainty and with conflict. These are both states that are more likely associated with negative than positive affect [47]. The introduction of a novel or complex stimulus poses a problem for an individual [48]. The individual has had limited prior experience with this low-probability stimulus. Therefore, they do not know how to respond to it. However, the stimulus does not leave the individual ambivalent, because it will have some resemblance to previously encountered stimuli. The individual has multiple possible responses to the stimulus. Many of these may be in competition. With increasing exposure to the stimulus, there is a reduction in response competition and tension. Too much exposure to the stimulus may again lead to boredom [49]. According to Sulmont-Rossé et al. [25], the balance between the effect of resolving uncertainty and the effect of boredom during repeated exposure may be dependent on the initial arousal potential of the product. They hypothesise that when someone is exposed to a stimulus with an arousal potential that is higher than their arousal optimum, the uncertainty-reducing effect may outweigh the boredom effect, and the appreciation of the individual for the stimulus will increase with exposure. To assess the arousal potential of soft drinks, Sulmont-Rossé et al. aggregated unfamiliarity (frequency of tasting, familiarity, and description), puzzlement (complexity and surprise), element (number of aromas and distinctiveness), and intensity into an index. They concluded that the more a drink was perceived as unfamiliar, the more the liking scores increased with exposure. They explained this by indicating that "new and complex stimuli may share some resemblance with stimuli encountered in the past and elicit antagonistic responses in the individual". In this way, extra information is brought into the experiment, i.e., the information obtained in the past. This effect naturally adds to the complexity of the phenomenon of sensory perception. When there are no negative consequences associated with a new stimulus, the individual becomes less uncertain about the food, and the product may be regarded as safe. This reduces tension and increases food acceptance and liking.

As indicated, there are problems with the use of the term "complexity" in the literature. The term is ambiguous, and the definition varies from author to author. Because complexity and perceived complexity are so poorly defined, it is possible that individuals will allocate different meanings to the term "complexity", not only based on their prior experiences, but also in relation to changing contexts [7]. The issue is complicated further because a stimulus may have different measures of complexity regarding each of its attributes. Some authors, such as Dember and Earl [50] and Smith and Dorfman [51], have proposed the use of other concepts such as variability, which can be defined in terms of a neutral metric derived from information theory. Complexity may be transformed into a psychological variable by defining it in terms of change in stimulation. Discrepancies/changes can arise during the observation of a stimulus. This indicates that there is uncertainty, or information, in the stimulus. In their study, Smith and Dorfman [51] used the Shannon measure of information, based on the works of Shannon [28] and Wiener [52], to classify their visual stimuli into different classes. Each class had the same level of uncertainty, measured in bits of information. Smith and Dorfman indicated that their information-theoretic concept of uncertainty was relatively unambiguous, contrary to complexity. This can be accommodated by using the premise that complexity is equal to information, and thus to the negative value of uncertainty, or entropy.

Adding further to the complexity of sensory perception of foods, one may combine different foods in a meal. When combining different foods, it may be of interest to describe the overall sensory impression of such combinations. Words such as harmony or complexity may be used for describing a meal or a dish, but it is sometimes difficult to relate specific sensory attributes to such descriptions. Such overall impressions can be described as composite or integrated attributes, meaning that they are composed of several sensory impressions, but that they are evaluated as a whole. Traditionally, in descriptive analysis, the aim is to use singular, non-redundant descriptors, with minimal overlap with other terms [53]. However, there have been several attempts with integrated approaches to describe overall impressions, such as total intensity of aroma and flavour, and the evaluation of balance or blend [54,55].

In this paper, we talk about stimulus complexity when it is related to the food product itself. In most cases, the number of ingredients is the key ingredient for a measure of complexity. As an example, according to Olabi et al. [11], more complex foods containing more spices were regarded as more flavourful and less bland. Some authors, however, challenge this view, due to the possibility of interaction between components, which has the potential to influence intensity or hedonistic values [8,26]. In addition to this point, it should be understood that when interactions between components are present, these add to the overall information of the system, and thus should be considered as a contributor to complexity.

In the context of perceived complexity, one can roughly discern two types of complexity measures: one based on the number of sensations that are selected (referred to as sensory complexity), and one that is measured on a scale (referred to as perceived complexity). An example of the first comes from Lévy et al. [12], who studied the complexity of orange juices to which they added different flavour extracts. They asked the panel to rate elaborateness, confusion, and homogeneity. They indicated that the perceived complexity was strongly correlated with the amount of selected flavours and novelty, but not much with the difficulty of describing the stimulus. However, it may be equally difficult to detect any one flavour. An example of perceived complexity is the use of the Visual Analogue Scale (VAS), or asking how difficult it is to describe a sensation.

Some publications combine two different forms of complexity. Weijzen et al. [18] studied the effects of intensity and complexity on sensory-specific satiety and food acceptance of various chicken soups and snacks, such as chocolate bars with incorporated nuts, with different flavours and textures. They assessed the complexity of the products by asking the panellists directly how complex the product was, how many different ingredients it contained, and how difficult it was to describe it. From their results, they concluded that complexity was strongly correlated with the number of perceived ingredients (sensory complexity), and that the difficulty of describing a product was inversely related to familiarity. They concluded that complexity decreased sensory-specific satiety. They also concluded that intensity increased sensory-specific satiety.

Ruijschop et al. [9] investigated the influence of complexity on satiation and food intake. Their samples consisted of two types of strawberry-aromatized sweetened yoghurts: one with a single-component strawberry aroma, and one with a multi-component strawberry aroma. They assessed the complexity by asking the panellists about the difficulty of describing the taste, the perceived number of ingredients (sensory complexity), and the perceived level of complexity. They observed significant differences in some of the complexity attributes (while the number of ingredients was not significantly different), but not in pleasantness. The multi-component aroma-flavoured yoghurt was able to enhance satiety because of increased sensory stimulation.

Paulsen et al. [32] indicated that complexity can be used to describe a meal or a dish, such as fish combined with a sauce, but that it is difficult to relate specific sensory attributes to such descriptions. They defined complexity as the sensation of many tastes and flavours.

As mentioned above, Lévy et al. [12] studied the complexity of orange juices to which they added different flavour extracts. Based on these attributes, they obtained the level of complexity for the different orange juices with added aromas and flavourings. They also asked the panellists about their initial liking for the products. The optimum level of complexity for each product and each panellist was obtained from the complexity level of the most-liked juice. Then, they exposed the panel to samples that were either more complex or less complex than their "complexity optimum", and looked at how the optimum level of complexity changed with repeated exposure.

Soerensen et al. [10], while predicting long-term consumer liking of chocolate, defined perceived complexity as the number of flavours, amounting to more than 800 different ones. The release of flavours depends on eating behaviour, making flavour perception, and thus the complexity of the perception, a dynamic property. The input/output of information during an experiment on panellists' perceptions while eating a food thus becomes dynamic. It is then impossible to talk about complexity. The annotation of perception complexity becomes even more difficult because of interpersonal differences in perception, and differences in previous experiences between panellists. Jellinek and

Köster [15] defined this problem of many receptors being involved at different times during perception as physiological complexity.

What all of the above-described measures of complexity have in common is their relation with the number of perceived attributes. We note that the observed relation may not necessarily have the correct functional dependence, and may deviate from a linear one, as is the case while using the expression of entropy. The relation between complexity and the number of perceived attributes, and the according measure of complexity in terms of entropy, may be linked to the concept of richness of sensation, and it may bear similarity to the concept of species richness in ecology [38].

5. Complexity, Degree of Dissimilarity, and Degree of Structure

The complexity of texture perception was studied by Larsen et al. [33–35] and Marcano et al. [6]. As already mentioned in Section 2, Larsen et al. defined "textural complexity" as determined during mastication, from the first bite through to the point of swallowing [34]. The number of texture attributes at any one time was established by asking the panel to name all the attributes they perceived while consuming the sample. It was concluded that for more texturally complex samples, there were more changes in textural attributes. One could rephrase this by saying that the more dissimilar the perceptions were, the more complex the samples were perceived as being. This rephrasing brings complexity in tune with the concept of uncertainty and its quantification in terms of the entropy measure.

In another experimental setting with dissimilarity measurements, Marcano et al. [6] studied the texture perception of different cheese pies with added visible particles. Rating of the cheese pies in terms of complexity was performed on a nine-point scale, from not complex at all to very complex. It was found that high complexity coincided with high heterogeneity, i.e., a high level of dissimilarity.

Observing differences is one way of assessing complexity, as we saw in Section 5. Another way is detecting differences in patterns, revealing the structural form of the information. This was, for example, performed in the work by Mielby et al. [21,22], which we discussed earlier, where pictures of fruits and vegetables were mixed in different ratios and cut in different patterns.

Regarding the complexity of taste perception, as studied by Bitnes et al. [27], the complexity of food products was investigated by means of mixtures of different tastes (sucrose, sodium chloride, citric acid, and caffeine) in water, iced tea, and tomato soup. Complexity, like in other cases, involved more than just the number of constituents and taste sensations, and also contained mixed stimuli, making the structure of the perception more diverse. Adding a new stimulus dimension, such as texture, made the product become more difficult to evaluate, adding a new element to the information structure, and thus causing it to be perceived as more complex.

Giacalone et al. [13] related complexity to the number of discernible elements within a stimulus and to the degree to which these elements coexist or conflict, and found that complexity was closely related to perceived ambiguity of a stimulus, as well as to the cognitive effort necessary for its interpretation. Ambiguity can be viewed as the contribution to the entropy, i.e., the negative value of the information, and the effort to retrieve the information to come to an interpretation can be viewed as the amount of information required for this. In other words, these authors also found evidence for the relevance of the concept of entropy, and of information, to the concept of complexity.

Herrmann et al. [17] also considered the effort of obtaining information as a measure for complexity, as obtained in their study of how scents influence cognition. More complex olfactory stimuli contain more information to be processed, thereby requiring more effort. This was proven by the amount of anagrams the panellists were able to solve.

6. Discussion and Future Directions

The measurement of complexity has been dealt with ambiguously in the sensory literature and beyond. Many studies show a relation between (perceived) complexity and the number of stimuli or ingredients. This is, likely correctly, debated, because interactions among the ingredients or stimuli also contribute to information, which is left out when only considering the number of ingredients. This is furthermore exemplified by means of a study by Ruijschop et al. [9], which showed that with a constant number of ingredients in products, the perceived complexity of these products still differed.

The most logical measure for complexity seems to be an information measure. Such a measure has also been used in some sensory food-related complexity studies. Notably, the functional dependency between information and number of ingredients is not a linear one, which constitutes another argument as to why complexity is most likely not related linearly to the number of ingredients in a food. The Shannon information measure, and its negative value, the uncertainty or entropy measure, may suffice for many occasions, although in this case, one needs to meticulously take into account the real number of possibilities of perceived signals, including those related to the context of the sensory experiment.

It is believed that re-analysing the appropriate sensory perception literature in terms of complexity as a Shannon information measure, instead of being simply equal to the number of ingredients or number of stimuli, will add to interpreting such sensory perception, and facilitate comparisons among different experiments. This was exemplified in reference [2].

Incorporating such a context-dependent element, however, poses challenges when using a measure of sensory profile complexity, particularly when complexity is defined as the effort required to retrieve the sensory profile. Such a measure relates back to the definition of complexity given by Kolmogorov, but seems hard to quantify unambiguously, due to the unclarity regarding how to embed context into the expression, which is in turn partially due to the fact that the measure will be dependent on the level of training of the individual that is perceiving the food.

Experiments probing dissimilarities can directly provide a measure of entropy, the negative value of the information, and as such, the negative value of complexity.

Overall, many experiments performed in the food sensory field point to the fact that complexity can be equated to the "amount of information obtained in the perception and interpretation process while consuming the food", and can be quantified in terms of an entropy measure by following the Shannon entropy measure. This explains the different possible experimental ways to determine (differences in) complexity, because the entropy measure can be obtained from experiments determining differences in either ambiguity, uncertainty, dissimilarity, or the amount of effort needed to obtain the information. This holds for food sensory perception in particular, and complex systems in general.

It is believed that using the Shannon entropy measure may enable better comparison among sensory perception outcomes determined in different laboratories and in different experimental settings, and that such use may also uncover possible differences in the contexts of sensory perception experiments. At the same time, considering sensory perception as the output of the complex system of "a human consuming food", and using the Shannon information to contribute towards its understanding, it is foreseeable that scientists who want to explore complex systems from a fundamental point of view may resort to the outcomes of food sensory perception experiments.

Author Contributions: Conceptualization, L.S., H.v.M., and E.v.d.L.; methodology, L.S., H.v.M. and E.v.d.L.; investigation, L.S., H.v.M. and E.v.d.L.; writing—original draft preparation, L.S.; writing—review and editing, E.v.d.L. and H.v.M.; supervision, H.v.M. and E.v.d.L.; project administration, E.v.d.L.; funding acquisition, E.v.d.L. All authors have read and agreed to the published version of the manuscript.

Funding: This research received no external funding.

Institutional Review Board Statement: Not applicable.

Data Availability Statement: No new data have been generated. All references used and referred to are publicly available.

Acknowledgments: We acknowledge the continued discussions with Theo Odijk on the topic of complexity. E.v.d.L. gratefully acknowledges Professor Signe Kjelstrup for pointing out reference [44].

Conflicts of Interest: The authors declare no conflicts of interest.

References

1. Köster, E.P. The psychology of food choice: Some often encountered fallacies. *Food Qual. Prefer.* **2003**, *14*, 359–373. [CrossRef]
2. Sturtewagen, L.; van Mil, H.; de Lavergne, M.D.; Stieger, M.; van der Linden, E.; Odijk, T. A quantitative information measure applied to texture perception attributes during mastication. *J. Texture Stud.* **2024**, *55*, e12816. [CrossRef]
3. Berlyne, D.E. Novelty, Uncertainty, Conflict, Complexity. In *Conflict, Arousal, and Curiosity*; American Psychological Association: Washington, DC, USA, 1960; pp. 18–44.
4. Berlyne, D.E. Complexity and incongruity variables as determinants of exploratory choice and evaluative ratings. *Can. J. Psychol.* **1963**, *17*, 274–290. [CrossRef]
5. Berlyne, D.E. Novelty, complexity, and hedonic value. *Percept. Psychophys.* **1970**, *8*, 279–286. [CrossRef]
6. Marcano, J.; Morales, D.; Vélez-Ruiz, J.F.; Fiszman, S. Does food complexity have a role in eliciting expectations of satiating capacity? *Food Res. Int.* **2015**, *75*, 225–232. [CrossRef]
7. Parr, W.V.; Mouret, M.; Blackmore, S.; Pelquest-Hunt, T.; Urdapilleta, I. Representation of complexity in wine: Influence of expertise. *Food Qual. Prefer.* **2011**, *22*, 647–660. [CrossRef]
8. Kroeze, J.H. The perception of complex taste stimuli. In *Psychological Basis of Sensory Evaluation*; Elsevier: London, UK, 1990; pp. 41–68.
9. Ruijschop, R.M.; Boelrijk, A.E.; Burgering, M.J.; de Graaf, C.; Westerterp-Plantenga, M.S. Acute effects of complexity in aroma composition on satiation and food intake. *Chem. Senses* **2010**, *35*, 91–100. [CrossRef]
10. Soerensen, J.G.; Waehrens, S.S.; Byrne, D.V. Predicting and Understanding Long-Term Consumer Liking of Standard Versus Novel Chocolate: A Repeated Exposure Study. *J. Sens. Stud.* **2015**, *30*, 370–380. [CrossRef]
11. Olabi, A.; Neuhaus, T.; Bustos, R.; Cook-Camacho, M.; Corvi, T.; Abdouni, L. An investigation of flavor complexity and food neophobia. *Food Qual. Prefer.* **2015**, *42*, 123–129. [CrossRef]
12. Levy, C.M.; MacRae, A.; Koster, E.P. Perceived stimulus complexity and food preference development. *Acta Psychol.* **2006**, *123*, 394–413. [CrossRef]
13. Giacalone, D.; Duerlund, M.; Bøegh-Petersen, J.; Bredie, W.L.; Frøst, M.B. Stimulus collative properties and consumers' flavor preferences. *Appetite* **2014**, *77*, 20–30. [CrossRef]
14. Melcher, J.M.; Schooler, J.W. The Misremembrance of Wines Past: Verbal and Perceptual Expertise Differentially Mediate Verbal Overshadowing of Taste Memory. *J. Mem. Lang.* **1996**, *35*, 231–245. [CrossRef]
15. Jellinek, J.S.; Koster, E.P. Perceived fragance complexity and its relationship to familiarity and pleasantness. *J. Soc. Cosmet. Chem.* **1979**, *34*, 83–97.
16. Matthias Laska, R.H. Ability to discriminate between related odor mixtures. *Chem. Senses* **1992**, *17*, 403–415. [CrossRef]
17. Herrmann, A.; Zidansek, M.; Sprott, D.E.; Spangenberg, E.R. The Power of Simplicity: Processing Fluency and the Effects of Olfactory Cues on Retail Sales. *J. Retail.* **2013**, *89*, 30–43. [CrossRef]
18. Weijzen, P.L.; Zandstra, E.H.; Alfieri, C.; de Graaf, C. Effects of complexity and intensity on sensory specific satiety and food acceptance after repeated consumption. *Food Qual. Prefer.* **2008**, *19*, 349–359. [CrossRef]
19. Frøst, M.B.; Mielby, L.H. Perceived complexity and its relationship to liking in a top-level gastronomic menu measured in a restaurant setting. In Proceedings of the Fourth European Conference on Sensory and Comsumer Research, Vitoria-Gasteiz, Spain, 5–8 September 2010.
20. Mielby, L.H.; Hopfer, H.; Jensen, S.; Thybo, A.K.; Heymann, H. Comparison of descriptive analysis, projective mapping and sorting performed on pictures of fruit and vegetable mixes. *Food Qual. Prefer.* **2014**, *35*, 86–94. [CrossRef]
21. Mielby, L.H.; Jensen, S.; Edelenbos, M.; Thybo, A.K. An Approach to Measuring Adolescents' Perception of Complexity for Pictures of Fruit and Vegetable Mixes. *J. Sens. Stud.* **2013**, *28*, 66–75. [CrossRef]
22. Mielby, L.H.; Kildegaard, H.; Gabrielsen, G.; Edelenbos, M.; Thybo, A.K. Adolescent and adult visual preferences for pictures of fruit and vegetable mixes—Effect of complexity. *Food Qual. Prefer.* **2012**, *26*, 188–195. [CrossRef]

23. Porcherot, C.I.; Issanchou, S. Dynamics of liking for flavoured crackers: Test of predictive value of a boredom test. *Food Qual. Prefer.* **1998**, *9*, 21–29. [CrossRef]
24. Reverdy, C.; Schlich, P.; Köster, E.; Ginon, E.; Lange, C. Effect of sensory education on food preferences in children. *Food Qual. Prefer.* **2010**, *21*, 794–804. [CrossRef]
25. Sulmont-Rossé, C.; Chabanet, C.; Issanchou, S.; Köster, E. Impact of the arousal potential of uncommon drinks on the repeated exposure effect. *Food Qual. Prefer.* **2008**, *19*, 412–420. [CrossRef]
26. Schifferstein, H.N.; Frijters, J.E. Contextual and sequential effects on judgments of sweetness intensity. *Percept. Psychophys.* **1992**, *52*, 243–255. [CrossRef]
27. Bitnes, J.; Ueland, Ø.; Møller, P.E.R.; Martens, M. Reliability of sensory assessors: Issues of complexity. *J. Sens. Stud.* **2009**, *24*, 25–40. [CrossRef]
28. Shannon, C.E. A Mathematical Theory of Communication. *Bell Syst. Tech. J.* **1948**, *27*, 379–423. [CrossRef]
29. Luce, R.D. Whatever Happened to Information Theory in Psychology? *Rev. Gen. Psychol.* **2003**, *7*, 183–188. [CrossRef]
30. Laming, D. Statistical Information and Uncertainty: A Critique of Applications in Experimental Psychology. *Entropy* **2010**, *12*, 720–771. [CrossRef]
31. Laming, D. Statistical Information, Uncertainty, and Bayes' Theorem: Some Applications in Experimental Psychology. In Proceedings of the European Conference on Symbolic and Quantitative Approaches to Reasoning and Uncertainty, Toulouse, France, 19–21 September 2001; Volume 2143, pp. 635–646.
32. Paulsen, M.T.; Ueland, Ø.; Nilsen, A.N.; Öström, Å.; Hersleth, M. Sensory perception of salmon and culinary sauces—An interdisciplinary approach. *Food Qual. Prefer.* **2012**, *23*, 99–109. [CrossRef]
33. Larsen, D.S.; Tang, J.; Ferguson, L.R.; James, B.J. Increased textural complexity in food enhances satiation. *Appetite* **2016**, *105*, 189–194. [CrossRef]
34. Larsen, D.S.; Tang, J.; Ferguson, L.; Morgenstern, M.P.; James, B.J. Textural Complexity is a Food Property—Shown Using Model Foods. *Int. J. Food Prop.* **2015**, *19*, 1544–1555. [CrossRef]
35. Larsen, D.S.; Tang, J.; Ferguson, L.; Morgenstern, M.P.; James, B.J. Oral Breakdown of Texturally Complex Gel-Based Model Food. *J. Texture Stud.* **2015**, *47*, 169–180. [CrossRef]
36. Tang, J.; Larsen, D.S.; Ferguson, L.; James, B.J. Textural Complexity Model Foods Assessed with Instrumental and Sensory Measurements. *J. Texture Stud.* **2016**, *48*, 9–22. [CrossRef]
37. Kolmogorov, A.N. Three approaches to the quantitative definition of information. *Int. J. Comput. Math.* **1968**, *2*, 157–168. [CrossRef]
38. Jost, L. Mismeasuring biological diversity: Response to Hoffmann and Hoffmann (2008). *Ecol. Econ.* **2009**, *68*, 925–928. [CrossRef]
39. Tsallis, C. Nonextensive statistics: Theoretical, experimental and computational evidences and connections. *Braz. J. Phys.* **1999**, *29*, 1–35. [CrossRef]
40. Odijk, T. (Lorentz Institute for Theoretical Physics, University of Leiden, Leiden, The Netherlands). Personal communication, 2024.
41. West, G.B.; Brown, J.H.; Enquist, B.J. The fourth dimension of life: Fractal geometry and allometric scaling of organisms. *Science* **1999**, *284*, 1677–1679. [CrossRef]
42. Bettencourt, L.M.A. The Origins of Scaling in Cities. *Science* **2013**, *340*, 1438–1441. [CrossRef]
43. Coppens, M.-O.; Froment, G.F. Diffusion and reaction in a fractal catalyst pore—I. Geometrical aspects. *Chem. Eng. Sci.* **1995**, *50*, 1013–1026. [CrossRef]
44. Magnanelli, E.; Wilhelmsen, Ø.; Acquarone, M.; Folkow, L.P.; Kjelstrup, S. The Nasal Geometry of the Reindeer Gives Energy-Efficient Respiration. *J. Non-Equilib. Thermodyn.* **2017**, *42*, 59–78. [CrossRef]
45. Leinster, T.; Cobbold, C.A. Measuring diversity: The importance of species similarity. *Ecology* **2012**, *93*, 477–489. [CrossRef]
46. Atick, J.J. Could information theory provide an ecological theory of sensory processing? *Network* **1992**, *3*, 213–251. [CrossRef]
47. Zajonc, R.B. Attidunal effects of mere response. *J. Personal. Soc. Psychol.* **1968**, *9*, 1–27. [CrossRef]
48. Harrison, A.B. Response competition, frequency, exploratory behaviour, and liking. *J. Personal. Soc. Psychol.* **1968**, *9*, 363–368. [CrossRef]
49. Zandstra, E.H.; Weegels, M.; Van Spronsen, A.; Klerk, M. Scoring or boring? Predicting boredom through repeated in-home consumption. *Food Qual. Prefer.* **2004**, *15*, 549–557. [CrossRef]
50. Dember, W.N.; Earl, R.W. Analysis of exploratory, manipulatory, and curiosity behaviors. *Psychol. Rev.* **1957**, *64*, 91–96. [CrossRef] [PubMed]
51. Smith, G.F.; Dorfman, D.D. The effect of stimulus uncertainty on the relationship between frequency of exposure and liking. *J. Personal. Soc. Psychol.* **1975**, *31*, 150–155. [CrossRef]
52. Wiener, N. *Cybernetics or Control and Communication in the Animal and the Machine*; MIT Press: Cambridge, MA, USA, 1949.
53. Lawless, H.T.; Heymann, H. *Sensory Evaluation of Food Principles and Practices*; Chapman & Hall: New York, NY, USA, 1998; 819p.

54. Giboreau, A.; Dacremont, C.; Egoroff, C.; Guerrand, S.; Urdapilleta, I.; Candel, D.; Dubois, D. Defining sensory descriptors: Towards writing guidelines based on terminology. *Food Qual. Prefer.* **2007**, *18*, 265–274. [CrossRef]
55. Meilgaard, M.C.; Carr, B.T.; Civille, G.V. *Sensory Evaluation Techniques*; CRC Press: Boca Raton, FL, USA, 1999.

entropy

MDPI

Article

Entropy Production and Filling Time in Hydrogen Refueling Stations: An Economic Assessment

Bruno F. Santoro [1], David Rincón [2] and Diego F. Mendoza [3,*,†]

1 Op2B—Optimization to Business, Av. Pompéia, 723, São Paulo 05023-000, Brazil; bruno.santoro@op2b.com.br
2 Department of Chemical and Biomolecular Engineering, University of California, Los Angeles, CA 90095, USA; fdrinconc@g.ucla.edu
3 Department of Chemical Engineering, Universidad de Antioquia, Medellín 050010, Colombia
* Correspondence: dfernando.mendoza@udea.edu.co
† Current address: Grupo de Ingeniería de Procesos Integrados, Universidad de Antioquia Medellin, Calle 70 No. 52-21, Antioquia 050010, Colombia.

Abstract: A multi-objective optimization is performed to obtain fueling conditions in hydrogen stations leading to improved filling times and thermodynamic efficiency (entropy production) of the de facto standard of operation, which is defined by the protocol SAE J2601. After finding the Pareto frontier between filling time and total entropy production, it was found that SAE J2601 is suboptimal in terms of these process variables. Specifically, reductions of filling time from 47 to 77% are possible in the analyzed range of ambient temperatures (from 10 to 40 °C) with higher saving potential the hotter the weather conditions. Maximum entropy production savings with respect to SAE J2601 (7% for 10 °C, 1% for 40 °C) demand a longer filling time that increases with ambient temperature (264% for 10 °C, 350% for 40 °C). Considering average electricity prices in California, USA, the operating cost of the filling process can be reduced between 8 and 28% without increasing the expected filling time.

Keywords: dynamic optimization; energy efficiency; vehicle fueling; irreversibility analysis

1. Introduction

Transportation using hydrogen fuel cells has brought new research opportunities for sustainable energy sources for personal and mass transit systems. For example, the Cal State LA Hydrogen Research and Fueling Facility has pioneered hydrogen on-site production for vehicles using electrolysis since 2014 [1]. The University of São Paulo projects for 2024 include a fueling station for renewable hydrogen from ethanol for buses and vehicles [2]. Similarly, Hynion and Toyota Norway have agreed to cooperate in expanding the hydrogen refueling station at Høvik [3]. To be economically competitive with traditional systems, fueling facilities for hydrogen transportation must be optimized from the energy expenditure point of view and user experience, which is affected among other factors by the filling time.

Hydrogen refueling stations (HRSs) technology has promised similar filling time as what is achieved with usual combustion engine vehicles. To attain such a goal, protocol SAE J2601 has been embedded in the HRSs system. In general lines, it simply specifies a pressure ramp to be provided to the system, whose slope is a function of the initial state of charge of the vehicle tank and the atmospheric temperature [4]. In addition, the SAE J2601 design is based only on the first law of thermodynamics. In order to avoid the static approach of SAE J2601 and improve filling time, first-law approaches have been proposed for HRSs using optimal control [5,6]. Still, entropy production for HRSs has been explored only recently, when minimal entropy production paths were identified to outperform SAE J2601 with 20 to 27% in energy savings, for a fixed filling time of 400 s for both approaches [7]. Due to space constraints, more details about SAE J2601 and other first-law contributions in the literature of HRSs are found elsewhere [7].

For other applications, there is a vast literature considering the second law of thermodynamics in process design and operation. By using a combination of entropy production minimization and nature-inspired designs, it was found that cylinder-shaped tubular reactors have less uniform entropy production than reactors with varying radius along the tube [8]. Energy (i.e., 50%) and cooling (i.e., 30%) savings were found when implementing entropy production minimization in an adiabatic distillation column for a binary system using rate-based and equilibrium-stage models [9]. Multiple entropy production minimization methods such as the equipartition of forces, equipartition of entropy production, and equal thermodynamic speed were compared in the same system in order to understand the effect of their assumptions on the optimal solution [10]. In heat transfer processes, a constant local entropy production rate showed lower entropy production than equipartition of thermal driving force [11]. Many other studies on entropy production have shown the benefits of this approach when optimizing classical chemical processes [12–15].

In this contribution, the filling time of HRSs is optimized by using an entropy balance together with a validated first-principles model. The proposed approach is compared with the protocol SAE J2601 under different scenarios of ambient temperatures. A multi-objective optimization is performed to obtain fueling conditions leading to optimal filling times with maximal thermodynamic efficiency quantified by the entropy production. As a result, a utopia-tracking approach representation is presented in terms of a Pareto frontier. Finally, the operating cost per filling cycle (i.e., economic analysis) of HRSs under the proposed approach and protocol SAE J2601 are compared under different scenarios.

There are three main contributions of this work:

- To include entropy production in a high-fidelity thermodynamic model of a hydrogen refueling station.
- To minimize the filling time of the HRS system under entropy production using optimal control and to compare it with SAE J2601 in a multi-objective approach.
- To analyze the operating cost of the HRS for different operating conditions.

2. Materials and Methods

2.1. Station Layout

A simplified version of an HRS is shown in Figure 1. The refueling process starts when hydrogen is dispensed from the high-pressure supply tanks (stream 6) with the hydrogen flow regulated by the flow control valve. Subsequently, the hydrogen exiting the control valve (stream 5) is thermally conditioned in a cooler that supplies the required cooling load through a refrigeration system. Once the hydrogen flow rate and temperature are adequate, hydrogen enters the vehicle tank through the inlet valve.

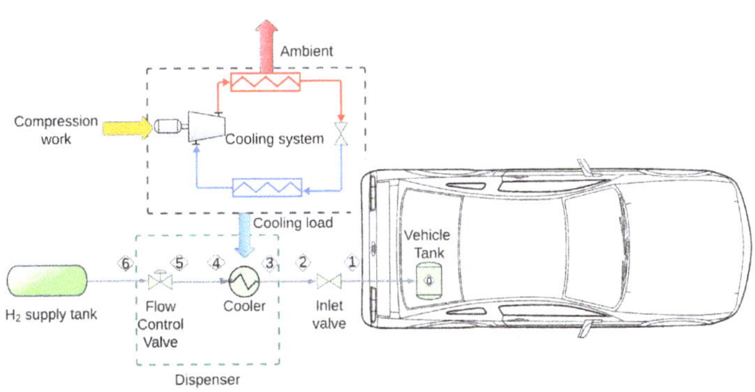

Figure 1. Hydrogen station layout.

2.2. Mathematical Model

Olmos and Manousiouthakis [5] presented a mathematical model for an HRS similar to the one presented in Figure 1, using mass and energy balances for each component in the system (i.e., tank, cooler, etc.). Later, such a model was expanded by Mendoza et al. [7] by including a high-accuracy thermodynamic model, valve pressure drop equations, and entropy balances for the main components of the system. In the mathematical representation of the refueling station, it is assumed that the dynamics of the cooler and valves are significantly faster than those of the tank. Consequently, the valves and cooler dynamics are represented by a set of algebraic equations, while the tank is modeled using differential equations.

In the current proposal, mass and energy balances for the hydrogen in the vehicle tank are given by Equations (1) and (2), respectively.

$$V_0 \frac{d\rho_0(t)}{dt} = \dot{m}(t) \tag{1}$$

where V_0 is the vehicle tank capacity, ρ_0 is the density of the hydrogen confined in the tank, and $\dot{m}$ is the hydrogen mass flow rate entering the tank. The left-hand side of the mass balance, Equation (1), represents the accumulation rate of hydrogen inside the tank due to the hydrogen supplied through the inlet valve, *iv*.

The energy balance for the gas confined in the vehicle tank is

$$V_0\rho_0(t)\frac{du_0(t)}{dt} = \dot{m}(t)\left[u_1(t) - u_0(t) + \frac{p_1(t)}{\rho_1(t)}\right] - \dot{Q}_{w0}(t) \tag{2}$$

where u_0, u_1 represent the internal energy of hydrogen in the tank and in stream 1, respectively; p_1, ρ_1 are the pressure and density of hydrogen at conditions of stream 1, and $\dot{Q}_{w0}$ is the heat transfer rate between the gas and the vehicle tank wall. The left-hand side of Equation (2) denotes the internal energy accumulation rate due to the energy interactions with the wall and with the hydrogen entering the tank.

The tank-wall dynamics is described by an energy balance, Equation (3), where the rate of change of the wall temperature, T_w, is a consequence of the heat interactions with the hydrogen in the tank, $\dot{Q}_{wo}$, and with the environment, $\dot{Q}_{w\infty}$. These heat flow rates are calculated using Newton's cooling law, which is represented by Equations (4) and (5), respectively.

$$\rho_w V_w C p_w \frac{dT_w(t)}{dt} = \dot{Q}_{w0}(t) - \dot{Q}_{w\infty}(t) \tag{3}$$

$$\dot{Q}_{w0}(t) = h_{w0}A_{w0}[T_0(t) - T_w(t)] \tag{4}$$

$$\dot{Q}_{w\infty}(t) = h_{w\infty}A_{w\infty}[T_w(t) - T_\infty(t)] \tag{5}$$

where, ρ_w, V_w, Cp_w are the density, volume and heat capacity of the vehicle tank wall, respectively. h_{w0}, $h_{w\infty}$ are the hydrogen/wall and wall/air convective heat transfer coefficients; A_{w0}, $A_{w\infty}$ are the heat transfer areas, and T_∞ is the temperature of the environment.

The cooling system is described by the energy balance in the cooler, Equation (6), and by the coefficient of performance (*COP*) of the refrigeration unit as defined in Equation (7).

$$u_4(t) + \frac{p_4(t)}{\rho_4(t)} + q_c(t) = u_3(t) + \frac{p_3(t)}{\rho_3(t)} \tag{6}$$

$$COP(t) = \frac{w_c(t)}{q_c(t)} \tag{7}$$

where w_c is the compression work input in the cooling unit and q_c is the heat transfer duty in the cooler. The variation of the COP as a function of the ambient temperature [16] is given by Equation (8).

$$COP(t) = 1.6\exp(-0.018(T_\infty(t) - 273.15)) \tag{8}$$

The inlet valve, iv, is modeled using an energy balance, Equation (9), and a design equation, Equation (10). Similarly, the flow control valve, fv, is modeled using an energy balance and a design equation, which are given by Equations (11) and (12), respectively.

$$u_2(t) + \frac{p_2(t)}{\rho_2(t)} = u_1(t) + \frac{p_1(t)}{\rho_1(t)} \tag{9}$$

$$\frac{\dot{m}(t)}{\rho_N} = \frac{514}{3.6 \times 10^5} k_{v,iv} \sqrt{\frac{p_1(t)(p_2(t) - p_1(t))}{\rho_N T_2(t)}} \tag{10}$$

$$u_6(t) + \frac{p_6(t)}{\rho_6(t)} = u_5(t) + \frac{p_5(t)}{\rho_5(t)} \tag{11}$$

$$\frac{\dot{m}(t)}{\rho_N} = \frac{514}{3.6 \times 10^5} k_{v,fv}(t) \sqrt{\frac{p_5(t)(p_6 - p_5(t))}{\rho_N T_5(t)}} \tag{12}$$

where k_v is the valve coefficient and ρ_N is the density at normal conditions (273.15 K and 101.325 kPa). The main difference between the valves is that the inlet valve is assumed to be completely open during the filling process, while the opening fraction of the flow control may be changed, $0 \leq x_{fv}(t) \leq 1$. A detailed description of the variables and its units is provided in the Abbreviations section.

2.3. Entropy Production

The total entropy production at the end of the fueling process, $S_{irr}(t_f)$, is calculated from the overall entropy balances for each component of the station,

$$S_{irr}(t_f) = \sum_{k \in \mathcal{K}} \int_0^{t_f} \left(\frac{dS(t)}{dt}\right)_{irr,k} dt, \quad \mathcal{K} \in \{0, w, iv, fv\} \tag{13}$$

where $(dS/dt)_{irr,k}$ is the instantaneous entropy production rate in component k, namely: vehicle tank (0), tank wall w, inlet valve (iv) and flow control valve (fv).

Remark 1. *Flow in pipes is assumed to be adiabatic and reversible, and heat transfer in the cooler is assumed to be reversible; both assumptions are necessary simplifications due to the lack of details about the actual pipework and heat exchanger type and flow configuration. It is important to notice that in* (13)*, the final fueling time t_f could be replaced by a particular time instant t to obtain the cumulative entropy generation up to that time.*

The entropy production in the gas confined in the tank, Equation (14), is calculated as the difference between the entropy outlet and inlet flows. The first term on the right-hand side corresponds to the outlet entropy flow due to the heat interaction with the tank wall, while the second and third terms correspond to the entropy flow due to the inlet hydrogen flow and to the instantaneous accumulation rate of entropy by the mass confined in the tank, respectively.

$$\left(\frac{dS(t)}{dt}\right)_{irr,0} = \frac{\dot{Q}_{w0}(t)}{T_w(t)} - \dot{m}(t)\, s_1(t) + V_0 \frac{d\rho_0(t) s_0(t)}{dt} \tag{14}$$

The term $d\rho_0(t)s_0(t)/dt$ in the in the preceding equation is expanded as follows:

$$V_0\frac{d\rho_0(t)s_0(t)}{dt} = V_0\rho_0(t)\frac{ds_0(t)}{dt} + s_0(t)V_0\frac{d\rho_0(t)}{dt} \tag{15}$$

where $ds_0(t)/dt$ is calculated from the fundamental equation of thermodyamics for a simple compressible substance,

$$\frac{ds_0(t)}{dt} = \frac{1}{T_0(t)}\frac{du_0(t)}{dt} - \frac{p_0(t)}{T_0(t)\rho_0^2(t)}\frac{d\rho_0(t)}{dt} \tag{16}$$

combining Equations (1), (2), (15) and (16) into Equation (14) yields the following expression

$$\begin{aligned}\left(\frac{dS(t)}{dt}\right)_{irr,0} &= \dot{Q}_{w0}(t)\left(\frac{1}{T_w(t)} - \frac{1}{T_0(t)}\right) \\ &+ \dot{m}(t)\left(u_1(t) + \frac{p_1(t)}{\rho_1(t)}\right)\left(\frac{1}{T_0(t)} - \frac{1}{T_1(t)}\right) + \dot{m}(t)\left(\frac{\mu_1(t)}{T_1(t)} - \frac{\mu_0(t)}{T_0(t)}\right)\end{aligned} \tag{17}$$

The terms on the right-hand side of Equation (17) show the entropy production as the sum of flows and thermodynamic forces related to heat transfer between the hydrogen confined in the tank and the wall (first term), with the temperature difference between the inlet gas and the hydrogen in the tank (second term), and with the chemical potential, μ, difference between the hydrogen entering the tank and the hydrogen confined in the tank (third term).

Entropy production within the wall arises from thermal interactions with both the confined gas and the environment,

$$\left(\frac{dS(t)}{dt}\right)_{irr,w} = \frac{\dot{Q}_{w\infty}(t)}{T_{w\infty}} - \frac{\dot{Q}_{w0}(t)}{T_w(t)} + \frac{\rho_w V_w C p_w}{T_w(t)}\frac{dT_w(T)}{dt} \tag{18}$$

The last term on the right-hand side of Equation (18) accounts for the entropy accumulation rate in the wall. Substituting the last term on the right-hand side of Equation (18) by the energy balance on the wall, Equation (3), we obtain the entropy production in the wall, Equation (19), as a product of the heat flow and the heat transfer thermodynamic force, i.e., the temperature difference between the wall and the environment.

$$\left(\frac{dS(t)}{dt}\right)_{irr,w} = \dot{Q}_{w\infty}(t)\left(\frac{1}{T_\infty} - \frac{1}{T_w(t)}\right) \tag{19}$$

Entropy production within the adiabatic valves is attributed to the difference between entropy flows carried by the outlet and inlet mass streams.

$$\left(\frac{dS(t)}{dt}\right)_{irr,iv} = \dot{m}(t)s_1(t) - \dot{m}(t)s_2(t) \tag{20}$$

$$\left(\frac{dS(t)}{dt}\right)_{irr,fv} = \dot{m}(t)s_5(t) - \dot{m}(t)s_6(t) \tag{21}$$

An insight about the nature of the irreversibility associated with the expansion in the adiabatic valve can be obtained using differential expressions for the entropy and enthalpy in the valves, as shown in Equations (22) and (23), respectively.

$$ds = -\frac{\partial}{\partial T}\left(\frac{1}{\rho}\right)_p dp + \frac{C_p}{T}dT \tag{22}$$

$$dh = \left[\frac{1}{\rho} - T\frac{\partial}{\partial T}\left(\frac{1}{\rho}\right)_p\right] + C_p dT = 0 \tag{23}$$

The constant pressure heat capacity, C_p, obtained from Equation (23) is replaced in Equation (22) providing an expression for the entropy change in the adiabatic valve as

$$ds = -\frac{dp}{\rho T} \tag{24}$$

From this expression, we obtain the entropy production in the valves

$$\left(\frac{dS(t)}{dt}\right)_{irr,iv} = -\dot{m}(t)\int_{p_2(t)}^{p_1(t)} \frac{dp}{\rho(t)T(t)} \tag{25}$$

$$\left(\frac{dS(t)}{dt}\right)_{irr,fv} = -\dot{m}(t)\int_{p_6(t)}^{p_5(t)} \frac{dp}{\rho(t)T(t)} \tag{26}$$

Equations (25) and (26) show that the entropy production in the valves is proportional to the instantaneous pressure drop in them.

2.4. Thermophysical Properties

The thermophysical and geometry parameters for the vehicle tank presented in Table 1 were obtained from the work of Olmos and Manousiouthakis [5].

Table 1. Vehicle tank wall properties [5].

Thermophysical		Geometry	
h_{wi} (kW m^{-2} K^{-1})	0.06	V_0 (m^3)	0.12
h_{w0} (kW m^{-2} K^{-1})	0.006	V_w (m^3)	0.0576
ρ_w (kg m^{-3})	1550	A_{wi} (m^2)	2.34
c_w (kJ kg^{-1} K^{-1})	1.374	A_{wi} (m^2)	2.88

All the thermodynamic properties of hydrogen were determined from a fundamental Helmholtz equation [17]. Detailed expressions for these properties are provided in Appendix A.

2.5. Optimization Problem

The minimization of filling time corresponds to an optimal control problem, where the inlet flow rate and the cooling load can be manipulated to achieve the fastest refueling while satisfying process safety constraints. Before posing an optimization problem for such a goal, it is necessary to manipulate the differential equations of the model to allow the simulation horizon to be taken as a variable. This can be accomplished through a scaling factor t_f, which is a decision variable of the problem corresponding to the final time of filling. Then, let $\tau = t/t_f$ be a dimensionless time to replace t in the model equations. From its definition, it is clear that the domain of integration is between 0 and 1, i.e., $0 \le \tau \le 1$. With this consideration, Equation (1) is rewritten as

$$V_0 \frac{1}{t_f}\frac{d\rho_0(\tau)}{d\tau} = \dot{m}(\tau) \tag{27}$$

After this change in variables in all equations, the optimization problem is

$$\min \int_0^{t_f} 1\, dt = t_f \tag{28}$$

s.t.

Equations (1) to (12)

$$0 \leq \dot{m}(\tau) \leq \dot{m}_{max} \tag{29}$$

$$0 \leq x_{fv}(\tau) \leq 1 \tag{30}$$

$$q_{c,max} \leq q_c(\tau) \leq 0 \tag{31}$$

$$SOC(\tau = 1) \geq SOC_{des} \tag{32}$$

$$S_{irr}(\tau = 1) \leq S_{irr}^{max} \tag{33}$$

Equations (29)–(31) are bounds on the manipulated variables, accounting for the physical limitations of the filling station. The constraint imposed by Equation (32) forces the state of charge (equivalent to the total amount of hydrogen in the tank) at the end of simulation horizon to achieve the desired value, SOC_{des}. In the results outlined in Section 3, the dynamic profile of all variables is converted back from τ to the dimensional value t for easier visualization and comparison with other techniques in the literature.

Entropy production is a secondary objective of the formulation. To deal with this multi-objective situation, Equation (33), an upper bound on the total generated entropy, is used in an ε-constraint method. More explicitly, S_{irr}^{max} is initially set as a sufficiently large value in order for the constraint defined by Equation (33) to be inactive, allowing the calculation of the absolute minimum filling time. Progressively, lower values of S_{irr}^{max} are chosen and lead to new points in the Pareto frontier (PF) of optimal solutions with minimal time and entropy production.

2.6. Economic Assessment

The model presented in Section 2.2 allows the calculation of energy consumption per filling cycle. In order to convert this expenditure into economic values, the commercial electric rates of Irvine, California (USA) in 2023 were considered as a benchmark [18]. In many regions, it is common for energy prices to change accordingly to the seasons of the year or even during a single day. In this study, only the monthly variation was considered by taking three data points: the highest (0.276 $/kWh, in July), the lowest (0.186 $/kWh, in February) and a typical value (0.204 $/kWh, in November).

2.7. Optimization Implementation and Solution Method

The problem was formulated using Pyomo [19], in particular with pyomo.dae [20] as a tool to convert the differential equations into an algebraic equivalent. Discretization has been made with 30 finite elements and two Radau collocation points per element. The resulting non-linear program was solved using CONOPT4. The typical computational time is around 20–30 s per simulation in a Windows 11 environment, using an Intel® Core™ i7-11800 CPU 2.30 GHz processor and 32 GB of RAM memory. The implemented model was validated in comparison with experimental data, achieving maximum deviations of around 1% in terms of state of charge and 7% of hydrogen temperature. Further details on model validation are provided in Ref. [7].

3. Results

3.1. Filling Time and Entropy Production

In order to evaluate the trade-off between filling time and thermodynamic efficiency, several simulations were conducted considering different ambient temperatures: 10, 25 and 40 °C. The simulations in this work consider that the vehicle tank is of type IV, starting at 5 MPa and the same temperature as the ambient. The filling station is considered to be H70-T40 (dispenser at 70 MPa and −40 °C), and the end goal is a state of charge of 99.1%.

A comparison of the optimization approach with the benchmark protocol SAE J2601 is shown in Figure 2. The squares correspond to specific optimizations after choosing particular values for S_{irr}^{max}, while the continuous lines joining them are a piece-wise linear approximation of the underlying Pareto frontier.

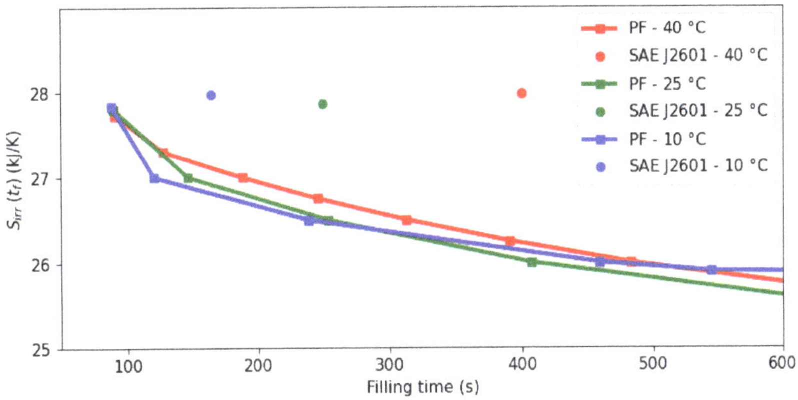

Figure 2. Pareto frontier of the total entropy production in the filling process, $S_{irr}(t_f)$, for an allowed filling time at different ambient temperatures.

The pressure ramp specified by SAE J2601 indirectly defines the corresponding filling time given the initial and final pressures. However, Figure 2 shows that such an operation (circular points) lies outside the Pareto frontier and is therefore suboptimal in terms of entropy minimization and filling time. For instance, when the ambient temperature is 40 °C, the protocol takes 400 s and generates 28 kJ/K, while the optimal profile achieves the same state of charge in only 91 s (77% lower) with an entropy production of 27.7 kJ/K (1% lower). The effect is more manifested the greater the ambient temperature, with possible time reductions of 160 s (64%) and 77 s (47%) at 25 °C and 10 °C, respectively.

Minimal entropy production can be achieved with a system operation approaching the thermodynamic limit of reversibility, which implies a slower process and smaller flow rate. Figure 2 shows that to move from the protocol toward a condition of minimal entropy has a significant cost in terms of filling time. For instance, when the ambient temperature is 10 °C, it would be necessary to spend around 600 s (264% increase) to achieve an entropy reduction of only 2 kJ/K (7%). For the 40 °C scenario, the minimum entropy is not shown in Figure 2 for scaling reasons, but it would take around 2000 s (350%) to reduce the entropy production by 3.5 kJ/K (12%). However, small decreases in entropy production may correspond to significant savings in operating costs due to there being smaller work loads in the cooling system.

System dynamics during the filling process can be observed at different points of the Pareto frontier in Figures 3–5 with a fixed ambient temperature of 40 °C. The evolution of the manipulated variables of the control system, flow rate and the cooling load is given in Figures 3 and 4, respectively, while Figure 5 shows the evolution of entropy production.

Figures 3–5 illustrate the optimal filling profile at different points of the Pareto frontier. For instance, when time is minimized with a virtually unbounded entropy production (S_{irr}^{max}= 50 kJ/K, red curves), the flow rate quickly achieves its maximum allowed value of $\dot{m}_{max}$ = 0.06 kg/s. Accordingly, the cooling load is more intense when there is a higher flow rate (S_{irr}^{max}= 50 kJ/K) or when there is a stricter requirement of inlet temperature (Protocol SAE J2601). As the requirement of maximal entropy production becomes more stringent, it is no longer feasible to operate at a maximum flow rate, dispersing the filling operation over time. Figure 5 shows that an excessively long filling time is necessary for the system to achieve its practical minimal point of entropy production at 25 kJ/K.

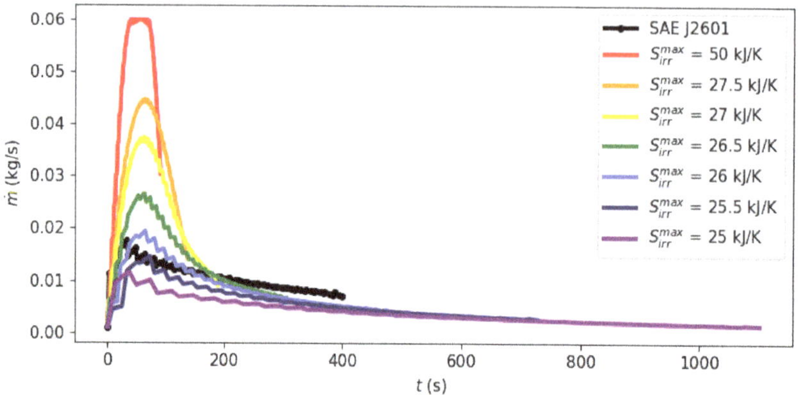

Figure 3. Evolution of the filling flow rate during the refueling process at different points of the Pareto frontier.

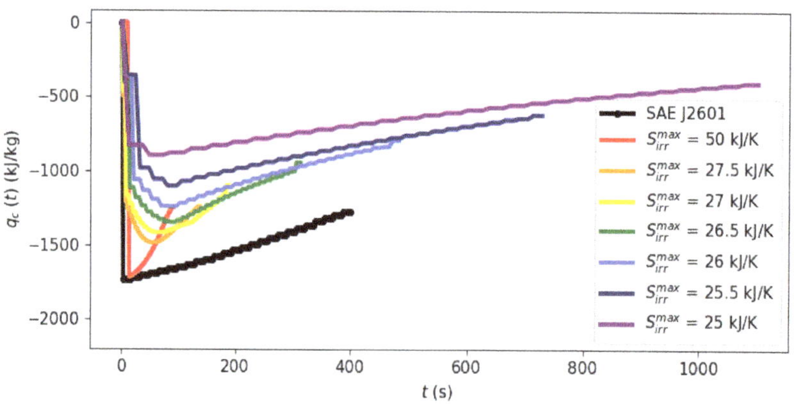

Figure 4. Evolution of the cooling load during the refueling process at different points of the Pareto frontier.

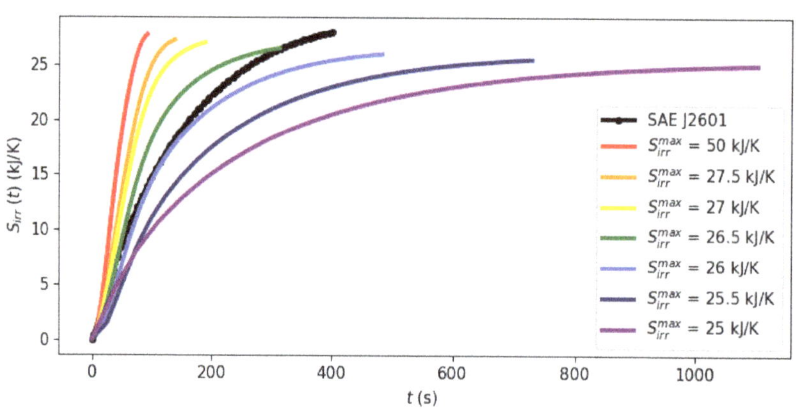

Figure 5. Evolution of the cumulative generated entropy up to time t, $S_{irr}(t)$, at different points of the Pareto frontier.

Remark 2. *Figures 3–5 illustrate a dynamic analysis of the system, meaning that the x-axis corresponds to time as the independent variable of the DAE system, as shown in Equations* (1)–(12). *On the other hand, Figure 2 is a static overview, because its x-axis, "Filling time", corresponds to t_f in the model formulation, i.e., the final time value. In other words, the static information of Figure 2 is contained in the dynamic plot of Figure 5, which shows the cumulative entropy generation up to each possible time t, including t_f. Notice that the end points of the entropy production trajectories in Figure 5 are exactly the Pareto frontier for the system at 40 °C that was presented in Figure 2.*

3.2. Economic Analysis

The trade-off between operating costs and filling time has been analyzed for three ambient temperatures (10, 25 and 40 °C) and typical seasonal electricity costs in Irvine (California, USA), in the Los Angeles metropolitan area. Results are shown in Figure 6, with the continuous line corresponding to usual electricity rates, while the shadowed zone represents the range between minimum and maximum values registered in 2023. The circles correspond to costs associated with filling according to the protocol SAE J2601, which once again lie outside the Pareto frontier.

Given that the vehicle tanks considered in this study have a capacity of 4.8 kg, to fill them from an initial pressure of 5 MPa and at ambient temperature to an *SOC* of 99.1% corresponds to a total mass of 4.3 kg to be refueled. Therefore, the total operational costs associated with a refueling of this magnitude when using the protocol SAE J2601 range from $0.19 (best-case scenario, with 10 °C ambient temperature and lowest electricity rate) to $0.66 (40 °C, highest electricity rate) per filling.

The order of magnitude of the operational cost agrees with the ones reported in other studies. For instance, [21] analyzed costs of HRSs in Germany and found compressor costs between 0.08 and 0.18 €/kg. In comparison with total hydrogen prices, ref. [22] placed a value around 10 €/kg considering the whole supply chain, which indicates that the bottleneck still lies in how to obtain the fuel in a cost-effective manner.

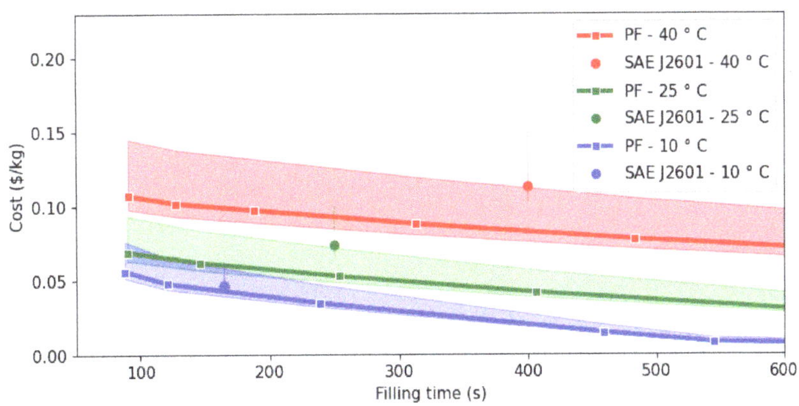

Figure 6. Pareto frontier for cost as a function of ambient temperature.

Ambient temperature plays a relevant role in the operational costs, since these increase with temperature, either using protocol SAE J2601 or the thermodynamic optimized filling process. This observation agrees with the fact that higher ambient temperatures demand more work from the cooling system regardless of the filling process followed, as shown in Table 2. However, the thermodynamic optimized process outperforms the SAE J2601 in all assessed scenarios, such that for the same filling time, the operation costs could be reduced by 8% at 10 °C, 28% at 25 °C, and 27% at 40 °C with respect to the SAE J2601 protocol.

Table 2. Compression work at different ambient temperatures.

Ambient Temperature (°C)	Compression Work (kJ)		
	SAE J2601	Min. time ($S_{irr}^{max}(t_f)$= 50 kJ/K)	Min. time ($S_{irr}^{max}(t_f)$= 26 kJ/K)
10 °C	3572 kJ	3225 kJ	748 kJ
25 °C	5664 kJ	5277 kJ	3117 kJ
40 °C	8625 kJ	8185 kJ	5930 kJ

In contrast to SAE J2601, the thermodynamic optimized operation allows the possibility of devising flexible operation with variable fueling times along the Pareto frontier. One case of interest would be having fast refueling with the same operation cost as SAE J2601: under this scenario, the filling times could be reduced between 40 s (10 °C) and 300 s (40 °C), allowing the station to achieve a higher throughput during the day when required (e.g., peak hours).

Considering a time delay of two minutes between vehicles and opening hours for the HRS between 8 AM and 8 PM, then it is possible to service between 83 (with outside temperature equal to 40 °C) and 151 vehicles (at 10 °C) if the SAE J2601 filling protocol is applied. If the operating costs are kept at the same level, the filling time could be reduced to around 90 s for temperatures above 10 °C, which in turn would allow around 200 vehicles to be serviced (increase between 54 and 120%).

On the other hand, if more emphasis is put on reducing costs while keeping the filling time of the protocol, it is possible to achieve higher savings. For instance, if a filling time equal to 500 s is considered acceptable for all temperatures, which is less than 10% slower than the SAE J2601 protocol for 40 °C, then savings of around 0.036 $/kg are possible for all temperatures. If 100 vehicles are serviced every day, the total savings amount to more than $5500 per HRS, which would compensate the additional investment required to operate under an optimal controller.

4. Conclusions

This study has evaluated the interdependency among filling time, entropy production and operating cost in HRSs. It highlights that the de facto standard of operation, defined by the protocol SAE J2601, is suboptimal in terms of these process variables, because it lies outside the Pareto frontier. Reductions of filling time from 47 to 77% are possible in the analyzed range of ambient temperature (from 10 to 40 °C) with higher savings potential the hotter the weather conditions.

Entropy production savings seem to demand a disproportionate increase in filling time; however, the economic analysis reveals that operating costs may be reduced by 30% with filling times 10% higher than what is obtained with the SAE J2601 protocol at 40 °C, which could be possible during off-peak hours.

Finally, the operating cost analysis shows possibilities for flexible operations to generate price policies which could be adapted to either fast, slow and normal fueling times with respect to the SAE J2601 protocol; this flexibility could facilitate the public adoption of hydrogen in the transportation sector, since both price and filling time can be adapted to customer needs. The identified economic and energy improvements provide a basis for developing more detailed mathematical models, particularly for the cooling system. A deeper economic analysis can subsequently enable a refined thermo-economic optimization of the refueling protocol. Further work could be conducted to enhance the safety aspects of SAE2601 by using constraint-based methodologies and/or barrier functions.

Author Contributions: Conceptualization, B.F.S., D.R. and D.F.M.; methodology, B.F.S., D.R. and D.F.M.; software, B.F.S., D.R. and D.F.M.; validation, B.F.S., D.R. and D.F.M.; formal analysis, B.F.S., D.R. and D.F.M.; investigation, B.F.S., D.R. and D.F.M.; resources, B.F.S., D.R. and D.F.M.; data curation, B.F.S., D.R. and D.F.M.; writing—original draft preparation, B.F.S., D.R. and D.F.M.; writing—review and editing, B.F.S., D.R. and D.F.M.; visualization, B.F.S.; supervision, D.F.M.; project administration, D.R.; funding acquisition, B.F.S., D.R. and D.F.M. All authors have read and agreed to the published version of the manuscript.

Funding: This research received no external funding.

Institutional Review Board Statement: This study did not require ethical approval.

Informed Consent Statement: Not applicable.

Data Availability Statement: Dataset available on request from the authors.

Acknowledgments: D.F.M. wishes to express his appreciation to Signe Kjelstrup for her support during the internship at the Norwegian University of Science and Technology, NTNU.

Conflicts of Interest: Author Bruno F. Santoro was employed by the company Op2B—Optimization to Business. The remaining authors declare that the research was conducted in the absence of any commercial or financial relationships that could be construed as a potential conflict of interest.

Abbreviations

The following abbreviations are used in this manuscript:

Latin	
A	Area, m^2
a	Helmholtz equation parameter, -
b	Helmholtz equation parameter, -
D	Helmholtz equation parameter, -
DAE	Differential algebraic equations
$\mathcal{K}$	Index
k	Valve flow capacity coefficient, $m^3\ h^{-1}$
h	Heat transfer coefficient, $kW\ m^{-2}\ K^{-1}$
HRS	Hydrogen refueling station
$\dot{m}(t)$	Mass flow rate, $kg\ s^{-1}$
$p(t)$	Pressure, kPa
p_c	Critical pressure, kPa
PF	Pareto frontier
ps	Pipe segment, -
$q(t)$	Specific heat flow, $kJ\ kg^{-1}$
R	Ideal gas constant, $kJ\ kg^{-1}\ K^{-1}$
$S(t)$	Entropy, $kJ\ K^{-1}$
$S_{irr}(t)$	Total entropy production, $kJ\ K^{-1}$
$s(t)$	Specific entropy, $kJ\ kg^{-1}\ K^{-1}$
SOC	State of charge
$T(t)$	Temperature, K
t	Time, s
t_i	Helmholtz equation parameter, -
T_c	Critical temperature, K
$u(t)$	Specific internal energy, $kJ\ kg^{-1}$
V	Vehicle tank volume, m^3

Subscripts

Greek

α	Helmholtz equation parameter, -
δ	Helmholtz equation parameter, -
γ	Helmholtz equation parameter, -
ϕ	Helmholtz equation parameter, -
$\rho(t)$	Density, kg m^{-3}
ρ_N	Density at normal conditions (273.15 K, 101.325 kPa), kg m^{-3}
ρ_c	Critical density, kg m^{-3}
ρ_{full}	Maximum admissible density in the vehicle tank kg m^{-3}
τ	Helmholtz equation parameter, -
f	Final
fv	Flow control valve
i	Index
iv	Inlet valve
irr	Irreversible
k	Index
w	Wall
0	Tank
32	Pipe segment 32
54	Pipe segment 54
∞	Ambient
0	Ideal gas
r	Residual
ref	Reference state
k	Index

Appendix A. Thermodynamic Properties of Hydrogen

Leachman and coworkers [17] proposed a Helmholtz equation to describe the thermodynamic properties of hydrogen of the form

$$\frac{a}{RT} = \alpha^0(\tau,\delta) + \alpha^r(\tau,\delta) \tag{A1}$$

where α^0 and α^r are the ideal gas and residual contributions to the dimensionless Helmholtz energy; both are functions of the reduced density $\delta = \rho/\rho_c$ and the inverse reduced temperature $\tau = T_c/T$.

$$\alpha^0(\tau,\delta) = \ln\delta + 1.5\ln\tau + a_1 + a_2\tau + \sum_{k=3}^{7} a_k \ln[1 - \exp(b_k\tau)] \tag{A2}$$

$$\alpha^r(\tau,\delta) = \sum_{i=1}^{7} N_i\delta^{d_i}\tau^{t_i} + \sum_{i=8}^{9} N_i\delta^{d_i}\tau^{t_i}\exp\left(-\delta^{p_i}\right) + \sum_{i=10}^{14} N_i\delta^{d_i}\tau^{t_i}\exp\left[\phi_i(\delta - D_i)^2 + \beta_i(\tau-\gamma_i)^2\right] \tag{A3}$$

Tables A1 and A2 provide the values of parameters needed in Equations (A2) and (A3).

From Equation (A1) and its derivatives, any other thermodynamic property of the system is calculated. In these contributions, it suffices to obtain a pressure–density–temperature relation (p–ρ–T), the internal energy u, the entropy s and the chemical potential, μ, of the substance.

$$\frac{p}{\rho RT} = -\frac{\rho}{RT}\left(\frac{\partial a}{\partial \rho}\right)_T \tag{A4}$$

$$\frac{u}{RT} = \frac{a}{RT} - \frac{1}{R}\left(\frac{\partial a}{\partial T}\right)_\rho \tag{A5}$$

$$\frac{s}{R} = -\frac{1}{R}\left(\frac{\partial a}{\partial T}\right)_{\rho} \tag{A6}$$

$$\frac{\mu}{RT} = \frac{a}{RT} + \frac{p}{\rho RT} \tag{A7}$$

The derivatives needed in Equations (A4)–(A6) follow the expressions:

$$\begin{aligned}\left(\frac{\partial \alpha^r}{\partial \delta}\right)_{\tau} = &\sum_{i=1}^{7} N_i d_i \delta^{d_i-1}\tau^{t_i} + \sum_{i=8}^{9} N_i \delta^{d_i}\tau^{t_i}\exp\left(-\delta^{p_i}\right)\left[\frac{d_i}{\delta} - p_i\frac{\delta^{p_i}}{\delta}\right] \\ &+ \sum_{i=10}^{14} N_i \delta_i^{d}\tau^{t_i}\left[\frac{d_i}{\delta} + 2\phi_i(\delta - D_i)\right]\exp\left[\phi_i(\delta - D_i)^2 + \beta_i(\tau-\gamma_i)^2\right]\end{aligned} \tag{A8}$$

$$\left(\frac{\partial \alpha^0}{\partial \tau}\right)_{\delta} = \frac{1.5}{\tau} + a_2 - \sum_{k=3}^{7}\frac{a_k b_k \exp\left(b_k\tau\right)}{1-\exp\left(b_k\tau\right)} \tag{A9}$$

$$\begin{aligned}\left(\frac{\partial \alpha^r}{\partial \tau}\right)_{\delta} = &\sum_{i=1}^{7} N_i \delta^{d_i} t_i \tau^{t_i-1} + \sum_{i=8}^{9} N_i \delta^{d_i} t_i \tau^{t_i-1}\exp\left(-\delta^{p_i}\right) \\ &+ \sum_{i=10}^{14} N_i \delta^{d_i}\tau^{t_i}\left[\frac{t_i}{\tau} + 2\beta_i(\tau-\gamma_i)\right]\exp\left[\phi_i(\delta - D_i)^2 + \beta_i(\tau-\gamma_i)^2\right]\end{aligned} \tag{A10}$$

Table A1. Coefficients used in α^0.

k	a_k	b_k
1	−1.4579856475	-
2	1.888076782	-
3	1.616	−16.0205159149
4	−0.4117	−22.6580178006
5	−0.792	−60.0090511389
6	0.758	−74.9434303817
7	1.217	−206.9392065168

Table A2. Coefficients used in α^r.

i	N_i	t_i	d_i	p_i	ϕ_i	β_i	γ_i	D_i
1	−6.93643	0.6844	1	-	-	-	-	-
2	0.01	1	4	-	-	-	-	-
3	2.1101	0.989	1	-	-	-	-	-
4	4.52059	0.489	1	-	-	-	-	-
5	0.732564	0.803	2	-	-	-	-	-
6	−1.34086	1.1444	2	-	-	-	-	-
7	0.130985	1.409	3	-	-	-	-	-
8	−0.777414	1.754	1	1	-	-	-	-
9	0.351944	1.311	3	1	-	-	-	-
10	−0.0211716	4.187	2	-	−1.685	−0.1710	0.7164	1.506
11	0.0226312	5.646	1	-	−0.489	−0.2245	1.3444	0.156
12	0.032187	0.791	3	-	−0.103	−0.1304	1.4517	1.736
13	−0.0231752	7.249	1	-	−2.506	−0.2785	0.7204	0.67
14	0.0557346	2.986	1	-	−1.607	−0.3967	1.5445	1.662

References

1. Genovese, M.; Blekhman, D.; Dray, M.; Fragiacomo, P. Hydrogen station in situ back-to-back fueling data for design and modeling. *J. Clean. Prod.* **2021**, *329*, 129737. [CrossRef]
2. Shell and USP Pioneer World's First Ethanol-Based Hydrogen Fuel Station. Available online: https://www.fuelsandlubes.com/shell-and-usp-pioneer-worlds-first-ethanol-based-hydrogen-fuel-station/ (accessed on 30 June 2024).

3. Hynion. Hynion and Toyota Norway Enter into a Cooperation Agreement to Drive the Infrastructure for Hydrogen in Norway. Available online: https://www.hynion.com/news/hynion-and-toyota-norway-enter-into-a-cooperation-agreement-to-drive-the-infrastructure-for-hydrogen-in-norway#:~:text=Currently%2C%20there%20is%20only%20one,station%20at%20HÃÿvik%20outside%20Oslo (accessed on 29 June 2024).
4. Reddi, K.; Elgowainy, A.; Rustagi, N.; Gupta, E. Impact of hydrogen SAE J2601 fueling methods on fueling time of light-duty fuel cell electric vehicles. *Int. J. Hydrogen Energy* **2017**, *42*, 16675–16685. [CrossRef]
5. Olmos, F.; Manousiouthakis, V.I. Hydrogen car fill-up process modeling and simulation. *Int. J. Hydrogen Energy* **2013**, *38*, 3401–3418. [CrossRef]
6. Olmos, F.; Manousiouthakis, V.I. Gas tank fill-up in globally minimum time: Theory and application to hydrogen. *Int. J. Hydrogen Energy* **2014**, 39, 12138–12157. [CrossRef]
7. Mendoza, D.; Rincon, D.; Santoro, B. Increasing energy efficiency of hydrogen refueling stations via optimal thermodynamic paths. *Int. J. Hydrogen Energy* **2024**, *50*, 1138–1151. [CrossRef]
8. Kizilova, N.; Shankar, A.; Kjelstrup, S. A Minimum Entropy Production Approach to Optimization of Tubular Chemical Reactors with Nature-Inspired Design. *Energies* **2024**, *17*, 432. [CrossRef]
9. Kingston, D.; Wilhelmsen, Ø.; Kjelstrup, S. Minimum entropy production in a distillation column for air separation described by a continuous non-equilibrium model. *Chem. Eng. Sci.* **2020**, *218*, 115539. [CrossRef]
10. Magnanelli, E.; Johannessen, E.; Kjelstrup, S. Entropy production minimization as design principle for membrane systems: Comparing equipartition results to numerical optima. *Ind. Eng. Chem. Res.* **2017**, *56*, 4856–4866. [CrossRef]
11. Johannessen, E.; Nummedal, L.; Kjelstrup, S. Minimizing the entropy production in heat exchange. *Int. J. Heat Mass Transf.* **2002**, *45*, 2649–2654. [CrossRef]
12. Johannessen, E.; Kjelstrup, S. Minimum entropy production rate in plug flow reactors: An optimal control problem solved for SO2 oxidation. *Energy* **2004**, *29*, 2403–2423. [CrossRef]
13. Nummedal, L.; Kjelstrup, S.; Costea, M. Minimizing the entropy production rate of an exothermic reactor with a constant heat-transfer coefficient: The ammonia reaction. *Ind. Eng. Chem. Res.* **2003**, *42*, 1044–1056. [CrossRef]
14. Kjelstrup, S.; Johannessen, E.; Rosjorde, A.; Nummedal, L.; Bedeaux, D. Minimizing the entropy production of the methanol producing reaction in a methanol reactor. *Int. J. Thermodyn.* **2000**, *3*, 147–153.
15. Røsjorde, A.; Kjelstrup, S.; Johannessen, E.; Hansen, R. Minimizing the entropy production in a chemical process for dehydrogenation of propane. *Energy* **2007**, *4*, 335–343. [CrossRef]
16. Elgowainy, A.; Reddi, K.; Lee, D.-Y.; Rustagi, N.; Gupta, E. Techno-economic and thermodynamic analysis of pre-cooling systems at gaseous hydrogen refueling stations. *Int. J. Hydrogen Energy* **2017**, *42*, 29067–29079. [CrossRef]
17. Leachman, J.W.; Jacobsen, R.T.; Penoncello, S.G.; Lemmon, E.W. Fundamental Equations of State for Parahydrogen, Normal Hydrogen, and Orthohydrogen. *J. Phys. Chem. Ref. Data* **2009**, *38*, 721–748. [CrossRef]
18. Orange County, California Electricity Rates & Statistics. Available online: https://findenergy.com/ca/orange-county-electricity/ (accessed on 9 July 2024).
19. Bynum, M.L.; Hackebeil, G.A.; Hart, W.E.; Laird, C.D.; Nicholson, B.L.; Siirola, J.D.; Watson, J.P.; Woodruff, D.L. *Pyomo-Optimization Modeling in Python*, 3rd ed.; Springer: Berlin/Heidelberg, Germany, 2021.
20. Nicholson, B.; Siirola, J.D.; Watson, J.P.; Zavala, V.M.; Biegler, L.T. pyomo.dae: A modeling and automatic discretization framework for optimization with differential and algebraic equations. *Math. Program. Comput.* **2018**, *10*, 187e223. [CrossRef]
21. Eißler, T.; Schumacher, G.; Behrens, J.; Mendler, F.; Voglstätter, C. Detailed Cost Analysis of Hydrogen Refueling Costs for Fleets. *Chem. Ing. Tech.* **2024**, *96*, 86–99. [CrossRef]
22. Simunović, J.; Pivac, I.; Barbir, F. Techno-economic assessment of hydrogen refueling station: A case study in Croatia. *Int. J. Hydrogen Energy* **2022**, *47*, 24155–24168. [CrossRef]

entropy

MDPI

Article

On the Elimination of Fast Variables from the Langevin Equation

Dick Bedeaux

PoreLab, Department of Chemistry, Norwegian University of Science and Technology, NO-7491 Trondheim, Norway; dick.bedeaux@ntnu.no

Abstract: In a multivariable system, there are usually a number of relaxation times. When some of the relaxation times are shorter than others, the corresponding variables will decay to their equilibrium value faster than the others. After the fast variables have decayed, the system can be described with a smaller number of variables. From the theory of nonequilibrium thermodynamics, as formulated by Onsager, we know that the coefficients in the linear flux–force relations satisfy the so-called Onsager symmetry relations. The question we will address in this paper is how to eliminate the fast variables in such a way that the coefficients in the reduced description for the slow variables still satisfy the Onsager relations. As the proof that Onsager gave of the symmetry relations does not depend on the choice of the variables, it is equally valid for the subset of slow variables. Elimination procedures that lead to symmetry breaking are possible, but do not consider systems that satisfy the laws of nonequilibrium thermodynamics.

Keywords: Langevin equation; fluctuation–dissipation theorems; fast and slow variables; Onsager symmetry relations; correlation functions

1. Introduction

In 1931, Onsager formulated [1–4] the classical theory of nonequilibrium thermodynamics. He showed that the coefficients in the linear flux–force relations for a multivariable system satisfy the so-called Onsager symmetry relations. When one describes the system with the Langevin equation, one adds random fluxes, the averages of which are zero. The correlation functions of the random fluxes are given by fluctuation–dissipation theorems [5–7]. These correlation functions are also symmetric for the interchange of the random fluxes.

When some of the variables decay to their equilibrium value faster than the others, one may after this decay describe the system with a smaller number of slow variables. An example of this is a multi-component fluid. The pressure distribution is the first to reach equilibrium. After this, the fluid can be treated as incompressible. Subsequently, the relaxation of the temperature is usually faster than the diffusion of the components. The advantage is that one may simplify the description to a smaller number of variables on longer time scales.

There are many ways in which one may eliminate fast variables. The proper methodology depends on the particular system considered. Some authors [8–11] claim that in the reduced description, the symmetry relations are no longer valid. For the systems they consider, and with their methodology to reduce the description, this is indeed what they find. Van Kampen [9] gives a very clear explanation of how to eliminate fast variables. In the systems, he considers that it is possible to identify a small parameter. The solution is then expanded in this parameter. The reduced description may no longer satisfy Onsager symmetry [8,9].

The unfortunate (or fortunate) fact is that the proof that Onsager gave of the symmetry relations does not depend on the choice of the variables, and it is equally valid for the subset of slow variables as for the full set. It is therefore clear that the systems some authors [8–11] consider, after eliminating the fast variables, no longer belong to the well established field of nonequilibrium thermodynamics. It is known that in the non-linear case the symmetry relations are no longer valid [12], but that is not the issue now.

The question addressed in this paper is how to reduce the description to slow variables such that the Onsager coefficients and the corresponding correlation functions still satisfy the symmetry relations. In the Section 2, I discuss the Langevin equation for n variables. It contains a $n \times n$ matrix of linear coefficients and n thermal random forces. The distribution of the variables around their equilibrium value is given. The average of the random forces is zero. Their auto-correlation functions are given by fluctuation–dissipation theorems in terms of the symmetric Onsager coefficients. In the Section 3, I discuss how to distinguish the fast and the slow variables. The equilibrium distribution of the slow variables is discussed in the Section 4. The derivation of the Langevin equation for the slow variables, including the verification of the symmetry of the Onsager coefficients and the correlation functions, is given in Section 5. The choice of physical variables is shortly discussed in Section 6, and this ends with a conclusion.

2. The Langevin Equation

Consider a system described by the variables $\mathbf{a} = (a_1, ..., a_n)$. These variables are chosen such that their equilibrium value is equal to zero, $\mathbf{a}_{eq} = 0$. I restrict myself to the case that the variables are either all symmetric or all asymmetric for time reversal. Given the entropy production, one has a set of conjugate fluxes and forces. Choosing all variables odd or even is equivalent to using a mixed set. As such, the choice of all even or all odd is no restriction. The velocities of a collection of Brownian particles are a good example of a set of odd variables. The extension to have mixed variables that are both symmetric and asymmetric for time reversal is straightforward [13]. The time dependence is given by the Langevin equation [13]:

$$\frac{d\mathbf{a}(t)}{dt} = -\mathbf{L}.\mathbf{a}(t) + \mathbf{f}(t) \tag{1}$$

Here, $\mathbf{L}$ is a $n \times n$ matrix of linear coefficients and $\mathbf{f}$ is the thermal random force, which is Gaussian and white. The average of the random force is zero

$$\langle \mathbf{f}(t) \rangle = 0 \tag{2}$$

The correlations are given by

$$\langle \mathbf{f}(t)\mathbf{f}(t') \rangle = 2\ell\delta(t - t') \tag{3}$$

where ℓ is the $n \times n$ Onsager matrix. Given the Gaussian nature of the random force, one may also calculate the higher order correlation functions. The Onsager relations imply that the matrix ℓ is symmetric.

The equilibrium distribution is given by

$$P_{eq}(\mathbf{a}) = \exp(S(\mathbf{a})/k_\mathrm{B}) / \int d\mathbf{a}' \exp(S(\mathbf{a}')/k_\mathrm{B}) \tag{4}$$

where $S(\mathbf{a})$ is the entropy and k_B is Boltzmann's constant. Expanding the entropy to second order in $\mathbf{a}$, one has

$$S(\mathbf{a}) = S_{eq} - \frac{1}{2}k_\mathrm{B}\mathbf{a}.\mathbf{g}.\mathbf{a} \tag{5}$$

where it is used so that the entropy has a maximum in equilibrium. The matrix $\mathbf{g}$ is symmetric and is given by

$$\mathbf{g} = -\frac{1}{k_\mathrm{B}} \frac{\partial^2}{\partial \mathbf{a}^2} S(\mathbf{a}) \tag{6}$$

Equation (5) together with Equation (4) gives

$$P_{eq}(\mathbf{a}) = (2\pi)^{-n/2} \sqrt{\det \mathbf{g}} \exp\left(-\frac{1}{2}\mathbf{a}.\mathbf{g}.\mathbf{a}\right) \tag{7}$$

This gives
$$\langle \mathbf{a} \rangle_{eq} = 0 \tag{8}$$
and
$$\langle \mathbf{aa} \rangle_{eq} = \mathbf{g}^{-1} \tag{9}$$
Given the Gaussian nature of P_{eq}, one may also calculate the higher order correlation functions.

The random force in the Langevin equation must have the appropriate strength in order to generate the proper size of the fluctuations around the equilibrium. This is found to imply that
$$\mathbf{L} = \ell.\mathbf{g} \quad \text{or} \quad \ell = \mathbf{L}.\mathbf{g}^{-1} \tag{10}$$
It should be realised that, while both ℓ and $\mathbf{g}$ are symmetric, $\mathbf{L}$ is not necessarily symmetric. One has
$$\mathbf{L}^T = \mathbf{g}.\ell \tag{11}$$
for the transposed matrix. The superscript T indicates the transpose of a matrix.

The solution of the Langevin Equation (1) may be written in the following form
$$\mathbf{a}(t) = \exp(-\mathbf{L}t).\mathbf{a}(0) + \int_0^t dt' \exp(-\mathbf{L}(t-t')).\mathbf{f}(t') \tag{12}$$
where $\mathbf{a}(0)$ is the value of $\mathbf{a}$ at time zero. The average of this equation gives
$$\langle \mathbf{a}(t) \rangle = \exp(-\mathbf{L}t).\mathbf{a}(0) \tag{13}$$
The thermal fluctuations around the average are
$$\delta\mathbf{a}(t) \equiv \mathbf{a}(t) - \langle \mathbf{a}(t) \rangle = \int_0^t dt' \exp(-\mathbf{L}(t-t')).\mathbf{f}(t') \tag{14}$$
The average of this fluctuation is by definition zero. The correlation function is found to be
$$\langle \delta\mathbf{a}(t)\delta\mathbf{a}(t') \rangle = \exp(-\mathbf{L}(t-t')).\mathbf{g}^{-1} = \mathbf{g}^{-1}.\exp(-\mathbf{L}^T(t-t')) \tag{15}$$
The second equality follows by expanding the exponents together with Equations (10) and (11). Comparing Equations (13) and (15) one sees that, as is usual, the relaxation of the correlation function and of the average are directly related.

In order to obtain more information about the relaxation times, we calculate the eigenvalues
$$\mathbf{L}.\mathbf{e}^r_\nu = \lambda_\nu \mathbf{e}^r_\nu \quad \text{and} \quad \mathbf{e}^l_\nu.\mathbf{L} = \lambda_\nu \mathbf{e}^l_\nu \tag{16}$$
where $\mathbf{e}^r_\nu$ and $\mathbf{e}^l_\nu$ are the right and the left eigenvectors, respectively. These eigenvectors satisfy the following orthogonality condition:
$$\mathbf{e}^l_\nu.\mathbf{e}^r_{\nu'} = \delta_{\nu\nu'} \tag{17}$$
They furthermore satisfy the following completeness relations:
$$\sum_{\nu=1}^n \mathbf{e}^r_\nu \mathbf{e}^l_\nu = \mathbf{1} \quad \text{and} \quad \sum_{\nu=1}^n \mathbf{e}^l_\nu \mathbf{e}^r_\nu = \mathbf{1} \tag{18}$$
The two forms of the completeness relation follow on from each other by taking the transpose of the matrix.

3. Fast and Slow Variables

For convenience, the eigenvalues are ordered such that $\text{Re}\lambda_1$ is the smallest one, $\text{Re}\lambda_2$ is the next smallest one, etc., until $\text{Re}\lambda_n$, which is the largest one. The real part of the first eigenvalue gives the inverse relaxation time for the slowest decay, while the real part of the last eigenvalue gives the inverse relaxation time for the fastest decay. I want to consider in particular a case where a subset of eigenvalues is much smaller than the rest of the eigenvalues. One may then define the time τ such that

$$\tau \mathrm{Re}\lambda_\nu \ll 1 \quad \text{for } \nu \leq m \quad \text{and} \quad \tau \mathrm{Re}\lambda_\nu \gg 1 \quad \text{for } \nu > m \tag{19}$$

When $t \gg \tau$, the fast decaying modes are equilibrated. Only m independent slow modes are then sufficient to describe the system for $t \gg \tau$. It is this reduction to a smaller subset of variables which will be studied. Given the validity of Equation (19), one may choose $\tau = 10/\mathrm{Re}\lambda_m$.

For large enough times one has

$$\langle a_\nu(t)\rangle \equiv \mathbf{e}^l_\nu.\langle \mathbf{a}(t)\rangle = 0 \quad \text{for} \quad t \gg \tau \quad \text{and} \quad \nu > m \tag{20}$$

This gives $(n-m)$ relations between the average variables for $t \gg \tau$. For the correlation functions, one similarly finds

$$\begin{aligned} \langle \delta a_\nu(t)\delta\mathbf{a}(t')\rangle &= \langle \delta\mathbf{a}(t)\delta a_\nu(t')\rangle = \\ \mathbf{e}^l_\nu.\langle \delta\mathbf{a}(t)\delta\mathbf{a}(t')\rangle &= \langle \delta\mathbf{a}(t)\delta\mathbf{a}(t')\rangle.\mathbf{e}^l_\nu = 0 \quad \text{for} \quad |t-t'| \gg \tau \quad \text{and} \quad \nu > m \end{aligned} \tag{21}$$

The equalities follow from the symmetry of the correlation function. Equation (21) implies that for $|t-t'| \gg \tau$, only a $m \times m$ submatrix remains for the m remaining independent variables.

After the fast modes have decayed, the remaining variables and the corresponding correlation matrix appear to decay from a different initial condition given by

$$\langle \mathbf{a}(\tau=0)\rangle = \sum_{\nu=1}^{m} \mathbf{e}^r_\nu \mathbf{e}^l_\nu.\mathbf{a}(0) \tag{22}$$

for the average and

$$\langle \delta\mathbf{a}(t)\delta\mathbf{a}(t-\tau)\rangle_{\tau=0} = \sum_{\nu,\nu'=1}^{m} \mathbf{e}^r_\nu \mathbf{e}^l_\nu.\mathbf{g}^{-1}.\mathbf{e}^l_{\nu'}\mathbf{e}^r_{\nu'} \tag{23}$$

for the correlation matrix. While the modified initial condition for the average seems rather obvious, it has been considered to be surprising for the correlation function [9–13]. Let me illustrate this by an example. Consider the velocity distribution of a Brownian particle suspended in a fluid. From ensemble theory, it follows that the distribution of the velocity $\mathbf{v}$ is given by the Maxwell distribution

$$P_{eq}(\mathbf{v}) = (2\pi M k_B T)^{-3/2} \exp\left(-\frac{v^2}{2Mk_B T}\right) \tag{24}$$

where M is the mass of the Brownian particle. It is known that the compression modes of the surrounding fluid are fast. Consider a sound velocity of 1000 m.s^{-1} and a particle size of 1 micron; this defines a time scale of $\tau = 10^{-9}$ s. When one measures the velocity of the particle with a time resolution large compared to τ, one sees a particle which carries along a vortex with fluid. On this time scale, it instantaneously adjusts its velocity to the velocity of the particle. The resulting modified initial velocity distribution is then

$$P_{eq}(\mathbf{v}) = \left[2\pi\left(M + \frac{1}{2}M_{fl}\right)k_B T\right]^{-3/2} \exp\left[-\frac{v^2}{2\left(M+\frac{1}{2}M_{fl}\right)k_B T}\right] \tag{25}$$

where M_{fl} is the mass of the displaced fluid. This result is now generally accepted, but for a while it was controversial [14–18].

4. The Reduced Equilibrium Distribution of the Slow Amplitudes

Define the projection on the left eigenvectors by

$$a_\nu(t) \equiv \mathbf{e}^l_\nu.\mathbf{a}(t) \quad \text{for} \quad \nu = 1, ..., n \tag{26}$$

These amplitudes will be chosen as an alternative set of variables. For $t \gg \tau$, only the first m of these variables remain. The rest have equilibrated. The reduced equilibrium distribution for the slow variables is given by

$$\begin{aligned} P^{slow}_{eq}(a_\nu; \nu &= 1, ..., m) \equiv \int (\prod_{\nu=m+1}^{n} da_\nu) P_{eq}(\mathbf{a}) \\ &= (2\pi)^{-n/2}(\det \mathbf{g})^{1/2} \int (\prod_{\nu=m+1}^{n} da_\nu) \exp(-\frac{1}{2}\mathbf{a}.\mathbf{g}.\mathbf{a}) \\ &= (2\pi)^{-n/2}(\det \mathbf{g})^{1/2} \int (\prod_{\nu=m+1}^{n} da_\nu) \exp(-\frac{1}{2}\sum_{\nu',\nu''=1}^{n} a_{\nu'} g_{\nu'\nu''} a_{\nu''}) \end{aligned} \tag{27}$$

where

$$g_{\nu'\nu''} \equiv \mathbf{e}^l_{\nu'}.\mathbf{g}.\mathbf{e}^l_{\nu''} \tag{28}$$

From the nature of the integration, it is clear that the reduced equilibrium distribution is again Gaussian

$$P^{slow}_{eq}(a_\nu; \nu = 1, ..., m) = (2\pi)^{-m/2}(\det \mathbf{g}^{slow})^{1/2} \exp(-\frac{1}{2}\sum_{\nu',\nu''=1}^{m} a_{\nu'} g^{slow}_{\nu'\nu''} a_{\nu''}) \tag{29}$$

where $\mathbf{g}^{slow}$ is now an $m \times m$ matrix.

There are two methods to find this reduced matrix. The first and more simple one is that the equilibrium correlations in the reduced description should be given by the right hand side of Equation (23) for the correlation function. This results in

$$\left[(\mathbf{g}^{slow})^{-1}\right]_{\nu\nu'} = \left[\mathbf{g}^{-1}\right]_{\nu\nu'} \quad \text{for} \quad \nu, \nu' = 1, ..., m \tag{30}$$

The inverse of the reduced matrix is therefore equal to the $m \times m$ submatrix for the slow eigenvectors of the inverse of the original $\mathbf{g}$ matrix.

In order to verify that this result is consistent with the straightforward integration of Equation (27), I also verify Equation (30) via direct integration of Equation (27). Write

$$P^{slow}_{eq}(A_\nu; \nu = 1, ..., m) = (2\pi)^{-n/2}(\det \mathbf{g})^{1/2} \int d\mathbf{a} (\prod_{\nu=1}^{m} \delta(A_\nu - a_\nu) \exp(-\frac{1}{2}\mathbf{a}.\mathbf{g}.\mathbf{a})) \tag{31}$$

Next, use $\delta(x) = \frac{1}{2\pi}\int dk \exp(ikx)$ and write

$$P^{slow}_{eq}(A_\nu; \nu = 1, ..., m) = (2\pi)^{-m-n/2}(\det \mathbf{g})^{1/2} \int d^m\mathbf{k} \int d\mathbf{a} \exp(i\mathbf{k}.(\mathbf{a} - \mathbf{A}) - \frac{1}{2}\mathbf{a}.\mathbf{g}.\mathbf{a}) \tag{32}$$

where $\mathbf{k} \equiv (k_1, ..., k_m, 0, ..., 0)$ and $\mathbf{A} \equiv (A_1, ..., A_m, 0, ..., 0)$. The expression in the exponent may then be written as

$$i\mathbf{k}.(\mathbf{a} - \mathbf{A}) - \frac{1}{2}\mathbf{a}.\mathbf{g}.\mathbf{a} = -\frac{1}{2}(\mathbf{a} - i\mathbf{g}^{-1}.\mathbf{k}).\mathbf{g}.(\mathbf{a} - i\mathbf{g}^{-1}.\mathbf{k}) + i\mathbf{k}.\mathbf{A} - \frac{1}{2}\mathbf{k}.\mathbf{g}^{-1}.\mathbf{k}$$

Upon substitution into Equation (32), one may perform the integral over $\mathbf{a}$, which is now a simple Gaussian integral, and one obtains

$$P_{eq}^{slow}(A_\nu;\nu=1,...,m) = (2\pi)^{-m}\int d^m\mathbf{k}\exp(i\mathbf{k}.\mathbf{A}-\frac{1}{2}\mathbf{k}.\mathbf{g}^{-1}.\mathbf{k}) \tag{33}$$

The vectors $\mathbf{k}$ and $\mathbf{A}$ may now be reduced to their m dimensional form. Notice that the exponent now contains the $m\times m$ dimensional submatrix of $\mathbf{g}^{-1}$. This $m\times m$ dimensional submatrix is by definition equal to $(\mathbf{g}^{slow})^{-1}$. Writing

$$i\mathbf{k}.\mathbf{A}-\frac{1}{2}\mathbf{k}.(\mathbf{g}^{slow})^{-1}.\mathbf{k} = -\frac{1}{2}(\mathbf{k}-i\mathbf{g}^{slow}.\mathbf{A}).(\mathbf{g}^{slow})^{-1}.(\mathbf{k}-i\mathbf{g}^{slow}.\mathbf{A})-\frac{1}{2}\mathbf{A}.\mathbf{g}^{slow}.\mathbf{A}$$

one obtains upon substitution into Equation (33) and integration over $\mathbf{k}$

$$P_{eq}^{slow}(\mathbf{A}) = (2\pi)^{-m/2}(\det\mathbf{g}^{slow})^{1/2}\exp(-\frac{1}{2}\mathbf{A}.\mathbf{g}^{slow}.\mathbf{A}) \tag{34}$$

This is equivalent to Equation (29).

The importance of the analysis in the last § is that the equilibrium distribution of the slow variables in fact follows directly from the equilibrium ensemble. Whether one directly reduces the microscopic description to these slow variables or does it via an intermediate macroscopic description using a larger number of variables, some of which are fast compared to the final slow variables, this method leads to exactly the same "initial condition" of the correlation functions of the slow variables. This is an important step to check the internal consistency of the procedure.

5. The Langevin Equation for the Slow Amplitudes

Now, turn to the derivation of the Langevin equation for the slow variables

$$\frac{d}{dt}a_\nu(t) = -\sum_{\nu'=1}^{m} L_{\nu\nu'}^{slow}a_{\nu'}(t)+f_\nu^{slow}(t) \quad \text{for} \quad \nu=1,...,m \tag{35}$$

The most convenient method to obtain this equation is to first consider the average of this equation

$$\frac{d}{dt}\langle a_\nu(t)\rangle = -\sum_{\nu'=1}^{m} L_{\nu\nu'}^{slow}\langle a_{\nu'}(t)\rangle \quad \text{for} \quad \nu=1,...,m \tag{36}$$

where the fact that the average of the reduced random force is also zero is used

$$\left\langle f_\nu^{slow}(t)\right\rangle = 0 \tag{37}$$

Using Equation (20), it follows that

$$L_{\nu\nu'}^{slow} = \mathbf{e}_\nu^l.\mathbf{L}.\mathbf{e}_{\nu'}^r = \lambda_\nu\delta_{\nu\nu'} \quad \text{for} \quad \nu,\nu'=1,...,m \tag{38}$$

It then follows that

$$\left\langle f_\nu^{slow}(t)f_{\nu'}^{slow}(t')\right\rangle = 2\ell_{\nu\nu'}^{slow}\delta(t-t') \tag{39}$$

with

$$L_{\nu\nu'}^{slow} = \sum_{\nu''=1}^{m}\ell_{\nu\nu''}^{slow}g_{\nu''\nu'}^{slow} \quad \text{or} \quad \ell_{\nu\nu'}^{slow} = \sum_{\nu''=1}^{m}L_{\nu\nu''}^{slow}\left[\left(\mathbf{g}^{slow}\right)^{-1}\right]_{\nu''\nu'} \tag{40}$$

With Equations (30) and (38), the last equality reduces to m of the original variables. Choose for this purpose a subset which remains independent for $t\gg\tau$. Relabel the original variables such that this subset consists of the first m variables. Define

$$\ell_{\nu\nu'}^{slow} = \sum_{\nu''=1}^{m}\mathbf{e}_\nu^l.\mathbf{L}.\mathbf{e}_{\nu''}^{\mathbf{r}}\mathbf{e}_{\nu''}^l.\mathbf{g}^{-1}.\mathbf{e}_{\nu'}^{\mathbf{r}} = \lambda_\nu\mathbf{e}_\nu^l.\mathbf{g}^{-1}.\mathbf{e}_{\nu'}^{\mathbf{r}} \tag{41}$$

In view of the fact that the matrix g^{-1} is symmetric, it follows that the reduced Onsager matrix in Equation (41) is symmetric.

6. A Reduced Description Using the Original Variables

An aspect of the use of the slow amplitudes is that they may be unpleasant combinations of the original variables. One may alternatively use a subset

$$\mathbf{a}^{red}(t) \equiv (a_1(t), ..., a_m(t)) \tag{42}$$

When $t \gg \tau$, this reduced set of variables is a linear combination of the amplitudes of the slow eigenfunctions. This implies that the use of the reduced set corresponds to the choice of a new basis set in the slow variable space. The Langevin equation is valid for all choices of the basis, and therefore also for $\mathbf{a}^{red}$ using $\mathbf{L}^{red}$, ℓ^{red} and $\mathbf{g}^{red}$.

7. Conclusions

After this long analysis, one may draw some conclusions:

(i) At any time scale, one has a natural choice of variables for which the relaxation from an initial state and the fluctuations are described in the usual way.

(ii) If one reduces a description by elimination of fast variables, both the linear (Onsager) coefficients and the equilibrium distribution change their values.

(iii) The reduced Onsager matrix ℓ^{red} and the correlation matrix $\mathbf{g}^{red}$ are not trivially obtained from ℓ and $\mathbf{g}$. For this, one needs the eigenvectors of $\mathbf{L} = \ell.\mathbf{g}$.

(iv) The Onsager matrix ℓ^{red} for the slow variables and the correlation matrix $\mathbf{g}^{red}$ are symmetric.

(v) Except for the case where one can conduct experiments on the fast and the slow time scale, there is no need to carry out this reduction. One simply uses a set of independent variables which have not relaxed and constructs $\mathbf{L}^{red}$, ℓ^{red} and $\mathbf{g}^{red}$ on the basis of physical arguments alone.

(vi) The initial state, equilibrium distribution and entropy depend on the time scale one considers. On each time scale, it will be such that processes that decay on a faster time scale have equilibrated when the experiment or simulation is started.

(vii) The initial state, equilibrium distribution and entropy for the reduced description cannot be found by substituting the equilibrium values of the fast variables in the more extended description.

To sum up, I have shown that the reduction in slow variables does not affect the structure of the equations we use. It does not affect the validity of the Onsager relations and the symmetry of the correlation functions.

Finally, notice that a coordinate transformation does not change the eigenvectors and eigenvalues of a matrix. Therefore, $\mathbf{L}^{red}$ and $\mathbf{L}^{slow}$ have the same eigenvectors and eigenvalues. Accordingly, one finds that all the relaxation times remain the same.

Funding: This research was funded by the Research Council of Norway grant number 262644 PoreLab.

Data Availability Statement: Data are contained within the article.

Acknowledgments: This paper is dedicated to my collaborator and wife Signe Kjelstrup on the occasion of her 75th birthday. I am grateful to the Center of Excellence Funding Scheme, project no. 262644 PoreLab from the Research Council of Norway.

Conflicts of Interest: The authors declare no conflict of interest

References

1. Onsager, L. Reciprocal relations in irreversible processes. I. *Phys. Rev.* **1931**, *37*, 405. [CrossRef]
2. Onsager, L. Reciprocal relations in irreversible processes. II. *Phys. Rev.* **1931**, *38*, 2265. [CrossRef]
3. Casimir, H.B.G. On Onsager's principle of microscopic reversibility. *Rev. Mod. Phys.* **1945**, *17*, 343. [CrossRef]
4. Onsager, L.; Machlup, S. Fluctuations and irreversible processes. *Phys. Rev.* **1953**, *91*, 1505. [CrossRef]
5. Callen, H.B.; Welton, T.A. Irreversibility and generalized noise. *Phys. Rev.* **1951**, *83*, 34. [CrossRef]

6. Green, M.S. Markoff random processes and the statistical mechanics of time-dependent phenomena. II. Irreversible processes in fluids. *J. Chem. Phys.* **1954**, *22*, 398–413. [CrossRef]
7. Kubo, R. The fluctuation-dissipation theorem. *Rep. Prog. Phys.* **1966**, *29*, 255. [CrossRef]
8. Geigenmüller, U.; Titulaer, U.M.; Felderhof, B.U. Systematic elimination of fast variables in linear systems. *Phys. A Stat. Mech. Appl.* **1983**, *119*, 41–52. [CrossRef]
9. van Kampen, N.G. Elimination of fast variables. *Phys. Rep.* **1985**, *124*, 69–160. [CrossRef]
10. Hubmer, G.F.; Titulaer, U.M. The Onsager-Casimir relations revisited. *J. Stat. Phys.* **1987**, *49*, 331–346. [CrossRef]
11. Titulaer, U.M. On the microscopic background of the Onsager-Casimir reciprocity relations. *J. Stat. Phys.* **1988**, *53*, 61–75. [CrossRef]
12. Cimmelli, V.A.; Sellitto, A.; Jou, D. A nonlinear thermodynamic model for a breakdown of the Onsager symmetry and the efficiency of thermoelectric conversion in nanowires. *Proc. Roy. Soc.* **2014**, *470*, 0265. [CrossRef]
13. de Groot, S.R.; Mazur, P. *Nonequilibrium Thermodynamics*; Amsterdam, The Netherlands, 1962.
14. Hynes, J.T. On Hydrodynamic Models for Brownian Motion. *J. Chem. Phys.* **1972**, *57*, 3497. [CrossRef]
15. Burgess, R.E. Brownian motion and the equipartition theorem. *Phys. Lett. A* **1973**, *42*, 395–396. [CrossRef]
16. Widom, A. Velocity fluctuations of a hard-core Brownian particle. *Phys. Rev. A* **1971**, *3*, 1394. [CrossRef]
17. Mazo, R.M. Theory of Brownian motion. IV. A hydrodynamic model for the friction factor. *J. Chem. Phys.* **1971**, *54*, 3712–3713. [CrossRef]
18. Chow, T.S.; Hermans, J.J. Limitations of the Langevin equation based on hydrodynamic models. *Proc. Koninkl. Nederl. Akad. Wet. Ser. B* **1974**, *77*, 18–25.

entropy

MDPI

Article

Enhanced Model for the Analysis of Thermoelectric Effects at Nanoscale: Onsager's Method and Liu's Technique in Comparison

Maria Di Domenico *,† **and Antonio Sellitto** †

Department of Industrial Engineering, University of Salerno, 84084 Fisciano, Italy; asellitto@unisa.it
* Correspondence: madidomenico@unisa.it; Tel.: +39-327-009-7226
† These authors contributed equally to this work.

Abstract: The aim of this paper is twofold. From the practical point of view, an enhanced model for the description of thermoelectric effects at nanoscale is proposed. From the theoretical point of view, instead, in the particular case of the proposed model, the equivalence between two classical techniques for the exploitation of the second law of thermodynamics is shown, i.e., Onsager's method and Liu's technique. An analysis of the heat-wave propagation is performed as well.

Keywords: Liu's technique; Onsager's method; second law of thermodynamics; thermoelectric effects; heat waves

1. Introduction

Physics is the science which targets the study of all natural phenomena, namely, all the events that can be correctly described by means of suitable quantities displaying only clear meanings in such a way that they should be always quantifiable and/or measurable, in principle. The final goal of physics (which is not only merely academic), in particular, is the achievement of the laws (if they truly exist) which may be used to predict the evolutions (in time and in space) of the aforementioned occurrences.

Physical phenomena, indeed, may be also coupled in some circumstances: this is the case, for example, for thermoelectricity, wherein the thermal and electric phenomena are strictly related in such a way that heat can be converted into electricity, and vice versa. In this case, as well as in many others, one necessarily has to therefore couple different laws in order to obtain the so-called theoretical model, i.e., the field equations which should be employed to carry out the aimed predictions.

Generally speaking, in a theoretical model, the field equations may be made by the following three sets of equations:

1. Balance laws, i.e., the statements of the causes of the change that an extensive basic quantity (such as mass, energy, momentum, entropy and so on) undergoes over a certain domain;
2. Evolution equations, i.e., the statements of the causes of the change of intensive quantities (e.g., internal variables), and/or of *extended* physical quantities (the densities of the fluxes of the momentum and the energy, for example);
3. Constitutive equations which ensure the closure of the problem from the mathematical point of view.

The balance laws have their own clear physical roots. The same statement, however, does not always hold either for the evolution equations, or for the constitutive equations. To this end, we note that the recent decades have particularly been the witnesses of the interest in going beyond some classical formulations and/or theories of Continuum Mechanics; for example, Fourier's law for heat conduction, Fick's law for diffusion, and Newton's law for viscous flow. Among the several reasons for seeking the generalizations of those laws, it is surely important to cite the well-known theoretical consequence that the introduction of the aforementioned relations into the balance equations leads to parabolic partial–differential equations, which therefore imply that the perturbations propagate with infinite speeds. This behavior is obviously at odds with the experimental evidence, and it is also disturbing from a theoretical point of view, because collective molecular effects should be expected to propagate at finite velocity, not only in a relativistic framework, but even from a non-relativistic point of view. It is currently understood that this unpleasant property can be avoided by taking into account the finite non-vanishing relaxation time of the respective fluxes (e.g., the heat flux, the diffusion flux, the momentum flux) which are sometimes also generically called dissipative fluxes. By looking at this problem from a pure mathematical point of view, one may note that a theoretical model should ultimately display a hyperbolic structure so that the actions at distance are precluded, and a finite speed of the propagation of disturbances is guaranteed.

There is also another important reason for the surge of modern theories. One should be aware, in fact, that classical formulations of Continuum Mechanics usually refer to situations which are very close to the thermodynamic equilibrium of the system. The raise of nano-technologies and their wide employments, for example, require the correct understanding of situations which are usually *very* far from equilibrium, i.e., situations wherein some well-known classical laws (as those quoted above) break down. Nano-technologies have been a real breakthrough; in the case of thermoelectric effects, for example, several research groups have recently confirmed that both one-dimensional (1D) and two-dimensional (2D) materials significantly exhibit better performances than those of the usual three-dimensional (3D) materials [1–4]. In this regard, in the literature, several interesting theoretical models for the correct description of thermoelectric effects at nano-scale appeared in recent years [5–13]. Since there could naturally be some discussions and concerns about the theoretical models which involve evolution and/or constitutive equations beyond the classical ones, asking what the right benchmark for them would be is a very good question. Like all the good questions, unfortunately, it does not have a unique answer.

In developing a new theoretical model, the exploitation of the second law should be (one of) the first steps to take: if the model agrees with it, in fact, then one is sure that all the mathematical solutions of the field equations (i.e., the aimed predictions) are physically acceptable [14], in principle.

In the literature, there are several techniques for testing the agreement with the second law [15–20]. Roughly speaking, the main differences among the different techniques for the exploitation of the second law lie in the starting point and the consequent final goal. In the case of Onsager's method [15,16], for example, one initially assumes suitable constitutive relations both for the specific entropy, and for its flux: the final goal of this method, in fact, is the derivation of the (a priori unknown)evolution equations. In the case of the Coleman–Noll procedure [17] and Liu's technique [18], as well as in their gradient extensions [19,20], however, the starting point is usually the form of the evolution equations, and the final goals are the constitutive relations for the fluxes and/or the specific entropy.

Whether those techniques are equivalent, or not, is another interesting theoretical question. We incidentally note that in Ref. [21] the authors proved that the procedures proposed by Coleman and Noll (see Ref. [17]) and by Liu (see Ref. [18]), in the case of a rigid body undergoing a temperature gradient, are equivalent in essence: they, in fact, bring about the same restrictions on the constitutive relations.

In the present paper, by assuming that the state space is

$$\mathcal{Z} \equiv \left\{ z; u; \mathbf{q}; \overset{\circ}{\mathbf{Q}}; Q; \mathbf{i} \right\} \tag{1}$$

(refer to the Appendix A for the physical meanings of the different quantities involved), from the practical point of view, we aim to show the agreement between the following theoretical model for the analysis of thermoelectric effects at nanoscale in a rigid and isotropic body Ω

$$\partial_t z = -\nabla \cdot \mathbf{i} \qquad \forall \mathbf{X} \in \Omega, \forall t \geq 0 \tag{2a}$$

$$\partial_t u = -\nabla \cdot \mathbf{q} + \mathbf{i} \cdot \mathbf{E} \qquad \forall \mathbf{X} \in \Omega, \forall t \geq 0 \tag{2b}$$

$$\partial_t \mathbf{q} = -\frac{\mathbf{q}}{\tau} - \left(\frac{\lambda \mathbf{I}}{\tau} - \frac{2\mathbf{Q}}{\theta}\right) \cdot \nabla\theta - \nabla \cdot \overset{\circ}{\mathbf{Q}} - \nabla Q + \frac{\Pi}{\tau}\mathbf{i} \qquad \forall \mathbf{X} \in \Omega, \forall t \geq 0 \tag{2c}$$

$$\partial_t \overset{\circ}{\mathbf{Q}} = -\frac{\overset{\circ}{\mathbf{Q}}}{\tau_1} - \left(\frac{2\ell^2}{\tau\tau_1}\right) \nabla \overset{\circ}{\mathbf{q}}^{\mathrm{sym}} \qquad \forall \mathbf{X} \in \Omega, \forall t \geq 0 \tag{2d}$$

$$\partial_t Q = -\frac{Q}{\tau_0} - \left(\frac{5\ell^2}{3\tau_0\tau}\right) \nabla \cdot \mathbf{q} \qquad \forall \mathbf{X} \in \Omega, \forall t \geq 0 \tag{2e}$$

$$\partial_t \mathbf{i} = -\frac{\mathbf{i}}{\tau_e} + \frac{\sigma_e}{\tau_e}(\mathbf{E} - \epsilon\nabla\theta) - \frac{\epsilon\sigma_e}{\lambda\tau_e}\mathbf{q} \qquad \forall \mathbf{X} \in \Omega, \forall t \geq 0 \tag{2f}$$

and the second law of thermodynamics. Since that analysis will be performed both by Onsager's method, and by Liu's procedure, the above proof of compatibility will be also used here to claim that those two different techniques are equivalent in the special case of Equations (2). For the sake of clarity, in Appendix A, the physical meanings of the different quantities used throughout this paper are listed.

For the needs of this paper, we note that the form of the state space in Equation (1) is compatible with the Extended Irreversible Thermodynamics approach [22,23] which is based upon the assumption that the entropy may not only depend on the classical variables, but also on the fluxes [23–26]. As a consequence of Equation (1), in particular, in this paper, the Gibbs relation (which turns out the differential form of the state function s) will be generalized as

$$\mathrm{d}s = \left(\frac{\partial s}{\partial z}\right)\mathrm{d}z + \left(\frac{\partial s}{\partial u}\right)\mathrm{d}u + \left(\frac{\partial s}{\partial \mathbf{q}}\right) \cdot \mathrm{d}\mathbf{q} + \left(\frac{\partial s}{\partial \overset{\circ}{\mathbf{Q}}}\right) \cdot \mathrm{d}\overset{\circ}{\mathbf{Q}} + \left(\frac{\partial s}{\partial Q}\right)\mathrm{d}Q + \left(\frac{\partial s}{\partial \mathbf{i}}\right) \cdot \mathrm{d}\mathbf{i} \qquad \forall \mathbf{X} \in \Omega, \forall t \geq 0 \tag{3}$$

For the sake of clarity, we note that Equations (2a) and (2b) are (the consequences of) balance laws, whereas Equations (2c)–(2f) have to be evolution equations. We also note that the net current into every volume $V \subset \Omega$ is given by $-\int_{\partial V} \mathbf{i} \cdot d\mathbf{S}$, with $d\mathbf{S}$ as the outward pointing normal of the boundary of V. Furthermore, the net current is also defined as the derivative in time of the total charge, which can be described as $\int_V z\,dV$. Applying the divergence theorem, Equation (2a) is obtained. In particular, Equations (2c) and (2f) can be seen as possible generalizations (see, for example, Ref. [27]) of the first thermoelectric model proposed by William Thomson (Lord Kelvin) in the XIX century [28].

The layout of the paper is as follows. In Section 2, the compatibility between Equations (2) and the second law is analyzed by means of Onsager's method [15,16]. In Section 3, the compatibility between Equations (2) and the second law is analyzed by means of Liu's procedure [18]. In Section 4, it is shown that Equations (2) constitute a hyperbolic system. In Section 5, both final remarks are given, and the physical quantities involved in this paper are summarized in Table A1.

2. Onsager's Method

In this section, we test the agreement of the theoretical model in Equations (2) with the second law of thermodynamics by means of Onsager's method [15,16]. As previously observed, the final goal of that method is the derivation of the evolution equations, i.e., the field equations which allow us to predict the evolutions of the extended field variables of a theoretical model: for this reason, therefore, in this section, we will properly derive Equations (2c)–(2f) in agreement with the second law, whereas Equations (2a) and (2b) simply arise from the local balance of the charge, and of the energy, respectively.

The first steps to take in Onsager's method for the exploitation of the second law are the assumptions both about the forms of the different coefficients of Equation (3), and about the constitutive equation for the specific-entropy flux; therefore, here, we start by assuming the following constitutive relations

$$\frac{\partial s}{\partial z} = \epsilon \qquad \forall \mathbf{X} \in \Omega,\ \forall t \geq 0 \tag{4a}$$

$$\frac{\partial s}{\partial u} = \frac{1}{\theta} \qquad \forall \mathbf{X} \in \Omega,\ \forall t \geq 0 \tag{4b}$$

$$\frac{\partial s}{\partial \mathbf{q}} = -\left(\frac{\tau}{\lambda \theta^2}\right)\mathbf{q} \qquad \forall \mathbf{X} \in \Omega,\ \forall t \geq 0 \tag{4c}$$

$$\frac{\partial s}{\partial \overset{\circ}{\mathbf{Q}}} = -\left(\frac{\tau_1 \tau^2}{2\lambda \ell^2 \theta^2}\right)\overset{\circ}{\mathbf{Q}} \qquad \forall \mathbf{X} \in \Omega,\ \forall t \geq 0 \tag{4d}$$

$$\frac{\partial s}{\partial Q} = -\left(\frac{3\tau_0 \tau^2}{5\lambda \ell^2 \theta^2}\right)Q \qquad \forall \mathbf{X} \in \Omega,\ \forall t \geq 0 \tag{4e}$$

$$\frac{\partial s}{\partial \mathbf{i}} = -\left(\frac{\tau_e}{\sigma_e \theta}\right)\mathbf{i} \qquad \forall \mathbf{X} \in \Omega,\ \forall t \geq 0 \tag{4f}$$

which involve the form of the specific entropy, and

$$\mathbf{J}^{(s)} = \frac{\mathbf{q}}{\theta} - \frac{\tau}{\lambda \theta^2}\left(\overset{\circ}{\mathbf{Q}} + Q\mathbf{I}\right) \cdot \mathbf{q} + \frac{\Pi}{\theta}\mathbf{i} \qquad \forall \mathbf{X} \in \Omega,\ \forall t \geq 0 \tag{5}$$

which instead gives the form of the specific-entropy flux. Recalling that the local balance of the specific entropy reads

$$\partial_t s + \nabla \cdot \mathbf{J}^{(s)} = \sigma^{(s)} \qquad \forall \mathbf{X} \in \Omega,\ \forall t \geq 0 \tag{6}$$

then the use of the other two balance laws of our model, i.e., Equations (2a) and (2b), allows us to straightforwardly obtain

$$\begin{aligned} \sigma^{(s)} = &- \frac{\tau}{\lambda \theta^2}\left[\partial_t \mathbf{q} + \frac{\lambda}{\tau}\nabla\theta - \frac{2}{\theta}\left(Q\mathbf{I} + \overset{\circ}{\mathbf{Q}}\right) \cdot \nabla\theta + \nabla \cdot \overset{\circ}{\mathbf{Q}} + \nabla Q - \frac{\Pi}{\tau}\mathbf{i}\right] \cdot \mathbf{q} \\ &\frac{\tau^2 \tau_1}{2\lambda \ell^2 \theta^2}\left(\partial_t \overset{\circ}{\mathbf{Q}} + \frac{2\ell^2}{\tau \tau_1}\nabla \overset{\circ}{\mathbf{q}}^{\mathrm{sym}}\right) \cdot \overset{\circ}{\mathbf{Q}} - \frac{3\tau^2 \tau_0}{5\lambda \ell^2 \theta^2}\left(\partial_t Q + \frac{5\ell^2}{3\tau \tau_0}\nabla \cdot \mathbf{q}\right)Q \\ &- \frac{\tau_e}{\sigma_e \theta}\left[\partial_t \mathbf{i} - \frac{\sigma_e}{\tau_e}\left(\mathbf{E} - \frac{\Pi}{\theta}\nabla\theta\right) + \frac{\Pi \sigma_e}{\lambda \tau_e \theta}\mathbf{q}\right] \cdot \mathbf{i} + \left(\frac{\Pi}{\theta} - \epsilon\right)\nabla \cdot \mathbf{i} \end{aligned} \tag{7}$$

$\forall \mathbf{X} \in \Omega$ and $\forall t \geq 0$, once the quantity $\left(\frac{\Pi}{\lambda \theta^2}\right)\mathbf{q} \cdot \mathbf{i}$ has been added, and subtracted, in the left-hand side of Equation (6) for our purposes.

From Equation (7) it can be argued that in our case the specific-entropy production is a linear combination of the thermodynamic fluxes $\mathbf{q}$, $\overset{\circ}{\mathbf{Q}}$, Q and $\mathbf{i}$, and the following thermodynamic forces

$$\mathbf{f}^{q} := \partial_t \mathbf{q} + \frac{\lambda}{\tau}\nabla\theta - \frac{2}{\theta}\left(Q\mathbf{I} + \overset{\circ}{\mathbf{Q}}\right) \cdot \nabla\theta + \nabla \cdot \overset{\circ}{\mathbf{Q}} + \nabla Q - \frac{\Pi}{\tau}\mathbf{i} \tag{8a}$$

$$\mathbf{F}^{Q} := \partial_t \overset{\circ}{\mathbf{Q}} + \frac{2\ell^2}{\tau \tau_1}\nabla \overset{\circ}{\mathbf{q}}^{\mathrm{sym}} \tag{8b}$$

$$F^{Q} := \partial_t Q + \frac{5\ell^2}{3\tau \tau_0}\nabla \cdot \mathbf{q} \tag{8c}$$

$$\mathbf{f}^{i} := \partial_t \mathbf{i} - \frac{\sigma_e}{\tau_e}\left(\mathbf{E} - \frac{\Pi}{\theta}\nabla\theta\right) + \frac{\Pi \sigma_e}{\lambda \tau_e \theta}\mathbf{q} \tag{8d}$$

which appear therein. In order that the thermodynamic unilateral constraint $\sigma^{(s)}$ is always (i.e., $\forall\, \mathbf{X} \in \Omega$ and $\forall\, t \geq 0$) fulfilled, it is sufficient to assume

$$\frac{\mathbf{q}}{\tau} = -\mathbf{f}^{q} \qquad \forall\, \mathbf{X} \in \Omega,\ \forall\, t \geq 0 \tag{9a}$$

$$\frac{\overset{\circ}{\mathbf{Q}}}{\tau_1} = -\mathbf{F}^{Q} \qquad \forall\, \mathbf{X} \in \Omega,\ \forall\, t \geq 0 \tag{9b}$$

$$\frac{Q}{\tau_0} = -F^{Q} \qquad \forall\, \mathbf{X} \in \Omega,\ \forall\, t \geq 0 \tag{9c}$$

$$\frac{\mathbf{i}}{\tau_e} = -\mathbf{f}^{i} \qquad \forall\, \mathbf{X} \in \Omega,\ \forall\, t \geq 0 \tag{9d}$$

together with the constitutive assumption

$$\epsilon = \frac{\Pi}{\theta} \tag{10}$$

which is sometimes referred to as the *second Kelvin relation*. The final insertion of Equations (9) into Equation (8) directly yields the evolution equations we were searching for here, namely Equations (2c)–(2f).

The above thermodynamic considerations, therefore, allow us to claim that the theoretical model in Equation (2) agrees with the second law of thermodynamics.

We note that Equations (9), referring to the linear relation between the fluxes and the forces [22], represent the simplest way to ensure that $\sigma^{(s)} \geq 0$. In this case, in fact, we have

$$\sigma^{(s)} = \left(\frac{1}{\lambda\theta^2}\right)\mathbf{q}\cdot\mathbf{q} + \left(\frac{\tau^2}{2\lambda\ell^2\theta^2}\right)\overset{\circ}{\mathbf{Q}}\cdot\overset{\circ}{\mathbf{Q}} + \left(\frac{3\tau^2}{5\lambda\ell^2\theta^2}\right)Q^2 + \left(\frac{1}{\sigma_e\theta}\right)\mathbf{i}\cdot\mathbf{i} \qquad \forall\, \mathbf{X} \in \Omega,\ \forall\, t \geq 0 \tag{11}$$

Other sufficient (and more refined) conditions could be obtained by allowing, for example, the thermodynamic forces to non-linearly depend on the fluxes [23].

In closing this section, we note that Equations (4) is tantamount to assume the following constitutive equation for s

$$s = s_0(u) + \epsilon z - \left(\frac{\tau}{2\lambda\theta^2}\right)\mathbf{q}\cdot\mathbf{q} - \left(\frac{\tau_1\tau^2}{4\lambda\ell^2\theta^2}\right)\overset{\circ}{\mathbf{Q}}\cdot\overset{\circ}{\mathbf{Q}} - \left(\frac{3\tau_0\tau^2}{10\lambda\ell^2\theta^2}\right)Q^2 - \left(\frac{\tau_e}{2\sigma_e\theta}\right)\mathbf{i}\cdot\mathbf{i} \qquad \forall\, \mathbf{X} \in \Omega,\ \forall\, t \geq 0 \tag{12}$$

wherein $s_0(u)$ stands for the part of the local-equilibrium specific entropy which only depends on the internal energy (or the temperature).

3. Liu's Procedure

In this section, we test the agreement of the theoretical model in Equations (2) with the second law of thermodynamics by means of Liu's procedure [18].

From a practical point of view, we initially admit here the validity of the model Equations (2), and look at the constitutive relations for s and $\mathbf{J}^{(s)}$, which guarantee that $\sigma^{(s)} \geq 0$, $\forall\, \mathbf{X} \in \Omega$ and $\forall\, t \geq 0$. Liu's technique [18], in fact, is based on the method of the Lagrange multipliers and allows us to derive necessary and sufficient conditions (that is, the aforementioned constitutive relations) ensuring that the linear combination between the specific entropy production and the field equations is always compliant with the second law of thermodynamics [29]. The coefficients of that linear combination are exactly the Lagrange multipliers. Recalling that the specific entropy production is given by the left-hand side of Equation (6), the above aimed constitutive relations have to be derived in such a way that the following constraint for the *extended* entropy inequality

$$\begin{aligned}
&\partial_t s + \nabla \cdot \mathbf{J}^{(s)} - \Lambda^z(\partial_t z + \nabla \cdot \mathbf{i}) - \Lambda^u(\partial_t u + \nabla \cdot \mathbf{q} - \mathbf{i} \cdot \mathbf{E}) \\
&- \mathbf{\Lambda}^q \cdot \left[\partial_t \mathbf{q} + \frac{\mathbf{q}}{\tau} + \left(\frac{\lambda \mathbf{I}}{\tau} - \frac{2\mathbf{Q}}{\theta}\right) \cdot \nabla\theta + \nabla \cdot \overset{\circ}{\mathbf{Q}} + \nabla Q - \frac{\Pi}{\tau}\mathbf{i}\right] \\
&- \mathbf{\Lambda}^Q \cdot \left[\partial_t \overset{\circ}{\mathbf{Q}} + \frac{\overset{\circ}{\mathbf{Q}}}{\tau_1} + \left(\frac{2\ell^2}{\tau\tau_1}\right) \nabla \overset{\circ}{\mathbf{q}}^{\mathrm{sym}}\right] - \Lambda^Q\left[\partial_t Q + \frac{Q}{\tau_0} + \left(\frac{5\ell^2}{3\tau_0\tau}\right)\nabla \cdot \mathbf{q}\right] \\
&- \mathbf{\Lambda}^i \cdot \left[\partial_t \mathbf{i} + \frac{\mathbf{i}}{\tau_e} - \frac{\sigma_e}{\tau_e}(\mathbf{E} - \epsilon\nabla\theta) + \frac{\epsilon\sigma_e}{\lambda\tau_e}\mathbf{q}\right] \geq 0
\end{aligned} \tag{13}$$

is fulfilled $\forall \mathbf{X} \in \Omega$ and $\forall t \geq 0$. For the sake of clarity, we note that in the above inequality, the parameters Λ^z, Λ^u, $\mathbf{\Lambda}^q$, $\mathbf{\Lambda}^Q$, Λ^Q and $\mathbf{\Lambda}^i$ are the so-called Lagrange multipliers which, in principle, may depend on the whole set of the state variables (1). The same conclusion also holds for the constitutive equations of s and $\mathbf{J}^{(s)}$.

Taking into account the usual relation $\mathrm{d}u = c_v \mathrm{d}\theta$, by writing the explicit forms both of $\partial_t s$, and of $\nabla \cdot \mathbf{J}^{(s)}$ in terms of the state variables, from inequality (13), one can reach the following two sets of necessary and sufficient conditions, which guarantee that second law of thermodynamics is always fulfilled

$$\frac{\partial s}{\partial z} = \Lambda^z \qquad \forall \mathbf{X} \in \Omega,\ \forall t \geq 0 \tag{14a}$$

$$\frac{\partial s}{\partial u} = \Lambda^u \qquad \forall \mathbf{X} \in \Omega,\ \forall t \geq 0 \tag{14b}$$

$$\frac{\partial s}{\partial \mathbf{q}} = \mathbf{\Lambda}^q \qquad \forall \mathbf{X} \in \Omega,\ \forall t \geq 0 \tag{14c}$$

$$\frac{\partial s}{\partial \overset{\circ}{\mathbf{Q}}} = \mathbf{\Lambda}^Q \qquad \forall \mathbf{X} \in \Omega,\ \forall t \geq 0 \tag{14d}$$

$$\frac{\partial s}{\partial Q} = \Lambda^Q \qquad \forall \mathbf{X} \in \Omega,\ \forall t \geq 0 \tag{14e}$$

$$\frac{\partial s}{\partial \mathbf{i}} = \mathbf{\Lambda}^i \qquad \forall \mathbf{X} \in \Omega,\ \forall t \geq 0 \tag{14f}$$

and

$$\frac{\partial \mathbf{J}^{(s)}}{\partial z} = \mathbf{0} \qquad \forall \mathbf{X} \in \Omega,\ \forall t \geq 0 \tag{15a}$$

$$\frac{\partial \mathbf{J}^{(s)}}{\partial \theta} = \left[\frac{\lambda}{\tau}\mathbf{I} - \frac{2}{\theta}\left(\overset{\circ}{\mathbf{Q}} + Q\mathbf{I}\right)\right] \cdot \mathbf{\Lambda}^q + \frac{\sigma_e\epsilon}{\tau_e}\mathbf{\Lambda}^i \qquad \forall \mathbf{X} \in \Omega,\ \forall t \geq 0 \tag{15b}$$

$$\frac{\partial \mathbf{J}^{(s)}}{\partial \mathbf{q}} = \Lambda^u\mathbf{I} + \left(\frac{2\ell^2}{\tau\tau_1}\right)\mathbf{\Lambda}^Q + \left(\frac{5\ell^2}{3\tau\tau_0}\right)\Lambda^Q\mathbf{I} \qquad \forall \mathbf{X} \in \Omega,\ \forall t \geq 0 \tag{15c}$$

$$\frac{\partial \mathbf{J}^{(s)}}{\partial \overset{\circ}{\mathbf{Q}}} = \mathbf{\Lambda}^q\mathbf{I} \qquad \forall \mathbf{X} \in \Omega,\ \forall t \geq 0 \tag{15d}$$

$$\frac{\partial \mathbf{J}^{(s)}}{\partial Q} = \mathbf{\Lambda}^q \qquad \forall \mathbf{X} \in \Omega,\ \forall t \geq 0 \tag{15e}$$

$$\frac{\partial \mathbf{J}^{(s)}}{\partial \mathbf{i}} = \Lambda^z\mathbf{I} \qquad \forall \mathbf{X} \in \Omega,\ \forall t \geq 0 \tag{15f}$$

together with the *reduced* entropy inequality

$$\sigma^{(s)} = \Lambda^u \mathbf{E} \cdot \mathbf{i} - \mathbf{\Lambda}^q \cdot \left(\frac{\mathbf{q}}{\tau} - \frac{\Pi}{\tau} \mathbf{i} \right) - \mathbf{\Lambda}^Q \cdot \frac{\overset{\circ}{\mathbf{Q}}}{\tau_1} - \Lambda^Q \frac{Q}{\tau_0} - \mathbf{\Lambda}^i \cdot \left(\frac{\mathbf{i}}{\tau_e} - \frac{\sigma_e}{\tau_e} \mathbf{E} + \frac{\sigma_e \epsilon}{\lambda \tau_e} \mathbf{q} \right) \geq 0 \qquad \forall \mathbf{X} \in \Omega, \forall t \geq 0 \tag{16}$$

If the following identifications are made

$$\Lambda^z := \frac{\Pi}{\theta} \tag{17a}$$

$$\Lambda^u := \frac{1}{\theta} \tag{17b}$$

$$\mathbf{\Lambda}^q := -\left(\frac{\tau}{\lambda \theta^2} \right) \mathbf{q} \tag{17c}$$

$$\mathbf{\Lambda}^Q := -\left(\frac{\tau_1 \tau^2}{2 \lambda \ell^2 \theta^2} \right) \overset{\circ}{\mathbf{Q}} \tag{17d}$$

$$\Lambda^Q := -\left(\frac{3 \tau_0 \tau^2}{5 \lambda \ell^2 \theta^2} \right) Q \tag{17e}$$

$$\mathbf{\Lambda}^i := -\left(\frac{\tau_e}{\sigma_e \theta} \right) \mathbf{i} \tag{17f}$$

then, based on the previous thermodynamic restrictions, it can be easily inferred that the theoretical model described by Equations (2) agrees with the second law if

- the specific entropy is given by Equation (12);
- the specific-entropy flux is given by Equation (5).

Equations (12) and (5) arise, respectively, from the integration of Equations (14) and (15) (details can be found in Appendix B).

By coupling Equations (16) and (17), it is also straightforward to obtain that the specific-entropy production is still given by Equation (11).

4. The Propagation Speed of Waves

We devote this section to pointing out the hyperbolic character of the theoretical model in Equations (2).

In order to find the speed of the propagation of the waves predicted by the equation, here, we employ the tool of the acceleration waves in particular; that is, we assume the presence in our system of a traveling surface $\mathcal{S}$ across which the state variables are continuous, whereas all their derivatives (both in time, and in space) display at most finite discontinuities. From the mathematical point of view, an acceleration wave may model, for example, the thermal perturbation generated by a periodic variation in time of the temperature in a point of the system with respect to its steady-state reference level.

For the sake of simplicity, here, we also suppose that $\mathcal{S}$ is moving into equilibrium zones, that is, we assume that the region $\Omega^\star$ ahead of $\mathcal{S}$ is such that

$$z = z_0 \qquad \theta = \theta_0 \qquad \mathbf{q} = \mathbf{q}_0 \qquad \mathbf{Q} = \mathbf{Q}_0 \qquad \mathbf{i} = \mathbf{i}_0 \qquad \forall \mathbf{X} \in \Omega^\star, \forall t \geq 0 \tag{18}$$

with z_0, θ_0, $\mathbf{q}_0$, $\mathbf{Q}_0$ and $\mathbf{i}_0$ being the stationary reference levels. If all the material functions are constant (i.e., if they only display vanishingly small variations when the temperature of the system changes), then from Equations (2) we may directly obtain

$$\llbracket \partial_t z \rrbracket = -\llbracket \nabla \cdot \mathbf{i} \rrbracket \tag{19a}$$
$$\llbracket \partial_t u \rrbracket = -\llbracket \nabla \cdot \mathbf{q} \rrbracket \tag{19b}$$
$$\llbracket \partial_t \mathbf{q} \rrbracket = -\left(\frac{\lambda \mathbf{I}}{\tau} - \frac{2\mathbf{Q}_0}{\theta_0} \right) \llbracket \nabla \theta \rrbracket - \llbracket \nabla \cdot \overset{\circ}{\mathbf{Q}} \rrbracket - \llbracket \nabla Q \rrbracket \tag{19c}$$
$$\llbracket \partial_t \overset{\circ}{\mathbf{Q}} \rrbracket = -\left(\frac{2\ell^2}{\tau \tau_1} \right) \llbracket \nabla \overset{\circ}{\mathbf{q}}^{\text{sym}} \rrbracket \tag{19d}$$
$$\llbracket \partial_t Q \rrbracket = -\left(\frac{5\ell^2}{3\tau_0 \tau} \right) \llbracket \nabla \cdot \mathbf{q} \rrbracket \tag{19e}$$
$$\llbracket \partial_t \mathbf{i} \rrbracket = -\frac{\sigma_e \epsilon}{\tau_e} \llbracket \nabla \theta \rrbracket \tag{19f}$$

Making use of the classical Hadamard relations [30,31]

$$\llbracket \partial_t z \rrbracket = -U_N \widehat{z} \tag{20a}$$
$$\llbracket \partial_t \theta \rrbracket = -U_N \widehat{\theta} \tag{20b}$$
$$\llbracket \partial_t \mathbf{q} \rrbracket = -U_N \widehat{\mathbf{q}} \tag{20c}$$
$$\llbracket \partial_t \overset{\circ}{\mathbf{Q}} \rrbracket = -U_N \widehat{\overset{\circ}{\mathbf{Q}}} \tag{20d}$$
$$\llbracket \partial_t Q \rrbracket = -U_N \widehat{Q} \tag{20e}$$
$$\llbracket \partial_t \mathbf{i} \rrbracket = -U_N \widehat{\mathbf{i}} \tag{20f}$$

wherein $\widehat{z} = \mathbf{n} \cdot \llbracket \nabla z \rrbracket$, $\widehat{\theta} = \mathbf{n} \cdot \llbracket \nabla \theta \rrbracket$, $\widehat{\mathbf{q}} = \mathbf{n} \cdot \llbracket \nabla \mathbf{q} \rrbracket$, $\widehat{\overset{\circ}{\mathbf{Q}}} = \mathbf{n} \cdot \llbracket \nabla \overset{\circ}{\mathbf{Q}} \rrbracket$, $\widehat{Q} = \mathbf{n} \cdot \llbracket \nabla Q \rrbracket$ and $\widehat{\mathbf{i}} = \mathbf{n} \cdot \llbracket \nabla \mathbf{i} \rrbracket$ are the wave amplitudes, with $\mathbf{n}$ being the normal unit to $\mathcal{S}$, and U_N the non-vanishing wave speed, by direct computations from Equations (19), the following system of homogeneous algebraic equations holds

$$U_N \widehat{z} - \mathbf{n} \cdot \widehat{\mathbf{i}} = 0 \tag{21a}$$
$$U_N \widehat{\theta} - \frac{1}{c_v} \mathbf{n} \cdot \widehat{\mathbf{q}} = 0 \tag{21b}$$
$$U_N \widehat{\mathbf{q}} - \left(\frac{\lambda}{\tau} \mathbf{n} - \frac{2}{\theta_0} \mathbf{Q}_0 \cdot \mathbf{n} \right) \widehat{\theta} - \widehat{\overset{\circ}{\mathbf{Q}}} \cdot \mathbf{n} - \widehat{Q} \mathbf{n} = \mathbf{0} \tag{21c}$$
$$U_N \widehat{\overset{\circ}{\mathbf{Q}}} - \frac{4\ell^2}{3\tau_1 \tau} \widehat{\mathbf{q}} \mathbf{n} = \mathbf{0} \tag{21d}$$
$$U_N \widehat{Q} - \frac{5\ell^2}{3\tau_0 \tau} \widehat{\mathbf{q}} \cdot \mathbf{n} = 0 \tag{21e}$$
$$U_N \widehat{\mathbf{i}} - \frac{\sigma_e \epsilon}{\tau_e} \widehat{\theta} \mathbf{n} = \mathbf{0} \tag{21f}$$

In order to avoid Sys. (21) only admitting the trivial solution for the wave amplitudes, the following relation has to hold

$$U_N^4 \left[U_N^2 - U_{\text{mc}}^2 (1 + \chi_1 + \chi_2 - \chi_3) \right] = 0 \tag{22}$$

wherein

$$U_{\text{mc}} = \sqrt{\frac{\lambda}{\tau c_v}} \tag{23}$$

is the second-sound speed in the Maxwell–Cattaneo theory, and

$$\chi_1 = \frac{4\ell^2 c_v}{3\tau_1 \lambda} \tag{24a}$$

$$\chi_2 = \frac{5\ell^2 c_v}{3\tau_0 \lambda} \tag{24b}$$

$$\chi_3 = \frac{2\tau\, \mathrm{tr}(\mathbf{Q}_0)}{\lambda \theta_0} \tag{24c}$$

are three dimensionless parameters accounting for the non-local effects (through ℓ^2), the relaxational effects (through the relaxation times of the second-order fluxes), and the stationary reference values for the temperature and the second-order flux, respectively. By avoiding the trivial solution $U_N = 0$, Equation (22) allows us to claim that the speed of propagation predicted by the theoretical model in Equations (2) is given by

$$U_N = U_{\mathrm{mc}} \sqrt{1 + \chi_1 + \chi_2 - \chi_3} \tag{25}$$

According to Equation (25), below we list some brief comments about the role played by non-local and non-linear effects.

- The role of the non-local effects is introduced by χ_1 and χ_2; since those parameters are always positive, it is possible to claim that non-local effects always enhance the speed of propagation with respect to U_{mc}.
- The role of the non-linear effects is introduced by χ_3; since it may attain, in principle, both positive and negative values, we can claim that non-linear effects can either enhance or reduce the speed of propagation with respect to U_{mc}.
- Whenever both the non-local effects, and the non-linear ones, are negligible in Equations (2), then the predicted speed of propagation (25) reduces to U_{mc}.

5. Final Remarks

The possibility of converting the heat into the electric current, and vice versa, is one of the best-known coupled transport phenomena since its discovery two centuries ago. Notwithstanding this, in the past, thermoelectric devices practically achieved a scant success; this was largely the consequence of their very low efficiencies. The rise of nano-technologies has been a real breakthrough for thermoelectricity, because they allow us to currently develop new thermoelectric materials [32–34]; by obtaining the device's sizes comparable to the mean-free path of the different heat carriers (phonons, electrons, holes, etc.), in fact, it is now possible to significantly enhance the thermoelectric features of a material.

The research regarding thermoelectricity, however, should not only be limited to the search of new materials, but it should also look at enhanced and even more refined theoretical models for the correct description of the thermoelectric phenomenon, in accordance with the essence of physics. For this reason, in the present paper, we proposed the model in Equations (2) for the description of thermoelectric effects at nano-scale. That model has the main merit of giving emphasis to non-local and non-linear effects, the influence of which has been especially pointed out here on the speed of propagation of thermal waves.

We note that the aforementioned model neither arises from a rigorous microscopic derivation, nor lies on experimental observations: it is therefore currently meant only as theoretical proposal. In order to give a physical value to Equations (2), however, in Sections 2 and 3 we scrutinized them in view of the second law of thermodynamics. That analysis has been performed by means of two different procedures: both of them turned out the compatibility of the theoretical model described by Equations (2) with the basic tenets of Continuum Mechanics. The study of the waves' propagation of Section 4 also indicates that the theoretical model (2) is hyperbolic, namely, it is compatible with the physical evidence that mechanical disturbances can only propagate with a finite speed.

By looking at a future analysis for the rigorous proof, we close this paper by observing that the perfect agreement between the results contained in Section 2 and those of Section 3 could be the first evidence of the possible equivalence between Onsager's method [15,16] and Liu technique [18].

Author Contributions: Conceptualization, A.S.; methodology, M.D.D. and A.S.; formal analysis, M.D.D.; writing—original draft preparation, M.D.D. and A.S. All authors have read and agreed to the published version of the manuscript.

Funding: This research received no external funding.

Institutional Review Board Statement: Not applicable.

Data Availability Statement: Data are contained within the article.

Acknowledgments: This work has been carried out under the auspices of GNFM (Italian National Group of Mathematical Physics, INdAM). The following research projects are also acknowledged: "Transport Phenomena in Low Dimensional Structures: Models, Simulations and Theoretical Aspects" (NextGenerationEU PRIN2022—CUP no. D53D23006000006); "Non Linear Models for Magma Transport and Volcanoes Generation" (National Recovery and Resilience Plan/NRRP—NextGenerationEU—CUP no. D53D23018760001).

Conflicts of Interest: The authors declare no conflicts of interest.

Appendix A

Table A1. List of the physical quantities.

Symbol	Meaning	Unit of Measurement
θ	non-equilibrium temperature	K
θ_0	reference value of temperature	K
u	specific internal energy	$\frac{J}{m^3}$
$\mathbf{q}$	local heat flux	$\frac{W}{m^2}$
$\mathbf{q}_0$	reference value of the local heat flux	$\frac{W}{m^2}$
$\mathbf{Q}$	flux of the local heat flux	$\frac{W}{m\,s}$
$\mathbf{Q}_0$	reference value of the flux of the local heat flux	$\frac{W}{m\,s}$
$\overset{\circ}{\mathbf{Q}}$	deviatoric (traceless) part of $\mathbf{Q}$	$\frac{W}{m\,s}$
Q	volumetric part of $\mathbf{Q}$	$\frac{W}{m\,s}$
z	charge per unit volume of electrons	$\frac{C}{m^3}$
z_0	reference value of the charge per unit volume of electrons	$\frac{C}{m^3}$
$\mathbf{i}$	electrical-current density	$\frac{C}{m^2 s}$
$\mathbf{i}_0$	reference value of the electrical-current density	$\frac{C}{m^2 s}$
$\mathbf{E}$	electric field	$\frac{N}{C}$
τ	relaxation time of the heat carriers	s
τ_e	relaxation time of electrons	s
τ_1	relaxation time of $\overset{\circ}{\mathbf{Q}}$	s
τ_0	relaxation time of Q	s
t	time	s
λ	thermal conductivity	$\frac{W}{m\,K}$
ℓ	mean-free path of the heat carriers	m
$\mathbf{x}$	position vector	m
c_v	specific heat at constant volume	$\frac{J}{m^3 K}$
ϵ	Seebeck coefficient	$\frac{J}{K\,C}$
Π	Peltier coefficient	$\frac{J}{C}$
σ_e	electrical conductivity	$\frac{C^2}{J\,m\,s}$
s	specific entropy	$\frac{J}{m^3 K}$

Table A1. *Cont.*

Symbol	Meaning	Unit of Measurement
$\mathbf{J}^s$	specific-entropy flux	$\frac{\mathrm{W}}{\mathrm{m^2K}}$
σ^s	specific-entropy production	$\frac{\mathrm{W}}{\mathrm{m^3K}}$
Λ^z	Lagrange multiplier	$\frac{\mathrm{J}}{\mathrm{KC}}$
Λ^u	Lagrange multiplier	$\frac{1}{\mathrm{K}}$
$\boldsymbol{\Lambda}^q$	Lagrange multiplier	$\frac{\mathrm{s}}{\mathrm{mK}}$
$\boldsymbol{\Lambda}^{\mathring{Q}}$	Lagrange multiplier	$\frac{\mathrm{s}}{\mathrm{m^2K}}$
Λ^Q	Lagrange multiplier	$\frac{\mathrm{s}}{\mathrm{m^2K}}$
$\boldsymbol{\Lambda}^i$	Lagrange multiplier	$\frac{\mathrm{C}}{\mathrm{m^2sK}}$
U_N	propagation speed	$\frac{\mathrm{m}}{\mathrm{s}}$
U_{mc}	propagation speed in the Maxwell–Cattaneo theory	$\frac{\mathrm{m}}{\mathrm{s}}$
$\mathbf{f}^q$	Thermodynamic force	$\frac{\mathrm{W}}{\mathrm{m^2s}}$
$\mathbf{F}^q$	Thermodynamic force	$\frac{\mathrm{W}}{\mathrm{ms^2}}$
F^Q	Thermodynamic force	$\frac{\mathrm{W}}{\mathrm{ms^2}}$
$\mathbf{f}^i$	Thermodynamic force	$\frac{\mathrm{C}}{\mathrm{m^2s^2}}$

Appendix B. Integration of Equations (15)

The aim of this section is to give details about how Equation (5) arises from the integration of Equations (15), together with the constitutive assumption (10).

Starting from Equation (15a), we infer that $\mathbf{J}^s = \mathbf{J}^s\left(\theta, \mathbf{q}, \mathring{\mathbf{Q}}, Q, \mathbf{i}\right)$. Assuming Equations (17), Equations (15) become

$$\frac{\partial \mathbf{J}^{(s)}}{\partial z} = \mathbf{0} \qquad \forall \mathbf{X} \in \Omega,\ \forall t \geq 0 \tag{A1a}$$

$$\frac{\partial \mathbf{J}^{(s)}}{\partial \theta} = \left(-\frac{\tau}{\lambda\theta^2}\right)\left[\frac{\lambda}{\tau}\mathbf{I} - \frac{2}{\theta}\left(\mathring{\mathbf{Q}} + Q\mathbf{I}\right)\right] \cdot \mathbf{q} - \frac{\epsilon}{\theta}\mathbf{i} \qquad \forall \mathbf{X} \in \Omega,\ \forall t \geq 0 \tag{A1b}$$

$$\frac{\partial \mathbf{J}^{(s)}}{\partial \mathbf{q}} = \frac{1}{\theta}\mathbf{I} - \left(\frac{\tau}{\lambda\theta^2}\right)\mathring{\mathbf{Q}} - \left(\frac{\tau}{\lambda\theta^2}\right)Q\mathbf{I} \qquad \forall \mathbf{X} \in \Omega,\ \forall t \geq 0 \tag{A1c}$$

$$\frac{\partial \mathbf{J}^{(s)}}{\partial \mathring{\mathbf{Q}}} = -\frac{\tau}{\lambda\theta^2}\mathbf{q}\mathbf{I} \qquad \forall \mathbf{X} \in \Omega,\ \forall t \geq 0 \tag{A1d}$$

$$\frac{\partial \mathbf{J}^{(s)}}{\partial Q} = -\frac{\tau}{\lambda\theta^2}\mathbf{q} \qquad \forall \mathbf{X} \in \Omega,\ \forall t \geq 0 \tag{A1e}$$

$$\frac{\partial \mathbf{J}^{(s)}}{\partial \mathbf{i}} = \frac{\Pi}{\theta}\mathbf{I} \qquad \forall \mathbf{X} \in \Omega,\ \forall t \geq 0 \tag{A1f}$$

Equation (A1f) leads to

$$\mathbf{J}^s = \frac{\Pi}{\theta}\mathbf{i} + \alpha_1\left(\theta, \mathbf{q}, \mathring{\mathbf{Q}}, Q\right) \tag{A2}$$

with α_1 as a suitable vectorial function of the aforementioned variables. Substituting the previous equation into Equation (A1e), one can reach

$$\alpha_1 = -\frac{\tau}{\lambda\theta^2}Q\mathbf{q} + \alpha_2\left(\theta, \mathbf{q}, \mathring{\mathbf{Q}}\right) \tag{A3}$$

The coupling between Equations (A2) and (A3), together with Equation (A1d), furnishes

$$\mathbf{J}^s = \frac{\Pi}{\theta}\mathbf{i} - \frac{\tau}{\lambda\theta^2}\left(\mathring{\mathbf{Q}} + Q\mathbf{I}\right) \cdot \mathbf{q} + \alpha_3(\theta, \mathbf{q}) \tag{A4}$$

Substituting Equation (A4) into Equation (A1c) one obtains

$$\alpha_3 = \frac{\mathbf{q}}{\theta} + \alpha_4(\theta) \tag{A5}$$

Finally, taking into account Equations (A5), (A4) and (A1b), the requested Expression (5) for the entropy flux arises.

References

1. Boukai, A.I.; Bunimovich, Y.; Tahir-Kheli, J.; Yu, J.K.; Goddard III, W.A.; Heath, J.R. Silicon nanowires as efficient thermoelectric materials. *Nature* **2008**, *451*, 168–171. [CrossRef] [PubMed]
2. Hochbaum, A.; Chen, R.; Delgado, R.D.; Liang, W.; Garnett, E.C.; Najarian, M.; Majumdar, A.; Yang, P. Enhanced thermoelectric performance of rough silicon nanowires. *Nature* **2008**, *451*, 163–167. [CrossRef] [PubMed]
3. Hung, N.; Nugraha, A.; Saito, R. Designing high-performance thermoelectrics in two-dimensional tetradymites. *Nano Energy* **2019**, *58*, 743–749. [CrossRef]
4. Patel, A.; Singh, D.; Sonvane, Y.; Thakor, P.; Ahuja, R. High Thermoelectric Performance in Two-Dimensional Janus Monolayer Material WS-X (X=Se and Te). *ACS Appl. Mater. Interfaces* **2020**, *12*, 46212–46219. [CrossRef] [PubMed]
5. Sun, X.; Zhang, Z.; Dresselhaus, M. Theoretical modeling of thermoelectricity in Bi nanowires. *Appl. Phys. Lett.* **1999**, *74*, 4005–4007. [CrossRef]
6. Cheng, C.H.; Huang, S.Y.; Cheng, T.C. A three-dimensional theoretical model for predicting transient thermal behavior of thermoelectric coolers. *Int. J. Heat Mass Transf.* **2010**, *53*, 2001–2011. [CrossRef]
7. Sellitto, A.; Cimmelli, V.A.; Jou, D. Thermoelectric effects and size dependency of the figure-of-merit in cylindrical nanowires. *Int. J. Heat Mass Transf.* **2013**, *57*, 109–116. [CrossRef]
8. Jou, D.; Sellitto, A.; Cimmelli, V.A. Multi-temperature mixture of phonons and electrons and nonlocal thermoelectric transport in thin layers. *Int. J. Heat Mass Transf.* **2014**, *71*, 459–468. [CrossRef]
9. Sellitto, A.; Cimmelli, V.A.; Jou, D. Heat Transport with Phonons and Electrons and Efficiency of Thermoelectric Generators. In *Mesoscopic Theories of Heat Transport in Nanosystems*; SEMA SIMAI Springer Series; Springer International Publishing: Cham, Switzerland, 2016; Volume 6, pp. 133–166.
10. Lee, H. A Theoretical Model of Thermoelectric Transport Properties for Electrons and Phonons. *J. Electron. Mater.* **2016**, *45*, 1115–1141. [CrossRef]
11. Shen, Z.G.; Wu, S.Y.; Xiao, L.; Yin, G. Theoretical modeling of thermoelectric generator with particular emphasis on the effect of side surface heat transfer. *Energy* **2016**, *95*, 367–379. [CrossRef]
12. Gupta, R.; Kumar, N.; Kaur, P.; Bera, C. Theoretical model for predicting thermoelectric properties of tin chalcogenides. *Phys. Chem. Chem. Phys.* **2020**, *22*, 18989–19008. [CrossRef] [PubMed]
13. Zeng, Y.J.; Wu, D.; Cao, X.H.; Zhou, W.X.; Tang, L.M.; Chen, K.Q. Nanoscale Organic Thermoelectric Materials: Measurement, Theoretical Models, and Optimization Strategies. *Adv. Funct. Mater.* **2020**, *30*, 1903873. [CrossRef]
14. Muschik, W.; Ehrentraut, H. An amendment to the Second Law. *J. Non-Equilib. Thermodyn.* **1996**, *21*, 175–192. [CrossRef]
15. Onsager, L. Reciprocal relations in irreversible processes I. *Phys. Rev.* **1931**, *37*, 405–426. [CrossRef]
16. Onsager, L. Reciprocal relations in irreversible processes II. *Phys. Rev.* **1931**, *38*, 2265–2279. [CrossRef]
17. Coleman, B.D.; Noll, W. The thermodynamics of elastic materials with heat conduction and viscosity. *Arch. Ration. Mech. Anal.* **1963**, *13*, 167–178. [CrossRef]
18. Liu, I.S. Method of Lagrange multipliers for exploitation of entropy principle. *Arch. Rational Mech. Anal.* **1972**, *46*, 131–138. [CrossRef]
19. Cimmelli, V.A.; Sellitto, A.; Triani, V. A new thermodynamic framework for second-grade Korteweg-type viscous fluids. *J. Math. Phys.* **2009**, *50*, 053101. [CrossRef]
20. Cimmelli, V.A.; Sellitto, A.; Triani, V. A generalized Coleman-Noll procedure for the exploitation of the entropy principle. *Proc. R. Soc. A* **2010**, *466*, 911–925. [CrossRef]
21. Triani, V.; Papenfuss, C.; Cimmelli, V.A.; Muschik, W. Exploitation of the Second Law: Coleman-Noll and Liu procedure in comparison. *J. Non-Equilib. Thermodyn.* **2008**, *33*, 47–60. [CrossRef]
22. Lebon, G.; Jou, D.; Casas-Vázquez, J. *Understanding Non-Equilibrium Thermodynamics*; Springer: Berlin, Germany, 2008.
23. Jou, D.; Casas-Vázquez, J.; Lebon, G. *Extended Irreversible Thermodynamics*, fourth revised ed.; Springer: Berlin, Germany, 2010.
24. Machlup, S.; Onsager, L. Fluctuations and Irreversible Process. II. Systems with Kinetic Energy. *Phys. Rev.* **1953**, *91*, 1512–1515. [CrossRef]
25. Gyarmati, I. *Nonequilibrium Thermodynamics*; Springer: Berlin, Germany, 1970.
26. Jou, D.; Zakari, M. Information theory and heat transport in relativistic gases. *J. Phys. A* **1995**, *28*, 1585–1592. [CrossRef]
27. Di Domenico, M.; Carlomagno, I.; Sellitto, A. Wave propagation at nano-scale in coupled transport phenomena: application to thermoelectricity. *Meccanica* **2024**. [CrossRef]
28. Mahan, G.; Sales, B.; Sharp, J. Thermoelectric materials: New approaches to an old problem. *Phys. Today* **1997**, *50*, 42–47. [CrossRef]

29. Lebon, G.; Jou, D.; Casas-Vázquez, J.; Muschik, W. Weakly nonlocal and nonlinear heat transport in rigid solids. *J. Non-Equilib. Thermodyn.* **1998**, *23*, 176–191. [CrossRef]
30. Straughan, B. *Heat Waves*; Springer: Berlin, Germany, 2011.
31. Carlomagno, I.; Di Domenico, M.; Sellitto, A. High-order fluxes in heat transfer with phonons and electrons: application to wave propagation. *Proc. R. Soc. A* **2021**, *477*, 20210392. [CrossRef]
32. Jaziri, N.; Boughamoura, A.; Müller, J.; Mezghani, B.; Tounsi, F.; Ismail, M. A comprehensive review of Thermoelectric Generators: Technologies and common applications. *Energy Rep.* **2020**, *6*, 264–287. [CrossRef]
33. Xing, T.; Song, Q.; Qiu, P.; Zhang, Q.; Gu, M.; Xia, X.; Liao, J.; Shi, X.; Chen, L. High efficiency GeTe-based materials and modules for thermoelectric power generation. *Energy Environ. Sci.* **2021**, *14*, 995–1003. [CrossRef]
34. Bu, Z.; Zhang, X.; Hu, Y.; Chen, Z.; Lin, S.; Li, W.; Xiao, C.; Pei, Y. A record thermoelectric efficiency in tellurium-free modules for low-grade waste heat recovery. *Nat. Commun.* **2022**, *13*, 237. [CrossRef]

entropy

MDPI

Article

Transported Entropy of Ions and Peltier Coefficients in 8YSZ and 10Sc1CeSZ Electrolytes for Solid Oxide Cells

Aydan Gedik * and Stephan Kabelac

Institute of Thermodynamics, Leibniz University Hannover, Welfengarten 1, D-30167 Hannover, Germany; kabelac@ift.uni-hannover.de

* Correspondence: gedik@ift.uni-hannover.de; Tel.: +49-511-762-2155

Abstract: In this study, the transported entropy of ions for 8YSZ and 10Sc1CeSZ electrolytes was experimentally determined to enable precise modeling of heat transport in solid oxide cells (SOCs). The Peltier coefficient, crucial for thermal management, was directly calculated, highlighting reversible heat transport effects in the cell. While data for 8YSZ are available in the literature, providing a basis for comparison, the results for 10Sc1CeSZ show slightly smaller Seebeck coefficients but higher transported ion entropies. Specifically, at 700 °C and an oxygen partial pressure of $p_{O_2} = 0.21$ bar, values of $S^*_{O^{2-}} = 52 \pm 10$ J/K·F for 10Sc1CeSZ and $S^*_{O^{2-}} = 48 \pm 9$ J/K·F for 8YSZ were obtained. The transported entropy was also validated through theoretical calculations and showed minimal deviations when comparing different cell operation modes ($O_2 \mid\mid O^{2-} \mid\mid O_2$ and $H_2, H_2O \mid\mid O^{2-} \mid\mid O_2$). The influence of the transported entropy of the ions on the total heat generation and the partial heat generation at the electrodes is shown. The temperature has the greatest influence on heat generation, whereby the ion entropy also plays a role. Finally, the Peltier coefficients of 8YSZ for all homogeneous phases agree with the literature values.

Keywords: solid oxide cell (SOC); non-equilibrium thermodynamics (NET); Seebeck coefficient; Peltier coefficient; transported entropy of ions; electrodes heat generation; reversible heat effects

1. Introduction

For cell development and for the optimisation of operating strategies in solid oxide cells (SOCs), a reliable understanding of the heat, mass and charge transport mechanisms is of great importance. In particular, the performance and lifetime of SOCs are influenced by the temperature distribution. Uncontrolled temperature changes can lead to local overheating or undercooling and thus to high-temperature gradients within a cell. A comprehensive literature review on thermal management in SOFC stacks shows that excessive temperature gradients can lead to delamination and cracks in the electrolyte and electrodes [1]. For this reason, it is necessary to recognise not only the amount of total heat in the cell but also the local generation and/or local absorption of heat in all parts of the cell. A systematic establishment of heat balance equations and assessment of transport mechanisms can provide information.

The heat is divided into two types. On the one hand, *irreversible* heat is caused by dissipated energy (ohmic resistances and overvoltages at the electrodes among other sources). On the other hand, *reversible* heat is due to the reaction entropies at the electrodes. The electrochemical reaction at the three-phase interface leads to a Peltier effect, in which Peltier heat is reversibly released or absorbed at each electrode. Using the non-equilibrium thermodynamics (NET) approach, the influence of the Peltier heat can be clearly determined. According to NET, the following equation applies to the heat transport of a homogeneous phase i [2]:

$$J_q^i = -\lambda^i \cdot \frac{dT}{dy} + \frac{\pi^i}{F} \cdot j, \tag{1}$$

where λ^{i} is the thermal conductivity, π^{i} is the Peltier coefficient of the homogen phase i and j is the current density. According to the monocausal approach, the first part of the right-hand side of the equation describes the heat transport induced by a temperature gradient (Fourier heat). The second part of the right-hand side of the equation describes the Peltier heat induced by a charge transport. The importance of the Peltier effect is emphasised by means of a 1D SOFC model with an 8YSZ electrolyte that a heat flux flows in the direction of the temperature maximum near the cathode reaction layer [3]. In comparison, another investigation using a 2D SOEC model with an 8YSZ electrolyte shows that the Peltier effect is more extensive and more pronounced in SOEC operation than in SOFC operation [4]. In both studies, the data for the Peltier coefficients for the homogeneous phases are taken from the literature. Precise knowledge of the Peltier coefficients is essential for a realistic representation of heat transport and thus successful heat management. Compared to the homogeneous phase of the electrolyte, however, the Peltier coefficients for the homogeneous phases of the anode and cathode can be calculated simply.

Peltier coefficients are not directly measurable variables. However, they can be determined via the direct measurement of Seebeck coefficients. If an electrical charge transport occurs due to an imposed temperature gradient, this effect is referred to as the thermoelectric effect or the Seebeck effect. The type of non-isothermal electrochemical cells are referred to as thermocells. A direct relationship between the Peltier coefficient π and the Seebeck coefficient α_S can be established via Onsager's reciprocity relationship [2]:

$$\alpha_S = \left(\frac{\mathrm{d}\phi}{\mathrm{d}T}\right)_{\mathrm{j}\to 0,\, \mathrm{d}T\to 0} = -\frac{\pi}{F \cdot T}. \tag{2}$$

The measuring method is based on recording the change in cell potential in an open circuit when a temperature gradient is applied. An electrochemical cell with identical electrodes is considered, in each of which the back-and-forth reaction of an electrochemical reaction takes place.

The determination of Seebeck coefficients of electrochemical converters is common in the literature. Investigations into the Seebeck coefficient are carried out primarily in the field of thermoelectric power generators. Thermogalvanic cells can be used to generate electricity directly from waste heat. In the works [5–9], examples of possible electrochemical converters that can be used for such an objective are considered. Also, with the aim of achieving optimum thermal management, reversible thermal effects for relevant electrodes of different compositions and lithiation levels with different electrolytes are being investigated in the field of lithium-ion batteries. A comprehensive overview can be found in the publication by Gunnarshaug et al. [10]. Supported by previous investigations, the significance of the Peltier heat of a single electrode with regard to the entire cell is also clarified there. With the same aim, the works [11–17] investigate the reversible heat effects of other electrochemical half-cell reactions that are used in fuel cell and electrolysis technology and classified in the literature.

There are several studies in the literature on experimental work on oxygen-concentration cells with solid oxide electrolytes in the high-temperature range. Kiukkola and Wagner [18] showed by electromotive force (EMF) measurements that ZrO_2-CaO electrolytes exhibit stable ionic conduction at a temperature of 870 °C over a wide range of atmospheric oxygen partial pressures (10^{-5} to 1 atm), with electronic conduction contributions being minimal. Further studies were then carried out on the ionic and electronic conductivity of solid oxide electrolytes in other oxygen partial pressure and temperature ranges [19–24]. By adjusting the H_2/H_2O and CO/CO_2 mixtures, the oxygen activity could be precisely regulated, thus ensuring a reducing atmosphere. Park et al. [25] established empirical correlations for the conductivity of ions, electrons and holes for 8YSZ as a function of temperature and oxygen partial pressure. The ionic conductivity of oxide ceramics is strongly dependent on the temperature, the doping and the dopant. For example, Kosacki et al. [26] show that ionic conduction for ScSZ predominates more in other oxygen partial pressure ranges than it does for 8YSZ.

Fundamental theoretical insights into the behavior of the thermocells and important phenomenological equations for describing their properties are provided by the work of Holtan et al. [27]. With the knowledge gained from the studies on the stability of oxide ceramics over wide temperature and oxygen partial pressure ranges, numerous investigations were undertaken with regard to the thermopower of oxide ceramics with the aim of obtaining more precise information on the mobility of the oxygen ion and its dependencies. Table 1 summarises some values for the Seebeck coefficients of YSZ and ScSZ [28]. Further studies with CSZ can be found, for example, in [24,29–31]. The results show that the Seebeck coefficient is clearly dependent on the oxygen partial pressure of the surrounding atmosphere and the concentration of oxygen vacancies in the mixed oxide. Furthermore, there is no clear dependence on temperature [32]. The NET approach makes it possible to calculate the transported ion entropy and the heat of transfer coupled to it; so, it can be concluded from this information that the transport of the ions does not take place via interstitial sites but via vacancies. From this, it was possible to show that the transported ion entropy is independent of the oxygen partial pressure. Ratkje et al. describe the temperature dependence via the Thomson coefficient of the ion [33].

Table 1. Seebeck coefficients for Y_2O_3 and Sc_2O_3 ceramics from the literature.

T [K]	Dopant	mol%	p_{O_2} [bar]	$-\alpha_S$ [mV/K]	Reference
960–1310	Y_2O_3	9	$-\alpha_S = 0.46 + 0.0219 \cdot \ln(159/(p_{O_2} \cdot 750))$		[29]
1448	Y_2O_3	10	0.21	0.47	from [28]
		20		0.44	
		40		0.42	
1873	Y_2O_3	4.5	1	0.503	from [28]
		5.8		0.498	
		6.8		0.486	
		7.6		0.485	
		8.4		0.471	
1273	Y_2O_3	8	0.21	0.303	[34]
1186	Y_2O_3	8	0.21	0.46	[35]
1273	Y_2O_3	8	0.21	0.423	[36]
1186	Y_2O_3	8	0.21	0.43	[32]
			1	0.46	
1273	Y_2O_3	3	0.21	0.498	[33]
		5		0.487	
		8		0.495	
		12		0.478	
1273	Y_2O_3	2	0.21	0.533	[28]
		8		0.486	
		18		0.405	
1273	Sc_2O_3	12.5	1	0.4	[28]
1873	Sc_2O_3	4.4	1	0.468	[28]
		6.9		0.467	
		8		0.461	
		8.7		0.46	

In the work of Takekara et al. [34] and Kanamura et al. [36], the measurement of the Seebeck coefficient of 8YSZ is used for the entropy changes and the associated heat generation of the individual electrodes. Using measurements and empirical correlations, they show how the entropy change in the half-cell reactions behaves as a function of the partial pressure of the hydrogen and oxygen in order to ensure optimisation of the energy efficiency in solid oxide fuel cells through targeted control of this entropy change. The work of Ratkje et al. [33] shows slightly different results, which determines a greater asymmetry in the heat production of the half-cell reactions. The discrepancy is explained by the value of the transported ion entropy, which can have a major influence on the amount of heat produced by the individual electrodes and can even determine whether the anode-side half-cell reaction is exothermic or endothermic. Based on these results, Fischer et al. [37]

use a spatially discretised stationary model of a tubular SOFC to show how relevant the influence of the separate consideration of the half-cell reactions is on the heat distribution. This illustrates the significance of the transported ion entropy and how important it is to know this material-specific value.

10Sc1CeSZ is considered a promising electrolyte material for SOCs due to its stable cubic phase and high ionic conductivity [38,39]. Studies on the entropy change in the half-cell reactions, the transported ion entropy and the resulting information on the Peltier coefficient are not yet known. Therefore, the results of this work should close these data gaps. These data are determined from Seebeck coefficient measurements as a function of temperature, oxygen and hydrogen partial pressures. In addition, the anode-side Peltier heat is determined from the cathode-side Seebeck coefficient and the entropy change in the entire cell. The measurements are first carried out with an 8YSZ electrolyte to validate the procedure by comparing the results with the literature. Subsequently, the measurements are repeated with a 10SC1CeSZ electrolyte and compared with the previously obtained data.

2. Materials and Methods

2.1. Theory

The following O_2 concentration cell is analysed:

$$\mathrm{Pt}(T^{s,a})|\mathrm{O_2}\,(p^{s,a},T^{s,a})||\mathrm{O^{2-}}||\mathrm{O_2}\,(p^{s,c},T^{s,c})|\mathrm{Pt}(T^{s,c}), \tag{3}$$

with the anode-side electrochemical reaction equation:

$$1/2\,\mathrm{O^{2-}} \rightarrow 1/4\,\mathrm{O_2}\,(\mathrm{g}) + \mathrm{e^-} \tag{4}$$

and its back reaction at the cathode. The system is divided into five subsystems (I–V); see Figure 1. The EMF is measured between the Pt wires (I and II) at the hot ambient temperature T_0 in an open circuit. The temperatures at the contact points between the Pt wire and the cell housing correspond to the ambient temperature ($T_a = T_c = T_0$). The temperature of the anode (IV) is kept constant at $T_{s,a}$, while the temperature of the cathode (V) $T_{s,c}$ is changed so that a temperature difference $\Delta T = T_{s,c} - T_{s,a} > 0$ is set between the electrodes. The temperature difference ΔT leads to an electrical potential $\Delta\phi$, which is made up of the individual partial potentials. For small ΔT and $j \rightarrow 0$, the potential difference $\Delta\phi$ is proportional to the temperature difference ΔT, with the resulting proportionality constant being the Seebeck coefficient according to [40]:

$$\alpha_S = \left(\frac{\mathrm{d}\phi}{\mathrm{d}T}\right)_{j\rightarrow 0,\,\mathrm{d}T\rightarrow 0} = \frac{\Delta_a\phi + \Delta_c\phi}{\mathrm{d}T} + \frac{\Delta_e\phi}{\mathrm{d}T} + \frac{\Delta_{a,e}\phi + \Delta_{e,c}\phi}{\mathrm{d}T}. \tag{5}$$

with $\Delta_i\phi$ as the potential difference in the homogeneous phase i (a: anode, c: cathode, e: electrolyte). Two indices are used to describe transport variables over the surface (s). The first subscript indicates the phase under consideration and the second subscript denotes the nearest phase. For example, $\Delta_{a,e}\phi$ is the difference in potential between the electrolyte (e) and the anode (a). The negative sign in the relationship between the Seebeck coefficient α_S and the Peltier coefficient π in Equation (2) results from the definition of Peltier heat. According to this definition, the reversible Peltier heat is absorbed at the anode when one Faraday positive charge is transported from the anode to the cathode. This also results in a positive definition of the electric current [40,41]. The transported entropy S^* of the charge carriers follows from the entropic balance equation and $\pi = \mathrm{F}\,[J_q^i/j]_{\mathrm{d}T=0,\mathrm{d}\mu=0}$ [42]:

$$S^* = \frac{\pi}{T} = -F \cdot \alpha_S. \tag{6}$$

To establish the reversible entropy balance based on Equation (6) for the individual subsystems according to this definition, the green arrows shown in Figure 1 are used.

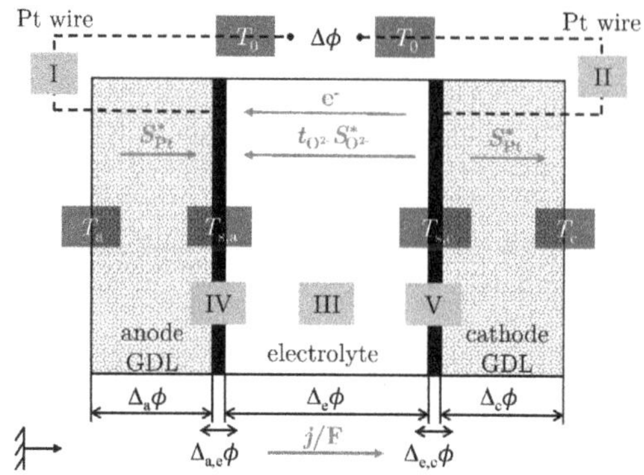

Figure 1. Schematic illustration of the oxygen concentration cell divided into the individual subsystems (I–V).

2.1.1. Electric Leads

For the anode-side and cathode-side gas diffusion layer (GDL), oxygen is chosen as the reference frame so that $J_{O_2} = 0$. The reversible entropy balances for the transport of electrons through the Pt wires are considered when one Faraday positive charge is transported from the anode to the cathode [41]:

$$\Delta_a\phi = -S^*_{Pt}(T_{s,a} - T_0), \tag{7}$$

$$\Delta_c\phi = -S^*_{Pt}(T_0 - T_{s,c}), \tag{8}$$

so

$$\frac{\Delta_a\phi + \Delta_c\phi}{dT} = S^*_{Pt}(T). \tag{9}$$

With S^*_{Pt} as the transported entropy of the electrons in Pt, which can be determined from the data of Moore et al. [43]. T is the mean temperature of the both electrodes.

2.1.2. Surfaces of the Electrodes

The Gibbs "Dividing Surface" approach is used to describe thermodynamic properties of surfaces between two homogeneous phases. The excess quantities introduced there serve to quantify the distribution and quantity of the respective property in comparison to the homogeneous phases. The approach, which was originally set up for a global equilibrium of the system, can be applied to the NET by taking into account time-dependent changes and defining a local equilibrium within a small volume element. For the complete anode-side, the excess entropy production rate $\dot{\sigma}^{s,a}$ is described according to this approach [40] by:

$$\begin{aligned}\dot{\sigma}^{s,a} = J_q^{a,e}\left(-\frac{1}{T^{a,e}}\frac{T^{s,a} - T^{a,e}}{T^{s,a}}\right) + J_q^{e,a}\left(-\frac{1}{T^{e,a}}\frac{T^{e,a} - T^{s,a}}{T^{s,a}}\right) + \ldots \\ \ldots + J_{O_2}^{e,a}\left[-\left(\frac{\mu^s_{O_2}}{T^{s,a}} - \frac{\mu^{e,a}_{O_2}}{T^{e,a}}\right)\right] + j\left[-\frac{1}{T^{s,a}}(\phi^{e,a} - \phi^{a,e})\right] + r^{s,a}\left(-\frac{1}{T^{s,a}}\Delta_n G^{s,a}\right).\end{aligned} \tag{10}$$

In contrast to the fluxes, the excess entropy production rate is independent of the reference frame. According to [44], the surface reference frame in which the surface is at rest is therefore selected for the fluxes. The description of the indices can be taken from Figure 1. The first two terms of the equation represent the contributions to the heat flow. This is followed by the term for the chemical potential of the oxygen, which describes the outgoing mass flow. This is followed by the contribution of the electrical work of the surface. The last part of the equation represents the term for the reaction rate. Equation (10) can be simplified by the following assumptions, cf. [40]:

- constant temperatures due to the surfaces: $T^{a,e} = T^{e,a} = T^{s,a}$
- chemical equilibrium for the absorbed oxygen: $\mu^{a,e}_{O_2}(T^{s,a}) = \mu^{e,a}_{O_2}(T^{s,a}) = \mu^{s}_{O_2}(T^{s,a})$
- constant current: $r^{s,a} = -j/\mathrm{F}$.

If the same assumptions are also made for the surface on the cathode side, the electrical potential differences in the surfaces in the reversible case follow:

$$\Delta_{a,e}\phi = -\,\Delta_n G^{s,a}/\mathrm{F} = -\frac{1}{\mathrm{F}}\left(\frac{1}{4}\mu^{s,a}_{O_2}(T^{s,a})\right), \tag{11}$$

$$\Delta_{e,c}\phi = -\,\Delta_n G^{s,c}/\mathrm{F} = -\frac{1}{\mathrm{F}}\left(-\frac{1}{4}\mu^{s,c}_{O_2}(T^{s,c})\right). \tag{12}$$

The relationship $(\delta\mu_{O_2}/\delta T)_p = -S_{O_2}$ results:

$$\frac{\Delta_{a,e}\phi + \Delta_{e,c}\phi}{dT} = -\frac{1}{\mathrm{F}}\left(\frac{1}{4}S_{O_2}(T)\right). \tag{13}$$

The molar entropy $S_{m,k}$ of component k at temperature T is calculated by the entropic equation of state of ideal gases with the approach according to Kabelac et al. [45] for the molar isobaric heat capacities.

2.1.3. Electrolyte

As in [3], the positive ion lattice is chosen as the homogeneous phase in the electrolyte as a reference frame so that $t_{O^{2-}} = 1$ applies to the transport number of ions. When a positive charge of a Faraday passes from the anode to the cathode, the reversible entropy balance follows [41]:

$$\frac{\Delta_e\phi}{dT} = -\frac{\pi_e}{T^{s,a}} = \frac{1}{2}S^*_{O^{2-}}(T^{s,a}). \tag{14}$$

$S^*_{O^{2-}}$ is the transported entropy of the ions. As already mentioned in the introduction, there are some studies on the transported entropy of the ions, which result in a determined value range of $S^*_{O^{2-}} = 42 - 72\,\mathrm{J/KF}$ according to [33].

2.1.4. Summary of the O_2 Concentration Cell

From the consideration of the individual subsystems, the Seebeck coefficient of the oxygen concentration cell can be calculated according to Equation (5). Thus, the Seebeck coefficient of the oxygen concentration cell results in:

$$\alpha_{S,O_2} = \left(\frac{d\phi}{dT}\right)_{j\to 0,\, dT\to 0} = \frac{1}{\mathrm{F}}\left(-\frac{1}{4}S_{O_2} + S^*_{Pt} + \frac{1}{2}S^*_{O^{2-}}\right), \tag{15}$$

from which the transported entropy of the ions $S^*_{O^{2-}}$ can be calculated by direct measurement of the Seebeck coefficient (as S_{O_2} and S^*_{Pt} are known). Equation (14) can then be used to determine the Peltier coefficient of the electrolyte π_e. The Peltier coefficient for the oxygen reaction then follows from the relationship from Equation (6), cf. [3]:

$$\frac{\pi_c}{T^{s,c}} = -\frac{1}{4}S_{O_2} + S^*_{Pt}. \tag{16}$$

2.1.5. Extension to $H_2/H_2O - O_2$ Cell

To determine the Peltier coefficient for the hydrogen reaction, a H_2/H_2O concentration cell is now considered according to the following equation:

$$\mathrm{Pt}(T^{s,a})\,|\,H_2\,(p^{s,a},T^{s,a}), H_2O\,(p^{s,a},T^{s,a})\,||\,O^{2-}\,||\,H_2\,(p^{s,c},T^{s,c}), H_2O\,(p^{s,a},T^{s,a})\,|\,\mathrm{Pt}(T^{s,c}). \tag{17}$$

with the anode-side electrochemical reaction equation:

$$1/2\,H_2 + 1/2\,O^{2-} \rightarrow e^- + 1/2\,H_2O \tag{18}$$

and its back reaction at the cathode. Analogous to the O_2 concentration cell, the same procedure can also be applied to the H_2/H_2O concentration cell so that the Seebeck coefficient α_{S,H_2} applies:

$$\alpha_{S,H_2} = \left(\frac{d\phi}{dT}\right)_{j\to 0,\, dT\to 0} = \frac{1}{F}\left(-\frac{1}{2}S_{H_2O} + \frac{1}{2}S_{H_2} + S^*_{Pt} + \frac{1}{2}S^*_{O^{2-}}\right), \tag{19}$$

so that the Peltier coefficient of the hydrogen reaction results according to [3]:

$$\frac{\pi_a}{T^{s,a}} = -\frac{1}{2}S_{H_2O} + \frac{1}{2}S_{H_2} + S^*_{Pt}. \tag{20}$$

From the divided consideration of the individual half-cell reactions, the entropy change ΔS or reaction entropy of the $H_2/H_2O - O_2$ reaction for the isothermal case [41] follows by combination:

$$F\left(\frac{d\phi_{O_2}}{dT}\right) - F\left(\frac{d\phi_{H_2}}{dT}\right) = F\left(\frac{d\phi_{tot}}{dT}\right) = \frac{1}{2}S_{H_2O} - \frac{1}{4}S_{O_2} - \frac{1}{2}S_{H_2} = \Delta S, \tag{21}$$

while $\alpha_{S,O_2} = (d\phi_{O_2}/dT)$ corresponds to the Seebeck coefficient in Equation (15). If the potential difference in the case of the oxygen concentration cell $\alpha_{S,O_2} = (d\phi_{O_2}/dT)$ and the potential difference in the $H_2/H_2O - O_2$ cell $\alpha_{S,tot} = (d\phi_{tot}/dT)$ are measured, the Seebeck coefficient $\alpha_{S,H_2} = (d\phi_{H_2}/dT)$ of an H_2/H_2O concentration cell can be calculated according to Equation (21). A plausibility check of the measured Seebeck coefficient $d\phi_{tot}/dT$ can be performed from the calculation of the total reaction entropy ΔS. In addition, the transported entropy of the ions can then be calculated using Equation (19) to determine the transported entropy of the ions and compare it with the calculated values from Equation (15), which must be identical.

2.2. Experiments

Test Environment

The Seebeck coefficients according to Equations (15) and (21) are measured using two different test environments; see Figure 2.

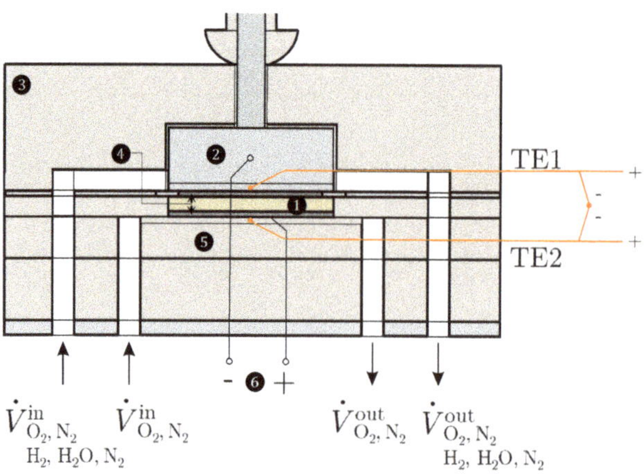

Figure 2. Measuring chamber: (1) sample, (2) inconel block with channels, (3) housing, (4) Pt mesh, (5) ceramic plate with channels and heater, (6) Pt wires.

For this purpose, an SOFC/SOEC test bench (Evaluator C1000-HT) from HORIBA FuelCon (Magdeburg-Barleben, Germany) is available, which is suitable for analysing individual components. For the realisation of the oxygen concentration cell, only oxygen (O_2) and nitrogen (N_2) are available as gases for both electrodes, whereby different partial pressures of the oxygen can be set. In contrast, in the $H_2/H_2O - O_2$ cell, hydrogen (H_2), water (H_2O) and nitrogen (N_2) are supplied to the anode-side electrode. The volume flows of the gases can be precisely controlled in the range from 0 to 0.5 Nl/min using mass flow controllers. The test specimens (1) are 8YSZ and 10Sc1CeSZ oxide ceramic specimens with the dimensions $40 \times 40 \times 10$ mm from CerPoTech (Heimdal, Norway). The surfaces of both test specimens are coated with a uniform 300 µm thick Pt coating by sputtering, which makes them electrochemically active. Pt nets (4) lie on these surfaces, which are in contact with Pt wires (6), via which the potential difference is measured. The Solartron ModuLab XM ECS electrochemical test system from AMETEK SI (Meerbusch, Germany) is available for this purpose. The ceramic base (5) contains flow channels for the cathode-side gases. A NiCr heating wire (Nikrothal® 80) from Kanthal GmbH (Mörfelden-Walldorf, Germany), which is controlled by an external voltage source, is placed inside these channels. The heating wire is electrically insulated from the Pt network using a ceramic paste. The temperature difference $\Delta T = T^{s,c} - T^{s,a} > 0$ is measured directly by connecting two type-S thermocouples (TE1 and TE2) with opposite poles. These thermocouples are attached to the top and bottom of the sample and are calibrated using the Pegasus 4853 high-temperature calibrator from ISOTECH (Merseyside, England). The test environment is closed off by an oven that is heated to a temperature of ϑ_{amb} = 1000 °C. The manufacturers give the uncertainties of all measuring instruments used. The GUM [46] and DIN EN 60584-1 [47] are used to determine the measurement uncertainty and the limiting deviation.

3. Experimental Calculation

After heating the measuring chamber (mc), the gas tightness of the set-up was determined by comparing the measured open circuit voltage (OCV) and the calculated Nernst voltage for different partial pressure ratios $x_{O_2} = \ln(p^c_{O_2}/p^a_{O_2})$, also $x_{H_2/H_2O-O_2} = \ln(p_{H_2O}/p_{H_2} \cdot p^{0.5}_{O_2})$ was checked. This electrochemical leak test was carried out for measuring chamber temperatures of ϑ_{mc} = 700 °C, 800 °C and 900 °C. To additionally check the influence of the total volume flow rates at the electrodes, the measurements were repeated once with $\dot{V}_{total} = 0.2$ Nl/min and $\dot{V}_{total} = 0.5$ Nl/min. All measurements showed good agreement with the calculated Nernst voltage. During the operation of the oxygen concentration cell, maximum deviations of 15% were found for high partial pressure ratios of $x_{O_2} = \pm 2.3$, which can be attributed to the limited accuracy of the mass flow control at low oxygen volume flows. In $H_2/H_2O - O_2$ operation, the maximum deviations were 1.5 %.

The gas tightness was rated as sufficient, so an impairment of the Pt coatings on the electrochemical reaction could be ruled out.

3.1. O_2 Cell

Three different experiments are carried out in the operation of the oxygen cell. To investigate the influence of temperature dependence, these experiments are repeated at temperatures $\vartheta_{mc} = 700\,°C, 800\,°C$ and $900\,°C$. The total volume flow rates are identical on both electrodes, $\dot{V}^a_{total} = \dot{V}^c_{total}$. As described above, the surface temperature at the cathode T^c is heated evenly by a heating wire. Temperature differences of $\Delta T = 0 - 15$ K are approached. The three experiments are carried out as follows:

(1) Identical oxygen composition at both electrodes, $x^a_{O_2} = x^c_{O_2} = 0.1, 0.2, 0.5, 0.8, 1$,
(2) Variation in the anode-side oxygen composition $x^a_{O_2} = 0.1, 0.2, 0.5, 0.8, 1$ with a constant cathode-side oxygen composition of $x^c_{O_2} = 1$,
(3) Variation in the cathode-side oxygen composition $x^c_{O_2} = 0.1, 0.2, 0.5, 0.8, 1$ with a constant anode-side oxygen composition of $x^a_{O_2} = 1$.

3.2. $H_2/H_2O - O_2$ Cell

Three different experiments are also carried out for this operation. The framework conditions with regard to the volume flows and the temperature difference correspond to those used in the operation of the oxygen concentration cell. The three experiments are carried out as follows:

(4) Variation in the cathode-side composition $x^c_{O_2} = 0.1, 0.21, 0.4, 1$ with an anode-side composition according to $x^a_{H_2} + x^a_{H_2O} = 1$ at $\vartheta_{mc} = 700\,°C$,

(5) Variation in the anode-side composition according to $x^a_{H_2} + x^a_{H_2O} + x^a_{N_2} = 1$ with a cathode-side composition of $x^c_{O_2} = 0.21$ for $\vartheta_{mc} = 700\,°C$,

(6) Variation in the measuring chamber temperature $\vartheta_{mc} = 700\,°C, 800\,°C, 900\,°C$ with an anode-side composition of $x^a_{H_2} + x^a_{H_2O} = 1$ and a cathode-side composition of $x^c_{O_2} = 0.21$.

4. Results and Discussion

4.1. O_2 Cell

4.1.1. Seebeck Coefficient

Figure 3 shows the measured Seebeck coefficients $\alpha_{S,O_2} = dE/dT$ for an oxygen concentration cell with identical oxygen composition at both electrodes for the temperatures $\vartheta_{mc} = 700\,°C, 800\,°C$ and $900\,°C$ for 8YSZ and 10Sc1CeSZ.

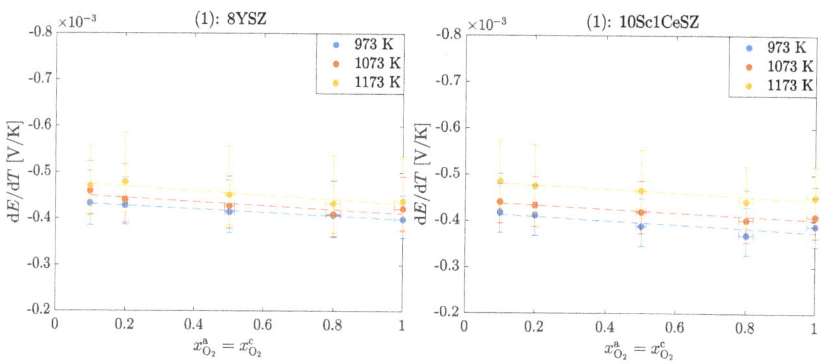

Figure 3. Seebeck coefficients α_{S,O_2} in the operation of an oxygen concentration cell for experiment (1): 8YSZ and 10Sc1CeSZ.

The Seebeck coefficients of 8YSZ measured in this experiment are slightly below the values given in the literature; see Table 1. However, since there are also large scatters between the literature values and the measured values are also subject to some errors, the results are assessed as plausible. The large vertical error bars result from the temperature measurement at the temperatures under consideration. The horizontal error bars result from the uncertainties of the mass flow controllers. This should explain that there is no true value in the literature, but that rather all measured values can be considered valid. The negative signs indicate that heat is generated on the anode-side surface and is reversibly transported inside the cell from the cold surface to the hot surface.

The following applies to the dependence of the Seebeck coefficient on the oxygen partial pressure $\alpha_{S,O_2}(p_{O_2})$:

$$\left(\frac{\alpha_{S,O_2}}{d \ln p_{O_2}/p_0}\right) = \frac{R}{4F} = 0.0215\,\text{mV/K}, \tag{22}$$

in which the following dependencies can be determined from the available results; see Table 2.

Table 2. Slope of the linear relationship between the Seebeck coefficient α_{S,O_2} and the partial pressure of oxygen p_{O_2} for 8YSZ and 10Sc1CeSZ for $\vartheta_{mc} = 700\,°C, 800\,°C, 900\,°C$; see Figure 3.

T [K]	Dopant	Slope [mV/K]
973.15	Y_2O_3	0.014 ± 0.022
973.15	Sc_2O_3	0.018 ± 0.017
1073.15	Y_2O_3	0.019 ± 0.023
1073.15	Sc_2O_3	0.016 ± 0.010
1173.15	Y_2O_3	0.018 ± 0.029
1173.15	Sc_2O_3	0.017 ± 0.014

Figure 3 shows a slight temperature dependence of the Seebeck coefficient $\alpha_{S,O_2}(T)$, according to [33]:

$$\left(\frac{\alpha_{S,O_2}}{\mathrm{d}\ln T/T_0}\right) = \frac{1}{\mathrm{F}}\left(-\frac{1}{4}C_{m,p,O_2} + \tau^*_{Pt} + \frac{1}{2}\tau^*_{O^{2-}}\right). \tag{23}$$

For 8YSZ, this results in a negative slope of $-0.00022 \pm 0.0003\,\mathrm{mV/K^2}$ and for 10Sc1CeSZ of $-0.00031 \pm 0.00026\,\mathrm{mV/K^2}$. There is no explicit temperature dependence addressed in the literature [32]. Some studies claim that a temperature dependence arises only from the molar isobaric heat capacity of oxygen C_{m,p,O_2} and the Thomson coefficient of the electron τ^*_{Pt} so that these two terms cancel each other out and there is no temperature dependence of the transported entropy of the oxygen ion results [48]. Other studies show temperature independence of the Seebeck coefficient, neglect the Thomson coefficient of the electron and equate the Thomson coefficient of the ion $\tau^*_{O^{2-}}$ with the molar isobaric heat capacity [33]. From all considerations, as well as from the results here, a very low temperature dependence (of α_{S,O_2} or $S^*_{O^{2-}}$) results. A comparison of the Seebeck coefficients of the two materials shows that the relative deviation of 10Sc1CeSZ to 8YSZ at 700 °C is about −5.5%, at 800 °C about −2.55% and at 900 °C is even 2.02%. These values show that the Seebeck coefficient of 8YSZ is slightly higher than that of 10Sc1CeSZ at higher temperatures, with differences decreasing with increasing temperature. This correlation is established by the absolute measurement results. Due to the presence of large error bars, this statement is only qualitative. The studies by Ahlgreen et al. [28] and Pizzini et al. [31] show that the Seebeck coefficient decreases with increasing doping and increasing vacancies. However, the differences here are too small to give a valid conclusion.

Figure 4 shows the results of experiments (2) and (3) for 8YSZ and 10Sc1CeSZ. The fundamental differences in terms of temperature and material dependence can also be seen in these experiments, as in the previous experiment. A clear difference to experiment (1) can be seen in the slopes of the Seebeck coefficients, which are more pronounced in these experiments; see Table 3.

Table 3. Slope of the linear relationship between the Seebeck coefficient α_{S,O_2} and the partial pressure of oxygen $p^i_{O_2}$ for 8YSZ and 10Sc1CeSZ for $\vartheta_{mc} = 700\,°C, 800\,°C, 900\,°C$; see Figure 4.

T [K]	Dopant	Slope for $x^a_{O_2} = 1$ [mV/K]	Slope for $x^c_{O_2} = 1$ [mV/K]
973.15	Y_2O_3	0.047 ± 0.022	−0.04 ± 0.02
973.15	Sc_2O_3	0.035 ± 0.021	−0.036 ± 0.019
1073.15	Y_2O_3	0.06 ± 0.029	−0.04 ± 0.027
1073.15	Sc_2O_3	0.048 ± 0.026	−0.025 ± 0.023
1173.15	Y_2O_3	0.06 ± 0.04	−0.027 ± 0.034
1173.15	Sc_2O_3	0.034 ± 0.027	−0.024 ± 0.023

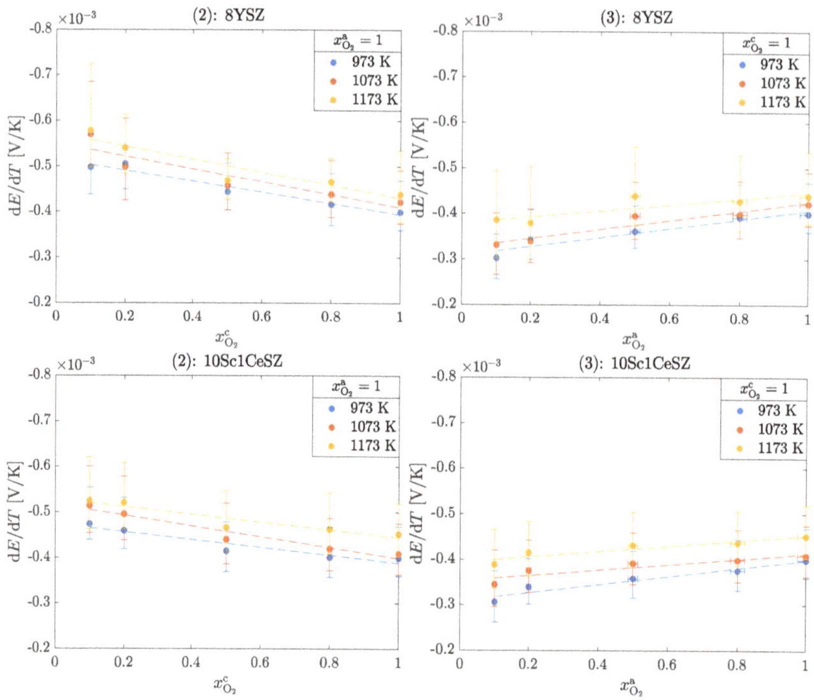

Figure 4. Seebeck coefficients α_{S,O_2} in the operation of an oxygen cell for experiments (2) and (3): 8YSZ and 10Sc1CeSZ.

As in experiment (1), it can be seen in experiment (2) that the Seebeck coefficients increase with the ln of the decreasing oxygen partial pressure. This effect is further enhanced by the difference in composition. Equation (15) cannot be used for these experiments because there are different oxygen partial pressures at the electrodes; see Equation (12). If experiment (2) is reversed, the results from experiment (3) are obtained. The behavior of the Seebeck coefficients is reversed, so they decrease with the ln of the decreasing oxygen partial pressure.

4.1.2. Transported Ion Entropy and Peltier Coefficient

The transported entropy of the oxygen ions $S^*_{O^{2-}}$ and the Peltier coefficients of the cathode π_c and the electrolyte π_e are calculated from the results of experiment (1) using Equations (14) and (16). In contrast to the Peltier coefficient of the cathode, the mean values $\hat{S^*_{O^{2-}}}$ and $\hat{\pi}_e$ are calculated due to the independence of the oxygen partial pressure. Measurement inaccuracies are taken into account when calculating the mean values. For reasons of clarity, the uncertainties of all individual values have been omitted. The results are summarized in Table 4.

Due to the direct dependence of the transported entropy of the ions $S^*_{O^{2-}}(\alpha_{S,O_2})$ on the Seebeck coefficient, this shows a corresponding behavior with changes in the oxygen partial pressure, the temperature and the material. The values determined are slightly above the previously assumed value of $S^*_{O^{2-}}$ = 42 J/KF at 1273 K according to [33]. This also results in slightly higher Peltier coefficients for the electrolyte $\pi_e(S^*_{O^{2-}})$. The influence of the transported entropy of the ions is examined in more detail in the last part of the results. If the Peltier coefficients of the two homogeneous phases are compared, it can also be stated here that the effect is more pronounced at the cathode than in the electrolyte, which has already been shown in the modeling results [3,4].

Table 4. Transported entropy of the ions $S^*_{O^{2-}}$ and Peltier coefficient for the electrolyte π_e and cathode π_c from the results in Figure 3.

T [K]	Dopant	p_{O_2} [bar]	$S^*_{O^{2-}}$ [J/KF]	π_e [J/C]	π_c [J/C]
973.15	Y_2O_3	0.1	50	−0.25	−0.66
		0.21	48	−0.24	−0.65
		0.5	47	−0.24	−0.63
		0.8	47	−0.23	−0.62
		1	47	−0.23	−0.62
			$\hat{S}^*_{O^{2-}} = \mathbf{48 \pm 9}$	$\hat{\pi}_e = \mathbf{-0.24 \pm 0.04}$	
1073.15		0.1	47	−0.26	−0.74
		0.21	48	−0.26	−0.73
		0.5	47	−0.26	−0.71
		0.8	48	−0.26	−0.70
		1	45	−0.25	−0.69
			$\hat{S}^*_{O^{2-}} = \mathbf{47 \pm 11}$	$\hat{\pi}_e = \mathbf{-0.26 \pm 0.06}$	
1173.15		0.1	47	−0.28	−0.83
		0.21	43	−0.26	−0.81
		0.5	44	−0.26	−0.79
		0.8	46	−0.27	−0.77
		1	44	−0.26	−0.77
			$\hat{S}^*_{O^{2-}} = \mathbf{45 \pm 14}$	$\hat{\pi}_e = \mathbf{-0.27 \pm 0.09}$	
973.15	Sc_2O_3	0.1	53	−0.26	−0.66
		0.21	51	−0.25	−0.65
		0.5	52	−0.26	−0.63
		0.8	54	−0.27	−0.62
		1	49	−0.24	−0.62
			$\hat{S}^*_{O^{2-}} = \mathbf{52 \pm 10}$	$\hat{\pi}_e = \mathbf{-0.26 \pm 0.05}$	
1073.15		0.1	51	−0.26	−0.74
		0.21	49	−0.26	−0.73
		0.5	48	−0.26	−0.71
		0.8	49	−0.26	−0.70
		1	47	−0.25	−0.69
			$\hat{S}^*_{O^{2-}} = \mathbf{49 \pm 10}$	$\hat{\pi}_e = \mathbf{-0.27 \pm 0.06}$	
1173.15		0.1	44	−0.28	−0.83
		0.21	43	−0.26	−0.81
		0.5	41	−0.26	−0.79
		0.8	43	−0.27	−0.77
		1	40	−0.26	−0.77
			$\hat{S}^*_{O^{2-}} = \mathbf{42 \pm 11}$	$\hat{\pi}_e = \mathbf{-0.25 \pm 0.07}$	

4.2. $H_2/H_2O - O_2$ Cell

Figure 5 shows the measured Seebeck coefficients $\alpha_{S,tot} = dE/dT$ for an $H_2/H_2O - O_2$ cell for experiment (4). With the help of the following results, the transported entropy of the ions $S^*_{O^{2-}}$ can also be determined; these must then correspond to the values from Table 4. Furthermore, the Peltier coefficient of the homogeneous phase on the anode side π_a is calculated according to Equation (20).

According to Equation (21), the overall reaction is exothermic at every operating point ($\alpha_{S,tot} < 0$). It can be seen that with increasing oxygen partial pressure at the cathode and with increasing hydrogen partial pressure at the anode, the Seebeck coefficient decreases and thus less heat is released. There is no clear difference between the materials, which is to be expected and seems to be correct, as the same reaction entropy ΔS is released in both materials. It is remarkable that the order of magnitude of the Seebeck coefficients here is similar to that in Figure 3. At an oxygen partial pressure of $p_{O_2} = 0.21$ bar, experiment (1) results in a value of -0.43 ± 0.05 mV/K for 8YSZ and here with increasing hydrogen partial pressure p_{H_2} from -0.4 ± 0.3 to -0.2 ± 0.2 mV/K. From this observation, it can be concluded that heat production on the cathode-side surface dominates the overall reaction. The following Figure 6 shows the measured Seebeck coefficients $\alpha_{S,tot} = \mathrm{d}E/\mathrm{d}T$ for experiment (5).

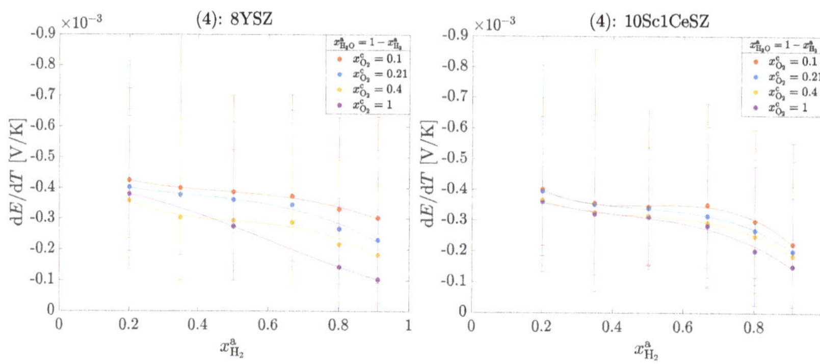

Figure 5. Seebeck coefficients $\alpha_{S,tot}$ in the operation of a $H_2/H_2O - O_2$ cell for experiment (4): 8YSZ and 10Sc1CeSZ. The solid lines within a measurement series are used for clear visualization.

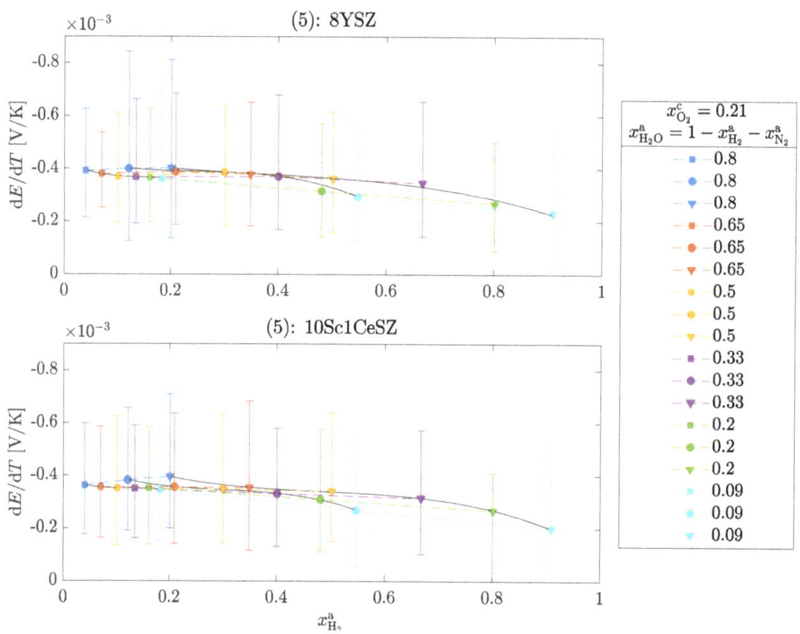

Figure 6. Seebeck coefficients $\alpha_{S,tot}$ in the operation of a $H_2/H_2O - O_2$ cell for experiment (5): 8YSZ and 10Sc1CeSZ. The solid lines within a measurement series are used for clear visualization.

Operating points of the same color mean that the supplied volume flow of water $\dot{V}_{H_2O}$ was kept constant and the volume flows of hydrogen and nitrogen were varied so that $\dot{V}_{total}$ = const. This resulted in a constant water partial pressure p_{H_2O} with a varrying p_{H_2}/p_{N_2} ratio. Continuous operating points typically have a constant p_{H_2}/p_{N_2} ratio with varying water partial pressure p_{H_2O}. Here, too, the Seebeck coefficients decrease with increasing hydrogen partial pressure. However, the Seebeck coefficients show deviations from this behavior within a constant water partial pressure; here, $p_{H_2O} = 0.8$ and $p_{H_2O} = 0.65$. With increasing hydrogen partial pressure, slight increases can be seen in these points, which does not correspond to theory. One possible reason for the deviation is the inaccurate control of the hydrogen supply quantity at these low partial pressures. Finally, the results of experiment (6) are shown in Figure 7.

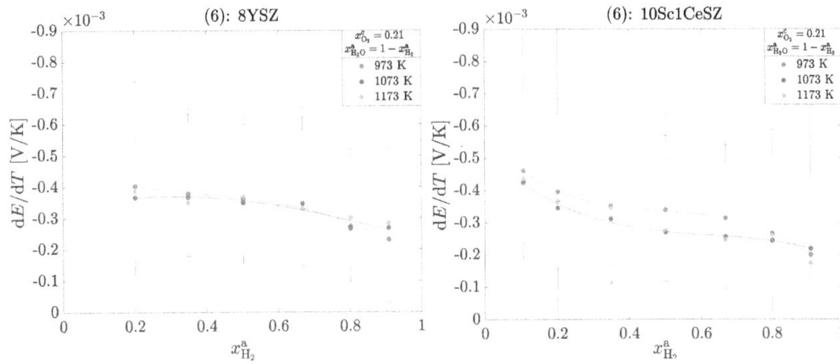

Figure 7. Seebeck coefficients $\alpha_{S,tot}$ in the operation of a $H_2/H_2O - O_2$ cell for experiment (6): 8YSZ and 10Sc1CeSZ. The solid lines within a measurement series are used for clear visualization.

As in the operation of the oxygen concentration cell, the temperature dependence should be determined here. There are no clear dependencies for either material. According to theory, the following applies to an electron $F \cdot \Delta^R S_m(700\,°C, x_{H_2O} = x_{H_2} = 0.5, x_{O_2} = 0.21) = 0.318\,V/K$ and $F \cdot \Delta^R S_m(900\,°C, x_{H_2O} = x_{H_2} = 0.5, x_{O_2} = 0.21) = 0.324\,V/K$ for 8YSZ. Accordingly, the total reaction entropy $\Delta^R S_m$ and thus also the Seebeck coefficient increase with increasing temperature. However, this change is so small that it is not clearly seen within the measurement inaccuracies.

4.2.1. Transported Ion Entropy and Peltier Coefficient

Table 5 summarizes the transported entropies of the ions $S^*_{O^{2-}}$, the Peltier coefficients of the electrolyte π_e and the anode π_a from the results of Figure 5 according to Equations (19), (20) and (14). First, the mean values for $S^*_{O^{2-}}$ and π_e were calculated at p_{O_2} = const. and then the total mean value was calculated from these individual mean values. These values are compared with the values from Table 4 and it can be seen that the values match with a slight inaccuracy, which corresponds to the theory. It can also be seen that the Peltier coefficients of the homogeneous phase on the anode side are about half as large as the Peltier coefficients on the cathode side, which illustrates the stronger manifestation of the Peltier effect at the cathode.

Experiment (5) also shows very good agreement with the results from the oxygen concentration cell; see Table 6. A closer look at the transported ion entropy for the two materials shows higher values for 8YSZ than for 10Sc1CeSZ in $H_2/H_2O - O_2$ mode. However, based on the previously obtained results and the findings from the literature, this behavior should be reversed. The deviations are due to the measurement in accuracies. The same behavior can also be seen in the results of experiment (6); see Table 7. Furthermore, a clear temperature dependence of the transported ion entropy can also be seen here. The significance of the results obtained for the operational behavior of an SOFC/SOEC is discussed below.

Table 5. Transported entropy of the ions $S^*_{O^{2-}}$ and Peltier coefficient for the electrolyte π_e and anode π_a for $T = 973$ K from the results in Figure 5.

Dopant	p_{H_2} [bar]	p_{O_2} [bar]	$S^*_{O^{2-}}$ [J/KF]	π_e [J/C]	π_a [J/C]
Y_2O_3	0.2	0.21	53	−0.26	−0.28
	0.35		54	−0.27	−0.32
	0.5		56	−0.28	−0.34
	0.67		59	−0.29	−0.37
	0.8		50	−0.25	−0.4
	0.91		50	−0.25	−0.44
			$\hat{S^*_{O^{2-}}} = 54 \pm 37$	$\hat{\pi_e} = -0.27 \pm 0.2$	
Y_2O_3	0.2	1	54	−0.27	−0.28
	0.35		55	−0.27	−0.32
	0.5		45	−0.22	−0.34
	0.67		45	−0.22	−0.37
	0.8		31	−0.16	−0.4
	0.91		31	−0.16	−0.44
			$\hat{S^*_{O^{2-}}} = 44 \pm 39$	$\hat{\pi_e} = -0.22 \pm 0.2$	
Y_2O_3	"	0.1	$\hat{S^*_{O^{2-}}} = 60 \pm 47$	$\hat{\pi_e} = -0.3 \pm 0.2$	
Y_2O_3	"	0.4	$\hat{S^*_{O^{2-}}} = 46 \pm 38$	$\hat{\pi_e} = -0.23 \pm 0.2$	
			$\hat{S^*_{O^{2-}}} = 51 \pm 40$	$\hat{\pi_e} = -0.25 \pm 0.2$	
Table 4			$\hat{S^*_{O^{2-}}} = 48 \pm 9$	$\hat{\pi_e} = -0.24 \pm 0.043$	
Sc_2O_3	0.2	0.21	51	−0.26	−0.28
	0.35		49	−0.25	−0.32
	0.5		52	−0.26	−0.34
	0.67		53	−0.26	−0.37
	0.8		50	−0.25	−0.4
	0.91		44	−0.22	−0.44
			$\hat{S^*_{O^{2-}}} = 50 \pm 38$	$\hat{\pi_e} = -0.25 \pm 0.2$	
Sc_2O_3	0.2	1	53	−0.26	−0.28
	0.35		51	−0.25	−0.32
	0.5		54	−0.27	−0.34
	0.67		55	−0.27	−0.37
	0.8		52	−0.26	−0.4
	0.91		48	−0.24	−0.44
			$\hat{S^*_{O^{2-}}} = 52 \pm 36$	$\hat{\pi_e} = -0.26 \pm 0.2$	
Sc_2O_3	"	0.1	$\hat{S^*_{O^{2-}}} = 50 \pm 40$	$\hat{\pi_e} = -0.25 \pm 0.2$	
Sc_2O_3	"	0.4	$\hat{S^*_{O^{2-}}} = 50 \pm 36$	$\hat{\pi_e} = -0.25 \pm 0.2$	
			$\hat{S^*_{O^{2-}}} = 51 \pm 38$	$\hat{\pi_e} = -0.25 \pm 0.2$	
Table 4			$\hat{S^*_{O^{2-}}} = 52 \pm 10$	$\hat{\pi_e} = -0.26 \pm 0.050$	

Table 6. Transported entropy of the ions $S^*_{O^{2-}}$ and Peltier coefficient for the electrolyte π_e from the results in Figure 6.

T [K]	Dopant	p_{H_2O} [bar]	p_{H_2} [bar]	$\hat{S}^*_{O^{2-}}$ [J/KF]	$\hat{\pi}_e$ [J/C]
973.15	Y_2O_3	0.8	0.04, 0.12, 0.2	46	−0.23
		0.65	0.07, 0.21, 0.35	49	−0.25
		0.5	0.1, 0.3, 0.5	53	−0.26
		0.33	0.13, 0.4, 0.67	56	−0.28
		0.2	0.16, 0.48, 0.8	53	−0.26
		0.09	0.18, 0.55, 0.91	57	−0.28
				$\hat{S}^*_{O^{2-}} = \mathbf{52 \pm 40}$	$\hat{\pi}_e = \mathbf{-0.26 \pm 0.2}$
Table 4				$\hat{S}^*_{O^{2-}} = \mathbf{48 \pm 9}$	$\hat{\pi}_e = \mathbf{-0.24 \pm 0.043}$
973.15	Sc_2O_3	0.8	0.04, 0.12, 0.2	46	−0.23
		0.65	0.07, 0.21, 0.35	47	−0.23
		0.5	0.1, 0.3, 0.5	51	−0.25
		0.33	0.13, 0.4, 0.67	54	−0.27
		0.2	0.16, 0.48, 0.8	55	−0.27
		0.09	0.18, 0.55, 0.91	56	−0.28
				$\hat{S}^*_{O^{2-}} = \mathbf{51 \pm 41}$	$\hat{\pi}_e = \mathbf{-0.25 \pm 0.2}$
Table 4				$\hat{S}^*_{O^{2-}} = \mathbf{52 \pm 10}$	$\hat{\pi}_e = \mathbf{-0.26 \pm 0.05}$

Table 7. Transported entropy of the ions $S^*_{O^{2-}}$ and Peltier coefficient for the electrolyte π_e from the results in Figure 7.

T [K]	Dopant	$\hat{S}^*_{O^{2-}}$ [J/KF]	$\hat{\pi}_e$ [J/C]
1073.15	Y_2O_3	51 ± 40	−0.27 ± 0.2
Table 4		47 ± 11	−0.26 ± 0.06
1173.15		48 ± 40	−0.29 ± 0.2
Table 4		44 ± 14	−0.27 ± 0.09
1073.15	Sc_2O_3	47 ± 40	−0.26 ± 0.2
Table 4		49 ± 10	−0.27 ± 0.06
1173.15		39 ± 39	−0.23 ± 0.2
Table 4		42 ± 11	−0.25 ± 0.07

4.2.2. Heat Generation of the Electrodes of an SOC

The results for the transported entropy of the ions $S^*_{O^{2-}}$ from experiments (1), (4), (5) and (6) can be used to calculate the heat generation of the half-cell reactions of an SOC. Figure 8 illustrates the total heat production $T \cdot \Delta S$ and the partial heat production at the electrodes $T \cdot \Delta S_i$ for four different scenarios based on the results obtained.

The overall reaction in all scenarios is exothermic and consists of an exothermic reaction at the cathode and a predominantly endothermic reaction at the anode. The reference scenario (a) describes an operation at a temperature of $T = 973.15\,\text{K}$ and a constant oxygen partial pressure on the cathode side of $p_{O_2} = \text{const.} = 0.21$ bar. The transported ion entropy is $S^*_{O^{2-}} = 52 \pm 10\,\text{J/KF}$ for 10Sc1CeSZ and $S^*_{O^{2-}} = 48 \pm 9\,\text{J/KF}$ for 8YSZ. It can be seen that there are minimal differences in the partial heat production of the individual electrodes between the two materials. In general, it can be seen that 10Sc1CeSZ produces slightly less heat at the cathode and absorbs slightly less heat at the anode than 8YSZ.

Taking the same operating conditions, the influence of the previously assumed transported ion entropy of $S^*_{O^{2-}} = 37\,\text{J/KF}$ is also determined in addition to the reference scenario (b). As this value is significantly lower, this results in higher overall amounts of heat that are absorbed or released at the electrodes. It can also be seen that an endothermic anode reaction occurs at low hydrogen partial pressures. This finding is also shown by Fischer et al. [37].

If oxygen is supplied instead of air on the cathode side, only slight differences in the total heat production and the heat production on the cathode side are recognizable (c). Increasing the oxygen partial pressure leads to less heat being released at the cathode, which results in a slight reduction in total heat production.

The last scenario examined is the influence of temperature (d). An increase in temperature by $T = 200\,\mathrm{K}$ leads to a significant increase in heat production in the overall reaction and at the cathode. Although more heat is also absorbed in the anode, the exothermic part of the cathode reaction dominates.

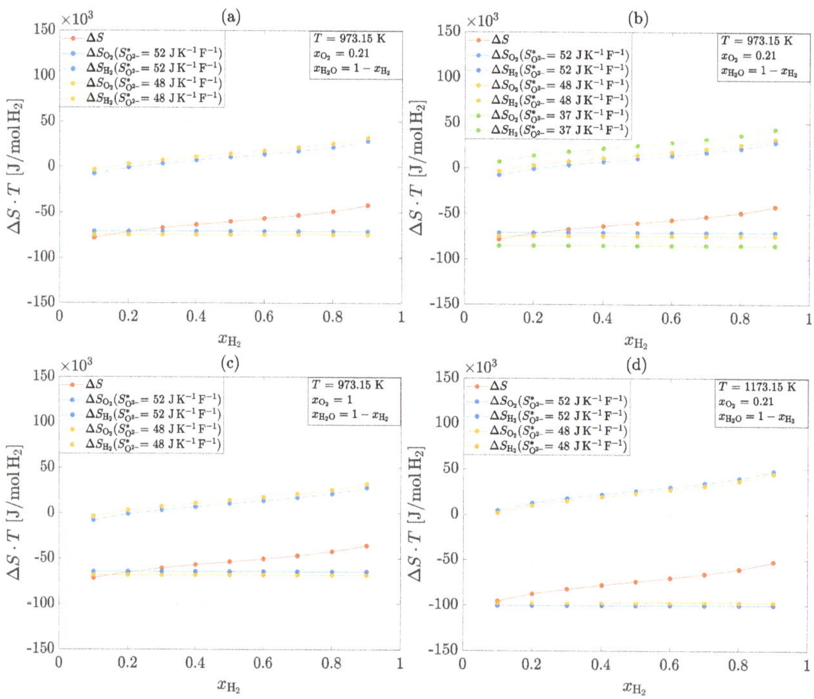

Figure 8. Total heat production and the partial heat production at the electrodes for four different scenarios: 1. Reference scenario (**a**). 2. Variation in the transported entropy of the ion (**b**). 3. Change in oxygen patial pressure from $p_{O_2} = 0.21$ bar to $p_{O_2} = 1$ bar (**c**). 4. Change in temperature from $T = 973.15$ K to $T = 1173.15$ K (**d**).

5. Conclusions

For exact modeling and calculation of the heat transport in SOCs, the transported entropy of the ions for an 8YSZ and a 10Sc1CeSZ electrolyte was determined experimentally in this work, from which, in turn, the Peltier coefficient could be calculated directly. The Peltier effect leads to the transport of reversible heat in the cell. In our previous studies and also in the literature, it has already been clearly shown that this effect must be taken into account for successful thermal management. In contrast to 10Sc1CeSZ, some data exist for 8YSZ in the literature, so the plausibility of these results could be shown. The transported entropy of the ions was determined by direct measurement of the Seebeck coefficient. This was measured for a O_2 concentration cell, a $O_2 - O_2$ cell and a $H_2/H_2O - O_2$ cell in the temperature range from 700 to 900 °C. The different operating modes were used for methodological variation but should provide and do provide consistent results.

The results from the O_2 concentration cell for 8YSZ show good agreement with the results from the literature. It was found that the measured Seebeck coefficients of 10Sc1CeSZ are slightly smaller than those of 8YSZ. This results in a higher transported entropy of the ions. For a temperature of 700 °C and an oxygen partial pressure of $p_{O_2} = 0.21$ bar in a value of $S^*_{O^{2-}} = 52 \pm 10$ J/KF for 10Sc1CeSZ and $S^*_{O^{2-}} = 48 \pm 9$ J/KF for 8YSZ results. The determination of the transported entropy of the ions via the operation of a $H_2/H_2O - O_2$ cell has not been shown previously in the literature. The validity of the measured Seebeck coefficients could be demonstrated with a theoretical calculation of the total reaction entropy. The results for the transported entropy of the ions showed deviations of about 6% at 8YSZ and −2.6% at 10Sc1CeSZ compared to the results from the O_2 concentration cell operation. However, taking into account the measurement errors, these deviations were assessed as small. The influence of the transported entropy of the ions, the oxygen partial pressure and the temperature on the total heat generation and on the partial heat generation of the electrodes of a $H_2/H_2O - O_2$ cell was also investigated. As expected, the temperature has the greatest influence on the heat generation of the half-cell reactions and thus also on the total heat generation in the cell. The transported entropy of the ions has an effect of $T \cdot S^*_{O^{2-}}$ on the heat generation of both half-cell reactions.

Based on the successful measurement results, values for the Peltier coefficients were determined of the homogeneous phases (anode, cathode and electrolyte) that correspond to those in the literature. A negative sign of the coefficient was found for all phases, whereby the coefficient for the homogeneous phase of the cathode was twice as high as that for the anode and the electrolyte. Previous studies have already shown that the Peltier effect at the cathode has the greatest influence on the system.

Author Contributions: A.G. and S.K. conceived the idea and the theoretical background. A.G. carried out the measurements. A.G. analyzed the data and evaluated the computational data. S.K. supervised the work. A.G. wrote the paper. A.G. made the figures. A.G. and S.K. revides the paper. All authors have read and agreed to the published version of the manuscript.

Funding: The authors gratefully acknowledge the financial support by the Deutsche Forschungsgemeinschaft (DFG, funding code KA 1211/32-1) for financial support.

Institutional Review Board Statement: Not applicable.

Data Availability Statement: The original contributions presented in the study are included in the article, further inquiries can be directed to the corresponding author.

Acknowledgments: The Deutsche Forschungsgemeinschaft (DFG) provided the experimental setup and materials as part of the major research instrumentation programm with contract number INST 187/630-1 FUGG. The authors gratefully acknowledge the financial support by the Deutsche Forschungsgemeinschaft (DFG, funding code KA 1211/32-1 and KA 1211/32-2) for financial support. We would also like to thank the Institute of Micro Production Technology (IMPT) at Leibniz University in Hannover, who carried out the platinum coating of the electrolyte blocks using a sputtering process.

Conflicts of Interest: The authors declare no conflicts of interest.

Abbreviations

The following abbreviations are used in this manuscript:

8YSZ	8 mol% yttria stabilized zirconia
10Sc1CeSZ	10 mol% scandia and 1 mol% ceria stabilized zirconia
EMF	Electromotive force
GDL	Gas diffusion layer
NET	Non-equilibrium thermodynamics
OCV	Open circuit voltage
SOEC	Solid oxide fuel cell
SOFC	Solid oxide electrolyzer cell

References

1. Zeng, Z.; Qian, Y.; Zhang, Y.; Hao, C.; Dan, D.; Zhuge, W. A review of heat transfer and thermal management methods for temperature gradient reduction in solid oxide fuel cell (SOFC) stacks. *Appl. Energy* **2020**, *280*, 115899. [CrossRef]
2. Kjelstrup Ratkje, S.; Bedeaux, D.; Johannessen, E.; Gross, J. *Non-Equilibrium Thermodynamics for Engineers*, 2nd ed.; World Scientific: Hackensack, NJ, USA, 2017.
3. Gedik, A.; Lubos, N.; Kabelac, S. Coupled Transport Effects in Solid Oxide Fuel Cell Modeling. *Entropy* **2022**, *24*, 224. [CrossRef]
4. Gedik, A.; Wachtel, J.; Kabelac, S. Comparative Analysis of Loss Mechanism Localization in a Semi-2D SOEC Single Cell Modell: Non-Equilibrium Thermodynamics versus Monocausal-Based Approach. *ECS Adv.* **2024**, *3*, 014501. [CrossRef]
5. Børset, M.T.; Kang, X.; Burheim, O.S.; Haarberg, G.M.; Xu, Q.; Kjelstrup, S. Seebeck coefficients of cells with lithium carbonate and gas electrodes. *Electrochim. Acta* **2015**, *182*, 699–706. [CrossRef]
6. Kang, X.; Børset, M.T.; Burheim, O.S.; Haarberg, G.M.; Xu, Q.; Kjelstrup, S. Seebeck coefficients of cells with molten carbonates relevant for the metallurgical industry. *Electrochim. Acta* **2015**, *182*, 342–350. [CrossRef]
7. Barragán, V.M.; Kristiansen, K.R.; Kjelstrup, S. Perspectives on Thermoelectric Energy Conversion in Ion-Exchange Membranes. *Entropy* **2018**, *20*, 905. [CrossRef]
8. Kristiansen, K.R.; Barragán, V.M.; Kjelstrup, S. Thermoelectric Power of Ion Exchange Membrane Cells Relevant to Reverse Electrodialysis Plants. *Phys. Rev. Appl.* **2019**, *11*, 044037. [CrossRef]
9. Kjelstrup, S.; Kristiansen, K.R.; Gunnarshaug, A.F.; Bedeaux, D. Seebeck, Peltier, and Soret effects: On different formalisms for transport equations in thermogalvanic cells. *J. Chem. Phys.* **2023**, *158*, 020901. [CrossRef]
10. Gunnarshaug, A.F.; Vie, P.J.S.; Kjelstrup, S. Review—Reversible Heat Effects in Cells Relevant for Lithium-Ion Batteries. *J. Electrochem. Soc.* **2021**, *168*, 050522. [CrossRef]
11. Kamata, M.; Ito, Y.; Oishi, J. Single electrode peltier heat of a hydrogen electrode in H2SO4 and NaOH solutions. *Electrochim. Acta* **1987**, *32*, 1377–1381. [CrossRef]
12. Grimstvedt, A.; Ratkje, S.K.; Forland, T. Theory of Thermocells: Transported Entropies, and Heat of Transfer in Sulfate Mixtures. *J. Electrochem. Soc.* **1994**, *141*, 1236–1242. [CrossRef]
13. Ratkje, S.K.; Ottøy, M.; Halseid, R.; Strømgård, M. Thermoelectric power relevant for the solid-polymer-electrolyte fuel cell. *J. Membr. Sci.* **1995**, *107*, 219–228. [CrossRef]
14. Blinov, V.; Kjelstrup, S.; Bedeaux, D.; Sharivker, V. The Role of the Transported Entropy of Lead Ions in Partially Thermostated and Adiabatic Cells. *J. Electrochem. Soc.* **2001**, *148*, E364. [CrossRef]
15. Fang, Z.; Wang, S.; Zhang, Z.; Qiu, G. The electrochemical Peltier heat of the standard hydrogen electrode reaction. *Thermochim. Acta* **2008**, *473*, 40–44. [CrossRef]
16. Kjelstrup, S.; Vie, P.; Akyalcin, L.; Zefaniya, P.; Pharoah, J.G.; Burheim, O.S. The Seebeck coefficient and the Peltier effect in a polymer electrolyte membrane cell with two hydrogen electrodes. *Electrochim. Acta* **2013**, *99*, 166–175. [CrossRef]
17. Gunnarshaug, A.F.; Kjelstrup, S.; Bedeaux, D. The heat of transfer and the Peltier coefficient of electrolytes. *Chem. Phys. Lett.* **2020**, *738*, 100040. [CrossRef]
18. Kiukkola, K.; Wagner, C. Measurements on Galvanic Cells Involving Solid Electrolytes. *J. Electrochem. Soc.* **1957**, *104*, 379. [CrossRef]
19. Schmalzried, H. Über Zirkondioxyd als Elektrolyt für elektrochemische Untersuchungen bei höheren Temperaturen. *Z. Für Elektrochem. Berichte Der Bunsenges. Für Phys. Chem.* **1962**, *66*, 572–576. [CrossRef]
20. Bray, D.T.; Merten, U. Transport Numbers in Stabilized Zirconia. *J. Electrochem. Soc.* **1964**, *111*, 447. [CrossRef]
21. Steele, B.C.H.; Alcock, C.B. Factors influencing the performance of solid oxide electrolytes in high temperature thermodynamic measurements. *Trans. Met. Soc. AIME* **1965**, *233*, 1359–1367.
22. Baker, R.; West, J.M. Solid electrolytes for use at steelmaking temperatures. *J. Iron Steel Inst.* **1966**, *204*, 212–216.
23. Fischer, W.A.; Janke, D. (Eds.) *Die Anwendbarkeit von ZrO2-Y2O3-, ZrO2-CaO- und ThO2-Y2O3-Festelektrolyten bei 1000 bis 1600 °C*; VS Verlag für Sozialwissenschaften: Wiesbaden, Germany, 1972. [CrossRef]
24. Fischer, W.A.; Pieper, C. Die elektrische Leitfähigkeit und Thermokraft von reinem und mit Calciumoxid stabilisiertem Zirkonoxid bei Temperaturen zwischen 1000 und 1700 °C und Sauerstoffpartialdrücken zwischen 1 und 10–16 atm. *Arch. Für Das Eisenhüttenwesen* **1973**, *44*, 251–259. [CrossRef]
25. Park, J.H.; Blumenthal, R.N. Electronic Transport in 8 Mole Percent Y_2O_3–ZrO_2. *J. Electrochem. Soc.* **1989**, *136*, 2867–2876. [CrossRef]
26. Kosacki, I. Nonstoichiometry and electrical transport in Sc-doped zirconia. *Solid State Ionics* **2002**, *152–153*, 431–438. [CrossRef]
27. Holtan, H.; Mazur, P.; de Groot, S.R. On the theory of thermocouples and thermocells. *Physica* **1953**, *19*, 1109–1118. [CrossRef]
28. Ahlgren, E.O.; Poulsen, F.W. Thermoelectric power of stabilized zirconia. *Solid State Ionics* **1995**, *82*, 193–201. [CrossRef]
29. Fischer, W. Die Thermokraft von kubisch stabilisiertem Zirkonoxid zwischen Sauerstoffelektroden. *Z. Für Naturforschung A* **1967**, *22*, 1575–1581. [CrossRef]
30. Ruka, R.J.; Bauerle, J.E.; Dykstra, L. Seebeck Coefficient of a $(ZrO_2)_{0.85}$ $(CaO)_{0.15}$ Electrolyte Thermocell. *J. Electrochem. Soc.* **1968**, *115*, 497. [CrossRef]
31. Pizzini, S.; Riccardi, C.; Wagner, V.; Sinistri, C. On the Thermoelectric Power of Stabilized Zirconia. *Z. Für Naturforschung A* **1970**, *25*, 559–565. [CrossRef]

32. Yoo, H.I.; Hwang, J.H. Thermoelectric behavior of single crystalline ZrO2(+8mo Y2O3). *J. Phys. Chem. Solids* **1992**, *53*, 973–981. [CrossRef]
33. Ratkje, S.K.; Møller-Holst, S. Exergy effeciency and local heat production in solid oxide fuel cells. *Electrochim. Acta* **1993**, *38*, 447–453. [CrossRef]
34. Takehara, Z.i.; Kanamura, K.; Yoshioka, S. Thermal Energy Generated by Entropy Change in Solid Oxide Fuel Cell. *J. Electrochem. Soc.* **1989**, *136*, 2506–2511. [CrossRef]
35. Ratkje, S.K.; Forland, K.S. The Transported Entropy of Oxygen Ion in Yttria–Stabilized Zirconia. *J. Electrochem. Soc.* **1991**, *138*, 2374–2376. [CrossRef]
36. Kanamura, K.; Yoshioka, S.; Takehara, Z.i. Dependence of Entropy Change of Single Electrodes on Partial Pressure in Solid Oxide Fuel Cells. *J. Electrochem. Soc.* **1991**, *138*, 2165–2167. [CrossRef]
37. Fischer, K.; Seume, J.R. Location and Magnitude of Heat Sources in Solid Oxide Fuel Cells. *J. Fuel Cell Sci. Technol.* **2009**, *6*, 011002. [CrossRef]
38. Nguyen, Q.M.; Takahashi, T. *Science and Technology of Ceramic Fuel Cells*; Elsevier Science: Amsterdam, The Netherlands; New York, NY, USA, 1995.
39. Brodnikovskyi, Y.; McDonald, N.; Polishko, I.; Brodnikovskyi, D.; Brodnikovska, I.; Brychevskyi, M.; Kovalenko, L.; Vasylyev, O.; Belous, A.; Steinberger-Wilckens, R. Properties of 10Sc1CeSZ-3.5YSZ(33-, 40-, 50-wt.%) Composite Ceramics for SOFC Application. *Mater. Today Proc.* **2019**, *6*, 26–35. [CrossRef]
40. Kjelstrup, S.; Bedeaux, D. *Non-Equilibrium Thermodynamics of Heterogeneous Systems*; Series on advances in statistical mechanics; World Scientific: Hackensack, NJ, USA, 2008; Volume 16.
41. Forland, K.S.; Forland, T.; Ratkje, S.K. *Irreversible Thermodynamics: Theory and Applications*; John Wiley & Sons: Chichester, UK, 1988.
42. Kjelstrup, S.; Bedeaux, D.; Johannessen, E.; Gross, J. *Non-Equilibrium Thermodynamics for Engineers*; World Scientific: Hackensack, NJ, USA, 2010. [CrossRef]
43. Moore, J.P.; Graves, R.S. Absolute Seebeck coefficient of platinum from 80 to 340 K and the thermal and electrical conductivities of lead from 80 to 400 K. *J. Appl. Phys.* **1973**, *44*, 1174–1178. [CrossRef]
44. Bedeaux, D.; Kjelstrup, S. Heat Mass and Charge Transport and Chemical Reactions at Surfaces. *Int. J. Thermodyn.* **2005**, *8*, 25–41.
45. Kabelac, S.; Siemer, M.; Ahrendts, J. Thermodynamische Stoffdaten für Biogase. *Forsch. Ingenieurwesen* **2005**, *70*, 46–55. [CrossRef]
46. International Organization for Standardization. *ISO/IEC Guide 98-3:2008: Uncertainty of Measurement—Part 3: Guide to the Expression of Uncertainty in Measurement (GUM)*; ISO: Geneva, Switzerland, 2008. Available online: https://www.iso.org/standard/50461.html (accessed on 14 October 2024).
47. *DIN EN 60584-1:2014-07. Thermoelemente_- Teil_1: Thermospannungen und Grenzabweichungen (IEC_60584-1:2013)*; Deutsche Fassung EN_60584-1:2013. DIN Media GmbH: Berlin, Germany, 2014. [CrossRef]
48. Ahlgren, E.; Willy Poulsen, F. Thermoelectric power of YSZ. *Solid State Ionics* **1994**, *70–71*, 528–532. [CrossRef]

entropy

MDPI

Article

An Instructive CO_2 Adsorption Model for DAC: Wave Solutions and Optimal Processes

Emily Kay-Leighton and Henning Struchtrup *

Department of Mechanical Engineering, University of Victoria, Victoria, BC V8W 2Y2, Canada
* Correspondence: struchtr@uvic.ca

Abstract: We present and investigate a simple yet instructive model for the adsorption of CO_2 from air in porous media as used in direct air capture (DAC) processes. Mathematical analysis and non-dimensionalization reveal that the sorbent is characterized by the sorption timescale and capacity, while the adsorption process is effectively wavelike. The systematic evaluation shows that the overall adsorption rate and the recommended charging duration depend only on the wave parameter that is found as the ratio of capacity and dimensionless air flow velocity. Specifically, smaller wave parameters yield a larger overall charging rate, while larger wave parameters reduce the work required to move air through the sorbent. Thus, optimal process conditions must compromise between a large overall adsorption rate and low work requirements.

Keywords: direct air capture; CO_2; adsorption

1. Introduction

In 2015, at the UN Climate Change Conference (COP21), 196 parties signed the historic Paris Agreement, which set long-term goals for all nations to limit global warming to below 2 °C over pre-industrial levels [1]. More recently, however, the Intergovernmental Panel on Climate Change (IPCC) has stressed that in order to prevent the more severe impacts of climate change, the temperature increase must be limited to 1.5 °C above pre-industrial levels [2], requiring a reduction of greenhouse gas emissions by 45% from 2010 levels by 2030 and the achievement of net-zero by 2050 [3]. At COP28, in 2023, the first global stocktake was conducted to assess progress towards the Paris Agreement targets, and the key findings concluded that global emissions are not on target with the mitigation pathway consistent with the global temperature goal and that significantly more ambitious mitigation plans need to be developed and implemented to accelerate the reduction of emissions across all sectors [4].

While the path to limiting warming to 1.5 °C remains primarily driven by a reduction in emissions through behavioral changes and technological advancements and deployment, there is a growing acknowledgment of the critical role of negative emission technologies in achieving net-zero [5,6]. In particular, carbon dioxide (CO_2) removal strategies will be essential for offsetting the remaining emissions in sectors where a complete reduction is prohibitively challenging, such as in the aviation and heavy industries [5]. Furthermore, CO_2 removal strategies, which are still relatively new technologies, will be able to play a more significant role as the processes become more efficient and mitigation strategies begin to reach their maximum potential [6]. Thus, research, development, and the commercialization of carbon removal is of current importance so that the technology can be deployed at scale in the near term [6].

Carbon dioxide capture can take place in two forms: point source capture, which involves collecting carbon directly from the source of emission and direct air capture (DAC), whereby carbon is removed from ambient air. Generally, DAC is more energy-intensive and, therefore, more expensive than post-combustion point source capture, primarily due to the low concentration of CO_2 in the atmosphere compared to emission sources such as flue gases [7]. Despite this, DAC remains a promising and potentially vital technology. An analysis of the methods of decarbonization of US natural gas power revealed that DAC is more applicable than post-combustion capture for power plants, which have a lower capacity or barriers to retrofitting and is the cheaper capture method for about one-third of all power plant emissions [8]. DAC also has the significant advantage of being capable of off-setting both past and present emissions and emissions from mobile sources. Additionally, since DAC plants do not need to be built at the source of emissions, they can be strategically located in areas with widely available renewable energy sources or in close proximity with CO_2 storage or processing facilities. There is also some suggestion that DAC could be used in the development of an anthropogenic chemical carbon cycle, in which carbon emissions are captured and then recycled for use in fuels or other products [9,10].

As of 2022, there were 19 actively operating DAC plants collectively capturing approximately 10,000 tonnes of CO_2 per year with an average cost of between USD 250–USD 600 per tonne of CO_2 [5,11]. While these figures offer a promising start, they are well below the gigaton capacity and USD 100 per ton cost that will need to be realized in order for the technology to have a meaningful climate impact by 2050 [11].

There are a variety of methods of DAC, including collection using physical and chemical adsorption, electrochemical methods, electrodialysis, membranes, mineral carbonation, cryogenics, and photocatalytic CO_2 conversion. Each method has technical, economic, and environmental considerations [12], however, technologies using reversible sorbents are the most widely investigated and the only method currently used in commercial plants [13].

The adsorption technologies utilize a cyclical process with two steps. First, the CO_2 molecules in the air are collected on the surface of the sorbent material, which can be a liquid or solid. Then, once the material has reached capacity, the CO_2 molecules are released through a regeneration process. The method of CO_2 recovery depends on the type of sorbent material with currently utilized methods, including changes in temperature and pressure [14] and a change in moisture [15,16] for solid sorbents, and a series of chemical processes for liquid sorbents [17]. The use of liquid sorbents has been pursued and proven successful [17], and has advantages such as low cost and continuous operation; however, the energy requirements and system complexity are typically higher due to the chemical separation process and higher regeneration temperatures than solid sorbents [18,19].

Climeworks is a global leader in DAC technology. They are currently operating the world's largest direct air capture and storage plant, which has a capacity of up to 4000 tonnes per year, and are in the process of launching a new plant with a capacity of up to 36,000 tonnes per year [20]. The Climeworks technology uses a temperature vacuum swing adsorption process. During the adsorption process, fans are used to pull air through the collectors, which contain a porous solid sorbent material. Carbon dioxide in the air reacts chemically with the sorbent and binds to the material. If there is humidity in the air drawn into the collector, co-adsorption of water molecules will also occur. Once the material reaches saturation, the collector closes, and the desorption process begins. A vacuum system drops the pressure in the collector, and the sorbent material is heated to 100 °C. At these conditions, the equilibrium capacity of the material is significantly reduced, and thus, the CO_2, and if present, the water molecules, are released from the material and removed from the collector by the vacuum system. The outflowing gas steam is cooled in order to induce condensation of any water that is present, separating it from the CO_2. When the desorption phase finishes, the sorbent material is cooled, and the collector reopens to the environment to begin another adsorption phase [20]. The collected concentrated CO_2 is sequestered using a mineralization process conducted by Carbfix, a partner organization [21]. In the available research, the sorbent material used is APDES-NFC-FD, which is a chemisorbent composed of amine-functionalized nanofibrillated cellulose [22]; however, other materials have also been investigated for use in TVS DAC applications, such as amine-functionalized silica or alumina, carbonate on silica, and anionic resin [18].

Direct air capture technologies are relatively well-examined in the literature, and there exists a variety of sources detailing specific processes [23,24], analyzing the effectiveness of different sorbents materials [13,25], and reviewing the technical and economic characteristics of different methods [12,26–28]; however, the literature on the specific TVS process that is used by Climeworks is less comprehensive. The available work is primarily limited to research published by individuals associated with Climeworks and a handful of papers that are written by non-associated authors but are in reference to the Climeworks data. The experimental work includes validation and analysis of the TVS process using two different sorbent materials, analysis and modeling of the adsorption isotherms, study of stability of the adsorption material, and an evaluation of the co-adsorption of CO_2 and H_2O [29–32].

In complement to the experimental research, there is also literature on the development of mathematical models of the adsorption and desorption processes and theoretical analysis of the TVS method. Wurzbacher et al. [33] developed a comprehensive heat and mass transfer model of the desorption process and Deschamps et al. [34] used a mass and energy balance model in Aspen Adsorption software to evaluate the performance of the TVS technology at an industrial scale. Leonzio et al. [35] developed a mathematical model describing the adsorption and desorption processes, using it to carry out a comparative analysis of the energy requirements, removal capacities, and costs for several sorbent materials.

The current research on the TVS adsorption process for DAC includes limited but detailed experimental results and complex and comprehensive theoretical models; however, there exists a gap in the research to understand the process on a more general and fundamental level. Much of the existing work is specific to particular process set-ups and often, due to the proprietary nature of the technology does not provide full transparency of the input parameters and equations.

A full thermodynamic evaluation of adsorption-based DAC must account for the desorption stage of the process, which relies on thermal energy supply. However, in this contribution, we are interested only in the adsorption phase.

In the following, we present a simple and easy-to-apply adsorption model that provides insight into the general adsorption process without the need for detailed and accurate information about the specifics of a particular sorbent material. Specifically, we present an instructive model for the CO_2 adsorption step in TVS DAC systems, where non-dimensionalization identifies a handful of key parameters for evaluation, namely, as material parameters for the characteristic timescale for adsorption τ and the dimensionless adsorption capacity ϕ; as a physical parameter, the sorbent layer thickness L; and as process parameters, the air flow velocity v and the charging duration t_{ch}.

Mathematical analysis of the set of adsorption equations reveals traveling wave solutions, which are characterized by the wave parameter $\lambda = \frac{\phi+1}{v}$, that is the ratio of capacity parameter ϕ and dimensionless flow speed $v = \frac{Lv}{\tau}$. The same parameter λ governs the initial boundary value problem describing the charging of the sorbent. Thus, the rather general question for the best charging process reduces to two specific questions, namely, which value of λ and which charging duration t_{ch} to chose. While large flow speeds, that is small λ, yield faster charging, they require significantly more pump work; hence, one must consider the trade-off between the charging rate and work requirements.

The present model and its discussion provide a concise understanding of adsorption-based DAC processes, as it allows a systematic evaluation with few well-defined parameters. Specifically, the evaluation of the characteristic material properties τ, ϕ of the sorbent, together with the choices for process parameters λ and t_{ch} provides insight into the challenges and possibilities of these processes.

The remainder of this contribution proceeds as follows: In Section 2 we present the adsorption model, which is akin to a combustion model for reactive flow. Non-dimensionalization allows us to identify the relevant scales and corresponding dimensionless parameters. Their values are identified from available data [36]. Traveling wave solutions are found in Section 3, which identifies the wave parameter λ as the main process parameter. The charging processes are discussed in detail in Section 4 and are based on numerical solutions as well as wave solutions. The charging rate and duration and the corresponding work requirement are analyzed in detail. In Section 5, we explore the optimal system and process conditions, including an estimate of the required size of impactful adsorption-based DAC facilities. The paper ends with our conclusions.

The results presented below extend and refine preliminary work in the honours thesis of EKL [37].

2. Transport Equations

2.1. The Model

We describe the adsorption of CO_2 from air as it is forced through a porous adsorbing material. Detailed adsorption models are widely available and discussed in the literature for DAC processes, e.g., in Refs. [28,33–35]. Our interest focuses on the overall behavior of DAC adsorption systems, for which the simple model presented below provides a meaningful foundation.

For model development, we first look at the flow through a duct, with a molar adsorption capacity at the walls denoted as $\beta(x,t)$ $\left[\text{unit: } \frac{\text{mol}}{\text{m}^2}\right]$. The (temperature dependent) saturation capacity β_0 defines the maximum occupation of the surface, that is $\beta = \beta_0$ when all available sites are occupied. We only consider the adsorption at low temperature, as a strong process going mainly forward, similar to the simplified description of combustion processes [38]; see Appendix A for a discussion in the context of linear irreversible thermodynamics [39,40]. While the process is exothermic, the temperature does not noticeably change due to a large amount of air accompanying the CO_2, to which the adsorption heat is dissipated; see [19] for a discussion of this for a chemical DAC process.

As CO_2 carrying air is passing through the duct, some of the CO_2 is absorbed and some is desorbed, which we express in the equation

$$\frac{\partial \beta}{\partial t} = \frac{p_{CO_2}}{\sqrt{2\pi M_{CO_2} \bar{R} T}} \varpi \left(1 - \frac{\beta}{\beta_0}\right) - K\beta, \tag{1}$$

where the adsorption rate is a product of the following three terms: $\frac{p_{CO_2}}{\sqrt{2\pi M_{CO_2} \bar{R} T}}$ $\left[\frac{\text{mol}}{\text{m}^2\text{s}}\right]$ is the rate of the molecular collisions per area due to the thermal motion of particles as determined from kinetic theory of gases [41], ϖ [1] is an overall probability for hitting an adsorption site with adsorption occurring, and $\left(1 - \frac{\beta}{\beta_0}\right)$ is the probability that a site is unoccupied. Further, p_{CO_2} is the partial pressure of CO_2 in air, T is the constant temperature of air and adsorption material, $\bar{R}$ is the universal gas constant, and M_{CO_2} is the molar mass of CO_2. The desorption rate is proportional to occupation by means of the material parameter K, which is a function of temperature.

With the mole fraction $\chi = \frac{p_{CO_2}}{p_{\text{air}}}$ of CO_2 and the ideal gas law $p_{\text{air}} = \bar{\rho}_{\text{air}} \bar{R} T$, the equation assumes the form

$$\frac{\partial \beta}{\partial t} = \bar{\rho}_{\text{air}} \sqrt{\frac{R_{CO_2} T}{2\pi}} \varpi \left(1 - \frac{\beta}{\beta_0}\right) \chi - K\beta \tag{2}$$

where $R_{CO_2} = \frac{\bar{R}}{M_{CO_2}}$ is the gas constant of CO_2, and $\bar{\rho}_{\text{air}}$ is the mole density of air, which we consider to be a constant; indeed, temperature and pressure changes are ignored throughout our discussion.

We assume that the transport of CO_2 is due to convective flow with the air (velocity v) and diffusion in air (diffusion coefficient D), while molecules are adsorbed at the walls. Thus, the balance law for CO_2 in terms of the mole fraction reads

$$\bar{\rho}_{\rm air}\left(\frac{\partial\chi}{\partial t}+v\frac{\partial\chi}{\partial x}-\mathsf{D}\frac{\partial^2\chi}{\partial x^2}\right)=-\frac{\partial\beta}{\partial t}\frac{A_{\rm duct}}{V_{\rm duct}},\tag{3}$$

with the duct surface-to-volume ratio $\frac{A_{\rm duct}}{V_{\rm duct}}$ as a property of the material. For a cylindrical channel of radius r and length dx, we have

$$\frac{A_{\rm duct}}{V_{\rm duct}}=\frac{2\pi r dx}{\pi r^2 dx}=\frac{2}{r}\tag{4}$$

while for a general porous material, one typically finds the active area $\mathsf{a}_{\rm act}\left[\frac{\rm m^2}{\rm kg}\right]$ and the accessible volume $\mathsf{v}_{\rm acc}\left[\frac{\rm m^3}{\rm kg}\right]$ per mass of material, hence

$$\frac{A_{\rm duct}}{V_{\rm duct}}=\frac{\mathsf{a}_{\rm act}}{\mathsf{v}_{\rm acc}}\tag{5}$$

Collecting the above results, we have the simple transport model

$$\frac{\partial\beta}{\partial t}=\bar{\rho}_{\rm air}\sqrt{\frac{R_{\rm CO_2}T}{2\pi}}\varpi\left(1-\frac{\beta}{\beta_0}\right)\chi-K\beta\tag{6}$$

$$\frac{\partial\chi}{\partial t}+v\frac{\partial\chi}{\partial x}-\mathsf{D}\frac{\partial^2\chi}{\partial x^2}=-\frac{\mathsf{a}_{\rm act}}{\mathsf{v}_{\rm acc}}\sqrt{\frac{R_{\rm CO_2}T}{2\pi}}\varpi\left(1-\frac{\beta}{\beta_0}\right)\chi-\frac{1}{\bar{\rho}_{\rm air}}\frac{\mathsf{a}_{\rm act}}{\mathsf{v}_{\rm acc}}K\beta\tag{7}$$

This model is in agreement with the second law of thermodynamics as discussed in Appendix A for the non-dimensionalized equations.

2.2. Non-Dimensionalization

We now introduce dimensionless properties in order to simplify the notation, and to identify key parameters of the model. Specifically, we refer the lengthscale to the thickness of the adsorbent L, introduce a timescale τ that will be identified below, and consider the number of unoccupied sites β relative to available sites β_0, and the CO_2 load relative to that of incoming air, that is, we set

$$\tilde{x}=\frac{x}{L},\quad\tilde{t}=\frac{t}{\tau},\quad\tilde{\beta}=\frac{\beta}{\beta_0},\quad\tilde{\chi}=\frac{\chi}{\chi_{\rm atm}}\tag{8}$$

to find at first

$$\frac{\partial\tilde{\beta}}{\partial\tilde{t}}=\tau\frac{\bar{\rho}_{\rm air}\chi_{\rm atm}}{\beta_0}\sqrt{\frac{R_{\rm CO_2}T}{2\pi}}\varpi(1-\tilde{\beta})\tilde{\chi}-\tau K\tilde{\beta}\tag{9}$$

$$\frac{\partial\tilde{\chi}}{\partial\tilde{t}}+\frac{\tau v}{L}\frac{\partial\tilde{\chi}}{\partial\tilde{x}}-\frac{\tau\mathsf{D}}{L^2}\frac{\partial^2\tilde{\chi}}{\partial\tilde{x}^2}=-\tau\frac{\mathsf{a}_{\rm act}}{\mathsf{v}_{\rm acc}}\sqrt{\frac{R_{\rm CO_2}T}{2\pi}}\varpi(1-\tilde{\beta})\tilde{\chi}+\frac{\tau}{\bar{\rho}_{\rm air}}\frac{\mathsf{a}_{\rm act}}{\mathsf{v}_{\rm acc}}\frac{\beta_0}{\chi_{\rm atm}}K\tilde{\beta}\tag{10}$$

To proceed, we chose the timescale as

$$\tau=\frac{\sqrt{2\pi}\beta_0}{\sqrt{R_{\rm CO_2}T}\bar{\rho}_{\rm air}\chi_{\rm atm}\varpi}\tag{11}$$

and introduce the capacity parameter

$$\phi=\frac{\mathsf{a}_{\rm act}}{\mathsf{v}_{\rm acc}}\frac{\beta_0}{\bar{\rho}_{\rm air}\chi_{\rm atm}}.\tag{12}$$

as well as the desorption parameter

$$\mathsf{k} = \tau K = \frac{\sqrt{2\pi}\beta_0 K}{\sqrt{R_{CO_2}T}\bar{\rho}_{air}\chi_{atm}\varpi} \tag{13}$$

The definition of τ gives the timescale for adsorption from CO_2-carrying air (where $\tilde{\chi} = 1$). The capacity parameter ϕ is the ratio of the total available adsorption sites $\mathsf{a}_{act}\beta_0$ [unit: $\frac{m^2}{kg}\frac{mol}{m^2} = \frac{mol}{kg}$] and the number of CO_2 molecules in air-filled pores $\mathsf{v}_{acc}\bar{\rho}_{air}\chi_{atm}$ [unit: $\frac{m^3}{kg}\frac{mol}{m^3}1 = \frac{mol}{kg}$], both considered per mass of adsorbent.

Moreover, dimensionless velocity and the diffusion coefficient are

$$v = \frac{\tau v}{L}, \quad D = \frac{\tau \mathsf{D}}{L^2}. \tag{14}$$

To ease notation, we will in the following not write the tildes, that is, we write the dimensionless transport equations as

$$\frac{\partial \beta}{\partial t} = (1-\beta)\chi - \mathsf{k}\beta \tag{15}$$

$$\frac{\partial \chi}{\partial t} + v\frac{\partial \chi}{\partial x} - D\frac{\partial^2 \chi}{\partial x^2} = -\phi[(1-\beta)\chi - \mathsf{k}\beta] \tag{16}$$

Specifically, β is the relative amount of local adsorption, with $\beta = 0$ if all sites are free, and $\beta = 1$ when all sites are occupied; χ denotes the CO_2 load of the airflow, rescaled such that $\chi = 1$ for the incoming flow and $\chi = 0$ when all CO_2 is removed. The parameters of the system are the material dependent coefficients τ, ϕ, k, and D, which are considered to be constant, and the airflow velocity v is the only process parameter.

We point out that this model ignores the adsorption of water molecules that compete with CO_2 and thus reduce the overall adsorption of CO_2. Thus, the estimates below must be considered as a best case.

2.3. Typical Parameter Values

While we are interested in the principal behavior of the process rather than a detailed evaluation of realistic systems, it is worthwhile to have a look at the available data. Due to the proprietary nature of materials in development for industrial applications, detailed material data to relate to and fit our model is scarce. Thus, while a comprehensive examination of data for different materials would be interesting and desirable, we consider only one specific material presented in the patent by Climeworks [36], from which we extracted the values listed in Table 1.

Table 1. Data for the material described in [36].

Parameter	Value	
$\bar{\rho}_{air}$	$42.5\ \frac{mol}{m^3}$	standard air
χ_{atm}	4.2×10^{-4}	mole fraction of CO_2 in atmosphere
β_0	$4.6 \times 10^{-6}\ \frac{mol}{m^2}$	adsorption site density
a_{act}	$3.04 \times 10^{5}\ \frac{m^2}{kg}$	active area per mass
$\mathsf{v}_{acc} = \frac{1}{\rho_{mat}}$	$1.79 \times 10^{-3}\ \frac{m^3}{kg}$	accessible volume per mass
L	0.04 m	thickness of slab
A_{tot}	$1.25 \times 10^{-3}\ m^2$	cross section of slab
Q	$3.33 \times 10^{-5}\ \frac{m^3}{s}$	air mass flow rate
$v = \frac{Q}{A_{tot}}$	$2.67 \times 10^{-2}\ \frac{m}{s}$	flow speed
D	$1.6 \times 10^{-5}\ \frac{m^2}{s}$	CO_2 in air diffusion coefficient

For the fitting, we assumed that at the low temperatures during adsorption the desorption rate K is sufficiently small, so that the desorption processes can be ignored, that is, we set $\mathrm{k} = 0$. This implies that the equilibrium site density β_{eq} equals the adsorption site density β_0; all subsequent results rely on this simplifying assumption. See Appendix A for an additional discussion of the influence of the desorption parameter (i.e., the dimensionless equilibrium constant).

The timescale τ (11) is inversely dependent on the probability parameter ϖ, which, at best, is obtained from a comparison to the experimental results based on the solutions of the model. For the data from [36], our fitting resulted in the values

$$\begin{aligned} &\tau = \frac{\sqrt{2\pi}\beta_0}{\sqrt{R_{CO_2}T}\rho_{air}\chi_{atm}\varpi} && = 1.14 \times 10^4\ \mathrm{s} = 2.8\ \mathrm{h} && \text{characteristic timescale} \\ &\varpi && = 2.5 \times 10^{-10} && \text{probability parameter} \end{aligned}$$

For this data, at 2.8 h, the characteristic timescale is quite large. Obviously, materials with shorter characteristic times are advantageous for technical processes since the turnover time will be proportional to τ. The parameter ϖ in our simplified model is the probability for adsorption when a CO_2 molecule hits an adsorption site. For the data given, a molecule must hit the surface about 4×10^9 times before it is adsorbed.

For this data, the dimensionless parameters appearing in the model assume the values

$$\begin{aligned} &v = \frac{\tau v}{L} && = 7620 && \text{dimensionless air velocity} \\ &D = \frac{\tau \mathrm{D}}{L^2} && = 114 && \text{dimensionless diffusion coefficient} \\ &\phi = \frac{\mathrm{a_{act}}}{\mathrm{v_{acc}}}\frac{\beta_0}{\rho_{air}\chi_{atm}} && = 43770 && \text{capacity parameter} \\ &\mathrm{k} = \tau K && \ll 1 && \text{desorption parameter (ignored)} \end{aligned}$$

With velocity and diffusivity quite different in size, such that $\frac{D}{v} = \frac{\mathrm{D}}{vL} \ll 1$ processes are dominated by convection while diffusion can be ignored, we have set $D = 0$ from now on.

2.4. CO_2 Conservation

The conservation law for CO_2 molecules is obtained from Equations (15) and (16) by elimination of the adsorption rates as

$$\frac{\partial \chi + \phi\beta}{\partial t} + \frac{\partial v\chi}{\partial x} = 0 \tag{17}$$

We consider a domain of length $L = 1$ and a process of duration t_e.

Integration about the domain length yields

$$\frac{d}{dt}\int_0^1 (\chi(x,t) + \phi\beta(x,t))dx + v[\chi(1,t) - \chi(0,t)] = 0 \tag{18}$$

and subsequent integration over the process duration t_e yields

$$v\int_0^{t_e} [\chi(0,t) - \chi(1,t)]dt = \phi\int_0^1 [\beta(x,t_e) - \beta(x,0)]dx + \int_0^1 [\chi(x,t_e) - \chi(x,0)]dx. \tag{19}$$

The expression on the left of this equation is the difference between the overall inflow (at $x = 0$) and outflow (at $x = 1$), while the expression on the right is the overall amount of the CO_2 adsorbed ($\phi\beta$) and present as gas (χ) at t_e.

3. Wave Solution

When diffusion can be ignored, the transport model yields traveling wave solutions, which we explore in this section. As will be seen, the analytical travelling wave solutions describe the processes very well, apart from the initial stage of the process. With that, they are very helpful for the better understanding of process behavior.

For the computations, it is advantageous to use the relative number of unoccupied adsorption sites, defined as

$$\hat{\beta} = 1 - \beta \tag{20}$$

with $\hat{\beta} = 1$ when all sites are free, and $\hat{\beta} = 0$ when all sites are occupied. With this, and vanishing diffusivity D, the transport Equations (15) and (16) reduce to

$$\frac{\partial \hat{\beta}}{\partial t} = -\hat{\beta}\chi \tag{21}$$

$$\frac{\partial \chi}{\partial t} + v\frac{\partial \chi}{\partial x} = -\phi\hat{\beta}\chi \tag{22}$$

Note that transport is linear, but decay is non-linear.

3.1. Travelling Waves

We assume that both functions are traveling waves with an unknown wave speed αv proportional to the flow velocity v, so that traveling wave solutions are of the form

$$\hat{\beta}(x,t) = \hat{\beta}(x - \alpha v t) = \hat{\beta}(\xi) \tag{23}$$

$$\chi(x,t) = \chi(x - \alpha v t) = \chi(\xi) \tag{24}$$

where $\xi = x - \alpha v t$ is the variable for the traveling wave. The velocity factor α will be determined as we proceed.

We denote differentiation with respect to the wave variable ξ by prime, which gives the transport equations as

$$\alpha v \hat{\beta}' = \hat{\beta}\chi \tag{25}$$

$$-\alpha v \chi' + v\chi' = -\phi\hat{\beta}\chi \tag{26}$$

which combine to

$$\hat{\beta}' = -\frac{(1-\alpha)}{\phi\alpha}\chi'. \tag{27}$$

Integration yields

$$\hat{\beta} = -\frac{(1-\alpha)}{\phi\alpha}\chi + C \tag{28}$$

with the constant of integration C.

With this, the second equation becomes an ODE for χ that can be written in the compact form

$$\chi' = \frac{\chi^2}{\alpha v} - \lambda\chi, \tag{29}$$

where the wave parameter λ—which will play an important role in the following—is introduced as

$$\lambda = \frac{\phi C}{(1-\alpha)v}. \tag{30}$$

Integration yields, with the constant of integration Γ,

$$\left(\frac{1}{\chi} - \frac{1}{\lambda\alpha v}\right) = \Gamma \exp[\lambda\xi], \tag{31}$$

hence the general wave solution reads

$$\chi = \frac{\lambda\alpha v}{\Gamma\lambda\alpha v \exp[\lambda\xi] + 1}. \tag{32}$$

3.2. Determination of Constants of Integration

The constants α, λ (or C), and Γ are identified as follows:

With $v > 0$, the wave travels from left to right connecting states of CO_2-carrying air, with $\chi(\xi \to -\infty) = 1$, and fully depleted air, with $\chi(\xi \to \infty) = 0$. The solution (32) yields the second condition for all $\lambda > 0$, while the first condition demands that the wave parameter is the inverse wave speed

$$\lim_{\xi \to -\infty} \frac{\lambda \alpha v}{\lambda \alpha v \Gamma \exp[\lambda \xi] + 1} = \lambda \alpha v \overset{!!}{=} 1 \quad \Rightarrow \quad \lambda = \frac{1}{\alpha v} \tag{33}$$

With this, the solution for the relative CO_2 load reads

$$\chi(\xi) = \frac{1}{1 + \exp\left[\frac{1}{\alpha v}(\xi - \xi_0)\right]} \tag{34}$$

where the offset ξ_0 is defined through $\lambda \alpha v \Gamma = \exp[-\lambda \xi_0]$; note that negative values of Γ give divergent results, hence are not physical.

We are now left with one last problem, which is the question of the value of the velocity factor α. For this, we consider the solution for the site occupation $\hat{\beta} = 1 - \beta$, which from (28), (30), (33), and (34) becomes

$$\hat{\beta} = \frac{1-\alpha}{\phi \alpha} \left[\frac{1}{1 + \exp[-\lambda(\xi - \xi_0)]} \right] \tag{35}$$

In the limit $\xi \to \infty$, we have all unoccupied sites, hence $\hat{\beta} = 1$ and

$$1 = \hat{\beta}(\xi \to \infty) = \lim_{\xi \to -\infty} \left[\frac{1-\alpha}{\phi \alpha} \left[\frac{1}{1 + \exp[-\lambda(\xi - \xi_0)]} \right] \right] = \frac{1-\alpha}{\phi \alpha} \tag{36}$$

which gives the velocity factor and wave parameter as

$$\alpha = \frac{1}{\phi + 1}, \quad \lambda = \frac{\phi + 1}{v} \tag{37}$$

Note that with this

$$\hat{\beta} = \frac{1}{1 + \exp\left[-\frac{\phi+1}{v}(\xi - \xi_0)\right]} \quad \Longrightarrow \quad \beta = 1 - \hat{\beta} = \frac{1}{1 + \exp[\lambda(\xi - \xi_0)]} = \chi. \tag{38}$$

that is, for the wave solution $\beta(x,t) = \chi(x,t)$. This simple relation results from ignoring the desorption rate. For the non-zero desorption rate k wave parameter, the width of the wave and equilibrium conditions are affected by k, while the general behavior is the same as for the simpler case $\mathsf{k} = 0$.

3.3. Wave Structure

Combining the results from above, we find the traveling wave solution as

$$\chi(\xi) = \beta(\xi) = \frac{1}{1 + \exp[\lambda(\xi - \xi_0)]} = \frac{1}{1 + \exp[\lambda(x - \xi_0) - t]}; \tag{39}$$

that is, χ and β agree for the waves. Notably, the structure of the wave is determined only through the value of the wave parameter $\lambda = \frac{\phi+1}{v}$.

Figure 1 shows the structure of the wave for a variety of wave parameters $\lambda = \frac{\phi+1}{v} \in [0.5, 20]$, centered in the accessible domain

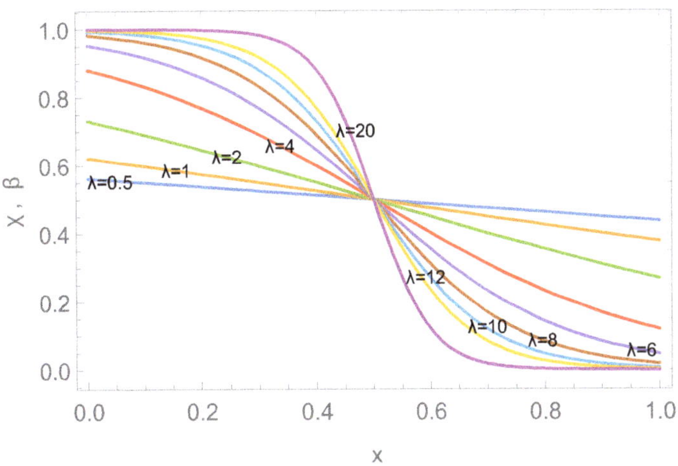

Figure 1. Wave solution centered at $x = 0.5$ for wave parameters $\lambda = \frac{\phi+1}{v}$ in $[0.5,\ 20]$.

Wave-like transport will be observed when the domain width L is larger than the effective width of the wave, which we approximate as the distance between points x_0, x_1 where the values of β are β_{th} and $1-\beta_{th}$, respectively, so that

$$\Delta x = x_0 - x_1 = \frac{1}{\lambda}\left[\ln\frac{1-\beta_{th}}{\beta_{th}} - \ln\frac{\beta_{th}}{1-\beta_{th}}\right] = \frac{2}{\lambda}\ln\frac{1-\beta_{th}}{\beta_{th}} \tag{40}$$

We note that the width is linear in the inverse wave parameter $\frac{1}{\lambda} = \frac{v}{\phi+1}$, hence small velocities v or large capacities ϕ give more compact waves, while the waves are wide for fast flows or low capacity.

Setting the threshold value to $\beta_{th} = 0.05$ gives an approximate dimensionless width $\Delta x = \frac{6}{\lambda}$. Hence, full wave structures will be observed in the domain ($L = 1$) for

$$\Delta x \lesssim 1 \quad \Longrightarrow \quad \lambda = \frac{\phi+1}{v} \gtrsim 6. \tag{41}$$

For the example data, $\lambda = \frac{\phi+1}{v} = 5.75$, hence wave-dominated transport is expected.

4. Charging Processes

4.1. Initial Boundary Value Problem

The charging process of the sorbent is the solution of our dimensionless model

$$\frac{\partial \beta}{\partial t} = (1-\beta)\chi \tag{42}$$

$$\frac{\partial \chi}{\partial t} + v\frac{\partial \chi}{\partial x} = -\phi(1-\beta)\chi \tag{43}$$

for the case that initially no CO_2 is in the domain, and all adsorption sites are free, that is

$$\text{initial conditions:} \quad \chi(x, t=0) = \beta(x, t=0) = 0 \tag{44}$$

Only one boundary condition is required, which is the (dimensionless) mole fraction of incoming air at $x = 0$:

$$\text{boundary condition:} \quad \chi(x=0, t) = 1. \tag{45}$$

Airflow velocity v is the only process parameter, while the capacity ϕ is a material parameter; note that the value of $v = \frac{\tau v}{L}$ depends on the time- and length-scale, which are properties of the material.

The numerical solution to this problem is straightforward: we used the NDSolve function of Wolfram Mathematica. Figure 2 shows, for $v = 10$, $\phi = 100$, or $\lambda = \frac{\phi+1}{v} = 10.1$, and the curves for $\chi(x,t)$ (orange) and $\beta(x,t)$ (green) over the domain at various times, as well as the wave solution (blue), where the wave shift ξ_0 was adjusted to provide agreement at larger times.

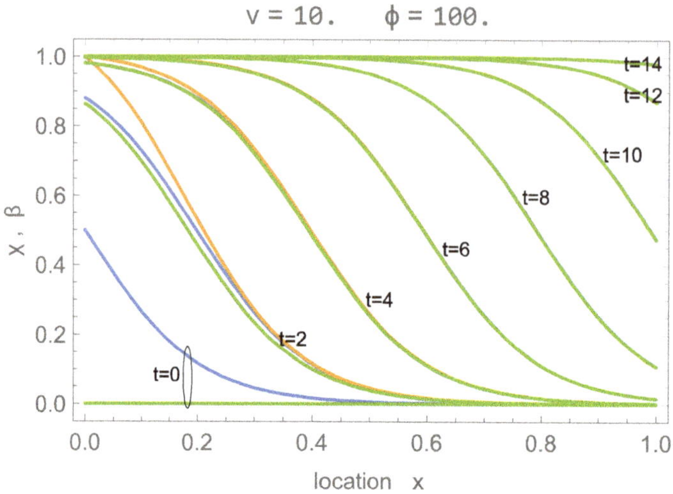

Figure 2. Solution of adsorption process with $v = 10$, $\phi = 100$, that is $\lambda = \frac{\phi+1}{v} = 10.1$, for times $t = 0, 2, 4, \ldots, 14$; green: $\beta(x,t)$; orange: $\chi(x,t)$; blue: wave solution for χ, β.

Figure 2 shows initial differences between χ and β, as well as their wave solutions. However, as the process proceeds, the numerical solution develops into the traveling wave solution, with all three curves agreeing well for times $t \geq 4$. The flow velocity v is relatively slow so that all CO_2 is adsorbed soon after entering. At time $\tau = 14$, the adsorption sites are almost fully occupied.

A further evaluation shows that the shift ξ_0 in the wave solution is independent of the parameter values. Specifically, we found that a shift of zero gives excellent long-time agreement with numerical solutions for all values at the parameters. With this, at the beginning of the process, the wave structure is centered at the inlet ($x = 0$),

$$\text{wave solution: } \chi(x,t) = \beta(x,t) = \frac{1}{1 + \exp[\lambda x - t]}. \tag{46}$$

Figure 3 shows the numerical and wave solutions for a fast charging process, at $\lambda = \frac{\phi+1}{v} = 1$, for which the signal width is considerably wider than the domain. Due to the large velocity, an abundance of CO_2 is provided so that adsorption occurs at all locations simultaneously, and all sites are filled at about $t = 4.5$. While the wave solution deviates in the early stages, from $t = 4$ forward, the agreement is quite good.

For the results above, we have given the values of velocity v and capacity ϕ separately. A notable outcome of our evaluation is that the solution behavior—numerical or wave solution—is determined only by the value of the wave parameter $\lambda = \frac{\phi+1}{v}$, which is inversely proportional to the width of fully developed waves. Thus, in the following, we will not refer to the individual values for v, ϕ but only to the values of the wave parameter λ, which is one of two process parameters to describe the charging of a sorbent; the other will be identified as the charging duration t_{ch}.

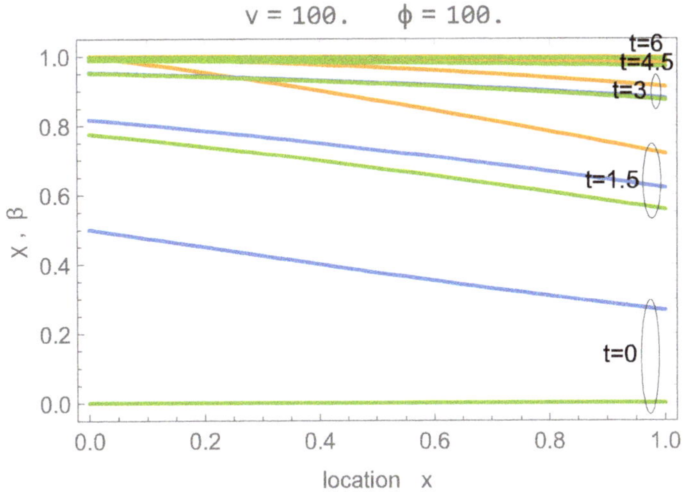

Figure 3. Solution of adsorption process with $\lambda = \frac{\phi+1}{v} = 1.01$ for $v = 100$, $\phi = 100$, for times $t = 0,\ 1.5,\ 3,\ 4.5,\ 6$; green: $\beta(x,t)$; orange: $\chi(x,t)$; blue: wave solution for χ, β.

4.2. Charging Duration

We proceed with the discussion of the required process duration to fill the adsorption sites to a certain level.

The CO_2 accumulation, which is the relative amount of CO_2 collected at time t, is the space average (with $L = 1$ in the dimensionless formulation),

$$\bar{\beta}(t) = \int_0^1 \beta(x,t)dx. \tag{47}$$

For the wave solution, the corresponding integral

$$\bar{\beta}(t) = \int_0^1 \frac{1}{1+\exp\left[\frac{\phi+1}{v}x - t\right]}dx \tag{48}$$

can be solved to give the CO_2 accumulation as an explicit function of time t and wave parameter $\lambda = \frac{\phi+1}{v}$,

$$\bar{\beta}(t) = 1 + \frac{1}{\lambda}\ln\left[\frac{1+\exp[-t]}{1+\exp[\lambda - t]}\right]. \tag{49}$$

As can be seen already in Figures 2 and 3, the wave solution does not properly predict the early stages of charging, as reflected in the non-zero values of $\bar{\beta}(t=0)$, which can be as large as 0.5 for $\lambda \to 0$.

Figure 4 shows the accumulation over time for four values of λ, comparing the numerical result (continuous) with the wave solution (dashed), again with good agreement for larger times as well as larger accumulations $\bar{\beta}$. For larger λ, the accumulation results from the fully developed wave traveling through the domain, which leads to the linear charging behavior that is clearly visible for $\lambda = 5, 10$.

An important question for applications is how long the process should run to make the best use of the sorbent material. We define the charging duration t_{ch} as the time required to reach a desired accumulation value $\bar{\beta}_{ch}$.

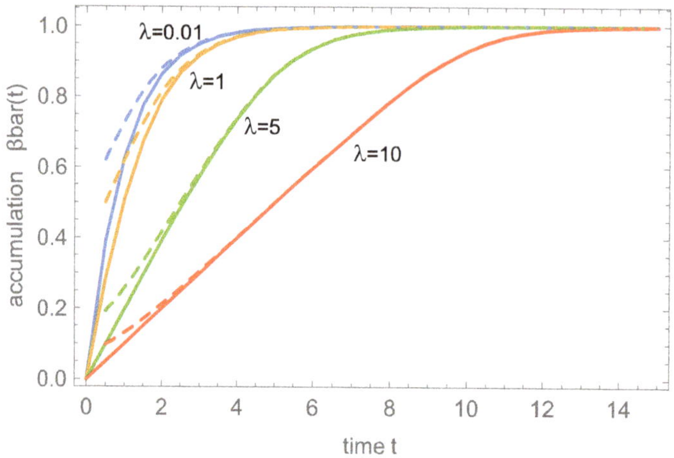

Figure 4. Accumulation $\bar{\beta}(t)$ over time for $\lambda = \frac{\phi+1}{v} = 0.01,\ 1,\ 5,\ 10$; continuous: numerical solution; dashed: wave solution.

The wave solution allows a quick evaluation of the charging time, which is found from the inversion of (49) as

$$t_{ch} = \ln \frac{\exp[\lambda \bar{\beta}_{ch}] - 1}{1 - \exp[-\lambda(1 - \bar{\beta}_{ch})]}. \tag{50}$$

Accordingly, the charging duration increases with $\lambda = \frac{\phi+1}{v}$ as well as with the desired value $\bar{\beta}_{ch}$, as shown in Figure 5 for the wave and numerical solution. Note that for the wave solution $\bar{\beta}_{ch}(0) \neq 0$ since the wave solution does not match the proper initial condition $\beta(x,0) = 0$. Thus the durations for small λ and small $\bar{\beta}_{ch}$ differ between wave and numerical solutions, while they agree well for larger λ and $\bar{\beta}_{ch}$. The plot gives a clear indication that charging to a large accumulation takes particularly long. Accordingly, in applications, one will not aim to charge to, say, 98%, but rather to a lower value and run the charge-desorption cycle more often.

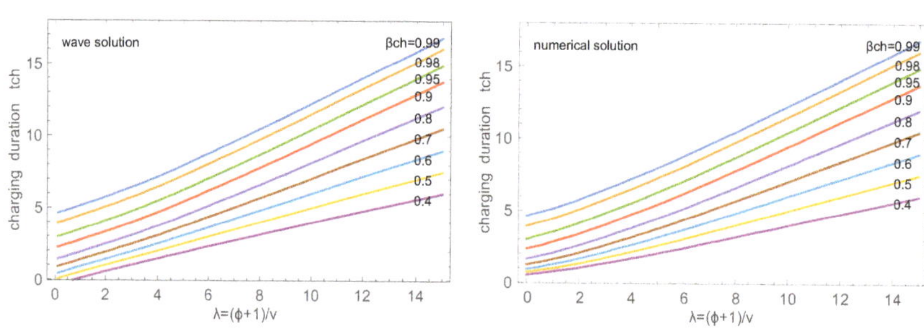

Figure 5. Charging duration t_{ch} in dependence of $\lambda = \frac{\phi+1}{v}$ for accumulations $\bar{\beta}_{ch} = 0.4, \ldots, 0.99$ for wave (left) and numerical (right) solutions.

4.3. Overall Charging Rate

For the best use of the sorbent in applications, one will be interested in large accumulations $\bar{\beta}_{ch}$ and short charging duration t_{ch}, that is, the large overall charging rates

$$\dot{\beta}_{ch} = \frac{\bar{\beta}_{ch}(t_{ch})}{t_{ch}}. \tag{51}$$

Figure 6 shows the charging rate $\dot{\beta}_{ch}$ as a function of charging time t_{ch} for the wave parameters $\lambda \in [0.5, 12]$ in a logarithmic scale. For larger durations, the overall rate $\dot{\beta}_{ch}$ decreases, hence it is advantageous to not fully charge the sorbent, but rather terminate charging earlier, and charge more often.

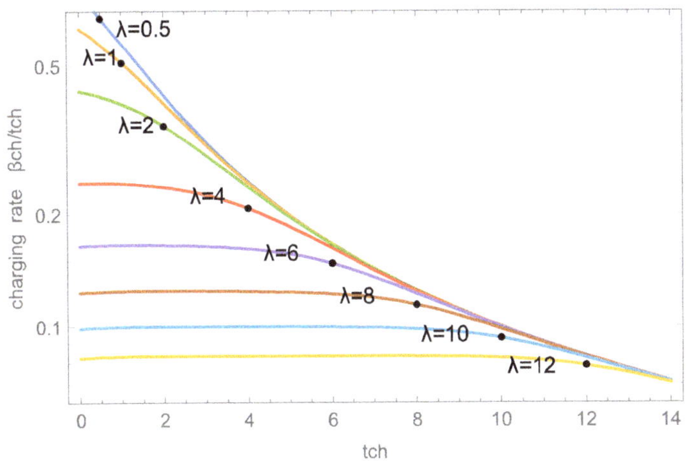

Figure 6. Charging rates $\dot{\beta} = \bar{\beta}_{ch}/t_{ch}$ as functions of charge duration t_{ch} for several $\lambda = \frac{\phi+1}{v}$ (numerical solution). The dots indicate the values at $t_{ch} = \lambda$. Note the logarithmic scale.

For smaller λ, that is large velocities, the wave width is larger than the domain, hence, as depicted in Figure 3, charging is distributed through the domain, with short charging durations yielding large rates.

For $\lambda \geq 4$, the charging rates are initially flat due to the wave-dominated charging process depicted in Figure 2. The exiting wave is centered at the domain exit ($x = 1$) for $t = \lambda$; hence the charging rate decreases around that time.

Not surprisingly, the figure indicates significantly higher charging rates for smaller $\lambda = \frac{\phi+1}{v}$, that is, a faster airflow. As will be discussed below, large velocities require considerably larger pump work and, thus, are not favorable for efficiency processes.

The dots in the figure indicate $t_{ch} = \lambda$, which appears to be a meaningful charging duration for all cases. Indeed, there would be little gain in charging with $\lambda = 0.5$ until $t_{ch} = 2$, which gives almost the same overall rate as charging with $\lambda = 1$ until $t_{ch} = 2$, but requires significantly more pump work to maintain the large flow velocity v.

Table 2 shows accumulations $\bar{\beta}_{ch}$ and overall rates $\dot{\beta}_{ch}$ for the case $t_{ch} = \lambda$ for the numerical and wave solutions, which agree well for $\lambda \geq 4$.

Notably, in particular for small λ, it is advantageous to not charge the sorbent to large accumulation, but rather to a lower accumulation (e.g., $\bar{\beta}_{ch} = 0.509$ for $\lambda = 1$) and more often.

Table 2. Charging conditions for $t_{ch} = \lambda$ from numerical solution; values in brackets are for the wave solution.

$\lambda = t_{ch}$	0.5	1	2	4	6	8	10	12
$\bar{\beta}_{ch}(t_{ch})$	0.335 (0.562)	0.509 (0.620)	0.688 (0.717)	0.829 (0.831)	0.886 (0.885)	0.913 (0.913)	0.931 (0.931)	0.942 (0.942)
$\dot{\beta}_{ch} = \frac{\bar{\beta}_{ch}(t_{ch})}{t_{ch}}$	0.67 (1.12)	0.509 (0.620)	0.344 (0.358)	0.207 (0.208)	0.148 (0.147)	0.114 (0.114)	0.093 (0.093)	0.079 (0.079)

4.4. Breakthrough Curve

We note that the accumulation $\bar{\beta}(t)$ is not accessible from direct measurements. Nevertheless, it can be determined from the breakthrough curve, i.e., the ratio of CO_2 mole fractions in the exiting and entering airflows [35], which both can be measured; in our dimensionless variables the breakthrough curve is given by

$$\frac{\chi(1,t)}{\chi(0,t)} = \chi(1,t), \tag{52}$$

since $\chi(0,t) = 1$, which is due to the definition of χ and the boundary condition (45). To proceed, we recall the space integrated conservation of CO_2 (18),

$$\frac{d}{dt}\int_0^1 (\chi(x,t) + \phi\beta(x,t))dx + v[\chi(1,t) - \chi(0,t)] = 0 \tag{53}$$

For any meaningful adsorption material, the capacity is large, $\phi \gg 1$, and we have seen that $\chi(x,t) \to \beta(x,t)$ for larger t. Hence, with $\chi(0,t) = 1$ and the definition (47) of the accumulation, we can approximate the conservation law as

$$\frac{d\bar{\beta}(t)}{dt} = \frac{v}{\phi+1}[1 - \chi(1,t)] = \frac{1}{\lambda}[1 - \chi(1,t)]. \tag{54}$$

Integration over the duration of the process, with $\bar{\beta}(t=0) = 0$, yields the accumulation at time t as

$$\bar{\beta}(t) = \frac{1}{\lambda}[t - \chi_{\text{out}}(t)], \tag{55}$$

where

$$\chi_{\text{out}}(t) = \int_0^t \chi(1,t')dt'. \tag{56}$$

Hence, the accumulation $\bar{\beta}(t)$ can be determined from the time integration of the breakthrough curve and the wave parameter λ, which must be determined from other measurements.

4.5. Excess Air

The capacity parameter ϕ depends only on the material; hence, it is unaffected by the process parameters. From the previous discussion, it is evident that smaller values of the wave parameter $\lambda = \frac{\phi+1}{v}$, that is, larger flow speeds v, are advantageous since they yield faster charging; hence, a better turn-around of charge–discharge processes and a better use of the sorbent. Next, we explore the relation between the faster flow and the total amount of air required, which, indeed, grows with increasing speed.

When $\chi_{\text{out}} \neq 0$, some CO_2 leaves the domain, which implies that excess airflow is required to provide the unused CO_2. Since work is required to force air through the porous sorbent, a larger air requirement is undesirable. To quantify excess air, we consider the amount of CO_2 collected relative to the CO_2 inflow with air, that is, the relative air usage over the charging duration t_{ch}, defined as the ratio of accumulated CO_2 to the inflow:

$$\Pi = \frac{v(\chi_{\text{in}}(t_{\text{ch}}) - \chi_{\text{out}}(t_{\text{ch}}))}{v\chi_{\text{in}}(t_{\text{ch}})} = 1 - \frac{\chi_{\text{out}}(t_{\text{ch}})}{\chi_{\text{in}}(t_{\text{ch}})} \tag{57}$$

where

$$\chi_{\text{in}}(t) = \int_0^t \chi(0,t')dt' = \int_0^t dt' = t. \tag{58}$$

is the time integral over the mole fraction at the inlet.

With (55) and (51), the air usage is directly related to the charging rate and the wave parameter,

$$\Pi = \lambda\frac{\bar{\beta}(t_{\text{ch}})}{t_{\text{ch}}} = \lambda\dot{\beta}_{\text{ch}}. \tag{59}$$

With the suggested charging duration $t_{ch} = \lambda$, the air usage equals dimensionless accumulation, $\Pi = \bar{\beta}(t_{ch})$. Depending on the value of λ, air usage can be as small as 33% for $\lambda = 0.5$, that is, for large velocities v.

4.6. Pump Work

Low air usage Π implies that considerable pump work must be used to push (or pull) excess air through the sorbent since not all CO_2 in the air can be collected. We proceed with determining the pump work required per mole of captured gas.

The pumping power to push a gas (i.e., air) through a porous medium of cross section A is given by the product of volume flow $\dot{V} = Av$ and pressure difference across the thickness L of the material:

$$\dot{W} = \dot{V}\Delta p = Av\Delta p. \tag{60}$$

The pressure difference is given by the Ergun equation [42]

$$\Delta p = \left(\frac{\mu_{\rm air}}{k} v_s + \frac{\rho_{\rm air}}{k_1} v_s^2\right) L, \tag{61}$$

where $\rho_{\rm air}$ and $\mu_{\rm air}$ are the mass density and viscosity of air, respectively,

$$v_s = \varepsilon v \tag{62}$$

The superficial velocity with the void fraction ε, and permeability k and inertial permeability k_1 are given as

$$k = \frac{d_p^2 \varepsilon^3}{150(1-\varepsilon)^2}, \quad k_1 = \frac{d_p \varepsilon^3}{1.75(1-\varepsilon)} \tag{63}$$

with d_p being the particle diameter of the sorbent.

The total work for pumping air through for the charging duration $\Delta t = \tau t_{ch}$ is, with $v = \frac{L}{\tau} v$,

$$W_{ch} = \dot{W}\Delta t = \varepsilon \frac{AL^3 \mu_{\rm air}}{\tau k}\left[t_{ch}\left(v^2 + \varepsilon \frac{L}{\tau}\frac{\rho_{\rm air} k}{\mu_{\rm air} k_1} v^3\right)\right]. \tag{64}$$

The total amount of CO_2 collected at time t_{ch} in the sorbent material of volume $V = AL$ is

$$n_{ch} = \bar{\beta}_{ch}[\beta_0 a_{act}][\rho_{mat} AL] \tag{65}$$

where $\beta_0 a_{act}$ is the number of adsorption sites per mass of material, and $\rho_{mat} AL$ is the mass in the volume AL.

With the reference work

$$\bar{w}_0 = \frac{\varepsilon \mu_{\rm air} L^2 (\phi+1)^2}{\tau k \beta_0 a_{act} \rho_{mat}} \tag{66}$$

the pump work per mole of CO_2 adsorbed assumes the compact form

$$\frac{W_{ch}}{n_{ch}} = \bar{w}_0 w_{ch}, \tag{67}$$

where w_{ch} is the dimensionless pump work required for charging the sorbent, which is inversely proportional to the overall charging rate $\dot{\beta}_{ch} = \frac{\bar{\beta}_{ch}}{t_{ch}}$:

$$w_{ch} = \frac{1}{\dot{\beta}_{ch}}\left[\frac{1}{\lambda^2} + \frac{\gamma}{\lambda^3}\right]; \tag{68}$$

with the dimensionless inertial parameter

$$\gamma = \varepsilon(\phi+1)\frac{L}{\tau}\frac{\rho_{\rm air} k}{\mu_{\rm air} k_1}. \tag{69}$$

Table 3 shows the data for the material considered by Climeworks [36], as well as the resulting values for reference work $\bar{w}_0$ and inertial parameter γ.

Table 3. Additional data for the material descibed in [36].

Parameter	Value	
ρ_{air}	$1.3\ \frac{\text{kg}}{\text{m}^3}$	mass density of air
μ_{air}	$1.81 \times 10^{-5}\ \frac{\text{kg}}{\text{ms}}$	viscosity of air
ρ_{mat}	$558\ \frac{\text{kg}}{\text{m}^3}$	mass density of sorbent
ε	0.43	void fraction
d_p	6.2×10^{-4} m	particle diameter
k	$6.3 \times 10^{-10}\ \text{m}^2$	permeability
k_1	4.9×10^{-5} m	inertial permeability
γ	0.06	inertial parameter
$\bar{w}_0$	$4.28\ \frac{\text{kJ}}{\text{mol}}$	reference work for pump

The dimensionless work w_{ch} depends on the charging rate $\dot{\beta}_{\text{ch}}$ and wave parameter λ. Table 4 shows the air usage factor and the dimensionless work for the charging rates of Table 2 (with $\gamma = 0.06$).

Table 4. Air usage factor Π and dimensionless work w_{ch} for charging duration $t_{\text{ch}} = \lambda$ (from numerical solution).

$\lambda = t_{\text{ch}}$	0.5	1	2	4	6	8	10	12
Π	0.335	0.509	0.688	0.829	0.886	0.913	0.931	0.942
w_{ch}	6.69	2.08	0.749	0.306	0.190	0.138	0.108	0.088

The work required for the reversible separation of CO_2 from air is in the order of $\bar{w}_{\text{rev}} \simeq 20\ \frac{\text{kJ}}{\text{mol}}$ [38]. The work for pumping is consumed by friction in the pores; hence, it must be considered as an irreversible loss [19,38,39]. For the example material, the reference value $\bar{w}_0 = 4.28\ \frac{\text{kJ}}{\text{mol}}$ is about one-fifth of the reversible separation work. With the data in Table 4, the actual pumping work $\bar{w} = \bar{w}_0 w_{\text{ch}}$ for $\lambda = 0.5$ is 1.43 times the overall reversible separation work. The behavior is strongly non-linear, with only 16% of the reversible work at $\lambda = t_{\text{ch}} = 2$.

Notably, the reference work (66) depends quadratically on sorbent thickness L; hence, it can be reduced significantly by using thinner layers of material.

5. Optimal System and Process Conditions

The discussion in the previous sections revealed the main parameters and process conditions to explore the fast and efficient adsorption of CO_2 from air in porous sorbents.

Specifically, one must distinguish between the material, physical, and process parameters, which will be discussed below.

External parameters that cannot be changed and thus will not be discussed further are as follows: environmental temperature T, air mole density $\bar{\rho}_{\text{air}}$, atmospheric CO_2 mole fraction χ_{atm}, air viscosity μ_{air}, and gas constant R_{CO_2}.

5.1. Material Parameters

One might state the goal of DAC as collecting CO_2 from the air in compact facilities at high adsorption rates.

Compact facilities demand sorbents with a high density of adsorption sites, that is, large values of $\rho_{\text{mat}} a_{\text{act}} \beta_0$ [unit: $\frac{\text{kg}}{\text{m}^3} \frac{\text{m}^2}{\text{kg}} \frac{\text{mol}}{\text{m}^2} = \frac{\text{mol}}{\text{m}^3}$], which can be separated into the demand for high material mass density ρ_{mat} and the high mass-specific number of adsorption sites

$\mathsf{a}_{act}\beta_0$. The latter is measured through the capacity parameter ϕ (12), which compares the number of adsorption sites relative to the number of CO_2 molecules in air-filled pores,

$$\phi = \frac{\mathsf{a}_{act}\beta_0}{\mathsf{v}_{acc}\tilde{\rho}_{air}\chi_{atm}}. \tag{70}$$

A high capacity is desirable for compactness; hence, the sorbent should have a large internal surface area a_{act} and large site density β_0.

The rate of adsorption is given through the reference time τ that sets the timescale for DAC processes. Revisiting our above line of arguments, this scale is effectively set in the equation for the number density of unoccupied adsorption sites $\hat{\beta} = 1 - \beta$ when permanently exposed to air with a full CO_2 load, which in dimensional form is the simple decay equation

$$\frac{d\hat{\beta}}{dt} = -\frac{\hat{\beta}}{\tau}. \tag{71}$$

This definition allows us to estimate the relevant timescale for any adsorption material or model.

For large adsorption rates, the reference time τ should be as small as possible. In our simple yet instructive model, we identified the timescale as (11)

$$\tau = \frac{\sqrt{2\pi}\beta_0}{\sqrt{R_{CO_2}T}\tilde{\rho}_{air}\chi_{atm}\varpi}. \tag{72}$$

With the site density β_0 being large for high capacity, small τ requires a large adsorption probability factor ϖ. Equivalent statements will be possible for more elaborate adsorption models, which lie outside the scope of this examination.

For typical materials, τ is measured in hours, and a typical charging duration is measured in multiples of τ; hence, the turnover times for adsorption–desorption cycles are rather long. Materials with shorter associated timescales are desperately needed.

5.2. Physical Parameters

The actual size of the system is given through the sorbent volume $V_{mat} = A_{mat}L$. With the volume flow of air entering as $\dot{V}_{air} = A_{mat}v$, the volume flow is proportional to the sorbent cross-section A_{mat}, which thus sets the overall size of the system.

The sorbent thickness L defines the length scale. In our evaluation, it appears to non-dimensionalize velocity v (see next section), and in the reference work (66)

$$\bar{w}_0 = \frac{\varepsilon\mu_{air}L^2(\phi+1)^2}{\tau k\beta_0\mathsf{a}_{act}\rho_{mat}}. \tag{73}$$

With work $\bar{w}_0$ proportional to L^2, a reduction of layer thickness has a strong influence on reducing pump work.

5.3. Process Parameters

With the material chosen and system geometry set, the only available process parameters are the dimensionless air velocity v or, alternatively, the wave parameter $\lambda = \frac{\phi+1}{v}$, and the charging rate $\dot{\beta}_{ch}$ or, alternatively, the charging duration t_{ch}.

As our discussion of charging processes has revealed, for efficient operation, these parameters are not independent. Specifically, when the wave parameter λ is chosen, a charging time not too different from $t_{ch} = \lambda$ yields good use of the material; see Figure 6 and Table 2.

With this, only the wave parameter λ remains to be chosen. With decreasing λ, both, the charging rate $\dot{\beta}_{ch}(\lambda)$ and the pump work $w_{ch}(\lambda)$ are increasing; see Tables 2 and 4, hence, there is no unique value of λ obtainable from an optimization. One has to choose an appropriate value to strike a balance between the contradictory demands of the large charging rate and small pump work.

While we refrained from a detailed discussion of the desorption part of the overall DAC process, a short comment is in order here. Desorption is a thermal process where the sorbent is heated to a suitable temperature (of the order of 100 °C). Heating and subsequent cooling of the sorbent is an inherently irreversible process, and the best use of the supplied heat will be made when the accumulation is relatively high.

With work increasing quadratically for small λ, the adsorption rate decreasing with large λ, and a desire for relatively large accumulation, we believe that the target value should lie in the range of

$$\lambda = 2 \cdots 6, \tag{74}$$

for which the charging rate assumes values in the range $[0.35 \cdots 0.15]$ and the accumulation is in the range of 70–90%. For the sample material and process of Ref. [36], we find $\lambda = 5.75$, which is at the upper end of our recommendation.

We emphasize that for small layer thickness L, the reference work $\bar{w}_0$ will be small as well; hence, the wave parameter can be smaller for large charging rates without demanding too much work.

Table 5 summarizes the recommendations in compact form.

Table 5. Material, physical, and process parameters, and their required range of values. See text for details.

Name	Symbol	Unit	Requirement
sorbent density	ρ_{mat}	$\frac{kg}{m^3}$	"large"
sorbent active area	a_{act}	$\frac{m^2}{kg}$	"large"
sorbent saturation capacity	β_0	$\frac{mol}{m^2}$	"large"
capacity factor	ϕ	1	$\gg 1$
reference time	τ	s	min, not h
probability factor	ϖ	1	$0 \ll \varpi \leq 1$
sorbent thickness	L	m	as small as possible
reference work	$\bar{w}_0$	$\frac{kJ}{mol}$	$\ll \bar{w}_{rev} = 20\frac{kJ}{mol}$
pump work	w_{ch}	1	"small"
charging duration	t_{ch}	1	$= \lambda$, "small"
charging rate	$\dot{\beta}_{ch}$	1	"large"
air velocity	$v = \frac{\tau}{L}v$	1	$= \frac{\phi+1}{\lambda}$
wave parameter	λ	1	$2 \cdots 6$ (balanced)

5.4. Physical Size

To close our arguments, we have a brief look at the required size for large-scale DAC facilities. Aiming for a real global impact, we consider CO_2 removal of $1\frac{t}{year}$ per person, which is $8\frac{Gt}{year} \simeq 10^6\frac{t}{h}$. Assuming the capacity of a single plant similar in size to CarbonEngineering's proposal [17,19] at $100\frac{t}{h}$, this requires 10,000 plants (which is one plant per 800,000 people).

We ask for the mass of sorbent m_{mat} required for one of these plants. The dimensional overall removal rate is

$$\dot{R}_{CO_2} = \frac{5}{6} m_{mat} M_{CO_2} a_{act} \frac{\beta_0}{\tau} \dot{\beta}_{ch} \tag{75}$$

where the factor $\frac{5}{6}$ results from the assumption that the time for desorption is $\frac{1}{5}$ of the time required for adsorption, so that five of six unit times are used for charging.

We evaluate this for the sample material from the Climeworks patent [36], as listed in Tables 1 and 3. Choosing $\lambda = 4$ with $\dot{\beta}_{ch} = 0.2$, we find the required mass of the sorbent as

$$m_{mat} = \frac{6}{5} \frac{\dot{R}_{CO_2} \tau}{M_{CO_2} a_{act} \beta_0 \dot{\beta}_{ch}} = 27{,}300 \text{ t}, \tag{76}$$

which fills the volume

$$V_{mat} = \frac{m_{mat}}{\rho_{mat}} = 49{,}000\ \mathrm{m}^3, \tag{77}$$

corresponding to a cube with an edge length of 36.6 m.

With a layer thickness of $L = 0.04$ m, the cross-section for the inflow is

$$A_{mat} = \frac{V_{mat}}{L} = 1.23 \times 10^6\ \mathrm{m}^2. \tag{78}$$

Obviously, all numbers are scaled with the characteristic time τ; hence, a decrease of τ, that is, using materials with a faster intake of CO_2 is essential.

Using the above numbers, we find a sorbent volume of 250 m^3 for Climeworks' 4000 t/year Orca plant, and 2250 m^3 for their 36,000 t/year Mammoth plant.

These estimates do not account for the adsorption of moisture from the air, which reduces the number of sites available for CO_2 molecules.

The data used here, as extracted from Ref. [36], are probably outdated, but we were not able to find data for newer materials for further evaluation. Certainly, the above numbers provide some idea of the significant physical size of the facilities that direct air capture of CO_2 demands.

6. Conclusions

A reduction in CO_2 emissions as well as direct air capture (DAC) of CO_2 are important tasks for humankind to reduce climate change. Above, we have studied the adsorption step in DAC systems by means of a simple but realistic model, where non-dimensionalization and mathematical evaluation identified the key parameters for these processes: the sorbent material is described through the adsorption timescale τ and its adsorption capacity ϕ, and through layer thickness L, while the running of the process depends on dimensionless air velocity v and charging duration t_{ch}.

Since the charging processes are governed only by the wave parameter $\lambda = \frac{\phi+1}{v}$, process control requires only to chose a suitable value for λ as well as the corresponding charging duration t_{ch}. Our model allows a systematic and quick evaluation of different choices that help with decision-making. While larger velocities, hence smaller values of λ, lead to faster charging rates, they imply considerably larger work demand, and as such, one will have to strike a compromise between the CO_2 collection rate and work requirements. Our evaluation leads to the recommendation that values in the range of $\lambda = 2 \cdots 6$ with $t_{ch} = \lambda$ should be considered for sorbent charging.

While the adsorption model used is relatively simple, it accounts for the proper physics of the problem. That is, more refined models will behave quite similarly, and, in particular, they will describe wave-like transport governed by a corresponding wave parameter (see Appendix A). Hence, the discussion of DAC performance measures and behavior will be rather similar to the above.

We hope that the instructive model and its evaluation will be helpful for further research into DAC adsorption processes, as well as being educational for those wishing to understand such processes better.

Author Contributions: Conceptualization, H.S.; Software, E.K.-L.; Formal analysis, E.K.-L.; Investigation, H.S.; Writing—original draft, E.K.-L. and H.S. All authors have read and agreed to the published version of the manuscript.

Funding: Support from the Natural Sciences and Engineering Research Council of Canada (NSERC) through Discovery Grant No. RGPIN-2022-03188 is gratefully acknowledged.

Institutional Review Board Statement: Not applicable.

Data Availability Statement: The original contributions presented in the study are included in the article, further inquiries can be directed to the corresponding author.

Acknowledgments: This paper is dedicated to the admirable Signe Kjelstrup, with congratulations and thanks for her outstanding contributions to research and education in nonequilibrium thermodynamics.

Conflicts of Interest: The authors declare no conflicts of interest.

Appendix A. Adsorption Rate and the Second Law

Appendix A.1. Transport Equations

We ask for the formulation of the second law of thermodynamics for the transport model (15) and (16). To this end, we consider the adsorption process akin to a chemical reaction where CO_2 (χ) and free adsorption sites ($\hat{\beta}$) "react" to form occupied adsorption sites (β). To proceed, we write the dimensionless reaction diffusion system as

$$\frac{\partial \phi \beta}{\partial t} = \Lambda \tag{A1}$$

$$\frac{\partial \phi \hat{\beta}}{\partial t} = -\Lambda \tag{A2}$$

$$\frac{\partial \chi}{\partial t} + v\frac{\partial \chi}{\partial x} - D\frac{\partial^2 \chi}{\partial x^2} = -\Lambda \tag{A3}$$

where $\Lambda(x,t)$ is the adsorption rate. In this dimensionless formulation the factor ϕ accounts for adjusting units between the variables, which in dimensional form are surface densities β, $\hat{\beta}$, and mole fraction χ. The number of adsorption sites are constant, hence

$$\beta + \hat{\beta} = 1. \tag{A4}$$

Appendix A.2. Linear Irreversible Thermodynamics

We proceed with the standard methods from nonequilibrium thermodynamics for the modeling of reactions, see, e.g., [40], which here simplify significantly since we consider only fixed temperatures and pressures. Then, it suffices to consider the dimensionless entropy density as

$$\eta = \chi(s_\chi - \ln \chi) + \phi\hat{\beta}\left(s_{\hat{\beta}} - \ln \hat{\beta}\right) + \phi\beta(s_\beta - \ln \beta) \tag{A5}$$

where the s_α are entropy constants. Taking the time derivative of η and replacing the time derivatives of the variables β, $\hat{\beta}$, χ by means of the transport equations yields after some standard manipulation of the second law of thermodynamics in the form [39,40]

$$\frac{\partial \eta}{\partial t} + \frac{\partial \varphi}{\partial x} = \sigma \tag{A6}$$

with the entropy flux

$$\varphi = v\chi(s_\chi - \ln \chi) - \left(D(s_\chi - \ln \chi - 1)\frac{\partial \chi}{\partial x}\right) \tag{A7}$$

and the entropy production rate

$$\sigma = \left[1 + s_\beta - s_\chi - s_{\hat{\beta}} + \ln \frac{\chi\hat{\beta}}{\beta}\right]\Lambda + \frac{\mathrm{D}}{\chi}\frac{\partial \chi}{\partial x}\frac{\partial \chi}{\partial x}. \tag{A8}$$

The entropy production consists of two contributions, which must be independently positive. The second term describes the entropy generation due to the diffusion of CO_2 in air, and is positive for the positive diffusion coefficient D. The first term describes entropy generation in the adsorption process and will now be used to model the adsorption rate Λ.

To proceed, we define the equilibrium constant k by writing

$$\ln \mathsf{k} = s_\chi + s_{\hat{\beta}} - s_\beta - 1, \tag{A9}$$

which gives the adsorption contribution to entropy generation as

$$\sigma_{\text{ad}} = \left[\ln \chi\hat{\beta} - \ln \mathsf{k}\beta\right]\Lambda \geqslant 0. \tag{A10}$$

Following the principles of classical linear irreversible thermodynamics [39,40], we interpret this expression as the product of the thermodynamic force $\mathcal{F} = \left[\ln \chi\hat{\beta} - \ln \mathsf{k}\beta\right]$ and the thermodynamic flux $\mathcal{J} = \Lambda$. For close to equilibrium processes, one tends to express the flux in a linear relation—a phenomenological equation—of the form $\mathcal{J} = \alpha\mathcal{F}$ with phenomenological coefficient α, but for chemical reactions and the adsorption process studied here, these are not useful. Instead, as is often done in reaction kinetics [40], we consider an Arrhenius-type ansatz by taking the exponentials of the two competing terms in the force as

$$\Lambda = \phi\left[\exp[\ln \chi\hat{\beta}] - \exp[\ln \mathsf{k}\beta]\right] = \phi\left[\chi\hat{\beta} - \mathsf{k}\beta\right], \tag{A11}$$

where the factor ϕ is chosen for agreement with the proposed model equations. The adsorption model studied in the bulk of this paper arises in the limit $\mathsf{k} \to 0$, which implies that the spontaneous detachment of CO_2 from an occupied site is effectively impossible. This is similar to the analysis of combustion processes, where only forward reactions are considered.

The chemical constant depends on temperature, $\mathsf{k}(T)$, and increases with temperature so that desorption occurs at larger temperatures.

Appendix A.3. Langmuir Isotherm

In equilibrium, the reaction rate must vanish, $\Lambda_{eq} = 0$, which gives the Langmuir isotherm as, with $\beta + \hat{\beta} = 1$,

$$\beta_{eq} = \frac{\chi}{\mathsf{k} + \chi}. \tag{A12}$$

In equilibrium with the atmosphere, in our dimensionless formulation, we have $\chi_{atm} = 1$, hence

$$\beta_{eq} = \frac{1}{1+\mathsf{k}}, \quad \hat{\beta}_{eq} = 1 - \beta_{eq} = \frac{\mathsf{k}}{1+\mathsf{k}}.$$

For the case $\mathsf{k} \to 0$, this reduces to the case studied in the main paper, where $\beta_{eq} = 1$, $\hat{\beta}_{eq} = 0$.

Appendix A.4. Wave Solution

Also, with $\mathsf{k} \neq 0$, the adsorption process is wave-like. For a wave connecting the equilibrium with air in the wake at $\chi_{-\infty} = 1$ and the post-desorption equilibrium state χ_∞, the wave solution is obtained from a straightforward calculation as

$$\chi(x,t) = \frac{1+\chi_\infty}{2} + \frac{1-\chi_\infty}{2}\tanh\left[-\frac{1-\chi_\infty}{2}(\lambda x - t - \xi_0)\right] \tag{A13}$$

$$\hat{\beta}(x,t) = \frac{\mathsf{k}}{1+\mathsf{k}}\left[1 + \frac{1-\chi(x,t)}{\mathsf{k}+\chi_\infty}\right] \tag{A14}$$

$$\beta(x,t) = \frac{1}{1+\mathsf{k}}\left[1 - \frac{\mathsf{k}(1-\chi(x,t))}{\mathsf{k}+\chi_\infty}\right] \tag{A15}$$

with the wave parameter

$$\lambda = \frac{1}{\alpha v} = \frac{\frac{\phi+1}{v}}{1 + \frac{\chi_\infty}{\mathsf{k}}} + \frac{\mathsf{k}^2\phi + \chi_\infty}{v(\mathsf{k}+\chi_\infty)}. \tag{A16}$$

We recognize the coefficient λ as the inverse wave speed, while the factor $(1-\chi_\infty)\lambda$ determines the width of the wave, which is wider for smaller values, i.e., larger χ_∞.

The two equilibrium states are, in the wake of the wave: $\xi \to -\infty$ with $\tanh(\infty) = 1$

$$\chi_{-\infty} = 1, \quad \hat{\beta}_{-\infty} = \frac{\mathsf{k}}{1+\mathsf{k}}, \quad \beta_{-\infty} = \frac{1}{1+\mathsf{k}}; \tag{A17}$$

and in front of the wave: $\xi \to \infty$ with $\tanh(-\infty) = -1$

$$\chi_\infty, \quad \hat{\beta}_\infty = \frac{\mathsf{k}}{\mathsf{k}+\chi_\infty}, \quad \beta_\infty = \frac{\chi_\infty}{\mathsf{k}+\chi_\infty}. \tag{A18}$$

The waves discussed in the bulk of the paper result in the limit $\mathsf{k} \to 0$, which also requires $\chi_\infty = 0$, so that in the limit $\mathsf{k} \to 0$, we find $\hat{\beta}_\infty = 1, \beta_\infty = 0$.

References

1. United Nations. Paris Agreement (Adopted 12 Dec. 2015, Entered into Force 4 Nov. 2016) ATS 2016 24. Available online: https://www.un.org/en/climatechange/paris-agreement (accessed on 11 September 2024).
2. Rogelj, J.; Shindell, D.; Jiang, K.; Fifita, S.; Forster, P.; Ginzburg, V.; Handa, C.; Kheshgi, H.; Kobayashi, S.; Kriegler, E.; et al. Global Warming of 1.5 °C. An IPCC Special Report on the Impacts of Global Warming of 1.5 °C above Pre-industrial Levels and Related Global Greenhouse Gas Emission Pathways, in the Context of Strengthening the Global Response to the Threat of Climate Change, Sustainable Development, and Efforts to Eradicate Poverty, IPCC Special Report (Intergovernmental Panel on Climate Change, Geneva). 2018. Available online: https://www.ipcc.ch/site/assets/uploads/sites/2/2019/05/SR15_Chapter2_High_Res.pdf (accessed on 11 September 2024).
3. United Nations. Net Zero Coalition. Available online: https://www.un.org/en/climatechange/net-zero-coalition (accessed on 11 September 2024).
4. UNFCCC. Technical Dialogue of the First Global Stocktake. Technical Report, August. 8 September 2023. Available online: https://unfccc.int/documents/631600 (accessed on 11 September 2024).
5. International Energy Agency. Net Zero by 2050—A Roadmap for the Global Energy Sector, Technical Report. October 2021. Available online: https://www.iea.org/reports/net-zero-by-2050 (accessed on 11 September 2024).
6. Environmental Defence Fund. A Revamped Cost Curve for Reaching Net-Zero Emissions. August 2021. Available online: https://www.edf.org/revamped-cost-curve-reaching-net-zero-emissions (accessed on 11 September 2024).
7. International Energy Agency. Direct Air Capture—Energy System. Available online: https://www.iea.org/energy-system/carbon-captureutilisation-and-storage/direct-air-capture (accessed on 11 September 2024).
8. Azarabadi, H.; Lackner, K.S. Postcombustion Capture or Direct Air Capture in Decarbonizing US Natural Gas Power? *Environ. Sci. Technol.* **2020**, *54*, 5102–5111. [CrossRef] [PubMed]
9. Olah, G.A.; Prakash, G.K.S.; Goeppert, A. Anthropogenic Chemical Carbon Cycle for a Sustainable Future. *Am. Chem. Soc.* **2011**, *133*, 12881–12898. [CrossRef] [PubMed]
10. Graves, C.; Ebbesen, S.D.; Mogensen, M.; Lackner, K.S. Sustainable hydrocarbon fuels by recycling CO_2 and H_2O with renewable or nuclear energy. *Renew. Sustain. Energy Rev.* **2011**, *15*, 1–23. [CrossRef]
11. Ozkan, M.; Nayak, S.P.; Ruiz, A.D.; Jiang, W. Current status and pillars of direct air capture technologies. *iScience* **2022**, *25*, 103990. [CrossRef] [PubMed]
12. Bouaboula, H.; Chaouki, J.; Belmabkhout, Y.; Zaabout A. Comparative review of Direct air capture technologies: From technical, commercial, economic, and environmental aspects. *Chem. Eng. J.* **2024**, *484*, 149411. [CrossRef]
13. Sanz-Perez, E.S.; Murdock, C.R.; Didas, S.A.; Jones, C.W. Direct Capture of CO_2 from Ambient Air. *Chem. Rev.* **2016**, *116*, 11840–11876. [CrossRef]
14. Global Thermostat. Solutions. Available online: https://www.globalthermostat.com/solutions (accessed on 11 September 2024).
15. Infinitree. Technology. Available online: http://www.infinitreellc.com/technology (accessed on 11 September 2024).
16. Carbon Collect. Mechanical Tree. Available online: https://carboncollect.com/mechanical-tree/ (accessed on 11 September 2024).
17. Keith, D.W.; Holmes, G.; Angelo, D.S.; Heidel, K. A process for capturing CO_2 from the atmosphere. *Joule* **2018**, *2*, 1573–1594. [CrossRef]
18. Deutz, S.; Bardow, A. Life-cycle assessment of an industrial direct air capture process based on temperature-vacuum swing adsorption. *Nat. Energy* **2021**, *6*, 203–213. [CrossRef]
19. Long-Innes, R.; Struchtrup, H. Thermodynamic loss analysis of a liquid-sorbent direct air carbon capture plant. *Cell Rep. Phys. Sci.* **2022**, *3*, 100791. [CrossRef]
20. Climeworks. Direct Air Capture. Available online: https://climeworks.com/direct-air-capture (accessed on 11 September 2024).
21. Carbfix. We Turn CO_2 into Stone. Available online: https://www.carbfix.com/ (accessed on 11 September 2024).

22. Wurzbacher, J.A. Development of a Temperature-Vacuum Swing Process for CO_2 Capture from Ambient Air. Doctoral Thesis, ETH Zürich, Zürich, Switzerland, 2015; Dissertation ETH No. 22172. [CrossRef]
23. Wijesiri, R.P.; Knowles, G.P.; Yeasmin, H.; Hoadley, A.F.A.; Chaffee, A.L. Technoeconomic Evaluation of a Process Capturing CO_2 Directly from Air. *Processes* **2019**, *7*, 503. [CrossRef]
24. Buijs, W.; de Flart, S. Direct Air Capture of CO_2 with an Amine Resin: A Molecular Modeling Study of the CO_2 Capturing Process. *Ind. Eng. Chem. Res.* **2017**, *56*, 12297–12304. [CrossRef] [PubMed]
25. Low, M.-Y.A.; Barton, L.V.; Pini, R.; Petit, C. Analytical review of the current state of knowledge of adsorption materials and processes for direct air capture. *Chem. Eng. Res. Des.* **2023**, *189*, 745–767. [CrossRef]
26. House, K.Z.; Baclig, A.C.; Ranjan, M.; Van Nierop, E.A.; Wilcox, J.; Herzog, H.J. Economic and energetic analysis of capturing CO_2 from ambient air. *Proc. Natl. Acad. Sci. USA* **2011**, *108*, 20428–20433. [CrossRef] [PubMed]
27. Zhang, Z.X.; Xu, H.J. Thermodynamic modeling on multi-stage vacuum-pressure swing adsorption (VPSA) for direct air carbon capture with extreme dilute carbon dioxide. *Energy* **2023**, *276*, 127550. [CrossRef]
28. Zhang, Z.X.; Xu, H.J.; Hua, W.S.; Zhao, C.Y. Thermodynamics analysis of multi-stage temperature swing adsorption cycle for dilute CO_2 capture, enrichment and purification. *Energy Convers. Manag.* **2022**, *265*, 115794. [CrossRef]
29. Wurzbacher, J.A.; Gebald, C.; Piatkowski, N.; Steinfeld, A. Concurrent Separation of CO_2 and H_2O from Air by a Temperature-Vacuum Swing Adsorption/Desorption Cycle. *Environ. Sci. Technol.* **2012**, *46*, 9191–9198. [CrossRef]
30. Gebald, C.; Wurzbacher, J.A.; Tingaut, P.; Zimmermann, T.; Steinfeld, A. Amine-Based Nanofibrillated Cellulose As Adsorbent for CO_2 Capture from Air. *Environ. Sci. Technol.* **2011**, *45*, 9101–9108. [CrossRef]
31. Gebald, C.; Wurzbacher, J.A.; Borgschulte, A.; Zimmermann, T.; Steinfeld, A. Single-Component and Binary CO_2 and H_2O Adsorption of Amine-Functionalized Cellulose. *Environ. Sci. Technol.* **2014**, *48*, 2497–2504. [CrossRef]
32. Gebald, C.; Wurzbacher, J.A.; Tingaut, P.; Steinfeld, A. Stability of Amine-Functionalized Cellulose during Temperature-Vacuum-Swing Cycling for CO_2 Capture from Air. *Environ. Sci. Technol.* **2013**, *47*, 10063–10070. [CrossRef]
33. Wurzbacher, J.A.; Gebald, C.; Brunner, S.; Steinfeld, A. Heat and mass transfer of temperature-vacuum swing desorption for CO_2 capture from air. *Chem. Eng. J.* **2016**, *283*, 1329–1338. [CrossRef]
34. Deschamps, T.; Kanniche, M.; Grandjean, L.; Authier, O. Modeling of Vacuum Temperature Swing Adsorption for Direct Air Capture Using Aspen Adsorption. *Clean Technol.* **2022**, *4*, 258–275. [CrossRef]
35. Leonzio, G.; Fennell, P.S.; Shah, N. A Comparative Study of Different Sorbents in the Context of Direct Air Capture (DAC): Evaluation of Key Performance Indicators and Comparisons. *Appl. Sci.* **2022**, *12*, 2618. [CrossRef]
36. Gebald, C.; Piatkowski, N.; Rüesch, T.; Wurzbacher, J.A. Low-Pressure Drop Structure of Particle Adsorbent Bed for Adsorption Gas Separation Process. Patent WO2014170184, 9 April 2014. Available online: https://patentscope.wipo.int/search/en/detail.jsf?docId=WO2014170184 (accessed on 11 September 2024).
37. Kay-Leighton, E. Thermodynamic Analysis of the Behaviour and Performance of the Adsorption Phase of Direct Air Capture Technology. Master's Thesis, University of Victoria, Victoria, BC, Canada, 2024
38. Struchtrup, H. *Thermodynamics and Energy Conversion*, 2nd ed.; Springer: Cham, Switzerland, 2024. [CrossRef]
39. Kjelstrup, S.; Bedeaux, D.; Johannessen, E.; Gross, J. *Non-Equilibrium Thermodynamics for Engineers*, 2nd ed.; World Scientific: Singapore, 2017. [CrossRef]
40. Struchtrup, H. *A Thermodynamic Introduction to Transport Phenomena*; Springer: Cham, Switzerland, 2024. [CrossRef]
41. Struchtrup, H. *Macroscopic Transport Equations for Rarefied Gas Flows*; Springer: Heidelberg, Germany, 2005. [CrossRef]
42. Ergun, S.; Orning, A.A. Fluid Flow through Randomly Packed Columns and Fluidized Beds. *Ind. Ind. Eng. Chem.* **1949**, *41*, 1179–1184. [CrossRef]

entropy

MDPI

Article

Internal Energy, Fundamental Thermodynamic Relation, and Gibbs' Ensemble Theory as Emergent Laws of Statistical Counting

Hong Qian

Department of Applied Mathematics, University of Washington, Seattle, WA 98195-3925, USA; hqian@uw.edu

Abstract: Statistical counting *ad infinitum* is the holographic observable to a statistical dynamics with finite states under independent and identically distributed N sampling. Entropy provides the infinitesimal probability for an observed empirical frequency $\hat{\nu}$ with respect to a probability prior $\mathbf{p}$, when $\hat{\nu} \neq \mathbf{p}$ as $N \to \infty$. Following Callen's postulate and through Legendre–Fenchel transform, without help from mechanics, we show that an internal energy $\mathbf{u}$ emerges; it provides a linear representation of real-valued observables with full or partial information. Gibbs' fundamental thermodynamic relation and theory of ensembles follow mathematically. $\mathbf{u}$ is to $\hat{\nu}$ what chemical potential μ is to particle number N in Gibbs' chemical thermodynamics, what $\beta = T^{-1}$ is to internal energy U in classical thermodynamics, and what ω is to t in Fourier analysis.

Keywords: emergent phenomenon; entropy; information; internal energy; probability theory; statistic; thermodynamics

1. Introduction

It is a pleasure to be a part of this celebration for Signe Kjelstrup. She has made significant contributions to nonequilibrium thermodynamics, in both theory and applications that include electrochemistry, transport in heterogeneous media, and T. L. Hill's small systems [1–3]. In this work we extend Gibbs' and Hill's approach to equilibrium thermodynamics [4] and show that the new logical path via the "crucial step" advocated in [5] is in fact a consequence of a limit theorem [6] in the mathematical theory of probability [7,8]. The results perfectly fit P. W. Anderson's notion of emergent phenomenon [9].

Sometimes a mathematical transform can provide a fundamental concept beyond just being a technique for solving a problem, and through which a new representation of a natural phenomenon emerges. A case in point is the Fourier transform (FT) that leads to the theory of harmonics in music instruments [10] and the very concept of optical spectrum. FT represents a function of time $f(t)$ in terms of $\tilde{f}(\omega)$, where ω is introduced as a novel notion, the temporal frequency of a sinusoidal oscillatory component in time [11]. The solutions to a large class of problems in differential calculus involving t can be very efficiently expressed through FT.

We show in the present paper that the fundamental notion of *internal energy* first appeared in the theory of thermodynamics in the 19th century, collectively developed by J. R. von Mayer, W. Rankine, R. Clausius, and W. Thomson among many others [12], is a concept that can be understood, and generalized, in statistical counting. The transformation in question is the Legendre–Fenchel transform (LFT) [13,14], a more refined mathematical formulation of the traditional Legendre transform [15].

When a simple statistical analysis is carried out on a set of data, correlated or not, it is usually supposed that they are from an identical probability distribution. One of the best understood systems that exhibit an invariant probability is the *ergodic dynamical system* [16]. The ergodic theory of classical Hamiltonian dynamics has been an intense research area in both physics and mathematics for more than a century [17,18]. Even when the data are from seemingly different "objects", say different individuals within a biological species, it is understood that an ergodic mating or mutational process is behind the statistical practice; and the conclusions drawn are most meaningful in this regard. Such an ergodic stochastic dynamic perspective has transformed cell biology through the notion of phenotypic switching in recent years [19].

2. Energetic Theory of Statistical Counting

Let us consider the repeated statistical samples *ad infinitum* of a system with finite state space $\mathcal{S} = \{0, 1, \cdots, n\}$. In the present work we shall restrict our discussion for independent and identically distributed (i.i.d.) samples. More general sampling of Markov data will be published elsewhere. The number counting $\boldsymbol{\nu} = (\nu_0, \cdots, \nu_n)$ with $\nu_0 + \cdots + \nu_n = N$ and counting frequency $\hat{\boldsymbol{\nu}} = \boldsymbol{\nu}/N$, not to be confused with the ω in FT above, has a homogeneous degree 1 neg-entropy function with respect to a given probability prior $\mathbf{p} = (p_0, \cdots, p_n)$ [8,20]:

$$\Phi(\boldsymbol{\nu}) = \sum_{i=0}^{n} \nu_i \ln\left(\frac{\nu_i}{p_i \sum_{j=0}^{n} \nu_j} \right). \tag{1}$$

The Appendix A provides the mathematical origin of the non-negative $\Phi(\boldsymbol{\nu})$ as a result of statistical counting. In information theory, it is interpreted as the "surprise" in observing the $\boldsymbol{\nu}$ under the assumption $\mathbf{p}$ [21,22]. It is a double-edged sword which tells the rareness of $\boldsymbol{\nu}$ (or $\hat{\boldsymbol{\nu}}$) with respect to $\mathbf{p}$ or erroneous model $\mathbf{p}$ with respect to empirical $\boldsymbol{\nu}$. The Kolmogorov probability $\mathbf{p}$ has two rather different roles in statistical inference and in statistical physics. In the former it has been identified as only half to the other "half of probability theory as it is needed in current applications—the principles for assigning probabilities by logical analysis of incomplete information—is not present at all in the Kolmogorov system" [20].

The application of modern probability to statistical physics involves the limit of sample *ad infinitum* represented by $N \to \infty$ [6]. In this case, the probability $\mathbf{p}$ is for all the systems with the same state space $\mathcal{S}$; it is not meant to be realistic for any particular system. It simply provides a "metric" under which each and every particular system has its own representation in terms of its complete information, $\boldsymbol{\nu}$. The $\Phi(\boldsymbol{\nu})$ is introduced to further gauge the differences among systems with different $\boldsymbol{\nu}$'s on the same $\mathcal{S}$; it becomes a "theory of everything." Because the $N \to \infty$ limit, there are no uncertainties in $\hat{\boldsymbol{\nu}}$; it is a definitive characterization of an i.i.d. statistical distribution with state space $\mathcal{S}$.

Therefore, statistical inference is about the mathematical model of a particular system, and statistical physics is about the mathematical representation of all systems with the same $\mathcal{S}$ under the supposition of i.i.d. data *ad infinitum*. The entropy in (1) is an emergent characterization in the limit of $N \to \infty$, with the starting point in terms of generative models [9]. It provides the *relationship* between $\boldsymbol{\nu}$ and $\mathbf{p}$ in the sampling process. It is an Eulerian degree 1 homogeneous function of $\boldsymbol{\nu}$: $\Phi(\lambda\boldsymbol{\nu}) = \lambda\Phi(\boldsymbol{\nu})$. This fits naturally to the fundamental thermodynamic postulate formulated by H. B. Callen [23]. The LFT of Φ as a function of the normalized $\hat{\boldsymbol{\nu}}$ then yields [13,14,24]:

$$\Psi(\mathbf{u}) = \inf_{\hat{\boldsymbol{\nu}}} \left\{ \sum_{i=0}^{n} \hat{\nu}_i u_i + \Phi(\hat{\boldsymbol{\nu}}) \right\} = -\ln \sum_{i=0}^{n} p_i e^{-u_i}, \tag{2a}$$

with corresponding optimal $\hat{\boldsymbol{\nu}}^*(\mathbf{u})$

$$\left(\hat{\boldsymbol{\nu}}^*\right)_i = \frac{p_i e^{-u_i}}{\sum_{\ell=0} p_\ell e^{-u_\ell}}, \text{ and } u_i = -\left(\frac{\partial \Phi(\hat{\boldsymbol{\nu}}^*)}{\partial \hat{\nu}_i} \right). \tag{2b}$$

Note that the second equation in (2b) is obtained when one uses calculus to solve the infimum in (2a); this recovers the traditional Legendre transform. Normalizing $\boldsymbol{\nu}$ to $\hat{\boldsymbol{\nu}}$ induces a gauge freedom in (2), an arbitrary additive constant to u_i. In statistical thermodynamics, the conjugate variable u_k introduced in Equation (2) has been interpreted as the *internal energy* of the state k, in k_BT unit [25]; then $\hat{\nu} \cdot \mathbf{u}$ is the mean internal energy of "the statistical system".

In a real-world laboratory working on a particular system, the ν tends to infinity as $N \to \infty$ but $\hat{\nu}$ converges to the intrinsic property of the statistical system. In statistical inference, the assumed $\mathbf{p}$, as a prior, then is expected to be replaced by the observed, real, posterior $\hat{\nu}$ according to conditional probability and/or Bayesian statistical logic [24,25]. This concludes the statistical investigation of the particular system with respect to the type of observations. The neg-entropy function in (1) actually provides a *meta-statistical* theory for all possible observed $\hat{\nu}$, assessing their respective infinitesimal probability (rate) with respect to the prior $\mathbf{p}$ (see Appendix A).

3. Maximization of Entropy $-\Phi$ Under Constraint by Empirical Mean Value

However, The complete counting for the entire state space $\mathcal{S}$ in terms of empirical frequencies $\hat{\nu}$ is only a *gedankenexperiment*. The significance of Gibbs' ensemble theory is in dealing with observations from a small set of real-valued observables $g_1(i), g_2(i), \cdots, g_J(i)$, where $i \in \mathcal{S}$ but $J \ll n$. These g's are random variables on the state space $\mathcal{S}$. In fact, their empirical mean values are linear combinations of the $\hat{\nu}$:

$$x_j = \sum_{i=0}^{n} \hat{\nu}_i g_j(i). \tag{3}$$

To fix mathematical notations, we append $g_0(i) = 1$ and $x_0 = 1$, which represent the fact that $\hat{\nu}$ is always normalized, and denote $(n+1) \times (J+1)$ matrix $\mathbf{G}_J$ with elements

$$(\mathbf{G}_J)_{ij} = \begin{cases} 1 & j = 0, \\ g_j(i) & j = 1, \cdots, J. \end{cases} \tag{4}$$

Equation (3) shows that if all the g's are linearly independent and $J = n$, then one can solve the normalized $\hat{\nu}$ uniquely from each set of x's: $\hat{\nu} = \mathbf{x}\mathbf{G}_n^{-1}$. We refer to such a set of observables as *holographic* with full information. In the following discussion, we shall always imagine the $(g_1, \cdots, g_J)$ as the first J component of a holographic observable $(g_0, g_1, \cdots, g_n)$. When $J < n$, there is missing information [20,22,25].

With a set of observed values $\mathbf{x}' = (x_1, \cdots, x_J)$ in hand where $J < n$, the *maximum entropy principle* (MEP) from classical thermodynamics [23] and the contraction principle from the mathematical theory of probability [8] assert that the most probable $\hat{\nu}^*$ that is consistent with the set of $\mathbf{x}'$ corresponds to minimum neg-entropy:

$$\hat{\nu}^* = \arg\inf_{\hat{\nu}} \left\{ \Phi(\hat{\nu}) \middle| \hat{\nu}\mathbf{G}_J = \mathbf{x}' \right\}. \tag{5}$$

The entire Gibbs' ensemble theory arises in solving the mathematical problem posed in Equation (5) through LFT. See Appendix A for its origin.

Entropy functions for different observables are different. First, for invertible $\mathbf{G}_n$, one has the entropy function for the holographic observable $\mathbf{x} = (1, x_1, \cdots, x_n)$:

$$\Phi_{\mathbf{x}}(\mathbf{x}) \equiv \Phi\big(\mathbf{x}\mathbf{G}_n^{-1}\big). \tag{6}$$

This is simply a change in the independent variables from ν to $\mathbf{x}$. Then in terms of this entropy function $\Phi_{\mathbf{x}}$, (5) becomes

$$\begin{aligned} \varphi(\mathbf{x}') &= \inf_{\hat{\nu}} \left\{ \Phi(\hat{\nu}) \middle| \hat{\nu}\mathbf{G}_J = \mathbf{x}' \right\} \\ &= \inf_{x_{J+1}, \cdots, x_n} \left\{ \Phi_{\mathbf{x}}(\mathbf{x}) \middle| x_1 = x_1', \cdots, x_J = x_J' \right\}. \end{aligned} \tag{7}$$

Intimately related to the generating function of a probability distribution, the LFT provides a powerful mathematical transform of the entropy functions $\Phi(\nu)$, $\Phi_{\mathbf{x}}(\mathbf{x})$, and $\varphi(\mathbf{x}')$ in terms of their conjugates in the energy representation: Parallel to the $\Psi(\mathbf{u})$ in (2) are,

$$\Psi_{\mathbf{y}}(\mathbf{y}) = \inf_{\mathbf{x}}\left\{\sum_{i=1}^{n} x_i y_i + \Phi_{\mathbf{x}}(\mathbf{x})\right\}, \tag{8}$$

$$\psi(\mathbf{y}') = \inf_{\mathbf{x}'}\left\{\sum_{j=1}^{J} x_j' y_j' + \varphi(\mathbf{x}')\right\}. \tag{9}$$

These psi's are now related through linear transformation:

$$\Psi_{\mathbf{y}}(\mathbf{y}) = \Psi(\mathbf{G}_n\mathbf{y}), \tag{10}$$

and projection:

$$\psi(\mathbf{y}') = \Psi_{\mathbf{y}}(y_1', \cdots, y_J', 0, \cdots, 0) = \Psi(\mathbf{G}_J\mathbf{y}') \tag{11a}$$

$$= -\ln \sum_{i=0}^{n} p_i \exp\left[-\sum_{j=1} g_j(i) y_j'\right]. \tag{11b}$$

And finally, since ψ is convex, the inverse LFT yields

$$-\varphi(\mathbf{x}') = \inf_{\mathbf{y}'}\left\{\sum_{i=1}^{J} x_i' y_i' - \psi(\mathbf{y}')\right\} = \left\{\begin{array}{lcl} -\varphi & = & \mathbf{y}' \cdot \nabla\psi(\mathbf{y}') - \psi(\mathbf{y}') \\ \mathbf{x}' & = & \nabla\psi(\mathbf{y}') \end{array}\right. \tag{12}$$

The optimization in (5) is completely "solved" in closed form, through LFT and its inverse, as a parametric function in terms of $\mathbf{y}'$ given in (12).

The equation $-\varphi = \mathbf{y}' \cdot \nabla\psi - \psi$ in (12) should be recognized as a generalization of the celebrated "entropy = mean internal energy − free energy", where

$$(\nabla\psi)_k = \frac{\sum_{i=0}^{n} g_k(i) p_i \exp \sum_{j=1}^{J} g_j(i) y_j}{\sum_{i=0}^{n} p_i \exp \sum_{j=1}^{J} g_j(i) y_j} \tag{13}$$

is the mean value of g_k following Equation (11b), whose conjugate variable is y_k'. The identification of $\mathbf{u} = \mathbf{G}_J\mathbf{y}'$ in (11a) with the first law of thermodynamics as formulated by Gibbs seems natural.

The $y_{J+1} = \cdots = y_n = 0$ in (11a) has a very clear thermodynamic interpretation: Since the conjugate variable $\mathbf{y}$ are the partial derivatives of the entropy function $\Phi_{\mathbf{x}}$ with respect to $\mathbf{x}$, finding x's with maximum entropy in Equation (7) is simply setting the corresponding $y = 0$, *e.g.*, letting the *entropic force* be zero. For each independent observable g_j, y_j is its "custom-designed" conjugate force and $y_j \times dx_j$ contributes a term to the internal energy as the "thermodynamic work" associated with g_j: The internal energy $\mathbf{u}$ is a highly flexible, adaptive representation of the $\hat{\nu}$. When $J = n$, $\mathbf{u} = \mathbf{G}_n\mathbf{y}$ and Equation (10) provides a complete "detailing" of the internal energy in terms of a set of holographic observables. MEP is for missing information [20].

4. Gibbs Distribution and Linear Algebraic Representation

There is a geometric picture associated with the above "thermodynamic analysis". As we have stated, counting frequency *ad infinitum* $\hat{\nu}$ is a fundamental, intrinsic property of an ergodic dynamical system. The space of all possible frequency distributions $\hat{\nu}$, with $\hat{\nu}_0 + \cdots + \hat{\nu}_n = 1$, is a n-dimensional hyper-plane in the positive quadrant of $\mathbb{R}^{n+1}$, known as a probability simplex $\mathscr{M}_n$. For a given set of observables $(g_1, \cdots, g_J)$, the $\mathscr{M}_n$ is foliated by $\hat{\nu}\mathbf{G}_j = \mathbf{x}'$ with different $\mathbf{x}'$. On each leave of the foliation, there is the most probable $\nu^*(\mathbf{x}')$, which is located at the tangent point between the $(n-J)$-dimensional leave and a $(n-1)$-dimensional level set of the $\Phi(\hat{\nu})$ function. At this point, $\nabla_\nu \Phi(\hat{\nu}) = -\mathbf{u}(\hat{\nu})$ is the normal vector to the $\mathbf{x}'$-leave in $\mathbb{R}^{n+1}$, and $\nabla\varphi(\mathbf{x}') = -\mathbf{y}'$ is its projection onto the J-manifold of $\mathbf{x}'$:

$$\nu^*(\mathbf{x}') = \begin{cases} \nu_i^* = \dfrac{1}{Z(\mathbf{y}')} p_i \exp\Big[-\sum\limits_{j=1}^{J} g_j(i) y_j'\Big] \\ x_j' = \dfrac{1}{Z(\mathbf{y}')} \sum\limits_{i=0}^{n} g_j(i) p_i \exp\Big[-\sum\limits_{j=1}^{J} g_j(i) y_j'\Big] \end{cases} \tag{14a}$$

in which

$$Z(\mathbf{y}') = \sum_{i=0}^{n} p_i \exp\left[-\sum_{i=1}^{J} g_j(i) y_j'\right]. \tag{14b}$$

All the other points on the same $\mathbf{x}'$-leave are no longer relevant: they are deemed statistically impossible under the prior $\mathbf{p}$ and observed $\mathbf{x}'$. The foliation therefore represents a partition of the $\mathscr{M}_n$ into macro- and micro-worlds: Transversing between different $\mathbf{x}'$-leaves are *macroscopic thermodynamic processes* that follow the $\mathbf{y}'(\mathbf{x}')$. According to the logic of Bayesian statistics, one should use the most suitable probability frequency distribution $\hat{\nu}^*(\mathbf{x}')$ to update the prior $\mathbf{p}$ for the particular system with observed $\mathbf{x}'$. The *microscopic world* is still random, due to missing information, but its prior is now updated. This is Gibbs' statistical ensemble.

With a given set of $(g_1, \cdots, g_J)$, the $\mathscr{M}_n$ is collapsed into J-manifold in $\mathbb{R}^{n+1}$, which is parametrized by the $\mathbf{x}'$, or equivalently $\mathbf{y}'$. There is no uncertainty in this "macroscopic" description. For a different set of g's and J', there will be a different J'-manifold. It will be desirable to treat different g's through transformations. We note that even though $\mathscr{M}_n$ is a "plane" in $\mathbb{R}^{n+1}$, it is not a linear Euclidean space since for any $c \neq 1$, $c\hat{\nu} \notin \mathscr{M}_n$, and neither are the $\mathbf{x}'$-leaves. They are affine manifolds [26]. The locating of $\hat{\nu}^*(\mathbf{x}')$ is a highly nonlinear procedure in the space of energies.

The LFT, in terms $\Psi(\mathbf{u})$, $\Psi_\mathbf{y}(\mathbf{y})$ and $\psi(\mathbf{y}')$, etc., enters as a powerful algebraic linear representation of the MEP procedure. The "collapse" of a holographic $\mathbf{y}$ to $\mathbf{y}'$ with missing information means simply neglecting all the extra dimensions: $y_{J+1} = \cdots = y_n = 0$. This is because due to the convexity of $\Phi(\hat{\nu})$, there is a one-to-one relation between $\hat{\nu}$ and $\mathbf{u} = -\nabla\Phi$ under a proper gauge fixing. And since the constrains to MEP in (5) are all linear due to the nature of observables being random variables, each g determines a 1-dimensional linear subspace in the space of $\mathbf{u}$.

5. Generalized Clausius Inequality

A combination of Equations (9) and (12a) yields a Clausius' inequality-like relation:

$$\varphi(\mathbf{x}') + \mathbf{x}' \cdot \mathbf{y}' - \psi(\mathbf{y}') \geq 0. \tag{15}$$

The thermodynamics equilibrium is between the observed mean value $\mathbf{x}'$ and its conjugate "force" $\mathbf{y}'$. When the equality holds, there is a relation between $\mathbf{x}'$ and $\mathbf{y}'$ which should be identified as a "the equation of state", with $\nabla_\mathbf{x}\varphi = -\mathbf{y}'$ and $\nabla_\mathbf{y}\psi = \mathbf{x}'$. When the $\mathbf{x}' \neq \nabla_\mathbf{y}\psi(\mathbf{y})$, the difference $\mathbf{x}' \cdot \mathbf{y}' - \psi(\mathbf{y}')$ can be interpreted as the nonequilibrium heat and φ again as the entropy; then, the inequality in (15) becomes the Clausius' inequality.

6. Generalized Gibbs–Duhem Equation

The celebrated Gibbs–Duhem equation in classical thermodynamics is a consequence of the entropy being a Eulerian degree 1 homogeneous function. Thus, for the $\Phi(\boldsymbol{\nu})$ in (1), we have

$$\Phi(\boldsymbol{\nu}) = \sum_{i=0}^{n} \nu_i \left(\frac{\partial \Phi}{\partial \nu_i}\right), \sum_{i=0}^{n} \nu_i \left(\frac{\partial^2 \Phi}{\partial \nu_i \partial \nu_j}\right) = 0,$$
$$\sum_{j=0}^{n} \mathrm{d}\nu_j \sum_{i=0}^{n} \nu_i \left(\frac{\partial^2 \Phi}{\partial \nu_i \partial \nu_j}\right) = \sum_{i=0}^{n} \nu_i \sum_{j=0}^{n} \left(\frac{\partial u_i}{\partial \nu_j}\right) \mathrm{d}\nu_j = 0,$$
$$\text{that is, } \sum_{i=0}^{n} \nu_i \mathrm{d}u_i = 0, \tag{16}$$

in which we have used (2b). We identify (16) as a generalized Gibbs–Duhem equation.

7. Conclusions

The mathematical theory of probability deals with a set of elementary events $\mathcal{S}$, on which the probability $\mathbf{p}$ and random variables g's are introduced. Applying this mathematics to the real world, each ergodic dynamical system with state space $\mathcal{S}$ has its own unique steady-state probability distribution which can be obtained as the $\hat{\boldsymbol{\nu}}$ from i.i.d. sampling *ad infinitum*.

Our present theory is to statistical inference obtaining particular $\hat{\boldsymbol{\nu}}$'s what dynamics is to kinematics in classical mechanics [27]. The entropy function in (1) arises in this context as a measure of the quantitative relationship between the assumed, "hypothesis" ($\mathbf{p}$) and the observed "data" ($\boldsymbol{\nu}$ and $\hat{\boldsymbol{\nu}}$), as "missing information" or "surprise" [21,22]. Motivated by the analogy to the Fourier analysis, our generalized Gibbs' theory suggests that the notion of thermo-energetics is a powerful mathematical transformation of the statistical description; $\hat{\boldsymbol{\nu}}$ and $\mathbf{u}$ are simply two *representations* of a same physical reality, the former being statistical while the latter thermo-energetic. $\mathbf{u}$ is to $\hat{\boldsymbol{\nu}}$ what chemical potential μ is to particle number N in J. W. Gibbs' chemical thermodynamics. With a fixed $\mathbf{p}$, the theory of probability [8] revealed a powerful, dual energetic representation for various different systems, with the same state space $\mathcal{S}$, in terms of their respective internal energy functions $\mathbf{u}$ [25]. This fundamental duality between counting frequency and internal energy of course has been recognized by L. Boltzmann already in 1880s, when he was developing the statistical mechanics as a foundation of classical thermodynamics under the principle of *equal probability a priori*. The present work shows that while the *probability and statistics* are fundamental as the foundation of thermodynamics, mechanics is not necessary. A similar conclusion was reached in the 1925 thesis of L. Szilard [28,29].

For sufficiently large N, the probability of observing a particular $\hat{\boldsymbol{\nu}}$ is asymptotically zero except $\hat{\boldsymbol{\nu}} = \mathbf{p}$. The significance of $\Phi(\hat{\boldsymbol{\nu}};\mathbf{p})$ is to provide a "high-resolution magnifying glass" for the asymptotically small

$$\exp\Big\{-N\Phi(\hat{\boldsymbol{\nu}};\mathbf{p})\Big\}. \tag{17}$$

This is known as the large deviations rate function in the modern theory of probability [8]. The entropy $-\Phi(\hat{\boldsymbol{\nu}},\mathbf{p})$ is a function of both $\hat{\boldsymbol{\nu}}$ and $\mathbf{p}$, $\Phi(\hat{\boldsymbol{\nu}};\mathbf{p}) \geq 0$ and $\Phi(\mathbf{p};\mathbf{p}) = 0$. For a given $\mathbf{p}$, it views each possible $\hat{\boldsymbol{\nu}}$ from a real system as a part of an entire class of systems under a common $\mathbf{p}$, a *meta-statistics*. If one chooses the true steady-state probability $\boldsymbol{\pi}$ of a particular system to replace $\mathbf{p}$, then Equation (17) gives the probability distribution of the uncertainties in the measurement $\hat{\boldsymbol{\nu}}$ from N samples. The second-order Taylor expansion near $\boldsymbol{\pi}$,

$$e^{-N\Phi(\hat{\boldsymbol{\nu}};\boldsymbol{\pi})} \simeq \exp\left[-N \sum_{i,j=0}^{n} \frac{(\hat{\nu}_i - \pi_i)(\delta_{ij} - \pi_i)(\hat{\nu}_j - \pi_j)}{\pi_i}\right],$$

is the central limit theorem for the statistics of counting frequency $\hat{\nu}$, with $\mathrm{Var}[\hat{\nu}_i] = \pi_i(1-\pi_i)/N$ and $\mathrm{Cov}[\hat{\nu}_i, \hat{\nu}_j] = -\pi_i\pi_j/N$. This is not the fluctuations within the $\boldsymbol{\pi}$ of the system itself. *Gibbs' theory of ensemble is about statistical measurements of a whole system; not about the individuals within.*

We choose to present our theory with finite state space $\mathcal{S}$ for mathematical simplicity. Formal generalization to continuous state space is straight forward if mathematical rigor is not required. Beyond the finite state space, it is known that modern probability and the theory of measures encounter challenges, *c.f.*, de Finetti's treatment of infinite sets and the axiom of choice of nonempity subsets [20]. In addition to continuous $\mathbb{R}^n$ [30], there are even larger Hilbert spaces of functions on $\mathbb{R}^n$ and/or von Neumann algebra of operators acting on a Hilbert space.

Funding: This research received no external funding.

Data Availability Statement: The original contributions presented in this study are included in the article. Further inquiries can be directed to the corresponding author.

Acknowledgments: I thank Jin Feng, Weishi Liu, Zhang-Ju Liu, Bing Miao, Zhongmin Shen, Xiang Tang, Yong-Shi Wu, and particularly Jun Zhang, for many helpful discussions, and the support from Olga Jung Wan Endowed Professorship.

Conflicts of Interest: The author declares no conflicts of interest.

Abbreviations

The following abbreviations are used in this manuscript:

LFT	Legendre–Fenchel transform
FT	Fourier transform

Appendix A. Statistical Counting *ad Infinitum*

In this section, we provide the mathematical reasoning for stating "entropy provides the infinitesimal probability for an observed frequency $\hat{\nu}$ with respect to a probability prior $\mathbf{p}$", "it characterizes the *relationship* between $\boldsymbol{\nu}$ and $\mathbf{p}$ in a sampling process", and the origin of Legendre–Fenchel transform in entropy analysis. The counting of independent and identically distributed samples with state space $\mathcal{S} = \{0, \cdots, n\}$ yields $\boldsymbol{\nu} = (\nu_0, \cdots, \nu_n)$, a $(n+1)$-tuple of non-negative integers. We call all the $\boldsymbol{\nu}$ with $N = \nu_0 + \cdots + \nu_n$ a *simplex for counting*. The simplex for counting grows with N, which we shall identify as "time". With a given prior probability $\mathbf{p} = (p_0, \cdots, p_n)$ on $\mathcal{S}$, statistical counting is a Markov process on a growing simplex, with probability:

$$P^{(N+1)}(\boldsymbol{\nu}) = \sum_{k=0}^{n} p_k P^{(N)}(\boldsymbol{\nu} - \boldsymbol{\delta}_k), \tag{A1}$$

in which $\boldsymbol{\delta}_k = (0, \cdots, 1, \cdots, 0)$ is the unit vector for the kth component. One can easily verify that

$$P^{(N)}(\boldsymbol{\nu}) = \frac{N!}{\nu_0! \cdots \nu_n!} p_0^{\nu_0} \cdots p_n^{\nu_n}$$

is a solution to (A1).

One is interested in the limit of counting *ad infinitum*, when all the ν_is are expected to tend to infinity as $N \to \infty$. On the increasing simplex for $\boldsymbol{\nu}$, the probability $P^{(N)}(\boldsymbol{\nu}) \to 0$. However, the properly normalized $\hat{\boldsymbol{\nu}} = \boldsymbol{\nu}/N$ converges, and $P^{(N)}$ as a function of the $\hat{\boldsymbol{\nu}}$ becomes sharper and sharper, concentrated around $\hat{\boldsymbol{\nu}}^* = \mathbf{p}$. To more precisely characterize this limiting situation, one introduces *counting frequency* $\hat{\boldsymbol{\nu}} = \boldsymbol{\nu}/N$. The space of $\hat{\boldsymbol{\nu}}$s then is called a *probability simplex* $\mathscr{M}_n$; Equation (A1) then becomes

$$\tilde{P}^{(N+1)}(\hat{\nu}) = \sum_{k=0}^{n} p_k \tilde{P}^{(N)}\Big(\big\{\tfrac{N+1}{N}\hat{\nu}_i - \tfrac{1}{N}\delta_{ik}\big\}\Big). \tag{A2}$$

Its limit is a Dirac-δ function: $\tilde{P}^{(\infty)} = 0$ for all $\hat{\nu} \neq \mathbf{p}$, and $\tilde{P}^{(\infty)} = \infty$ at $\hat{\nu} = \mathbf{p}$. However, "a higher order" infinitesimal analysis shows that [8]

$$\lim_{N\to\infty} \frac{1}{N} \ln \tilde{P}^{(N)}(\hat{\nu}) = -\sum_{i=0}^{n} \hat{\nu}_i \ln\left(\frac{\hat{\nu}_i}{p_i}\right) = -\Phi(\hat{\nu}). \tag{A3}$$

It is clear that entropy function $-\Phi(\nu)$ represents the infinitesimal prior probability $e^{-N\Phi(\nu)}$ on $\mathscr{M}_n$. For two ν's with different entropy values, $\Phi(\nu)$ and $\Phi(\nu')$, their probabilities $P^{\infty}(\nu)/P^{\infty}(\nu') = 0$ if $\Phi(\nu) > \Phi(\nu')$. This is the origin of the maximum entropy principle (MEP).

To understand the limit $P^{(N)}(\nu) \to 0$, one can also introduce the probability generating function [8]:

$$W^{(N)}(\mathbf{u}) = \sum_{\nu} P^{(N)}(\nu) e^{-\mathbf{u}\cdot\nu}, \tag{A4}$$

in which $\mathbf{u}\cdot\nu = u_0\nu_0 + \cdots + u_n\nu_n$. Then, Equation (A1) becomes

$$\begin{aligned} W^{(N+1)}(\mathbf{u}) &= \sum_{\nu} P^{(N+1)}(\nu) e^{-\mathbf{u}\cdot\nu} \\ &= \sum_{k=0}^{n} p_k e^{-\mathbf{u}\cdot\delta_k} \sum_{\nu} P^{(N)}(\nu - \delta_k) e^{-\mathbf{u}\cdot(\nu-\delta_k)} \\ &= W^{(N)}(\mathbf{u}) e^{-\Psi(\mathbf{u})}, \\ &\text{where } \Psi(\mathbf{u}) = -\ln \sum_{k=0}^{n} p_k e^{-u_k}. \end{aligned} \tag{A5}$$

The free energy function Ψ is meaningful for all finite N. This is why the partition function is valid even for small systems in Gibbs' theory of ensembles [13]. The Legendre–Fenchel transform of $\Psi(\mathbf{u})$ is precisely the the right-hand side of (A3):

$$\begin{aligned} & \inf_{\mathbf{u}} \Big\{ \mathbf{u}\cdot\nu - \Psi(\mathbf{u}) \Big\} \\ =\ & \inf_{\mathbf{u}} \left\{ \sum_{i=0}^{n} \nu_i \ln e^{u_i} + \ln \sum_{k=0}^{n} p_k e^{-u_k} \right\} \\ =\ & \inf_{\mathbf{u}} \left\{ -\sum_{i=0}^{n} \nu_i \ln\left[\frac{e^{-u_i}}{\sum_{k=0}^{n} p_k e^{-u_k}} \right] \right\} = -\sum_{i=0}^{n} \nu_i \ln \frac{\hat{\nu}_i}{p_i}, \end{aligned} \tag{A6}$$

in which the optimal $e^{-u_i} \propto \nu_i/p_i$. Legendre–Frenchel transform arises in the limit of $N \to \infty$ through the Laplace's method of evaluating asymptotic integrals, or the related Darwin–Fowler method of maximum term.

The analysis in this Appendix suggests that a proper interpretation of $\mathbf{p} = (p_0, \cdots, p_n)$ in Equation (A1) is not as an intrinsic property, for example, the generative model of data statistics, rather it should be interpreted as a choice of a "gauge" in terms of which a set of counting data is represented: each particular set of data *ad infinitum* is represented by the energy function $\mathbf{u}$, not $\mathbf{p}$, and the $\hat{\nu}$ is gauge invariant via the Boltzmann relation $\hat{\nu}_i \propto p_i e^{-u_i}$—this yields an i.i.d. generative model. Probability is *not* for generative models, it is for analyzing empirical measurements on random variables.

References

1. Førland, K.S.; Førland, T.; Kjelstrup, S. *Irreversible Thermodynamics: Theory and Applications*; John Wiley & Sons: Chichester, UK, 1988.
2. Kjelstrup, S.; Bedeaux, D. *Non-Equlibrium Thermodynamics of Heterogeneous Systems*; Series on Advances in Statistical Mechanics; World Scientific: Singapore, 2008; Volume 16. [CrossRef]

3. Bedeaux, D.; Kjelstrup, S.; Schnell, S.K. *Nanothermodynamics Theory and Applications;* World Scientific: Singapore, 2023. [CrossRef]
4. Guggenheim, E.A. *Modern Thermodynamics by the Methods of Willard Gibbs;* Methuen & Co.: New York, NY, USA, 1933.
5. Hill, T.L. A different approach to nanothermodynamics. *Nano Lett.* **2001**, *1*, 273–275. [CrossRef]
6. Khinchin, A.Y. *Mathematical Foundations of Statistical Mechanics;* Dover: New York, NY, USA, 1949.
7. Touchette, H. The large deviation approach to statistical mechanics. *Phys. Rep.* **2009**, *478*, 1–69. [CrossRef]
8. Dembo, A.; Zeitouni, O. *Large Deviations Techniques and Applications*, 2nd ed.; Springer: New York, NY, USA, 1998. [CrossRef]
9. Anderson, P.W. More is different: Broken symmetry and the nature of the hierarchical structure of science. *Science* **1972**, *177*, 393–396. [CrossRef] [PubMed]
10. Alm, J.F.; Walker, J.S. Time-frequency analysis of musical instruments. *SIAM Rev.* **2002**, *44*, 457–476. [CrossRef]
11. Fourier, J.B.J. *The Analytic Theory of Heat;* Freeman, A., Translator; Cambridge University Press: London, UK, 1878. [CrossRef]
12. Truesdell, C. *Rational Thermodynamics;* Springer: New York, NY, USA, 1984. [CrossRef]
13. Lu, Z.; Qian, H. Emergence and breaking of duality symmetry in thermodynamic behavior: Repeated measurements and macroscopic limit. *Phys. Rev. Lett.* **2022**, *128*, 150603. [CrossRef] [PubMed]
14. Galteland, O.; Bering, E.; Kristiansen, K.; Bedeaux, D.; Kjelstrup, S. Legendre-Fenchel transforms capture layering transitions in porous media. *Nanoscale Adv.* **2022**, *4*, 2660–2670. [CrossRef] [PubMed]
15. Rockafellar, R.T. *Convex Analysis;* Princeton University Press: Princeton, NJ, USA, 1970.
16. Qian, M.; Xie, J.S.; Zhu, S. *Smooth Ergodic Theory for Endomorphisms;* Lecture Notes in Mathematics; Springer: Berlin, Germany, 2009; Volume 1978. [CrossRef]
17. Dorfman, J.R. *An Introduction to Chaos in Nonequilibrium Statistical Mechanics;* Cambridge Lect. Notes in Phys.; Cambridge University Press: London, UK, 1999. [CrossRef]
18. Mackey, M.C. The dynamic origin of increasing entropy. *Rev. Mod. Phys.* **1989**, *61*, 981–1015. [CrossRef]
19. Qian, H.; Ge, H. *Stochastic Chemical Reaction Systems in Biology;* Lect. Notes on Math. Modelling in the Life Sci.; Springer Nature: Cham, Switzerland, 2021. [CrossRef]
20. Jaynes, E.T. *Probability Theory: The Logic of Science;* Cambridge University Press: London, UK, 2003.
21. Levine, R.D. Information theory approach to molecular reaction dynamics. *Annu. Rev. Phys. Chem.* **1978**, *29*, 59–92. [CrossRef]
22. Ben-Naim, A. *A Farewell to Entropy: Statistical Thermodynamics Based on Information;* World Scientific: Singapore, 2008. [CrossRef]
23. Callen, H.B. *Thermodynamics and an Introduction to Thermostatistics*, 2nd ed.; Wiley: New York, NY, USA, 1991.
24. Commons, J.; Yang, Y.J.; Qian, H. Duality symmetry, two entropy functions, and an eigenvalue problem in Gibbs' theory. *arXiv* **2021**. [CrossRef]
25. Qian, H. Statistical chemical thermodynamics and energetic behavior of counting: Gibbs' theory revisited. *J. Chem. Theory Comput.* **2022**, *18*, 6421–6436. [CrossRef]
26. Hong, L.; Qian, H.; Thompson, L.F. Representations and divergences in the space of probability measures and stochastic thermodynamics. *J. Comput. Appl. Math.* **2020**, *376*, 112842. [CrossRef]
27. Goldstein, H. *Classical Mechanics;* Addison-Wesley: New York, NY, USA, 1951.
28. Szilard, L. Über die ausdehnung der phänomenologschen thermodynamik auf die schwankungserscheinungen. *Z. Physik.* **1925**, *32*, 753–7888. [CrossRef]
29. Mandelbrot, B. On the derivation of statistical thermodynamics from purely phenomenological principles. *J. Math. Phys.* **1964**, *5*, 164–171. [CrossRef]
30. Miao, B.; Qian, H.; Wu, Y.S. Emergence of Newtonian deterministic causality from stochastic motions in continuous space and time. *arXiv* **2024**. [CrossRef]

entropy

MDPI

Article

Computing Entropy for Long-Chain Alkanes Using Linear Regression: Application to Hydroisomerization

Shrinjay Sharma [1], Richard Baur [2], Marcello Rigutto [2], Erik Zuidema [2], Umang Agarwal [3], Sofia Calero [4], David Dubbeldam [5] and Thijs J. H. Vlugt [1,*]

1 Engineering Thermodynamics, Process & Energy Department, Faculty of Mechanical Engineering, Delft University of Technology, Leeghwaterstraat 39, 2628 CB Delft, The Netherlands; s.sharma-6@tudelft.nl

2 Shell Global Solutions International B.V., Grasweg 39, 1031 HW Amsterdam, The Netherlands; richard.baur@shell.com (R.B.); marcello.rigutto@shell.com (M.R.); erik.zuidema@shell.com (E.Z.)

3 Shell Chemical LP, Monaca, PA 15061, USA; u.agarwal@shell.com

4 Department of Applied Physics, Eindhoven University of Technology, 5600 MB Eindhoven, The Netherlands; s.calero@tue.nl

5 Van 't Hoff Institute of Molecular Sciences, University of Amsterdam, Science Park 904, 1098 XH Amsterdam, The Netherlands; d.dubbeldam@uva.nl

* Correspondence: t.j.h.vlugt@tudelft.nl

Abstract: Entropies for alkane isomers longer than C_{10} are computed using our recently developed linear regression model for thermochemical properties which is based on second-order group contributions. The computed entropies show excellent agreement with experimental data and data from Scott's tables which are obtained from a statistical mechanics-based correlation. Entropy production and heat input are calculated for the hydroisomerization of C_7 isomers in various zeolites (FAU-, ITQ-29-, BEA-, MEL-, MFI-, MTW-, and MRE-types) at 500 K at chemical equilibrium. Small variations in these properties are observed because of the differences in reaction equilibrium distributions for these zeolites. The effect of chain length on heat input and entropy production is also studied for the hydroisomerization of C_7, C_8, C_{10}, and C_{14} isomers in MTW-type zeolite at 500 K. For longer chains, both heat input and entropy production increase. Enthalpies and absolute entropies of C_7 hydroisomerization reaction products in MTW-type zeolite increase with higher temperatures. These findings highlight the accuracy of our linear regression model in computing entropies for alkanes and provide insight for designing and optimizing zeolite-catalyzed hydroisomerization processes.

Keywords: entropy; linear regression; alkanes; hydroisomerization

1. Introduction

Entropy data are important for a wide range of hydrocarbon applications, from industrial processes [1] to environmental assessments [2]. Entropy changes during chemical reactions like hydroisomerization, combustion, and cracking have a large influence on the feasibility of such reactions [3]. Iso-alkanes with high branching are preferred over linear ones in sustainable aviation fuel and lubricants [4], making shape-selective zeolite-catalyzed hydroisomerization [5,6], also known as catalytic dewaxing, important for the production of these alkanes [7,8]. Understanding thermochemical properties, such as entropies and enthalpies, is essential for designing efficient processes and equipment for (iso)alkane handling and separation [9]. Process energy efficiency is best evaluated through the second law of thermodynamics [3,10]. The second law efficiency requires knowledge on exergy destruction which is the product of entropy production and the temperature of the environment [3,10]. Exergy destruction is the useful work that is destroyed due to irreversibilities in a process [11].

Thermochemical properties, including absolute entropies (S_0) for all isomers up to C_{10} at temperatures ranging from 0 to 1500 K, have been reported by Scott [12]. Experimental data for the thermochemical properties of various alkanes are available in the literature and can be accessed through the NIST Chemistry Webbook [13]. However, there are limited experimental data for alkanes longer than C_{10}. Several group contribution methods exist in the literature to compute thermochemical properties, which includes those by Benson et al. [14], Constantinou and Gani [15], Joback and Reid [16], Marero and Gani [17], Hukkerikar et al. [18], Albahri and Aljasmi [19], and Domalski and Hearing [20]. Yaw's Handbook [21] and the Design Institute for Physical Properties (DIPPR) database [22] also list thermochemical properties for many long-chain alkanes, which are either obtained from experiments or group contribution methods. Machine Learning (ML) models have emerged as an alternative for predicting thermochemical properties. Aldosari et al. successfully used Support Vector Regression (SVR), ν-SVR, and Random Forest Regression (RFR) algorithms to predict $\Delta_f S_0$, $\Delta_f H_0$, and $c_{p,0}$ for hydrocarbons [1]. For longer hydrocarbons, High-Dimensional Model Representation (HDMR) could be considered as another approach to predict these thermochemical properties [23].

Most of the group contribution methods in the literature are often inaccurate because these methods only consider first-order groups (CH_3, CH_2, CH, and C) or combine with a few second-order groups [15,24]. Second-order group contributions consider interactions between neighboring groups of atoms [25]. Increasing the number of second-order groups improves the predictions of thermochemical properties because these groups also account for interactions between the nearest neighboring groups of atoms [25]. Scott's correlation [26], based on statistical mechanics, provides accurate predictions till C_{10} isomers when compared to experiments. However, this correlation is complex and difficult to apply to long-chain alkanes (>C_{10}), as it requires numerous functions and fitting parameters. While Machine Learning models offer promising predictive power, these models may struggle with extrapolation for long-chain alkanes [1,27]. In such cases, linear regression (LR) can outperform ML models when the independent variables are accurate and the output has a linear relationship with these variables [27].

In our recent work, we developed a user-friendly linear regression (LR) model based on second-order group contributions to predict thermochemical properties like Gibbs free energies $(G_0 - H_0(0\ \mathrm{K}))$, enthalpies $(H_0 - H_0(0\ \mathrm{K}))$, Gibbs free energies of formation $\Delta_f G_0$, and enthalpies of formation $\Delta_f H_0$ for alkanes longer than C_{10} for temperatures in the range (0–1000) K [25]. The second-order groups consider the interactions between the central atom and the neighboring groups. These groups are solely determined based on the topology of the alkanes with SMILES strings as input. Therefore, no 3D structures of alkanes were used in our LR model. Here, we calculated absolute entropies S_0 from $(G_0 - H_0(0\ \mathrm{K}))$ and $(H_0 - H_0(0\ \mathrm{K}))$, predicted using the LR model. The entropy values are in excellent agreement with both Scott's data [12] and experimental values [26]. Our method effectively captures the variations in entropy in isomers due to differences in branching. The S_0 values are used to compute the entropy production for the hydroisomerization of alkanes in different zeolites at reaction equilibrium. Understanding the reaction product distribution at equilibrium is key to optimizing branched isomer yield in hydroisomerization [28]. Hydroisomerization involves the adsorption and subsequent dehydrogenation of linear alkanes at the metal sites of zeolites, forming alkenes [29]. These alkenes are protonated at the acid sites, resulting in the formation of alkyl carbenium ions [30]. Alkanes are regenerated from these alkylcarbenium ions via hydrogenation [30]. Hydroisomerization is accompanied by cracking reactions depending on the operating temperature of the reactor. This study is relevant for conditions where the hydroisomerization reaction approaches chemical equilibrium with a negligible amount of cracked products. Cracking reactions, being irreversible in nature [31], must be excluded from the reaction equilibrium distribution study. Calculating reaction equilibrium distributions for hydroisomerization reactions also provides insights into the shape selectivity effects of zeolites during these processes [32]. The calculation of entropy production in an equilibrium reactor involves steady-state mass and energy balances. The reaction equilibrium distribution provides the amount of isomers at the reactor outlet, while the energy balance determines the heat input required for the reactions. We analyzed the heat input and the entropy production for C_7 isomers in FAU-, ITQ-29, BEA-, MEL-, MFI-, MTW-, and MRE-type zeolites at 500 K, observing small variations in these zeolites due to differences in reaction equilibrium distributions. Both heat input and entropy production increase with longer alkanes and higher temperatures, as the enthalpies and the absolute entropies increase with increasing chain length and temperature.

This article is organized as follows: Section 2 presents the key equations for computing the absolute entropies of alkanes, the mass balance in the equilibrium reactor, the energy balance for calculating the heat input, and the entropy balance for determining the entropy production. Section 3 discusses the results, showing that the computed entropies are highly accurate. A minimal variation in entropy production and heat input is observed for different zeolites during hydroisomerization of linear alkanes. Both heat input and entropy production increase with longer alkane chains and higher temperatures. Section 4 provides conclusions on the accuracy and usefulness of the entropy data for alkanes and the effects of zeolites, chain length, and temperature on entropy production for hydroisomerization at reaction equilibrium. This article also includes Supporting Information SI.xlsx, which contains the computed entropy values derived from the enthalpies and Gibbs free energies predicted by our LR model for isomers ranging from C_1 to C_{14}. SI.xlsx also contains the comparison between the absolute entropies computed using our model [25] and those predicted using group contribution methods by Benson et al. [14] and Constantinou and Gani [15].

2. Theory

The absolute entropy S_0 is computed as follows [3]

$$S_0 = \left(\frac{1}{T}\right)[(H_0 - H_0(0\,\mathrm{K})) - (G_0 - H_0(0\,\mathrm{K}))] \tag{1}$$

where $(H_0 - H_0(0\,\mathrm{K}))$ is the enthalpy at temperature T relative to 0 K. 0 K is considered as the reference temperature in this study. $(G_0 - H_0(0\,\mathrm{K}))$ is the Gibbs free energy at temperature T relative to 0 K. These thermochemical properties are predicted using our LR model based on second-order group contributions for hydrocarbons [25]. The computed absolute entropies S_0 for alkane isomers $(C_1 - C_{14})$ at temperatures (0–1000) K are provided in the Supporting Information SI.xlsx. S_0 values are used to calculate the entropy production in a hydroisomerization reactor at reaction equilibrium (Figure 1), where ideal gas behavior is assumed. Calculating the entropy production requires molar flow rates at the inlet and the outlet of the reactor and the rate of heat input. Both heat input and entropy production are computed per unit moles of fluid mixture at the inlet. A steady-state molar balance is considered in the reactor as we assume that all isomerization reactions are at equilibrium:

$$\sum_{i=1}^{N_{\mathrm{comp}}} \dot{n}_{\mathrm{in},i} = \sum_{i=1}^{N_{\mathrm{comp}}} \dot{n}_{\mathrm{out},i} \tag{2}$$

where $\dot{n}_{\mathrm{in},i}$ and $\dot{n}_{\mathrm{out},i}$ are the molar flow rates of component i at the reactor column inlet and outlet, respectively. At the inlet, only the linear alkane is considered at 298.15 K. The composition at the outlet is obtained from the reaction equilibrium distribution for the hydroisomerization of linear alkanes at infinite dilution in the zeolite. The reaction equilibrium distribution is obtained by imposing a gas-phase reaction equilibrium for the alkane isomers and a simultaneous phase equilibrium between the gas and the adsorbed phase for each component [32]. This satisfies the reaction equilibrium distribution in the zeolites [33]. A similar approach was also adopted by Hansen et al. [34] to study the influence of silicalite-1 pores on the reaction equilibrium distribution of the propene metathesis reaction. The adsorbed phase loadings for these isomers are obtained using Henry's law. The Henry coefficients are computed using Widom's test particle insertion method [35,36] in the RASPA software [37–39]. The mole fractions in the adsorbed phase depends on the Henry coefficients and the gas-phase mole fractions. The total pressure cancels out from the equation [32]. For further details on this method, the reader is referred to Refs. [25,32]. The gas-phase reaction equilibrium distribution is computed using $(G_0 - H_0(0\ \text{K}))$ at the operating temperature T and $\Delta_f H_{0,i}(0\ \text{K})$ at 0 K [32]. For alkanes longer than C_{10} isomers, these properties are obtained using our LR model [25]. The rate of heat input $\dot{Q}_{\text{in}}$ to the reactor is computed from the steady-state energy balance for the reactor [3]:

$$\dot{Q}_{\text{in}} = \sum_{i=1}^{N_{\text{comp}}} \dot{n}_{\text{out},i} H_i - \sum_{i=1}^{N_{\text{comp}}} \dot{n}_{\text{in},i} H_i \tag{3}$$

where H_i is the enthalpy of component i, which is computed as follows [3]:

$$H_i(T) = \Delta_f H_{0,i}(0\ \text{K}) + (H_i(T) - H_i(0\ \text{K})) \tag{4}$$

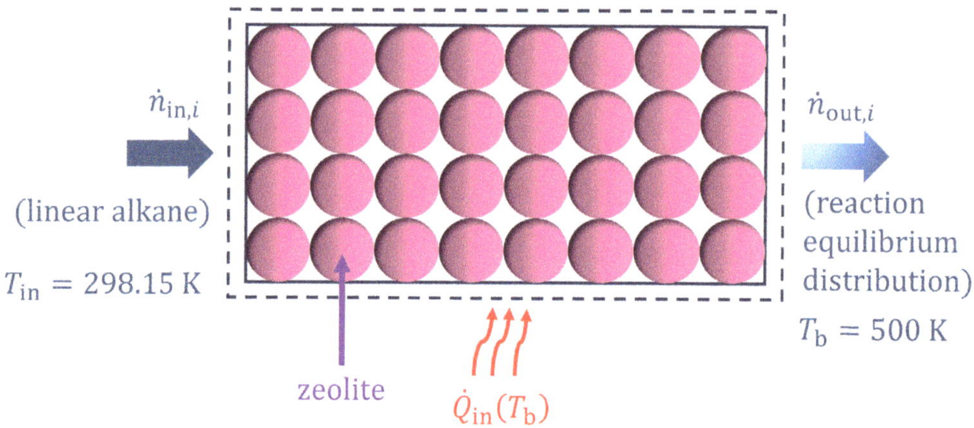

Figure 1. Schematic representation of a reactor for the hydroisomerization of linear alkanes at chemical equilibrium. The feed at the column inlet $(\dot{n}_{\text{in},i})$ is at 298.15 K. The heat input $(\dot{Q}_{\text{in},i})$ and the reactions take place at 500 K.

In Equation (4), $\Delta_f H_{0,i}(0\ \text{K})$ is the enthalpy of formation of component i at 0 K. The entropy production $\sigma_{\text{production}}$ is computed using the steady-state entropy balance for the reactor [3]:

$$\sigma_{\text{production}} = \sum_{i=1}^{N_{\text{comp}}} \dot{n}_{\text{out},i} S_{\text{out},i} - \sum_{i=1}^{N_{\text{comp}}} \dot{n}_{\text{in},i} S_{\text{in},i} - \frac{\dot{Q}_{\text{in}}}{T_{\text{b}}} \tag{5}$$

$S_{\text{in},i}$ and $S_{\text{out},i}$ in Equation (5) are the absolute entropies of component i at the inlet and the outlet of the reactor. $S_{\text{in/out},i}$ depends on the fluid composition at the inlet and outlet of the reactor [3]. The heat input and the reactions take place at temperature T_{b}.

$$S_{\text{in/out},i} = S_{0,i} - R \ln \left(\frac{x_{\text{in/out},i} P}{P_{\text{ref}}} \right) \tag{6}$$

where $S_{0,i}$ is the absolute entropy for an isolated molecule of component i, R is the universal gas constant, and $x_{\text{in/out},i}$ are the mole fractions of component i at the inlet or outlet of the reactor. P is the operating pressure of the reactor and P_{ref} is the reference pressure, which is considered as 1 bar in this study. The term $R \ln (P/P_{\text{ref}})$ drops out in the entropy balance (Equation (5)) because the total molar flow rates at the inlet and outlet are equal (Equation (2)). For all cases, the operating temperature T_{b} is 500 K.

3. Results and Discussion

Figure 2 compares the absolute entropies of C_7 (Figure 2a) and C_8 (Figure 2b) isomers at 298.15 K, calculated using our LR model [25], the group contribution methods of Benson et al. [14] and Constantinou and Gani [15], and data obtained from Scott's tables [12] and experimental results [26]. The data in the Scott's tables were obtained by fitting statistical mechanics-based correlations to experimental measurements [26]. The entropies predicted by our LR model [25] show excellent agreement with both Scott's tables and experimental data. Unlike the group additivity method by Benson et al. [14] and the group contribution approach by Constantinou and Gani [15], our LR model successfully distinguishes between different isomers based on the number, types and positions of the branches. The Mean Absolute Errors (MAEs) of $(H_0 - H_0(0\text{ K}))$ and $(G_0 - H_0(0\text{ K}))$, predicted by our linear regression model, are 1.012 kJ/mol and 0.181 kJ/mol, respectively [25]. The MAE for the product of temperature and absolute entropy S_0 is 1.03 kJ/mol, which exceeds the chemical accuracy of 4.184 kJ/mol. The dataset comparing the absolute entropies computed by our method and those predicted by group contribution methods of Benson et al. [14] and Constantinou and Gani [15] for C_{14} isomers at 298.15 K is included in the excel worksheet S0_comparison of the Supporting Information SI.xlsx.

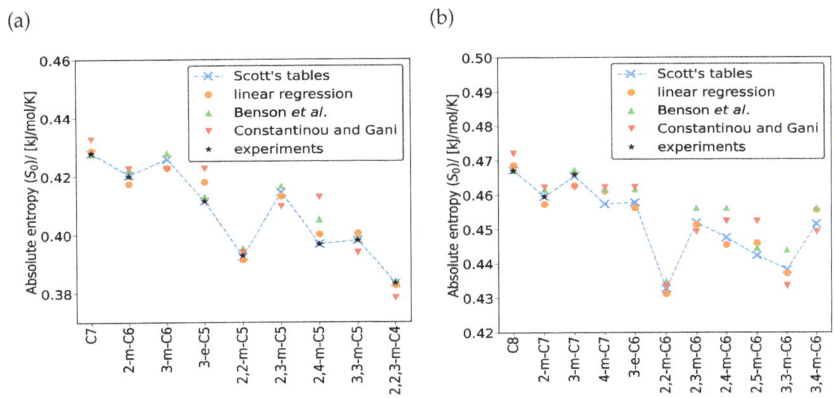

Figure 2. Computation of absolute entropies S_0 for (**a**) C_7 and (**b**) C_8 isomers at 298.15 K using our LR model, Benson et al.'s group additivity method [14], Constantinou and Gani's group contribution [15], Scott's tables [12], and the experimental data listed by Scott [26]. The predictions using the LR model are in excellent agreement with Scott's tables and the data from the experiments [26]. The dashed blue line through the data points obtained from Scott's tables is a guide to the eye.

The composition at the reactor column outlet represents the reaction equilibrium distribution for the hydroisomerization of alkanes in different zeolites at infinite dilution. Figure 3 shows the reaction equilibrium distributions for the hydroisomerization of C_7 isomers in FAU-, ITQ-29-, BEA-, MEL-, MFI-, MTW-, and MRE-type zeolites at 500 K. These zeolites are classified into three categories based on the shape and size of the pores, which show distinct selectivity for alkanes. FAU- and ITQ-29-type zeolites have cage-like pore structures [40]. BEA-, MEL-, and MFI-type zeolites feature three-dimensional channels with intersections [40]. MTW- and MRE-type zeolites contain one-dimensional pore structures [40]. The datasets for the reaction equilibrium distributions are included in the excel sheet xi_C7_500K of the Supporting Information SI.xlsx. In FAU- and ITQ-29-type zeolites (Figure 3a), the adsorbed phase selectivity is primarily influenced by the gas-phase thermochemical properties $\Delta_f H_0(0\text{ K})$ and $(G_0 - H_0(0\text{ K}))$. Dimethyl isomers (2,2-m-C_5, 2,3-m-C_5, and 2,4-m-C_5) are favored compared to n-C_7 inside these zeolites. Due to larger pore diameters, the influence of pore structure on selectivities is small. FAU- and ITQ-29-type zeolites can accommodate molecules as large as 11.24 Å and 11.05 Å, respectively [40]. BEA-, MEL-, and MFI-type zeolites with 3D channel-like pores connected via intersections show similar selectivities for n-C_7 and the monomethyl isomers (2-m-C_6 and 3-m-C_6), which are preferentially formed compared to dimethyl-, trimethyl-, and ethyl-branched isomers (Figure 3b). In BEA-, MEL-, and MFI-type zeolites, variations in selectivities are influenced by both gas-phase thermochemical properties and Henry coefficients. In MTW- and MRE-type zeolites, n-C_7 has the highest selectivity, followed by 2-m-C_6 and 3-m-C_6, with variations influenced largely by the Henry coefficients (Figure 3c) because of the smaller diameters and one-dimensional nature of the pores [32]. The overall differences in selectivities between the linear and the branched isomers are mainly determined by the pore diameters of the zeolites. For isomers with an identical degree of branching, the selectivities are determined by the distributions and the shapes of the pore structures.

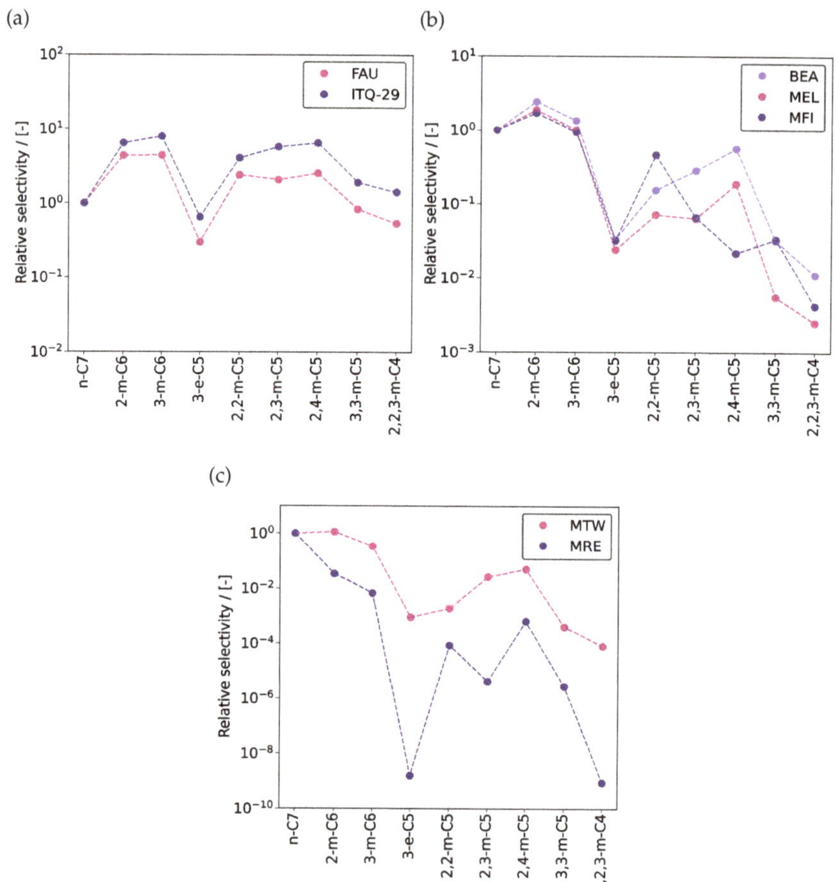

Figure 3. Selectivities of C_7 isomers relative to n-C_7 at reaction equilibrium in (**a**) FAU- and ITQ-29-type zeolites, (**b**) BEA-, MEL-, and MFI-type zeolites, and (**c**) MTW- and MRE-type zeolites at infinite dilution and 500 K. The absolute selectivity is defined as the mole fraction of a certain component divided by the sum of the mole fractions of all other components [41]. The relative selectivity refers to the ratio of the absolute selectivity of a specific isomer to that of the reference isomer, which is C_7 in this case. The dashed lines through the data points are a guide to the eye.

Figure 4 shows the variations in the rates of heat input (Figure 4a) and entropy production (Figure 4b) for the hydroisomerization of C_7 isomers in an equilibrium reactor for different zeolites. Both heat input and entropy production show small variations with the zeolites, influenced by the unique reaction equilibrium distributions in each type of these zeolites. The rate of heat input slightly increases from large pore zeolites (FAU- and ITQ-29-types) to narrower 1D zeolites (MTW- and MRE-types) (Figure 4a). This is because n-C_7, 2-m-C_6, and 3-m-C_6 isomers which have higher enthalpies compared to the multi-branched isomers are formed in large proportions in these 1D zeolites at reaction equilibrium. The entropy production decreases with decreasing pore size of the zeolites due to smaller values of entropies for these linear and mono-branched alkanes at reaction equilibrium in these zeolites. This may vary when cracking reactions are involved.

The hydroisomerization of linear alkanes n-C_7, n-C_8, n-C_{10}, and n-C_{14} is studied in MTW-type zeolite at 500 K and infinite dilution. With increasing chain length, both heat input to the reactor (Figure 5a) and entropy production (Figure 5b) increase because of the rapid increase in enthalpies and entropies with temperature for long-chain alkanes, as shown in Figure 5c,d, respectively. The enthalpies increase with chain length due to the larger number of carbon–hydrogen (C-H) and carbon–carbon (C-C) bonds, each adding to the total enthalpy of the molecule. Longer alkanes also have higher absolute entropies compared to the shorter ones because the larger number of atoms allows for more molecular arrangements [42]. This leads to an increase in entropy production during the hydroisomerization of longer alkanes.

Figure 6a shows an increase in the rate of heat input with increasing temperature. The internal energy increases at higher temperatures, which leads to higher enthalpy values. This increases the rate of heat requirement in the reactor column. The absolute entropies of alkanes increase with temperature because higher temperatures provide more energy to the molecules to occupy a larger number of configurations. This leads to an increase in the rate of entropy production with temperature (Figure 6b).

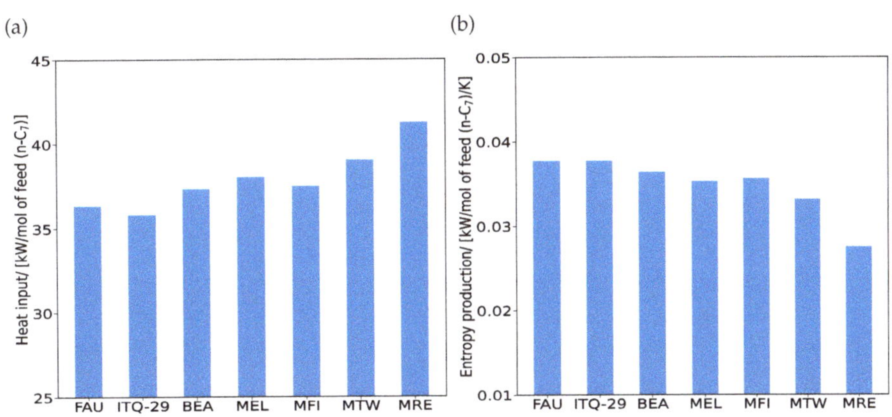

Figure 4. Variations of (**a**) heat input $\dot{Q}_{\mathrm{in}}$ and (**b**) entropy production $\sigma_{\mathrm{production}}$ for the hydroisomerization of C_7 isomers at reaction equilibrium in FAU-, ITQ-29-, BEA-, MEL-, MFI-, MTW-, and MRE-type zeolites. The reactions take place at 500 K and infinite dilution in the zeolite.

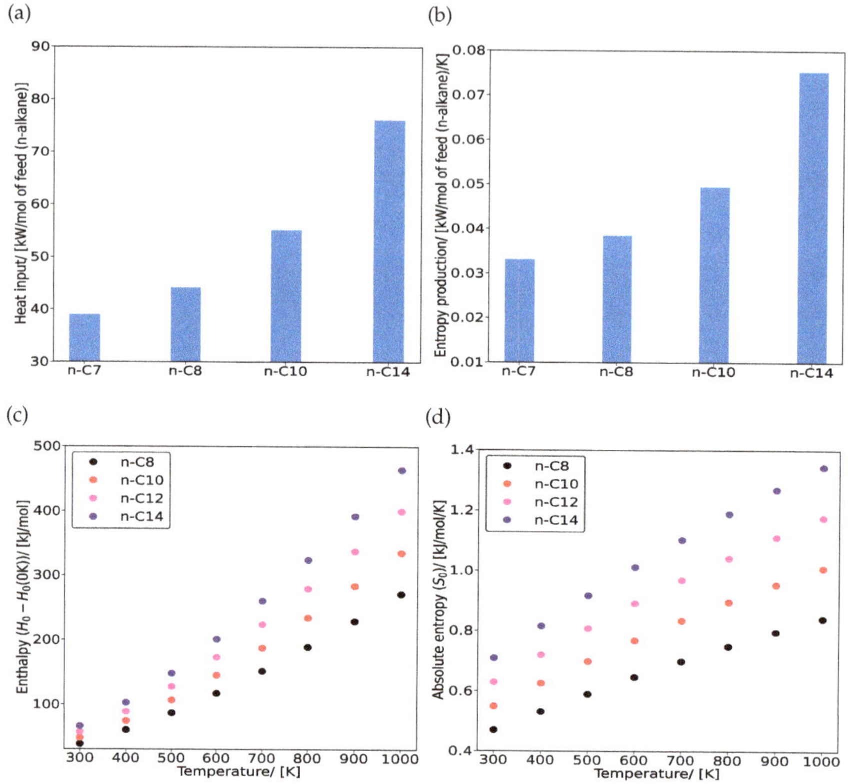

Figure 5. Effect of chain length on (**a**) heat input $\dot{Q}_{in}$ and (**b**) entropy production $\sigma_{production}$ for the hydroisomerization of C_7, C_8, C_{10}, and C_{14} isomers at reaction equilibrium in MTW-type zeolite. The reactions take place at 500 K and infinite dilution. Variations of (**c**) enthalpies $H_0 - H_0(0\text{ K})$ and (**d**) absolute entropies S_0 with temperatures in the range (300–1000) K for n-C_8, n-C_{10}, n-C_{12}, and n-C_{14} isomers.

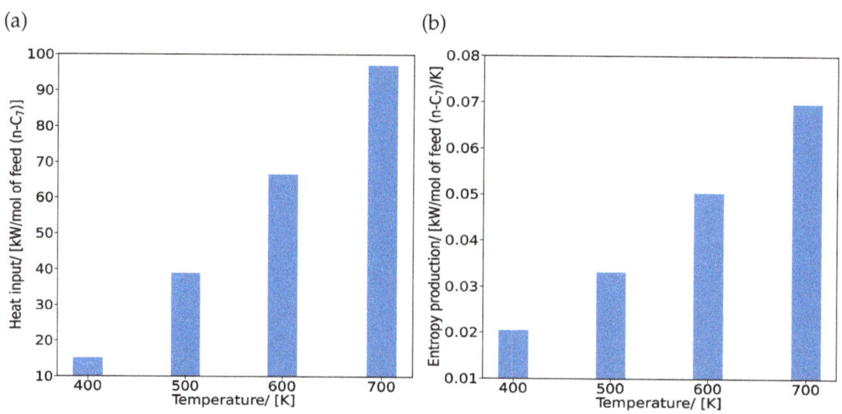

Figure 6. Effect of temperature on (**a**) heat input $\dot{Q}_{in}$ and (**b**) entropy production $\sigma_{production}$ for the hydroisomerization of C_7 isomers at reaction equilibrium in MTW-type zeolite at 400, 500, 600, and 700 K. The reactions take place in the zeolite at infinite dilution.

4. Conclusions

The absolute entropies (S_0) of alkanes computed from $(H_0 - H_0(0\ \mathrm{K}))$ and $(G_0 - H_0(0\ \mathrm{K}))$ as predicted by our LR model show excellent agreement with Scott's tables [12] and experimental data [26]. Our model effectively captures entropy variations based on the type, number, and positions of branches in isomers, outperforming group contribution methods like those by Benson et al. [14] and Constantinou and Gani [15]. The computed absolute entropies will be valuable for designing and optimizing alkane-based processes with a focus on second law efficiency. Unique reaction product distributions are observed in the hydroisomerization of C_7 isomers in different zeolites. Large-pore zeolites, such as FAU- and ITQ-29-type zeolites, favor dimethyl isomers compared to n-C_7, which is mainly due to $(G_0 - H_0(0\ \mathrm{K}))$ and $\Delta_f H_0(0K)$. In BEA-, MEL-, and MFI-type zeolites, normal alkanes and monomethyl isomers show similar selectivities with larger preferences compared to dimethyl and trimethyl isomers for C_7 and C_8 alkanes. In narrow-pore zeolites (MRE- and MTW-types), n-C_7 is preferred, followed by mono-branched isomers. Small variations in heat input and entropy production are observed for different zeolites because of the unique reaction product distributions, which are influenced by the shapes and sizes of the zeolite pores. These trends may shift when cracking reactions are considered. As the alkane chain length increases, both heat input and entropy production increase. This is due to the increase in enthalpies and entropies with temperature for longer chains. A larger number of C-H and C-C bonds in long-chain alkanes increases the internal energy, leading to a larger enthalpy and, consequently, higher heat input. The increased number of atoms in long-chain alkanes allows for more molecular conformations, leading to higher absolute entropies. At higher temperatures, both heat input and entropy production for the hydroisomerization of a certain alkane (C_7 in this case) increase. The rate of heat input increases because of the larger enthalpies of the isomers at higher temperatures. At higher temperatures, molecules gain more energy to occupy a larger number of configurations, causing an increase in entropy and therefore an increase in entropy production. This work demonstrates the accuracy of our LR model in predicting absolute entropies of alkanes and offers insights into heat requirements and entropy production during the formation of branched isomers in zeolite-catalyzed hydroisomerization, which will be useful for better design and optimization of this process. The computed entropies will also be relevant for improving energy efficiencies in other processes involving hydrocarbons.

Supplementary Materials: The following supporting information can be downloaded at: https://www.mdpi.com/article/10.3390/e26121120/s1.

Author Contributions: Conceptualization, S.S., R.B., M.R., E.Z., U.A., S.C., D.D. and T.J.H.V.; methodology, S.S., R.B., M.R., E.Z., U.A., S.C., D.D. and T.J.H.V.; validation, S.S., T.J.H.V. and D.D.; formal analysis, S.S.; investigation, S.S., R.B., M.R., E.Z., U.A., S.C., D.D. and T.J.H.V.; data curation, S.S., T.J.H.V. and D.D.; writing—S.S.; writing—review and editing, S.S., R.B., M.R., E.Z., U.A., S.C., D.D. and T.J.H.V.; supervision, R.B., M.R., E.Z., U.A., S.C., D.D. and T.J.H.V.; methodology, R.B., M.R., E.Z., U.A., S.C., D.D. and T.J.H.V.; project administration, M.R.; funding acquisition, T.J.H.V., D.D. and M.R. All authors have read and agreed to the published version of the manuscript.

Funding: This work was sponsored by NWO Domain Science for the use of supercomputer facilities. This work is part of the Advanced Research Center for Chemical Building Blocks, ARC-CBBC, which is co-funded and co-financed by the Netherlands Organization for Scientific Research (NWO) and the Netherlands Ministry of Economic Affairs and Climate Policy.

Institutional Review Board Statement: Ethical review and approval are not applicable as this study does not involve humans or animals.

Data Availability Statement: The raw data supporting the conclusions of this article are available in the Supporting Information SI.xlsx.

Acknowledgments: This work was sponsored by NWO Domain Science for the use of supercomputer facilities. This work is part of the Advanced Research Center for Chemical Building Blocks, ARC-CBBC, which is co-funded and co-financed by the Netherlands Organization for Scientific Research (NWO) and the Netherlands Ministry of Economic Affairs and Climate Policy. The authors acknowledge the use of computational resources of the DelftBlue supercomputer, provided by Delft High Performance Computing Center (https://www.tudelft.nl/dhpc) [43].

Conflicts of Interest: Authors Richard Baur, Marcello S. Rigutto and Erik Zuidema were employed by the company Shell Global Solutions International B.V. Author Umang Agarwal was employed by the company Shell Chemical LP. The funders had no role in the design of the study; in the collection, analyses, or interpretation of data; in the writing of the manuscript; or in the decision to publish the results.

References

1. Aldosari, M.N.; Yalamanchi, K.K.; Gao, X.; Sarathy, S.M. Predicting entropy and heat capacity of hydrocarbons using machine learning. *Energy AI* **2021**, *4*, 100054. [CrossRef]
2. Ruelle, P.; Kesselring, U.W. Aqueous solubility prediction of environmentally important chemicals from the mobile order thermodynamics. *Chemosphere* **1997**, *34*, 275–298. [CrossRef]
3. Moran, M.J.; Shapiro, H.N.; Boettner, D.D.; Bailey, M.B. *Fundamentals of engineering thermodynamics*, 7th ed.; John Wiley & Sons: Hoboken, NJ, USA, 2010.
4. Calis, H.; Lüke, W.; Drescher, I.; Schütze, A. *Handbook of Fuels: Energy Sources for Transportation*, 2nd ed.; Wiley Online Library: Weinheim, Germany, 2021.
5. Smit, B.; Maesen, T.L.M. Towards a molecular understanding of shape selectivity. *Nature* **2008**, *451*, 671–678. [CrossRef] [PubMed]
6. Smit, B.; Maesen, T.L. Molecular simulations of zeolites: Adsorption, diffusion, and shape selectivity. *Chem. Rev.* **2008**, *108*, 4125–4184. [CrossRef]
7. Rigutto, M.S.; van Veen, R.; Huve, L. Zeolites in hydrocarbon processing. In *Studies in Surface Science and Catalysis*; Elsevier: Amsterdam, The Netherlands, 2007; p. 855.
8. Mäki-Arvela, P.; Kaka khel, T.A.; Azkaar, M.; Engblom, S.; Murzin, D.Y. Catalytic hydroisomerization of long-chain hydrocarbons for the production of fuels. *Catalysts* **2018**, *8*, 534. [CrossRef]
9. Letcher, T.M. *Chemical Thermodynamics for Industry*, 1st ed.; Royal Society of Chemistry: Letchworth, UK, 2004.
10. Kanoğlu, M.; Çengel, Y.A.; Dinçer, İ. *Efficiency Evaluation of Energy Systems*, 1st ed.; Springer Science & Business Media: New York, NY, USA, 2012.
11. Kjelstrup, S.; Bedeaux, D.; Johannessen, E.; Gross, J. *Non-Equilibrium Thermodynamics for Engineers*, 2nd ed.; World Scientific Publishing Co. Pte. Ltd.: Singapore, 2010.
12. Scott, D.W. *Chemical Thermodynamic Properties of Hydrocarbons and Related Substances: Properties of the Alkane Hydrocarbons, C_1 Through C_{10}, in the Ideal Gas State from 0 to 1500 K*, 1st ed.; US Department of the Interior, Bureau of Mines: Washington, DC, USA, 1974.
13. Linstrom, P.J.; Mallard, W.G. The NIST Chemistry WebBook: A chemical data resource on the internet. *J. Chem. Eng. Data* **2001**, *46*, 1059–1063. [CrossRef]
14. Benson, S.W.; Cruickshank, F.; Golden, D.; Haugen, G.R.; O'neal, H.; Rodgers, A.; Shaw, R.; Walsh, R. Additivity rules for the estimation of thermochemical properties. *Chem. Rev.* **1969**, *69*, 279–324. [CrossRef]
15. Constantinou, L.; Gani, R. New group contribution method for estimating properties of pure compounds. *AIChE J.* **1994**, *40*, 1697–1710. [CrossRef]
16. Joback, K.G.; Reid, R.C. Estimation of pure-component properties from group-contributions. *Chem. Eng. Commun.* **1987**, *57*, 233–243. [CrossRef]
17. Marrero, J.; Gani, R. Group-contribution based estimation of pure component properties. *Fluid Phase Equilib.* **2001**, *183*, 183–208. [CrossRef]
18. Hukkerikar, A.S.; Meier, R.J.; Sin, G.; Gani, R. A method to estimate the enthalpy of formation of organic compounds with chemical accuracy. *Fluid Phase Equilib.* **2013**, *348*, 23–32. [CrossRef]
19. Albahri, T.A.; Aljasmi, A.F. SGC method for predicting the standard enthalpy of formation of pure compounds from their molecular structures. *Thermochim. Acta* **2013**, *568*, 46–60. [CrossRef]
20. Domalski, E.S.; Hearing, E.D. Estimation of the Thermodynamic Properties of Hydrocarbons at 298.15 K. *J. Phys. Chem. Ref. Data* **1988**, *17*, 1637–1678. [CrossRef]
21. Yaws, C.L. *Yaws' Handbook of Thermodynamic Properties for Hydrocarbons and Chemicals*, 1st ed.; Knovel: New York, NY, USA, 2009.
22. Bloxham, J.C.; Redd, M.E.; Giles, N.F.; Knotts IV, T.A.; Wilding, W.V. Proper use of the DIPPR 801 database for creation of models, methods, and processes. *J. Chem. Eng. Data* **2020**, *66*, 3–10. [CrossRef]
23. Hayes, M.Y.; Li, B.; Rabitz, H. Estimation of molecular properties by high-dimensional model representation. *J. Phys. Chem. A* **2006**, *110*, 264–272. [CrossRef]

24. Yalamanchi, K.K.; Van Oudenhoven, V.C.; Tutino, F.; Monge-Palacios, M.; Alshehri, A.; Gao, X.; Sarathy, S.M. Machine learning to predict standard enthalpy of formation of hydrocarbons. *J. Phys. Chem. A* **2019**, *123*, 8305–8313. [CrossRef]
25. Sharma, S.; Sleijfer, J.J.; op de Beek, J.; van der Zeeuw, S.; Zorzos, D.; Lasala, S.; Rigutto, M.S.; Zuidema, E.; Agarwal, U.; Baur, R.; et al. Prediction of Thermochemical Properties of Long-Chain Alkanes Using Linear Regression: Application to Hydroisomerization. *J. Phys. Chem. B* **2024**, *128*, 9619–9629. [CrossRef]
26. Scott, D.W. Correlation of the chemical thermodynamic properties of alkane hydrocarbons. *J. Chem. Phys.* **1974**, *60*, 3144–3165. [CrossRef]
27. Muckley, E.S.; Saal, J.E.; Meredig, B.; Roper, C.S.; Martin, J.H. Interpretable models for extrapolation in scientific machine learning. *Digit. Discov.* **2023**, *2*, 1425–1435. [CrossRef]
28. Estrada-Villagrana, A.D.; De La Paz-Zavala, C. Application of chemical equilibrium for hydrocarbon isomerization analysis. *Fuel* **2007**, *86*, 1325–1330. [CrossRef]
29. Gunawan, M.L.; Novita, T.H.; Aprialdi, F.; Aulia, D.; Nanda, A.S.F.; Rasrendra, C.B.; Addarojah, Z.; Mujahidin, D.; Kadja, G.T.M. Palm-oil transformation into green and clean biofuels: Recent advances in the zeolite-based catalytic technologies. *Bioresour. Technol. Rep.* **2023**, *23*, 101546. [CrossRef]
30. Poursaeidesfahani, A.; de Lange, M.F.; Khodadadian, F.; Dubbeldam, D.; Rigutto, M.; Nair, N.; Vlugt, T.J.H. Product shape selectivity of MFI-type, MEL-type, and BEA-type zeolites in the catalytic hydroconversion of heptane. *J. Catal.* **2017**, *353*, 54–62. [CrossRef]
31. Agarwal, U.; Rigutto, M.S.; Zuidema, E.; Jansen, A.P.J.; Poursaeidesfahani, A.; Sharma, S.; Dubbeldam, D.; Vlugt, T.J.H. Kinetics of zeolite-catalyzed heptane hydroisomerization and hydrocracking with CBMC-modeled adsorption terms: Zeolite Beta as a large pore base case. *J. Catal.* **2022**, *415*, 37–50. [CrossRef]
32. Sharma, S.; Rigutto, M.S.; Zuidema, E.; Agarwal, U.; Baur, R.; Dubbeldam, D.; Vlugt, T.J.H. Understanding shape selectivity effects of hydroisomerization using a reaction equilibrium model. *J. Chem. Phys.* **2024**, *160*, 214708. [CrossRef]
33. Matito-Martos, I.; Rahbari, A.; Martin-Calvo, A.; Dubbeldam, D.; Vlugt, T.J.H.; Calero, S. Adsorption equilibrium of nitrogen dioxide in porous materials. *Phys. Chem. Chem. Phys.* **2018**, *20*, 4189–4199. [CrossRef]
34. Hansen, N.; Jakobtorweihen, S.; Keil, F.J. Reactive Monte Carlo and grand-canonical Monte Carlo simulations of the propene metathesis reaction system. *J. Chem. Phys.* **2005**, *122*, 164705. [CrossRef]
35. Widom, B. Some topics in the theory of fluids. *J. Chem. Phys.* **1963**, *39*, 2808–2812. [CrossRef]
36. Frenkel, D.; Smit, B. *Understanding Molecular Simulation: From Algorithms to Applications*, 3rd ed.; Elsevier: Amsterdam, The Netherlands, 2023.
37. Dubbeldam, D.; Torres-Knoop, A.; Walton, K.S. On the inner workings of Monte Carlo codes. *Mol. Simul.* **2013**, *39*, 1253–1292. [CrossRef]
38. Dubbeldam, D.; Calero, S.; Ellis, D.E.; Snurr, R.Q. RASPA: Molecular simulation software for adsorption and diffusion in flexible nanoporous materials. *Mol. Simul.* **2016**, *42*, 81–101. [CrossRef]
39. Ran, Y.A.; Sharma, S.; Balestra, S.R.G.; Li, Z.; Calero, S.; Vlugt, T.J.H.; Snurr, R.Q.; Dubbeldam, D. RASPA3: A Monte Carlo code for computing adsorption and diffusion in nanoporous materials and thermodynamics properties of fluids. *J. Chem. Phys.* **2024**, *161*, 114106. [CrossRef]
40. Baerlocher, C.; McCusker, L.B.; Olson, D.H. *Atlas of Zeolite Framework Types*, 6th ed.; Elsevier: Amsterdam, The Netherlands, 2007.
41. Levenspiel, O. *Chemical Reaction Engineering*, 3rd ed.; John Wiley & Sons: New York, NY, USA, 1998.
42. Perrot, P. *A to Z of Thermodynamics*, 1st ed.; Oxford University Press: Guildford, UK, 1998.
43. Delft High Performance Computing Centre (DHPC). DelftBlue Supercomputer (Phase 2). Available online: https://www.tudelft.nl/dhpc/ark:/44463/DelftBluePhase2 (accessed on 26 July 2024).

entropy

MDPI

Article

Entropy Production in an Electro-Membrane Process at Underlimiting Currents—Influence of Temperature

Juan Carlos Maroto [1,2], Sagrario Muñoz [3] and Vicenta María Barragán [3,*]

1 Department of Electronics, Automation, and Communications, Comillas Pontifical University, 28049 Madrid, Spain; jcmaroto@comillas.edu
2 Department of Science and Aerospace, Universidad Europea de Madrid, 28670 Madrid, Spain
3 Department of Structure of Matter, Thermal Physics and Electronics, Faculty of Physics, Complutense University of Madrid, 28040 Madrid, Spain; smsm@ucm.es
* Correspondence: vmabarra@ucm.es

Abstract: The entropy production in the polarization phenomena occurring in the underlimiting regime, when an electric current circulates through a single cation-exchange membrane system, has been investigated in the 3–40 °C temperature range. From the analysis of the current–voltage curves and considering the electro-membrane system as a unidimensional heterogeneous system, the total entropy generation in the system has been estimated from the contribution of each part of the system. Classical polarization theory and the irreversible thermodynamics approach have been used to determine the total electric potential drop and the entropy generation, respectively, associated with the different transport mechanisms in each part of the system. The results show that part of the electric power input is dissipated as heat due to both electric migration and diffusion ion transports, while another part is converted into chemical energy stored in the saline concentration gradient. Considering the electro-membrane process as an energy conversion process, an efficiency has been defined as the ratio between stored power and electric power input. This efficiency increases as both applied electric current and temperature increase.

Keywords: cation-exchange membrane; limiting current; diffusion boundary layers; concentration polarization; entropy production; current–voltage curve; saline concentration gradient

1. Introduction

Concentration polarization at the surface of an ion-exchange membrane arises from the difference between the transport numbers of an ionic species in the membrane and in the free solution. At equilibrium, without electric current or other driving forces, the bulk concentration is uniform throughout the external solution, including the diffusion boundary layers. When an electrical current is applied, ions start to migrate to and through the ion exchange membrane. Since the electrical current is carried by both cations and anions in free solutions, but only by counterions through the ion-exchange membrane, the concentration profile across the boundary layer gives rise to a diffusional flux of ions to the membrane surface to balance the unequal migration fluxes; that is, polarization occurs. This effect is very important in applications involving ion-exchange membranes in electrolyte media, such as electrodialysis [1]. It is an essential limiting factor in the performance of electrically driven processes in general. Therefore, these phenomena have been widely studied in the literature [1–20].

Usually, polarization phenomena occurring with ion exchange membranes are studied using models based on the Nernst–Planck transport equation [3–5,7,9–12,14,15,17–20]. More recent works mainly focus on phenomena associated with overlimiting transport [4,7,11,13–15,20].

However, studies examining these systems from a thermodynamics point of view are rarer. Chehayeb and Lienhard [21] analyzed the entropy generation in electrodialysis, only considering the electric driving force. Generous et al. [22] also studied entropy generation of electrodialysis desalination using multi-component solutions. Ślezak et al. [6] used the non-equilibrium thermodynamics-based Kedem–Katchalsky equations to describe the influence of the concentration boundary layer on membrane potential in a single membrane system. Ślezak et al. [23] used the same formalism to evaluate S-entropy production in a single membrane system in concentration polarization conditions. However, in both cases, only a concentration gradient was initially imposed in the system.

Non-equilibrium thermodynamics gives a unified method to study coupled transport processes [24]. Kjelstrup and Bedeaux [25] have encouraged the use of the non-equilibrium thermodynamics formalisms to describe transport in heterogeneous media. It can describe coupled transport processes, such as those occurring under concentration polarization conditions. Entropy production in a system can be used to assess how energy resources are exploited within the system. Non-equilibrium thermodynamics formalism has been applied to study fuel cells [26], thermoelectric cells [27], or reverse electrodialysis processes [28]. The work of Manganelly et al. [29] showed that entropy production minimization can be used as a method to optimize the use of energy resources in a system.

However, in most cases, the formalism of the non-equilibrium thermodynamics is used to determine relationships between flows and forces present in the system rather than as a diagnostic tool for analyzing energy system.

The general purpose of this work is to analyze entropy production in an electrically driven process in a single membrane system under polarization concentration conditions in the underlimiting regime, considering the system as a heterogeneous medium. To this end, we combine classical polarization theory and irreversible thermodynamics formalism. The objective is to identify the different irreversible processes that contribute to entropy production in the system and evaluate the parameters that influence it.

2. Fundamentals

2.1. The System

The thermodynamic system considered is a heterogeneous system formed by different subsystems: electrodes, bulk solution, concentration boundary layers, and membrane.

Figure 1 (above) illustrates the system under study. It consists of a cation-exchange membrane separating two identical 1:1 electrolyte solutions of bulk concentration c_0 maintained at the same temperature and hydrostatic pressure. In this configuration, a constant electric current passes through the membrane system. As it is well known, due to the different mobilities of the counterions in membrane and in free solution phases, a current-induced concentration gradient is established within the solution and perpendicular to the membrane surface. The Nernst films model [30] is often used to analyze this phenomenon. This model assumes the formation of polarization layers adjoining the membrane at the two surfaces. Within these layers, termed diffusion boundary layers, convection is absent, and mass transfer is governed by diffusion and migration, resulting in a linear concentration profile (Figure 1, bottom) under steady-state conditions. The Nernst model overlooks the impact of convection on transport. Levich [31] refined this model, demonstrating that convective transport within the diffusion layers results in a smooth, monotonic concentration profile that asymptotically approaches the bulk concentration. For this study, we adopt the

Nernst model for simplicity. Linear concentration profiles are obtained in this model. Due to the small concentration gradient existing in the system, a significant convection effect is not expected.

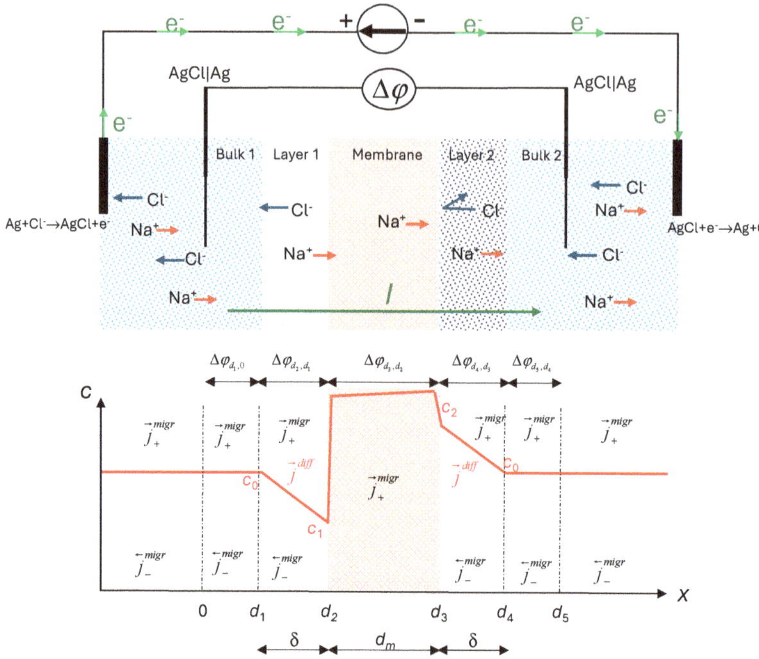

$$\Delta\varphi = \Delta\varphi_{d_1,0} + \Delta\varphi_{d_2,d_1} + \Delta\varphi_{d_3,d_2} + \Delta\varphi_{d_4,d_3} + \Delta\varphi_{d_5,d_4}$$

Figure 1. Sketch of the ion-exchange membrane system under study: An electric current passes through a cation exchange membrane separating two aqueous solutions of NaCl of equal concentration c_0 (**above**). The Ag | AgCl electrodes are reversible to Cl^- ion concentration profiles and different fluxes across the system (**bottom**).

This phenomenon, termed polarization concentration, significantly affects transport properties, particularly the cell voltage drop, $\Delta\varphi$, between bathing solutions on either side of the membrane.

The transport processes considered occur along the horizontal axis of the cell, referred to as the x-axis. Fluxes through the membrane are treated as scalar components of the vectorial flux in this direction. The membrane surfaces can be assumed to be in local equilibrium. The electric current is applied via reversible Ag | AgCl electrodes, and we also use reversible Ag | AgCl electrodes for measuring the electric potential difference. Treating the system as a multi-layer system, the total voltage drop comprises contributions of the different parts of the system: external electric circuit, electrodes, bulk solution, boundary solution layers, and membrane.

In the bulk solution, far from the diffusion boundary layers, ion transport occurs solely by migration, as there are no concentration gradients perpendicular to the membrane surface.

Within the highly permselective cation-exchange membrane, due to the anion exclusion, there is no significant concentration gradient inside it. Anions are more or less completely excluded from the membrane, and their flux by migration and diffusion in opposite directions is compared to that of the cations, which is generally small. Thus, it can be assumed, to a first approximation, that the transport of cations through the highly permselective membrane occurs primarily via migration driven by the electrical potential gradient [32].

In the diffusion boundary layers, ions are transported through a combination of migration and diffusion mechanisms (Figure 1, bottom). Reversible electrodes in the system are maintained under identical conditions of concentration, temperature, and hydrostatic pressure.

2.2. The Current–Voltage Curve of a Single Membrane System

According to the classical theory of concentration polarization [30], the electric current, I, should increase linearly with voltage, $\Delta\varphi$, at low voltages and then rise more slowly, eventually reaching a limiting value. This limiting value, termed limiting current, I_L, would correspond to the value of the current at which the concentration of the solution becomes zero at the membrane surface. At this stage, diffusion is unable to supply enough ions to compensate for the migration flux through the membrane, resulting in an ion depletion at this membrane surface. In a real system, however, the current–voltage curve exhibits a characteristic shape with three regions (Figure 2) clearly distinguished. After the linear relationship (I), the curve reaches a region in which the current remains nearly constant despite increasing voltage, a region known as the plateau region (II). Finally, a new region (III) of current increasing with voltage is observed.

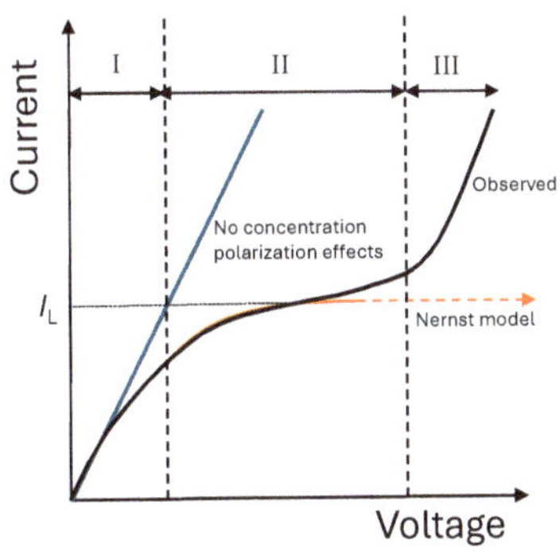

Figure 2. Sketch of a typical current–voltage curve of an ion-exchange membrane in an electrolyte solution (in black), showing the three distinct regions: (I) linear increase, (II) plateau, and (III) overlimiting transport. Profile predicted by the Nernst model (in red) and linear voltage–current curve (in blue) for a system without concentration polarization effects are also shown.

It indicates that the concentration of counterions does not become zero in the membrane surface due to the activation of overlimiting transport mechanisms, such as dissociation of water, gravitational convection, and electroconvection, which allow the increase in ionic transfer through the membrane [11,14]. Our study focuses on the underlimiting regime at currents below the limiting value.

Using the classical theory of concentration polarization, the Peers' equation [32] is obtained, which allows us to estimate the value of the limiting current, I_L, if the solute concentration, c_0, the diffusion coefficient of solute in water, D, the boundary layer thickness, δ, and the counter-ion transport numbers in the membrane, $\bar{t}_+$, and in free solution, t_+, are known:

$$I_L = \frac{AFDc_0}{\delta(\bar{t}_+ - t_+)} \tag{1}$$

where F is the Faraday constant and A is the membrane effective area. This equation is obtained by considering Nernst's linear concentration profile under steady-state conditions, where the concentration of the solution becomes zero at the membrane surface and the limiting current is achieved. The boundary layer thickness also depends on the hydrodynamic conditions.

For currents $I < I_L$, the concentrations at the dilute (c_1) and concentrated (c_2) solutions in the membrane/solution interfaces can be expressed as follows:

$$\begin{aligned} c_1 &= c_0\left(1 - \frac{I}{I_L}\right) \\ c_2 &= c_0\left(1 + \frac{I}{I_L}\right) \end{aligned} \tag{2}$$

In this approach, an equation can be derived for the electric potential difference $\Delta\varphi_{AB}$ between two fixed points placed on both sides of the membrane in the corresponding bulk solution when the stationary state has been reached after the injection of an electric current in the membrane system. This equation, obtained in a previous work [5], is expressed as follows:

$$\Delta\varphi_{AB} = \left(R_0 - \frac{RT}{F\Delta t_+ I_L}\right) I + \frac{RT}{F}\left[(2\Delta t_+) + (2\Delta t_+)^{-1}\right] \ln\left(\frac{1 + \frac{I}{I_L}}{1 - \frac{I}{I_L}}\right) \tag{3}$$

where T is the absolute temperature, R is the gas constant, $\Delta t_+ = \left(\bar{t}_+ - t_+\right)$ and R_0 indicate the total ohmic resistance of the system before the polarization layers are formed. This expression has two contributions, one due to the polarization potential established through the membrane system due to the concentration gradient and the other due to the change in the ohmic resistance of the system as a consequence of the formation of the layer.

Equation (3) is only applicable to electric currents such as $I < I_L$. It predicts that the electric potential difference tends to infinity as I becomes I_L and the polarization potential as well as the resistance of the depletion layer tend to infinity in the Nernst model.

2.3. *The Non-Equilibrium Thermodynamics Formalism*

The system in Figure 1 can be analyzed with the non-equilibrium thermodynamic formalism. As an isothermal electrochemical cell, it exhibits coupled transport of mass and charge. It consists of different parts that will contribute to the total entropy production. In the considered system, the only external force applied is an electric potential gradient.

According to the irreversible thermodynamics approach, for a homogeneous phase, the local entropy production, σ, which indicates the change in the entropy in a volume element, can be expressed as a function of the fluxes, J_i, and forced, X_i, existing in the system as:

$$\sigma = \sum_{i=1}^{n} J_i X_i \tag{4}$$

where n is the number of independent fluxes. Irreversible thermodynamics assumes linear relationships between fluxes and forces, and, in general:

$$J_i = \sum_{j=1}^{n} L_{ij} X_j \text{ for } j = 1, \ldots, n \tag{5}$$

where L_{ij} are the linear phenomenological coefficients. It is assumed that the linear relations are locally valid.

The total entropy production due to the irreversible processes occurring in the system, $\frac{dS_i}{dt}$, is obtained by integrating Equation (4) over the entire volume V of the system considered:

$$\frac{dS_i}{dt} = \int_V \sigma dV \tag{6}$$

Considering one-dimensional transport in the normal direction (x), the entropy production for the whole path, with a cross-section A, Equation (6) can be written as follows:

$$\frac{dS_i}{dt} = A\int \sigma(x)dx \quad (7)$$

The wasted or lost work per unit time is the energy dissipated as heat in the surrounding area and, in the isothermal system, can be expressed as [26]:

$$W_{dissipated} = T\frac{dS_i}{dt} \quad (8)$$

Systems with membranes are heterogeneous systems, consisting of bulk and surface parts, all of them with entropy production: electrode surfaces, bulk solution, polarization layers, and membranes. The total entropy production between two fixed points placed at both sides of the membrane in the corresponding bulk solution when the stationary state is reached can be calculated by applying Equation (7) to the different homogeneous parts of the system placed between these two points.

In our isothermal and isobaric system, only one force is externally applied, the electric potential gradient. However, as previously described, the polarization concentration effects caused by the pass of the electric current will result in a concentration gradient, and coupled transports of mass and charge will also occur in the cell.

The local entropy production is expressed as follows:

$$\sigma = \sum_j J_j\left(-\frac{1}{T}\frac{\partial \mu_{j,T}}{\partial x}\right) + J\left(-\frac{1}{T}\frac{\partial \varphi}{\partial x}\right) + \sum_r r\left(-\frac{\Delta_r G}{T}\right) \quad (9)$$

where J_i and J are, respectively, mass and electric charge flux densities, $\mu_{j,T}$ is the chemical potential at isothermal conditions of the j component, and r and $\Delta_r G$ are, respectively, the velocity and the reaction Gibbs energy of reaction r. If there are no reactions present in the system, the last term in Equation (9) is null.

Coupling between mass and electric charge fluxes can occur in the system. According to Equation (5), if there is only one reaction occurring in the system, the flux equations can be written as follows:

$$\begin{aligned} J_i &= \sum_j L_{ij}\left(-\frac{1}{T}\frac{\partial \mu_{j,T}}{\partial x}\right) + L_{i\varphi}\left(-\frac{1}{T}\frac{\partial \varphi}{\partial x}\right) \\ J &= \sum_j L_{\varphi j}\left(-\frac{1}{T}\frac{\partial \mu_{j,T}}{\partial x}\right) + L_{\varphi\varphi}\left(-\frac{1}{T}\frac{\partial \varphi}{\partial x}\right) \\ r &= L_{rr}\left(-\frac{1}{T}\Delta_r G\right) \end{aligned} \quad (10)$$

In the studied case, connecting leads and reversible electrodes are at the same temperature, pressure, and concentration. The only reaction presented in the system is the reversible reaction corresponding to the electrode surfaces required to change the charge carrier from chloride in solution to an electron in the external circuit. Thus, the main contribution to the entropy production will be due to the electrolyte solution, including boundary diffusion layers, and the membrane.

In analyzing membrane processes, it is more convenient to transform the previous equation into an expression involving the flux of neutral salt, J_s, more than the individual ion fluxes [33]. We choose the membrane as the natural frame of reference for the fluxes. Neglecting the water flux, we have two fluxes (salt and charge) and two involved forces (electric and chemical potential gradients)—Equation (9) can be written as a function of the salt flux as:

$$\sigma = -J_s \frac{1}{T}\frac{\partial \mu}{\partial x} - J\frac{1}{T}\frac{\partial \varphi}{\partial x} \tag{11}$$

and Equation (10) as:

$$\begin{aligned} J_s &= -L_{\mu\mu}\tfrac{1}{T}\tfrac{\partial \mu}{\partial x} - L_{\mu\varphi}\tfrac{1}{T}\tfrac{\partial \varphi}{\partial x} \\ J &= -L_{\varphi\mu}\tfrac{1}{T}\tfrac{\partial \mu}{\partial x} - L_{\varphi\varphi}\tfrac{1}{T}\tfrac{\partial \varphi}{\partial x} \end{aligned} \tag{12}$$

where we have dropped subscript T in the chemical potential gradient since the system is isothermal.

Introducing the salt transference coefficient, t_s, defined as the ratio of the salt flux and the electric current density at uniform composition, and the salt diffusion coefficient from Fick's law at zero current, D, the second expression in Equation (12) allows us to obtain a relation between both fluxes:

$$J_s = -D\frac{\partial c}{\partial x} + \frac{t_s}{F}J \tag{13}$$

where the relation:

$$\mu_{NaCl} = \mu^0_{NaCl} + 2RT\ln c_{NaCl} \tag{14}$$

has been used for relating the salt chemical potential and concentration; mean activities of ions of value 1 are considered. For Ag | AgCl electrodes reversible to the ion Cl^-, the salt transference number is equal to the transport number of the cation t_+, with:

$$t_+ + t_- = 1 \tag{15}$$

where t_- is the anion transport number.

3. Materials and Methods

The membrane used in this work was a commercial Nafion 117 membrane, a homogenous, highly selective cation-exchange membrane manufactured by Dupont Inc. According to the data provided by the manufacturer, this membrane has a nominal thickness of 183 μm and an IEC of 0.94 meq/g.

Sodium chloride (pro-analysis grade) and deionized, doubly distilled water were used to prepare the electrolyte solutions.

A sketch of the experimental setup used for measuring current–voltage curves, already described elsewhere [5], is shown in Figure 3. The membrane was vertically placed, separating two glass chambers containing two NaCl solutions of the same concentration (5 $molm^{-3}$). The exposed membrane surface area was 9.04×10^{-4} m^2. Both chambers were maintained at the same pressure, and all the experiments were carried out under isothermal conditions by immersing the cell in a thermostatic bath.

The solutions in the chambers were under natural convection conditions, without any stirring. A constant electric current was applied by means of a large active surface Ag | AgCl electrode, and Ag | AgCl probe electrodes were positioned on either side of the membrane to record the electric potential difference between them.

Current–voltage curves were obtained by measuring the electric potential difference at varying current levels. This process was repeated at different temperatures ranging from 3 to 40 °C.

At each temperature, the electric conductivity of the solution was measured with a conductivity meter, JENCO Model 1671. Closed vessels containing the solution were immersed in a thermal bath at the selected temperature. Once the thermal equilibrium was reached, the conductivity of the solutions was measured using a suitable conductivity probe. The accuracy of the conductivity measurements was 1 μScm^{-1}.

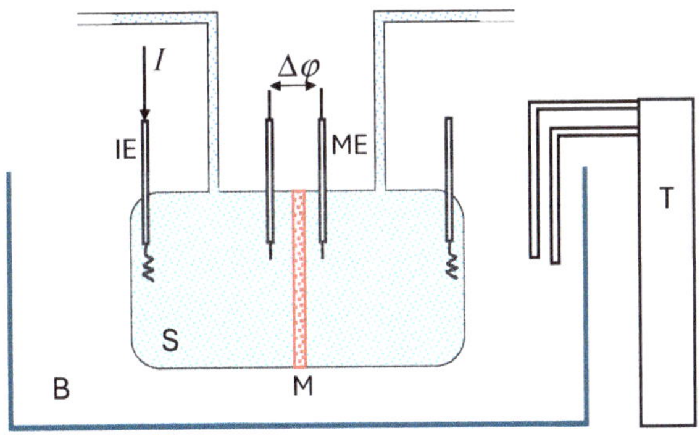

Figure 3. Sketch of the experimental setup for measuring current–voltage curves. I: electric current; B: bath; T: thermostat; S: solution; M: membrane; IE: Ag | AgCl injecting electrode; ME: voltage Ag | AgCl electrode.

4. Results and Discussion

4.1. Properties of the Electrolyte Solution

The electrical conductivity, κ, of the electrolyte solution has been measured at each temperature. The results are shown in Table 1. From these values, the diffusion coefficient, D, at each temperature was calculated using the following relationship [34]:

$$D = \frac{\Lambda RT}{2F^2} \tag{16}$$

where Λ is the specific conductance, defined by the following expression:

$$\Lambda\left(\text{Scm}^2\text{mol}^{-1}\right) = 100\frac{\kappa\left(\text{Scm}^{-1}\right)}{c(\text{M})} \tag{17}$$

and c is the molar concentration of the electrolyte. The results obtained are also shown in Table 1. Transport numbers of Na^+ at different temperatures have been estimated using information from literature [35,36].

Table 1. Values of the electrical conductivity, equivalent conductance, diffusion coefficient, and transport number for Na^+ in 5 mol m^{-3} NaCl aqueous solution.

T (K)	κ (μS cm^{-1}) *	Λ (Scm^{2}mol^{-1}) **	D (10^{-9} m^2 s^{-1}) **	t_+ **
276.15	350	71.62	0.884	0.360
279.15	376	76.94	0.959	0.386
283.15	415	84.92	1.074	0.389
288.15	474	96.99	1.248	0.392
293.15	535	109.47	1.433	0.394
298.15	608	119.91	1.596	0.395
303.15	654	131.98	1.787	0.397
313.15	753	154.08	2.155	0.399

* Measured. ** Estimated.

The data in Table 1 indicate that increasing temperature enhances the transport properties of the solution phase.

4.2. Current–Voltage Curves

Current–voltage curves were measured at different temperatures. They are presented in Figure 4a.

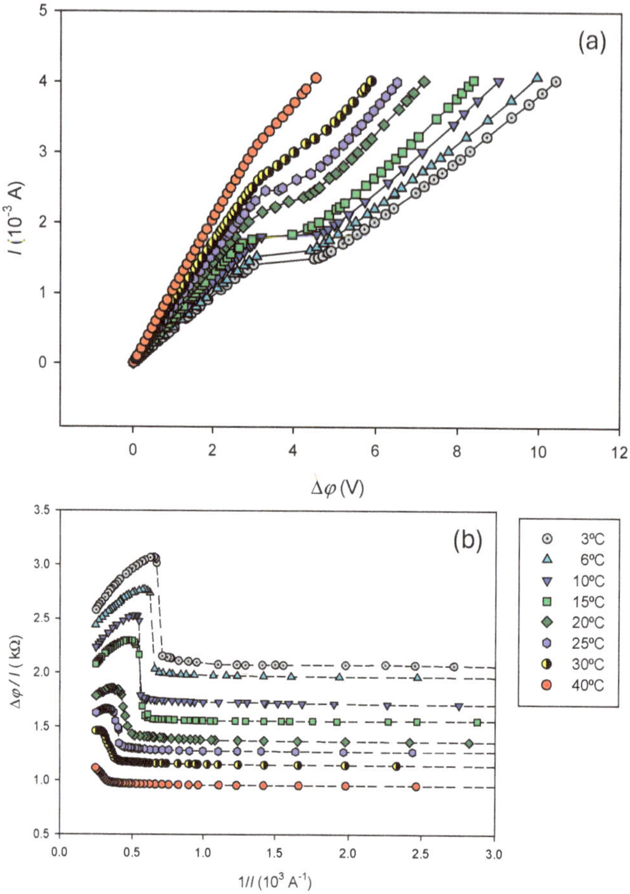

Figure 4. (**a**) Current–Voltage curves and (**b**) corresponding Cowan plots at different temperatures. Dashed lines are included as visual guides.

At temperatures below 30 °C, the obtained curves exhibit the typical profile for cation-exchange membranes immersed in electrolyte solutions, showing the three usual characteristic regions described earlier. Visual inspection of Figure 4a reveals that the second region, corresponding to the plateau, occurs at greater current and becomes less pronounced as temperature increases. Moreover, for 30 and 40 °C, a defined plateau is no longer clearly appreciated.

In the linear region (I) and in the approximately linear segment in region (III), the trend observed with increasing temperature may be attributed to a reduction in the system's electric resistance. The electric conductivity of an electrolyte solution increases with temperature [34–36], as do the ion transport number and the salt diffusion coefficient, facilitating ion transport. The electric conductivity of Nafion 117 is also temperature dependent. Results from the literature have shown that the electric conductivity of the H^+-form Nafion membrane is not a monotonic function of the temperature. Initially it increases with temperature, reaches a maximum value of about 60 °C, and then decreases [37]. In Na^+-form, an Arrhenius relationship has been observed for the electric conductivity with temperature [38].

In region II, the plateau width decreases with increasing temperature, and it becomes less defined, eventually disappearing entirely at 25 and 30 °C. At these temperatures, what is observed is rather an inflection on the current voltage curve. As the temperature increase facilitates ion transport in solution, it may indicate that cation concentration drop on the diluted surface of the membrane decreases with temperature, reducing the voltage drop, also affecting the different mechanisms involved in the overlimiting regime. Further studies are underway to better understand this behavior.

4.2.1. Determination of the Limiting Current Value

The limiting current I_L at each temperature was determined from the corresponding Cowan plots, presented in Figure 4b. The results are presented in Table 2. As was expected from the observation of plateaus in Figure 4a, I_L increases with temperature.

Table 2. Values of parameters R_0 and y $\bar{t}_+$, estimated from Equation (3), values of the limiting current estimated from the corresponding Cowan plots, $I_L{}^{\text{Cowan}}$, and diffusion layer thickness, δ, estimated from Peers' equation.

T (K)	R_0 (Ω)	$\bar{t}_+$	$I_L{}^{\text{Cowan}}$ (10^{-3} A)	δ (10^{-6} m)
276.15	1995.6 ± 1.3	0.88 ± 0.10	1.37	545
279.15	1896.17 ± 0.06	0.88 ± 0.06	1.52	555
283.15	1651.6 ± 0.3	0.89 ± 0.08	1.73	540
288.15	1495.00 ± 0.04	0.89 ± 0.03	1.85	592
293.15	1330.8 ± 1.2	0.88 ± 0.15	2.04	633
298.15	1247.0 ± 0.1	0.88 ± 0.08	2.31	629
303.15	1113.9 ± 0.2	0.88 ± 0.06	2.54	635
313.15	921.40 ± 0.04	0.90 ± 0.05	2.98	645

4.2.2. Determination of R_0 and Counterion Transport Number in the Membrane $\bar{t}_+$

Using the measured limiting currents, the experimental data ($\Delta\varphi$, I) in region I were fitted to Equation (3) using a minimization χ^2 method. It allowed us to estimate parameters R_0 and $\bar{t}_+$ at each temperature. Results are shown in Table 2. As expected, R_0 decreases with increasing temperature, likely due to the enhanced electrical conductivity of the electrolyte. As the electrical conductivity of Nafion membranes is high [39], the greater contribution to R_0 must be due to the electrolyte contribution. The results indicate that the Na^+ transport number in the membrane hardly varies with temperature.

By utilizing data from Tables 1 and 2, the thickness of the boundary diffusion layers at each temperature was estimated using the Peer's equation. Results are also shown in Table 2. The values obtained agree with typical values given in the literature for similar systems [8,10]. A slight increase in layer thickness was observed for temperatures above 20 °C. Experimental observations have confirmed that the Nernst model, which reasonably coincides with the theoretical diffusion boundary layer thickness calculated from the limiting current, effectively depicts the transport phenomena in ion-exchange membrane systems. By measuring the potential drop with a mobile micro-electrode at various distances from a cation-exchange membrane [40], it was observed that the diffusion boundary layers existed in the range of 300–350 µm. Thicknesses of a similar order of magnitude to those found in this work have been observed using laser interferometry [41].

4.3. *Entropy Production*

We apply Equations (7) and (11) to estimate the entropy production in each part of the system. The total entropy production is obtained by the following expression:

$$\frac{dS_i}{dt} = A\int_{\text{bulk}} \sigma(x)dx + A\int_{\text{layers}} \sigma(x)dx + A\int_{\text{membrane}} \sigma(x)dxA \quad (18)$$

In the bulk solutions, where no chemical potential gradient exists, ions move solely by migration. From Equation (11), the dissipation function, representing the local energy dissipated as heat due to irreversible processes, is expressed as:

$$T\sigma = -J\frac{d\varphi}{dx} \tag{19}$$

From Equation (7), and assuming steady-state conditions with constant fluxes, the entropy production in the bulk solution is given by:

$$T\frac{dS_i}{dt}\bigg|_{\text{bulk}} = A\int T\sigma(x)dx = -AJ\Delta\varphi_{\text{bulk}} \tag{20}$$

According to Figure 1 and the second expression in Equation (12), the electric potential difference across the bulk phase is given by:

$$\Delta\varphi_{bulk} = \Delta\varphi_{d_1,0} + \Delta\varphi_{d_5,d_4} = -J\frac{d_{\text{bulk}}T}{L_{\varphi\varphi}} \tag{21}$$

where d_{bulk} is the thickness of the two bulk phases in the cell. Considering the electric conductivity for a uniform distribution of salt,

$$\kappa \equiv \frac{L_{\varphi\varphi}}{T} \tag{22}$$

the total heat dissipation in the bulk solution can be expressed in terms of the total electric resistance in the bulk phase, R_{bulk}, and the electric current circulating along the system:

$$T\frac{dS_i}{dt}\bigg|_{\text{bulk}} = \frac{AJ^2 d_{\text{bulk}}}{\kappa_{\text{bulk}}} = I^2 R_{\text{bulk}} \tag{23}$$

where $I = J.A$. In the boundary diffusion layers, ions are transported by both migration and diffusion. Consequently, both terms in Equation (11) contribute to the entropy production. Using Equations (13) and (14), the total heat dissipation in layers 1 and 2 can be expressed as follows:

$$\begin{aligned} T\frac{dS_i}{dt}\bigg|_{\text{layer1}} &= 2ARTD\left(\frac{c_1-c_0}{\delta}\right)\ln\frac{c_1}{c_0} - AJ\Delta\varphi_{d_2,d_1} - \frac{AJRT2t_+}{F}\ln\frac{c_1}{c_0} \\ T\frac{dS_i}{dt}\bigg|_{\text{layer2}} &= 2ARTD\left(\frac{c_0-c_2}{\delta}\right)\ln\frac{c_0}{c_2} - AJ\Delta\varphi_{d_4,d_3} - \frac{AJRT2t_+}{F}\ln\frac{c_0}{c_2} \end{aligned} \tag{24}$$

where the electric potential difference in the boundary diffusion layers, $\Delta\varphi_{d_2,d_1}$ and $\Delta\varphi_{d_4,d_3}$ can be written as [16,33]:

$$\begin{aligned} \Delta\varphi_{d_2,d_1} &= -\frac{J\delta}{\kappa_{l_1}} - (1-2t_+)\frac{RT}{F}\ln\frac{c_1}{c_0} \\ \Delta\varphi_{d_4,d_3} &= -\frac{J\delta}{\kappa_{l_2}} - (1-2t_+)\frac{RT}{F}\ln\frac{c_0}{c_2} \end{aligned} \tag{25}$$

where κ_{l_1} and κ_{l_2} are the electric conductivity of dilute and concentrate layers, respectively, and the same thickness has been considered for both layers. By combining Equations (24), (25) and (22) for layer conductivity, the total heat dissipation in both layers can be written as follows:

$$T\frac{dS_i}{dt}\bigg|_{\text{layers}} = I^2 R_{layers} + 2ARTD\left(\frac{c_0 I}{\delta I_L}\right)\ln\frac{c_2}{c_1} - I(1-4t_+)\frac{RT}{F}\ln\frac{c_2}{c_1} \tag{26}$$

where R_{layers} indicates the ohmic resistance of both layers.

Entropy production within the membrane was evaluated considering its high selectivity. In this case, cations move predominantly due to migration under the influence of the electric potential gradient. Neglecting water flux, the expression for estimating the heat dissipation inside the membrane is similar to Equation (19) used for determining the bulk solution contribution:

$$T\frac{dS_i}{dt}\bigg|_{membrane} = A\int T\sigma(x)dx = -AJ\Delta\varphi_{membrane} = -AJ\Delta\varphi_{d_3,d_2} \tag{27}$$

but in this case, there are two contributions to the electric potential difference: one is the ohmic voltage drop across the membrane, and the other is the membrane potential [30]. Thus,

$$\Delta\varphi_{d_3,d_2} = -\frac{Jd_m}{\kappa_m} - (1-2\bar{t}_+)\frac{RT}{F}ln\frac{c_2}{c_1} \tag{28}$$

where κ_m is the membrane electric conductivity. Combining Equations (27) and (28) and using Equation (22) for the membrane conductivity, the total heat dissipation within the membrane is given by:

$$T\frac{dS_i}{dt}\bigg|_{membrane} = I^2R_m + I(1-2\bar{t}_+)\frac{RT}{F}ln\frac{c_2}{c_1} \tag{29}$$

The total heat dissipation is obtained by adding all contributions indicated in Equations (23), (26), and (29), resulting in:

$$T\frac{dS_i}{dt}\bigg|_{total} = I^2R_{total} + I\frac{2t_+RT}{F}ln\frac{c_2}{c_1} \tag{30}$$

Equation (30) indicates that the total dissipation of heat produced between the two reference electrodes placed in the system has a contribution of the ohmic resistances of the different parts of the system, but there is also a contribution of the diffusion transport in the solution due to the presence of the boundary diffusion layers, which contribute to the total electric potential drop with both ohmic and non-ohmic phenomena.

We can estimate the total ohmic resistance of the system using the model described in Section 2.2. According to this model, R_{total} can be expressed as the sum of the total ohmic system resistance in the absence of polarization layers, R_0, and the ohmic resistance change due to the formation of these layers, ΔR, expressed by:

$$\Delta R = -\frac{RT}{F\Delta t_+ I_L} + \frac{RT}{IF2\Delta t_+}\ln\left(\frac{1+\frac{I}{I_L}}{1-\frac{I}{I_L}}\right) \tag{31}$$

Using Equations (30) and (31), the total entropy production in the system can be estimated by the following equation:

$$\frac{dS_i}{dt}\bigg|_{total} = I^2\left(\frac{R_0}{T} - \frac{R}{I_LF\Delta t_+}\right) + \frac{IR}{F}\left(\frac{1}{2\Delta t_+} + 2t_+\right)\ln\left(\frac{1+\frac{I}{I_L}}{1-\frac{I}{I_L}}\right) \tag{32}$$

Figure 5 shows the values obtained for the total entropy production using Equation (32) and the data in Table 1 as a function of the current density at each temperature.

Visual inspection of Figure 5 shows that the total entropy production increases as the applied electric current increases and decreases as the system temperature increases.

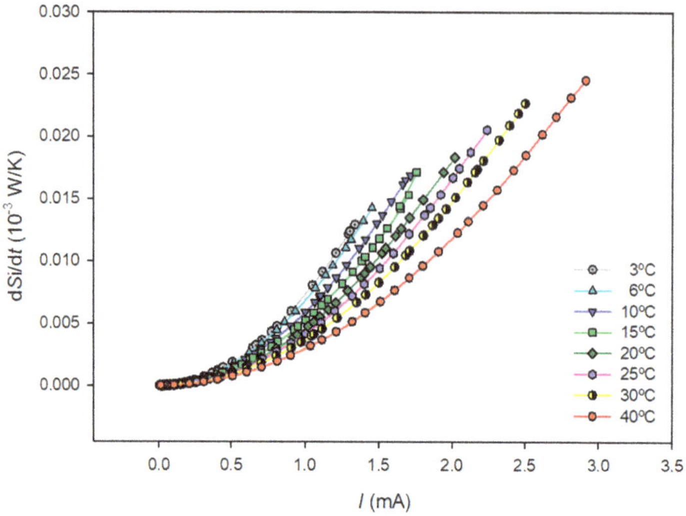

Figure 5. Estimated entropy production as a function of the applied electric current at different temperatures. Lines are provided as visual guides.

4.4. *Efficiency Analysis*

In the concentration polarization phenomena that occur in electrically driven membrane processes, electric work is partially dissipated as heat due to the irreversible processes occurring in the system. Equation (30) expresses this heat dissipation power, $\dot{W}_{\text{dissipated}}$. It is taken from the total electric power input. However, part of the energy is used to create the concentration gradient. Thus, this other part of the power input is converted into chemical energy stored in the saline gradient created in the boundary diffusion layers. In this sense, we can interpret the concentration polarization phenomenon in the system as an energy conversion process, and entropy generation can be used to analyze the energy efficiency of the process. The lower the dissipation, the higher the stored energy. The situation is similar to that in an electrodialysis process, where an electric current is passed through cation and anion exchange membranes placed in alternating order to create a concentration gradient [21].

From the model described in Section 2.2, we can write the electrical power input, $\dot{W}$, as follows:

$$\dot{W} = I \cdot \Delta\varphi_{AB} \tag{33}$$

By combining Equations (3), (32) and (33), the useful electric power is obtained as follows:

$$\dot{W}_{\text{stored}} = \dot{W} - T\frac{dS_i}{dt}\bigg|_{total} = \frac{IRT2\bar{t}_+}{F}\ln\left(\frac{1+\frac{I}{I_L}}{1-\frac{I}{I_L}}\right) \tag{34}$$

Equation (34) expresses that the power stored in the salt gradient will be maximum for an ideal selective membrane with $\bar{t}_+ = 1$. The dissipated power can also be calculated from:

$$\dot{W}_{\text{dissipated}} = \dot{W}_{\text{exp}} - \dot{W}_{\text{stored}} \tag{35}$$

where $\dot{W}_{\text{exp}} = I\Delta\varphi$ is the experimental power input. In the absence of a membrane, there would be no polarization layers, and the electric power input would be dissipated as Joule heat.

Figure 6 shows the stored and dissipated powers calculated from Equations (34) and (35), experimental current–voltage values, and the data given in Table 1.

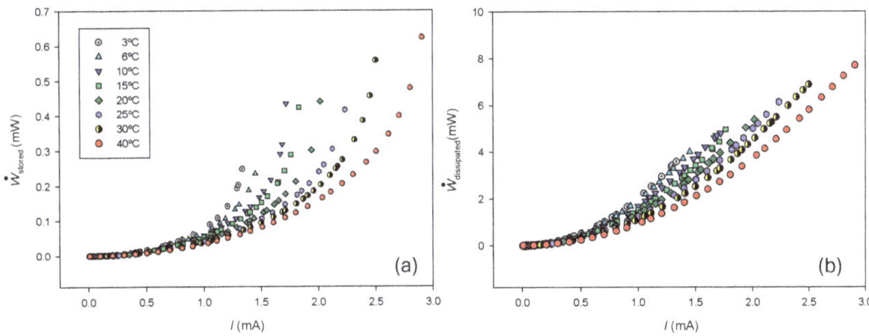

Figure 6. (**a**) Estimated stored and (**b**) dissipated powers as a function of the electric current at different temperatures.

As can be observed, both stored and dissipated powers increase with the electric current and decrease with temperature.

System efficiency was evaluated as the store-to-total-power ratio:

$$\eta(\%) = \frac{\dot{W}_{\text{stored}}}{\dot{W}_{\text{exp}}} 100 \tag{36}$$

The efficiency values are shown in Figure 7 as a function of $I_r = I/I_L$ at each temperature. Efficiency increases as the electric current input approaches the corresponding limit value, maybe due to an increase in the concentration difference established between both sides of the membrane. Although it increases both dissipated and stored powers, an increase in temperature generally results in better process efficiency.

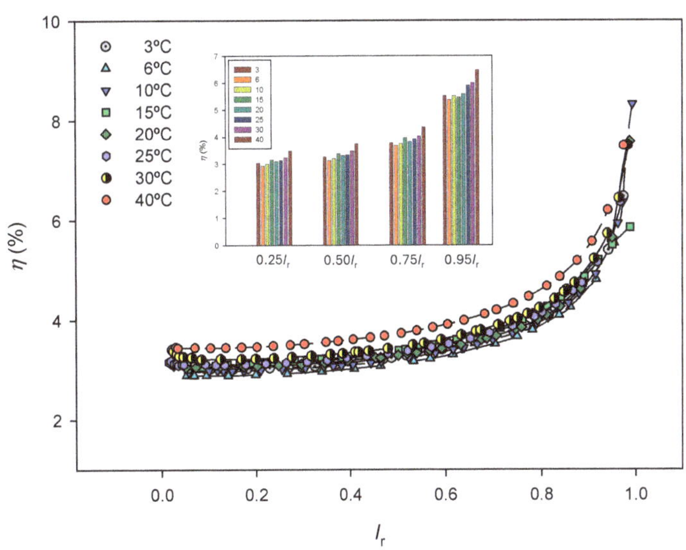

Figure 7. Estimated efficiencies as a function of $I_r = I/I_L$ at different temperatures. Dashed lines are only visual guides. The figure legend shows the efficiency values at different I_r percentages at the different temperatures for a better visualization.

It should be noted that the model does not consider the contribution of water transport to entropy production. However, it is expected that this contribution will only be relevant at low flow rates, as the relative importance of the osmotic contribution is reduced.

The model used is only valid for currents below the limiting value, and as the applied current approaches this limiting value, other mechanisms begin to emerge that have not been considered in this study. These mechanisms may contribute to a reduction in the thickness of the boundary diffusion layer and thus may strongly influence the overall entropy production of the system. The analysis of entropy production would be a useful tool to study dissipation effects in the overlimiting regime.

5. Conclusions

The entropy production in the polarization phenomena occurring in the underlimiting regime, when an electric current circulates through a single cation-exchange membrane system, has been estimated, and its variation with temperature has been studied.

To this purpose, classical polarization theory and irreversible thermodynamics formalism were used to analyze experimental current–voltage curves of the system at different temperatures.

The results obtained reveal that the total entropy production increases as the applied electric current increases and decreases as the system temperature increases.

The total dissipation of heat in the system has a contribution from the ohmic resistances of the different parts of the system, but there is also a contribution from the diffusion transport in the solution due to the presence of the boundary diffusion layers.

Considering the process as a conversion of electric work input into chemical energy stored in the saline gradient, the system efficiency has been evaluated as the ratio of stored to total power input. The results obtained show that efficiency increases with increasing electric current and temperature.

Author Contributions: Conceptualization, J.C.M., V.M.B. and S.M.; methodology, J.C.M. and V.M.B.; software, J.C.M.; validation, J.C.M., V.M.B. and S.M.; formal analysis, J.C.M.; investigation, J.C.M., V.M.B. and S.M.; resources, V.M.B.; data curation, J.C.M.; writing—original draft preparation, J.C.M. and V.M.B.; writing—review and editing, V.M.B. and S.M.; visualization, J.C.M., V.M.B. and S.M.; supervision, V.M.B. and S.M. All authors have read and agreed to the published version of the manuscript.

Funding: This scientific research was funded by Banco de Santander and Universidad Complutense of Madrid within the framework of Project PR108/20-02.

Institutional Review Board Statement: Not applicable.

Informed Consent Statement: Not applicable.

Data Availability Statement: The data are contained within this article.

Conflicts of Interest: The authors declare no conflicts of interest.

References

1. Campione, A.; Gurreri, L.; Ciofalo, M.; Micale, G.; Tamburini, A.; Cipollina, A. Electrodialysis for water desalination: A critical assessment of recent developments on process fundamentals, model and applications. *Desalination* **2018**, *434*, 121–160. [CrossRef]
2. Yamane, R.; Sata, T.; Mizutani, Y.; Onoue, Y. Concentration polarization phenomena in ion-exchange membrane electrodialysis. II. The effect of the condition of the diffusion-boundary layer on the limiting-current density and on the relative transport numbers of ions. *Bull. Chem. Soc. Jpn.* **1969**, *42*, 2741–2748. [CrossRef]
3. Tay, M.; Pourcelly, G.; Lebon, F.; Gavach, C. Polarization phenomena at the interfaces between an electrolyte solution and an ion exchange membrane. *J. Electroanal. Chem.* **1992**, *336*, 171–194.
4. Sistat, P.; Pourcelly, G. Chronopotentiometric response of an ion-exchange membrane in the underlimiting current-range. Transport phenomena within the diffusion layers. *J. Membr. Sci.* **1997**, *123*, 121–131. [CrossRef]

5. Barragán, V.M.; Ruiz-Bauzá, C. Current-voltage curves for ion-exchange membranes: A method for determining the liming current density. *J. Colloid Interface Sci.* **1998**, *205*, 365–373. [CrossRef] [PubMed]
6. Ślęzak, A.; Ślęzak, I.H.; Ślęzak, K.M. Influence of the concentration boundary layers on membrane potential in a single membrane system. *Desalination* **2005**, *184*, 113–123. [CrossRef]
7. Belova, E.I.; Lopatkova, G.Y.; Pismenskaya, N.D.; Nikonenko, V.; Larchet, C.; Pourcelly, G. Effect of anion-exchange membrane surface properties on mechanisms of overlimiting mass transfer. *J. Phys. Chem. B* **2006**, *110*, 13458–13469. [CrossRef]
8. Larchet, C.; Nouri, S.; Auclair, B.; Dammak, L.; Nikonenko, V. Application of chronopotentiometry to determine the thickness of diffusion layer adjacent to an ion-exchange membrane under natural convection. *Adv. Colloid Interface Sci.* **2008**, *139*, 45–61. [CrossRef] [PubMed]
9. Tanaka, Y. Mass transfer in a boundary layer and in an ion exchange membrane: Mechanism of concentration polarization and water dissociation. *Russ. J. Electrochem.* **2012**, *48*, 665–681. [CrossRef]
10. Moya, A.A.; Sistat, P. Chronoamperometric response of ion-exchange membrane systems. *J. Membr. Sci.* **2013**, *444*, 412–419. [CrossRef]
11. Nikonenko, V.; Kovalenko, A.; Urtenov, M.K.; Pismenskaya, N.D.; Han, J.; Sistat, P.; Pourcelly, G. Desalination at overlimiting currents: State-of-the-art and perspectives. *Desalination* **2014**, *342*, 85–106. [CrossRef]
12. Mareev, S.A.; Butylskii, D.Y.; Kovalenko, A.V.; Petukhova, A.V.; Pismenskaya, N.D.; Dammak, L.; Larchet, C.; Nikonenko, V. Accounting for the concentration dependence of electrolyte diffusion coefficient in the Sand and the Peers equation. *Electrochim. Acta* **2016**, *195*, 85–93. [CrossRef]
13. Wenten, I.G.; Khoiruddin, K.; Alkhadra, M.A.; Tian, H.; Bazant, M.Z. Novel ionic separation mechanisms in electrically driven membrane process. *Adv. Colloid Interface Sci.* **2020**, *284*, 102269. [CrossRef]
14. Barros, K.S.; Martí-Calatayud, M.C.; Scarazzato, T.; Bernardes, A.M.; Espinosa, D.C.R.; Pérez-Herranz, V. Investigation of ion-exchange membranes by means of chronopotentiometry: A compressive review on this highly informative and multipurpose technique. *Adv. Colloid Interface Sci.* **2021**, *293*, 102439. [CrossRef] [PubMed]
15. Stockmeier, F.; Schatz, M.; Habermann, M.; Linkhorst, J.; Mani, A. Direct 3D observation and unravelling of electroconvection phenomena during concentration polarization at in exchange membranes. *J. Membr. Sci.* **2021**, *640*, 119846. [CrossRef]
16. Batko, K.M.; Ślęzak-Prochazka, I.; Ślęzak, A.; Bajdur, W.M.; Ščurek, R. Modelling of the electrical membrane potential for concentration polarization conditions. *Entropy* **2022**, *24*, 138. [CrossRef]
17. Uzdenova, A. Ion transport in electromembrane system under the passage of direct current: 1D modelling approach. *Membranes* **2023**, *13*, 421. [CrossRef] [PubMed]
18. Uzdenova, A.M. Modeling of ion transport in a three-layer system with an ion-exchange membrane based on the Nernst Planck and displacement current equations. *Membr. Membr. Technol.* **2024**, *6*, 1–8. [CrossRef]
19. Nguyen, D.; Pham, V.-S. Modeling non-linear ion transport phenomena in ion-selective membranes: Three simplified models. *Sep. Purif. Technol.* **2024**, *333*, 125929. [CrossRef]
20. Zabolotsky, V.I.; Vasil'eva, V.I.; Lebedev, K.A.; Akberova, E.M.; Achoh, A.R.; Davidov, D.V.; Loza, S.A.; Dobryden, S.V. Towards optimized cation-exchange membranes for overlimiting current electrodialysis: Correlation between size of resin particles in membranes and mechanism of ion transport through them. *Chem. Eng. Sci.* **2024**, *295*, 120137. [CrossRef]
21. Chehayeb, K.M.; Lienhard, V.J.H. Entropy generation analysis of electrodialysis. *Desalination* **2017**, *414*, 184–198. [CrossRef]
22. Generous, M.M.; Qasen, N.A.A.; Zubair, S.M. Entropy generation analysis of electrodialysis desalination using multi-component groundwater. *Desalination* **2021**, *500*, 114858. [CrossRef]
23. Ślęzak, A.; Ślęzak-Prochazka, I.; Grzegorczyn, S.; Jasik-Ślęzak, J. Evaluation of S-entropy production in a single-membrane system in concentration polarization conditions. *Transp. Porous Med.* **2017**, *116*, 941–957. [CrossRef]
24. Kjelstrup, S.; Bedeaux, D.; Johannessen, E.; Gross, S. *Non-Equilibrium Thermodynamics for Engineers*, 2nd ed.; Word Scientific: Hackensack, NJ, USA, 2017.
25. Kjelstrup, S.; Bedeaux, D. *Non-Equilibrium Thermodynamics of Heterogeneous Systems*; Series on Advances in Statistical Mechanics-Volume 16; Word Scientific: Hackensack, NJ, USA, 2008.
26. Naterer, G.F.; Tokarz, C.D.; Avsec, J. Fuel cell entropy production with ohmic heating and diffusive polarization. *Int. J. Heat Mass Tranfer.* **2006**, *49*, 2673–2683. [CrossRef]
27. Kristiansen, K.; Barragán, V.M.; Kjelstrup, S. Thermoelectric power of ion exchange membrane cells relevant to reverse electrodialysis plants. *Phys. Rev. Appl.* **2019**, *11*, 044037. [CrossRef]
28. Kujawski, W.; Yaroshchuk, A.; Zholkovskiy, E.; Koter, I.; Koter, S. Analysis of membrane transport equations for reverse electrodialysis (RED) using irreversible thermodynamics. *Int. J. Mol. Sci.* **2020**, *21*, 6325. [CrossRef] [PubMed]
29. Magnanelli, E.; Wilhelmsen, Ø.; Johannessen, E.; Kjelstrup, S. Energy efficient design of membrane processes by use of entropy production minimization. *Comp. Chem. Eng.* **2018**, *117*, 105–116. [CrossRef]
30. Helfferich, F. *Ion Exchange*; Dover Publication, INC: Garden City, NY, USA, 1995.
31. Levich, V.G. *Physicochemical Hydrodynamics*; Prentice-Hall: New York, NY, USA, 1962.

32. Strathamann, H. *Ion-Exchange Membrane Separation Processes*; Membrane Science and Technology Series Volume 9; Elsevier: Amsterdam, The Netherlands, 2004.
33. Katchalsky, A.; Curran, P.F. *Thermodynamics in Biophysics*; Harvard University Press Cambridge: Cambridge, MA, USA, 1967.
34. Brockris, J.O.M.; Reddy, A.K.N. *Modern Electrochemistry*; Plenum Press: New York, NY, USA, 1970; Volume 1.
35. Lobo, V.M.; Quaresma, J.L. *Handbook of Electrolyte Solutions*; Parts A and B; Elsevier: Amsterdam, The Netherlands, 1989.
36. Robinson, R.A.; Stokes, R.H. *Electrolyte Solutions*, 2nd ed.; Dover Publication Inc.: Garden City, NY, USA, 2003.
37. Avramov, S.G.; Lefterova, E.; Penchev, H.; Sinigersky, V.; Slavcheva, E. Comparative study on the proton conductivity of perfluorosulfonic and polybenzimidazole based polymer electrolyte membranes. *Bulg. Chem. Commun.* **2016**, *48*, 43–50.
38. Hongsirikarn, K.; Goodwin, J.G.; Greenway, S.; Creager, S. Effect of cations (Na^+, Ca^{2+}, Fe^{3+}) on the conductivity of a Nafion membrane. *J. Power Sources* **2010**, *195*, 7213–7220. [CrossRef]
39. Slade, S.; Campbell, S.A.; Ralph, T.R.; Walsh, F.C. Ionic conductivity of an extruded Nafion 1100 EW series of membranes. *J. Electrochem. Soc.* **2002**, *149*, A1556–A1564. [CrossRef]
40. Choi, J.-H.; Park, J.-S.; Moon, S.-H. Direct measurement of concentration distribution within the boundary layer of an ion-exchange membrane. *J. Colloid Int. Sci.* **2002**, *251*, 311–317. [CrossRef]
41. Shaposhnik, V.A.; Grigorchuk, O.V.; Korzhov, E.N.; Vasil'Eva, V.I.; Klimov, V.Y. The effect of ion-conducting spacers on mass transfer-numerical analysis of laser interferometry. *J. Membr. Sci.* **1998**, *139*, 85–96. [CrossRef]

entropy

MDPI

Article

Transport Numbers and Electroosmosis in Cation-Exchange Membranes with Aqueous Electrolyte Solutions of HCl, LiCl, NaCl, KCl, $MgCl_2$, $CaCl_2$ and NH_4Cl

Simon B. B. Solberg *, Zelalem B. Deress, Marte H. Hvamstad and Odne S. Burheim

Department of Energy and Process Engineering, Norwegian University of Science and Technology, NO-7491 Trondheim, Norway; zelalem.b.deress@ntnu.no (Z.B.D.); odne.s.burheim@ntnu.no (O.S.B.)
* Correspondence: simon.b.b.solberg@ntnu.no

Abstract: Electroosmosis reduces the available energy from ion transport arising due to concentration gradients across ion-exchange membranes. This work builds on previous efforts to describe the electroosmosis, the permselectivity and the apparent transport number of a membrane, and we show new measurements of concentration cells with the Selemion CMVN cation-exchange membrane and single-salt solutions of HCl, LiCl, NaCl, $MgCl_2$, $CaCl_2$ and NH_4Cl. Ionic transport numbers and electroosmotic water transport relative to the membrane are efficiently obtained from a relatively new permselectivity analysis method. We find that the membrane can be described as perfectly selective towards the migration of the cation, and that Cl^- does not contribute to the net electric current. For the investigated salts, we obtained water transference coefficients, t_w, of 1.1 ± 0.8 for HCl, 9.2 ± 0.8 for LiCl, 4.9 ± 0.2 for NaCl, 3.7 ± 0.4 for KCl, 8.5 ± 0.5 for $MgCl_2$, 6.2 ± 0.6 for $CaCl_2$ and 3.8 ± 0.5 for NH_4Cl. However, as the test compartment concentrations of LiCl, $MgCl_2$ and $CaCl_2$ increased past 3.5, 1.3 and 1.4 mol kg^{-1}, respectively, the water transference coefficients appeared to decrease. The presented methods are generally useful for characterising concentration polarisation phenomena in electrochemistry, and may aid in the design of more efficient electrochemical cells.

Keywords: ion-exchange membrane; electrodialysis; permselectivity; electroosmosis; transport number

1. Introduction

Ion-exchange membranes are essential components of reverse electrodialysis power plants; they provide the means to mix fresh water and seawater in a controlled manner such that electric power can be extracted from the cell. The opposite process, electrodialysis, is also showing promise for use in wastewater treatment [1–3], water desalination [4], and demineralisation in food processing [5–13]. Cheap, selective membranes with low resistance are necessary for these processes [14,15]. As such, it is crucial to have accurate experimental methods and meaningful performance indicators for describing the transport phenomena in ion-exchange membranes.

Permselectivity is frequently used to describe a membrane's selectivity towards an ionic species, but several different definitions are used in the literature [16]. It has been shown that a particularly useful definition is the ratio of the measured electric potential across an ion-exchange membrane divided by the chemical potential difference of the salt in question, since it directly relates the permselectivity to the ionic transport numbers and the neutral component transference coefficients [17–20]. In aqueous electrolyte mixtures, the water transference coefficient is of particular interest, since it describes the electroosmotic transport of water dragged along as a hydration shell around charge-transporting ions [21,22]. The electroosmotic transport affects the amount of work involved in the movement of ions through the ion-exchange membrane, usually lowering the permselectivity and the reversible voltage difference across the membrane [19,23]. Both the transport numbers and the transference coefficients of the membrane are useful concepts for the description of ion-exchange membrane selectivity.

Water transference coefficients, t_w, and electroosmosis have been investigated for perfluorinated Nafion membranes in single-salt electrolyte mixtures, mainly by use of streaming potential measurements [24–29]. Examples of measured water transference coefficients for the Nafion 117 membrane are $t_w \approx 2.6$, ≈ 5 and ≈ 15 for chloride salts of H^+, K^+ and Mg^{2+}, respectively. This corresponds to 2.6, 5 and 30 moles of water per mole of cation transporting charge [21]. Clearly, the magnitude of electroosmotic transport can be large. Hydrophobic and hydrophilic polymers will likely facilitate different water transference coefficients, and these factors may be important for the design of efficient membranes for a particular process. Ionic transport numbers and water transference coefficients have also been extracted from electric potential differences across membranes through various methods of analysis [17,30]. We shall here further contribute to the discussion around strategies for analysing membrane potentials, with a focus on transport numbers and transference coefficients, by using a relatively novel permselectivity method.

We present new measurements of membrane potentials of the commercial cation-exchange membrane Selemion CMVN. These new electric potential measurements are for membranes in contact with aqueous single-salt solutions of HCl, LiCl, NaCl, $MgCl_2$, $CaCl_2$ and NH_4Cl. Furthermore, we shall use the novel permselectivity analysis method communicated by Solberg et al. [19] in order to determine ionic transport numbers and water transference coefficients for the Selemion CMVN equilibrated with these salts. The analysis method is derived from the theory of non-equilibrium thermodynamics, in which the fluxes and driving forces are obtained from the entropy production.

2. Thermodynamic Framework

2.1. The Measurable Electric Potential

The electric potential difference across the cation-exchange membrane was measured in a concentration cell, as illustrated in Figure 1. We used two Ag/AgCl electrodes, and the two electrolyte solution compartments with different salt concentrations were thoroughly stirred. The main contributions to the total cell voltage were therefore due to the salt concentration gradient across the membrane, as well as a small contribution from the differences in chemical potentials of the solid phases of the electrodes (the electrode bias) [21]. Measurements were taken under isobaric and isothermal conditions, and we aimed to achieve a pseudo-steady state where the membrane interfaces were in a Donnan equilibrium. Transport phenomena in the ion-exchange membrane of such a cell for a laboratory frame of reference were thoroughly described by Solberg et al. using non-equilibrium thermodynamics [20]. We repeat only the essential parts here.

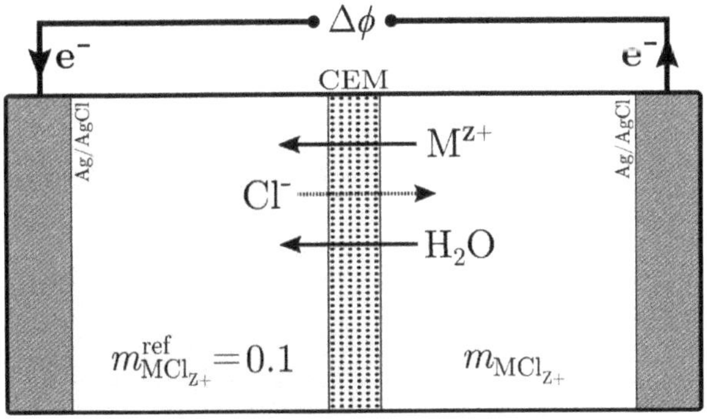

Figure 1. Sketch of the concentration cell used for the cation-exchange membrane (CEM) electric potential measurements, $\Delta\phi$. The composition is described by the molality, $m_{\mathrm{MCl}_{z+}}$, and the superscript "ref" denotes the reference compartment which always has a salt concentration of $m^{\mathrm{ref}}_{\mathrm{MCl}_{z+}} = 0.1$. Arrows show the ionic species transport directions that contribute to the electron direction displayed and the measured net electric potential.

Under open-circuit conditions, i.e., negligible electric current density, the measurable electric potential difference, $\Delta\phi$, across the ion-exchange membrane is [20,21]:

$$-F\Delta\phi = \frac{t_{\mathrm{M}^{z+}}}{z^{+}}\Delta\mu_{\mathrm{MCl}_{z+}} + t_w \Delta\mu_w \tag{1}$$

where F is the Faraday constant, $t_{\mathrm{M}^{z+}}$ and z^{+} are the transport number and charge of the cation and t_w is the water transference coefficient. Furthermore, $\Delta\mu_w$ is the chemical potential difference of water, and $\Delta\mu_{\mathrm{MCl}_{z+}}$ is the chemical potential difference of the neutral salt of the M^{z+} cation consisting of z^{+} chloride ions. The transport number and the water transference coefficient take on a mean value on the intervals $\Delta\mu_{\mathrm{MCl}_{z+}}$ and $\Delta\mu_w$, respectively. They are defined as:

$$t_{\mathrm{M}^{z+}} = z^{+}F\left(\frac{J_{\mathrm{M}^{z+}}}{j}\right)_{d\mu=0}, \qquad t_w = F\left(\frac{J_w}{j}\right)_{d\mu=0} \tag{2}$$

where $j = F(z^{+}J_{\mathrm{M}^{z+}} - J_{\mathrm{Cl}^{-}})$ is the current density and $J_{\mathrm{M}^{z+}}$, $J_{\mathrm{Cl}^{-}}$ and J_w are the molar fluxes of cations, chloride and water, respectively. The subscript $d\mu = 0$ refers to a uniform composition. It is essential for the definition of the transport numbers, but these quantities retain their physical meaning even outside of uniform composition. We make a distinction between the ionic transport number and the water transference coefficient according to the naming convention used by Kjelstrup and Bedeaux [21]. Transport numbers are here reserved for ions contributing to the electric current, while transference coefficients describe the amount of a neutral component that is indirectly transported by the current. Electroosmosis is the relevant example, describing the amount of water coupled to the movement of migrating ions.

We evaluate the two chemical potentials in terms of one state variable, i.e., the molality. It is related to the number of moles of salt, $n_{\mathrm{MCl}_{z+}}$, and the moles of water, n_w, through:

$$m_{\mathrm{MCl}_{z+}} = \frac{n_{\mathrm{MCl}_{z+}}}{M_w n_w} \tag{3}$$

where M_w is the molecular weight of water. With this state variable choice, in which the molality is the only independent variable, we define the chemical potentials as [31]:

$$\mu_{MCl_{z+}} = \mu^0_{MCl_{z+}} + RT \ln \left(\frac{m_{M^{z+}} m_{Cl^-}^{\nu+} \gamma^{\nu}_{MCl_{z+}}}{m^0_{M^{z+}} \left(m^0_{Cl^-}\right)^{\nu+}} \right)$$
$$\mu_w = \mu^0_w + RT \ln a_w \tag{4}$$

where R is the gas constant, T is the absolute temperature, $\gamma_{MCl_{z+}}$ is the mean molal activity coefficient, a_w is the water activity and the factors $\nu = \nu^+ + \nu^-$ represent the total number of ionic species, and the individual cationic and anionic species produced by the dissolution of a salt. Here, $\nu^- = 1$ since all investigated components are chloride salts. The superscript 0 denotes the reference state of the chemical potentials, here referring to infinitely dilute salt in water.

We shall now use a permselectivity definition consisting of the measured electric potential divided by the chemical potential difference of the salt. This is a common way to quantify a membrane "permselectivity" metric [18,19,32] that has the added benefit of scaling the measured electric potential such that large electric potentials do not introduce bias in the regression analysis. The resulting permselectivity can be understood in terms of either the apparent transport number or via the permselectivity analysis shown by Solberg et al. [19].

2.2. The Apparent Transport Number

The chemical potential differences of water and salt in the two-component mixture are related through the Gibbs–Duhem equation: $d\mu_w = -m_{M^{z+}} M_w d\mu_{MCl_{z+}}$ [21]. Here, $m_{MCl_{z+}} = m_{M^{z+}}$ denotes the cation molal concentration. This relation is frequently used to reduce the number of driving forces that need to be calculated [17], such that the measurable electric potential becomes:

$$-F\Delta\phi = \left(\frac{t_{M^{z+}}}{z^+} - \bar{m}_{M^{z+}} M_w t_w \right) \Delta\mu_{MCl_{z+}} \tag{5}$$

The overhead bar, $\bar{m}_{M^{z+}}$, signifies that the molal concentration takes on a mean value as the expression is integrated over the $\Delta\mu_{MCl_{z+}}$ interval. This mean value need not be the arithmetic mean of the reference and test concentrations used. The term wrapped in parentheses is frequently referred to as the apparent transport number, $t^{app}_{M^{z+}}$, and it is found by evaluating the permselectivity according to:

$$t^{app}_{M^{z+}} = -F \left(\frac{\Delta\phi}{\Delta\mu_{MCl_{z+}}} \right) = \left(\frac{t_{M^{z+}}}{z^+} - \bar{m}_{M^{z+}} M_w t_w \right) \tag{6}$$

This equation is sometimes also referred to as the Scatchard equation [33,34]. The method states that a decrease in the membrane permselectivity is due to a reduction in the apparent transport number, i.e., the ionic transport number corrected for water transport.

2.3. The Permselectivity

The permselectivity may also be evaluated without invoking the Gibbs–Duhem equation [19]. We have:

$$\alpha = -F \left(\frac{\Delta\phi}{\Delta\mu_{MCl_{z+}}} \right) = \frac{t_{M^{z+}}}{z^+} + t_w \left(\frac{\Delta\mu_w}{\Delta\mu_{MCl_{z+}}} \right) \tag{7}$$

where α is the permselectivity. The permselectivity and the apparent transport number are, in fact, the same quantity, but we use separate symbols to denote the two different analysis methods. For the permselectivity analysis, the transport number, $t_{MCl_{z+}}$, and the transference coefficient, t_w, are used as constant regression coefficients.

3. Data Reduction

The activity coefficient data necessary for the calculation of the chemical potentials of salt and water were gathered from the literature [31,35] and fitted to the Pitzer equations [36]. See the Supplementary Materials for a summary of the values used in the analysis.

3.1. The Apparent Transport Number

The apparent transport number of the cation-exchange membrane was found by first fitting an empirical function to the measured electric potential. The reported electric potentials are monotonically increasing functions of the chemical potential difference of the salt. Non-linear S-shaped functions such as the hyperbolic tangent or the error function are therefore useful. Other choices, such as polynomial functions, could introduce extrema in the apparent transport number function. We used the following function:

$$-F\Delta\phi = \frac{A\Delta\mu_{MCl_{z+}}}{\sqrt{1+B\left(\Delta\mu_{MCl_{z+}}\right)^2}} + \epsilon_\phi \tag{8}$$

where A and B are regression coefficients, and ϵ_ϕ is the regression residual error. This provides an empirical relation for the apparent transport number:

$$t^{app}_{M^{z+}} = -F\left(\frac{\Delta\phi}{\Delta\mu_{MCl_{z+}}}\right) = \frac{A}{\sqrt{1+B\left(\Delta\mu_{MCl_{z+}}\right)^2}} \tag{9}$$

3.2. The Permselectivity

The permselectivity was evaluated in terms of a linear function based on Equation (7) using the ionic transport number and the water transference coefficient as the regression coefficients.

$$\alpha = -F\left(\frac{\Delta\phi}{\Delta\mu_{MCl_{z+}}}\right) = C + D\left(\frac{\Delta\mu_w}{\Delta\mu_{MCl_{z+}}}\right) + \epsilon_\alpha \tag{10}$$

where ϵ_α is the regression residual, while $C = t_{M^{z+}}/z^+$ and $D = t_w$ are the regression coefficients. In the case of the ionic transport numbers or the water transference coefficients varying with the salt concentration, the regression model will be found lacking. This was the case for LiCl, $MgCl_2$ and $CaCl_2$, when the concentration of the salt in the test compartment increased past 3.5, 1.3 and 1.4 mol kg^{-1}, respectively. Therefore, only measurements with concentrations below these limits, corresponding to a driving force of $\Delta\mu_{MCl_{z+}} = 20$ kJ mol^{-1}, were used in the permselectivity analysis.

The resulting ionic transport numbers and water transference coefficients were then used together with Equation (9) to estimate the mean molal concentration across the two electrolyte solutions according to the following:

$$\bar{m}_{M^{z+}} = \frac{\frac{t_{M^{z+}}}{z^+} - t^{app}_{M^{z+}}}{M_w t_w} \tag{11}$$

The resulting mean concentration represents the contribution from the chemical potential difference of water, and it may deviate from the arithmetic mean of the test and reference solution concentrations.

4. Experimental Methods

Aqueous electrolyte solutions of chloride salts were prepared gravimetrically using demineralised water with an electrical conductivity of 5.5 $\mu S\ m^{-1}$. Of the alkali metal salts, LiCl (99%, Thermo Fisher Scientific, Waltham, MA, USA) and NaCl ($\geq$99%, Sigma-Aldrich, St. Louis, MO, USA) were investigated. The KCl results from the same concentration cell and cation-exchange membrane were obtained by Solberg [20], and they were included in this analysis. Of the alkaline earth metals, $MgCl_2 \cdot 6H_2O$ (AnalaR NORMAPUR, VWR Chemicals, Leicestershire, UK) and $CaCl_2 \cdot 2H_2O$ (AnalaR NORMAPUR, VWR Chemicals) were investigated. In addition, HCl (32%w/w EMSURE, Merck, Darmstadt, Germany) and NH_4Cl (ACS, Merck) were used. The molal concentrations used for the test solutions are shown in Table 1.

Table 1. The molal concentrations of the test solutions; the reference compartment molality is 0.1 mol kg^{-1} for all salts. Concentration cell measurements with KCl and the same cation-exchange membrane were taken from Solberg et al. [19]. Additional information for the relevant salts, such as thermodynamic data and calculated chemical potentials, is available in the Supplementary Materials.

HCl	LiCl	NaCl	KCl [19]	$MgCl_2$	$CaCl_2$	NH_4Cl
$m_{MCl_{z+}}$ (mol kg^{-1})						
0.2	0.5	0.5	0.5	0.2	0.2	0.2
0.5	1	1	1	0.5	0.5	0.5
0.9	2	2	2	1.0	1.0	1
1.7	4	4	3	1.9	2.0	2
4.1	6	6	4.7	4.8	5.8	6
-	8	-	-	-	-	-

The Selemion CMVN cation-exchange membrane (AGC Engineering Co., Ltd., Chiba, Japan) was delivered in the Na^+ form. Appropriately sized pieces of the membrane were stored in corresponding electrolyte solutions of 0.1 mol kg^{-1} for at least one week before use. The electrolyte solution was refreshed once during this week to ensure the absorption and equilibration of the target cation. A thickness of roughly 100 μm, a water uptake of 0.182 g $g_{polymer}^{-1}$ and an ion-exchange capacity of 1.40 meq $g_{polymer}^{-1}$ are typical for these commercial membranes [37].

Ag/AgCl reference electrodes were fabricated from two polished silver rods of suitable length ($\geq$99.95% Ag, 5 mm diameter, Thermo Fisher Scientific). The silver rods were electroplated in a 0.1 mol kg^{-1} diluted solution of HCl (32%w/w EMSURE, Merck) in order to form a silver chloride coating. An electric current of 1 mA was passed using Pt as the counter electrode over the course of 12 h. The Ag/AgCl electrodes were stored short-circuited in the aqueous 0.1 mol kg^{-1} HCl solution when not in use in order to minimise the bias potential between them. A Gamry Interface 5000E potentiostat was used for all logging of electric potentials and electroplating (with a declared standard deviation, σ, for the measured voltage in mV of $2\sigma = 1 + 0.003 \times \Delta\phi$). The bias potential was found as the lower limit of the open-circuit potential measured over a period of 300–600 s between the two reference electrodes exceeded 1 mV when situated in the same 0.1 mol kg^{-1} electrolyte solution. It was measured directly after measuring the electric potential of the concentration cell, and some (concentration-dependent) capacitance was evident as an initial quick drop in voltage and stabilisation around the true electrode bias potential. The silver chloride coating was refreshed if the measured electrode bias potential exceeded 1 mV.

The concentration cell used for the measurement of electric potentials across the cation-exchange membrane, illustrated in Figure 1, consisted of two compartments of roughly 300 mL. Each compartment was filled with 200 mL of electrolyte solution. The left-hand side of Figure 1 is denoted as the reference solution, while the right-hand side is referred to as the test solution. A cation-exchange membrane with an exposed area of 7.10 cm^2 separated the two compartments, preventing mixing of the two electrolyte solutions. The left-hand side of the cell always held a concentration of 0.1 mol kg^{-1} of the salt to be tested. The electrolyte solution concentrations on the right-hand side were varied. Both compartments were thoroughly stirred with magnetic stirrers to minimise the contribution from concentration polarisation in the boundary layers adjacent to the membrane. With one Ag/AgCl reference electrode in each compartment, the open-circuit potential of the concentration cell was measured over the course of at least 1800 s. Each concentration level was measured two times with fresh electrolyte solutions. These membrane potentials were measured under isobaric conditions (maintained by ensuring equal heights for the two solutions) at an ambient temperature of 295 $\pm$ 1 K, and the final value was taken as the mean of a stable reading of at least 900 s. A negative drift was generally not present in the measured potential, indicating that diffusive transport of both water (osmosis) and/or salt did not significantly alter the electrolyte solution concentrations over the course of the measurement. HCl, however, did show, in initial tests, a clearly visible drift that was more pronounced at high concentration gradients across the membrane. Therefore, 10 cation-exchange membranes (with a total thickness of roughly 1 mm) were used for the reported results HCl, while 1 membrane was sufficient for the other salts.

5. Results and Discussion

The measured electric potentials of the concentration cell, as illustrated in Figure 1, were dominated by the contribution from the concentration gradient across the cation-exchange membrane. In the observed electric potential, a small contribution was present due to differences in the chemical potentials of the solid phases of the electrodes (the electrode bias). This contribution was measured and subtracted from the reported electric potentials, but it was generally below 1 mV. The measured potentials are summarised together with thermodynamic data for the salts in the Supplementary Materials. Concentration gradients in the bulk electrolyte solutions were minimised by rigorous stirring, such that its contribution to the observed potential was negligible. The variance of each individual measurement was low, such that the double standard deviations fell within the scatter plot boxes and error bars were therefore left out of subsequent figures.

The methods applied in this work can be leveraged to measure particularly useful performance indicators for ion-exchange membranes; i.e., the transport numbers and electroosmosis. Such results may later be combined with measurements of the electric conductivity and diffusion coefficients to fully characterize the membrane transport phenomena in terms of the phenomenological coefficients [21]. This may then be compared with equilibrium properties for different membranes, such as water uptake, polymer type, ion-exchange capacity and the nature of the membrane fixed ion groups, in order to better understand these materials. Here, however, we focus on the determination of transport numbers and transference coefficients.

5.1. The Apparent Transport Number

The electric potential across the Selemion CMVN cation-exchange membrane is shown for HCl, LiCl, NaCl and $MgCl_2$ in Figure 2a. The other salts have been left out of this figure to enhance legibility, but equivalent trends are observed. The non-linear regression provides a good description of the electric potential as a function of the chemical potential, but there are still data trends that are unexplained by these non-linear models. This is illustrated in Figure 2b, where the regression residuals of NaCl and $MgCl_2$ are shown. The variance in repeated measurements is small, but residuals vary much more as the experimental conditions are varied. Errors in the calculation of the chemical potentials, where both activity coefficients and the molal concentration carry some uncertainty, could contribute to the model's inability to explain all trends. Nevertheless, the regression residuals are below 2.5% of the electric potential in all cases, such that the model provides a good estimate of the apparent transport numbers of the salts.

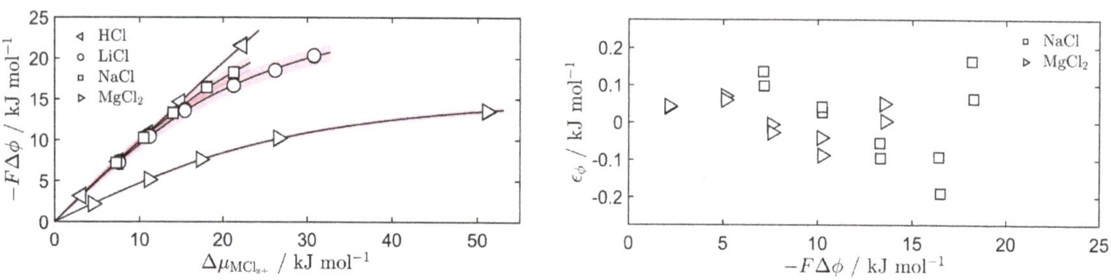

(**a**) Non-linear regression (**b**) Non-linear regression residuals

Figure 2. (**a**) Non-linear regression (solid lines) of the measured electric potential, $\Delta\phi$, as a function of the chemical potential difference across the membrane, $\Delta\mu_{MCl_{z+}}$, of the salts HCl (left-facing triangles), LiCl (circles), NaCl (squares) and $MgCl_2$ (right-facing triangles). Red shaded areas show the 95% confidence interval of the regression curves. (**b**) The regression residuals, ϵ_ϕ, of the non-linear curves of NaCl and $MgCl_2$ illustrating the difference in variance of the repeated measurements compared to between different measurements.

Individual salt apparent transport numbers, found from the non-linear regression models divided by the chemical potential difference of the salts, are shown in Figure 3. For salts of monovalent cations, they trend toward $t^{app}_{MCl_{z+}} = 1$ as the chemical potential difference across the membrane moves toward zero. They trend toward $t^{app}_{MCl_{z+}} = 0.5$ for divalent cation salts, since the cation valency is included in this apparent transport number definition. This indicates perfect cation selectivity for the membrane, $t_{M^{z+}} = 1$ and $t_{Cl^-} = 0$, in the case where the whole concentration cell has uniform composition. The subsequent decrease in the apparent transport number as the salt's driving force increases can be understood as either a reduction in the cation selectivity, an increased influence of water transference, or a combination of both.

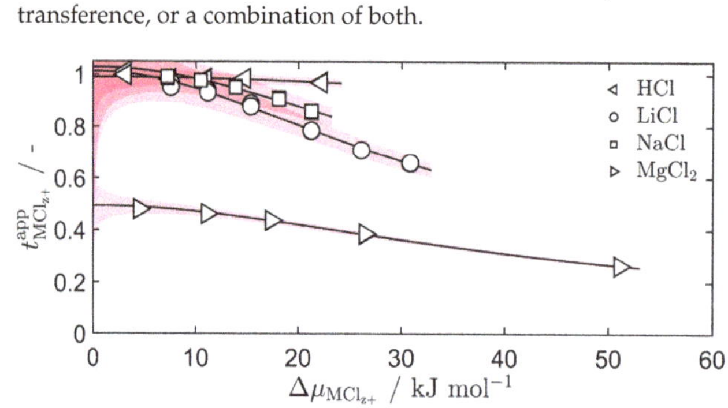

Figure 3. The apparent transport numbers, $t^{app}_{MCl_{z+}}$, as a function of the chemical potential difference across the membrane, $\Delta\mu_{MCl_{z+}}$, of the salts HCl (left-facing triangles), LiCl (circles), NaCl (squares) and $MgCl_2$ (right-facing triangles). Red shaded areas show the 95% confidence interval of the regression curves.

The physical significance of the water transference is understood as follows. Consider a spatial direction in which a concentration gradient of one aqueous salt is positive across the membrane, leading to a positive chemical potential gradient and a positive value of $-F\Delta\phi$. This means that the concentration cell can perform useful work in the external circuit, as long as the concentration difference persists. The chemical potential gradient of the water, however, is negative in this direction. Any water molecules transferred along with migrating cations—that is, electroosmosis—are transported in a direction that increases their chemical potential. This must decrease the available energy from the transport of cations from high to low concentration. By using Equation (1), this effect is fully captured by $t_w \Delta\mu_w$. By applying the Gibbs–Duhem equation, resulting in Equation (5), the water transference effect is instead captured by the apparent transport number, $t^{\mathrm{app}}_{\mathrm{MCl}_{z+}}$. When electric current is applied to the cell, desalinating one compartment by driving ions from low to high concentration, the effect is reversed.

It is important to note that the ionic transport numbers, $t_{\mathrm{M}^{z+}} + t_{\mathrm{Cl}^-} = 1$, are independent of any diffusive transport that is not accompanied by a net electric current in the external circuit. This means that it is possible for Cl^- to leak through the membrane in the same direction as M^{z+} even if $t_{\mathrm{M}^{z+}} = 1$, but such leakage must be accompanied by an equal amount of M^{z+}/z^+ such that there is no net charge transport.

The exact magnitude of the water transference coefficient is difficult to extract from the apparent transport number, since it requires knowledge of the mean cation concentration across the $\Delta\mu_{\mathrm{MCl}_{z+}}$ interval. This value may be different from the arithmetic mean of the two external solution concentrations. In the continuation of the analysis, it is therefore useful to apply the permselectivity model from Equation (7). This may allow for the direct identification of membrane water transference coefficients.

5.2. The Membrane Permselectivity

The permselectivity values of all of the investigated salts are shown in Figure 4. All monovalent cation salts, except for LiCl, show a linear permselectivity decrease as the ratio of chemical potentials is increased, and a lack-of-fit statistic [19,38] shows that statistically significant trends have been captured. It is possible that this decrease is a result of a decrease in cation selectivity; i.e., $t_{\mathrm{M}^{z+}} \neq 1$. However, a simpler explanation is that the permselectivity follows Equation (7) with constant transport numbers, where t_w is the slope of the linear regression and $t_{\mathrm{M}^{z+}}/z^+$ is the y-axis intercept. The obtained regression values are presented in Table 2. This model can fully explain the variation in the membrane permselectivity for most salts, but deviations from linearity arise for hygroscopic salts.

Table 2. Ionic transport numbers, $t_{\mathrm{M}^{z+}}$, and water transference coefficients, t_w, from the linear region of the permselectivity measured for the Selemion CMVN cation-exchange membrane. The reported values for LiCl, $\mathrm{MgCl_2}$ and $\mathrm{CaCl_2}$ are valid for test compartment salt concentrations below 3.5, 1.3 and 1.4 mol kg^{-1}, respectively.

	$t_{\mathrm{M}^{z+}}$	t_w
HCl	1.00 (±0.01)	1.1 (±0.8)
LiCl	1.00 (±0.01)	9.2 (±0.8)
NaCl	1.01 (±0.03)	4.9 (±0.2)
KCl	0.99 (±0.01)	3.7 (±0.4)
$\mathrm{MgCl_2}$	1.00 (±0.01)	8.5 (±0.5)
$\mathrm{CaCl_2}$	1.00 (±0.01)	6.2 (±0.6)
$\mathrm{NH_4Cl}$	1.00 (±0.01)	3.8 (±0.5)

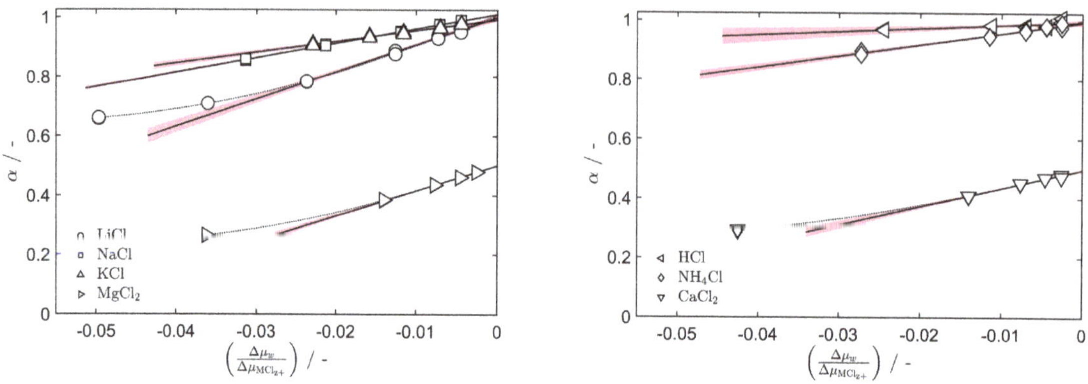

(**a**) LiCl (circles), NaCl (squares), KCl (triangles) and $MgCl_2$ (right-facing triangles).

(**b**) HCl (left-facing triangles), NH_4Cl (diamonds) and $CaCl_2$ (down-facing triangles).

Figure 4. Linear regression (solid lines) of the permselectivity, α, as a function of the chemical potential ratio, $\Delta\mu_w/\Delta\mu_{MCl_{z+}}$ in the linear (low-salt-concentration) region. Red shaded areas show the 95% confidence interval of the linear regressions, and non-linear curves (dotted lines) show the deviation from linearity at high salt concentrations for some salts.

For the hygroscopic salts LiCl, $MgCl_2$ and $CaCl_2$, there is a systematic deviation from the linear behaviour when the chemical potential difference of the salt exceeds $\Delta\mu_{MCl_{z+}} = 20$ kJ mol^{-1}. With a reference compartment molality of 0.1 mol kg^{-1}, this corresponds to the test molalities of around 3.5, 1.3 and 1.4 mol kg^{-1} for LiCl, $MgCl_2$ and $CaCl_2$, respectively. Nevertheless, we used the linear region below this in order to estimate the ionic transport numbers and water transference coefficients, as shown in Table 2. The ionic transport numbers are close to and have double standard deviation intervals that include unity, indicating that the Selemion CMVN cation-exchange membrane is adequately described by perfect cation selectivity. Any reduction in the permselectivity in the linear region can instead be explained by the magnitude of electroosmosis.

The non-linear region of the LiCl, $MgCl_2$ and $CaCl_2$ curves in Figure 4 is more complicated. The slopes become shallower, which could in part be due to $t_{M^{z+}} \neq 1$. In that case, Cl^- will carry charge and water in the opposite direction of M^{z+}, which leads to a reduction in the net water transference coefficient that is observed. On the other hand, the behaviour could also be explained exclusively by a reduction in the water transference coefficient while $t_{M^{z+}} = 1$. It may be necessary to measure the salt concentration over time during electrodialysis operation to fully characterise the transport coefficients in this non-linear regime. Measuring the permselectivity with both the test and reference compartment in the highly concentrated regime for these hygroscopic salts is a natural choice for the continuation of this work. Such measurements could verify if the ionic transport number decreases at high mean concentrations, and contribute to the construction of models for the concentration dependencies of the transport numbers and transference coefficients.

The water transference coefficient of H^+, $t_w = 1.1 \pm 0.8$, suggests that migrating protons generally enter and leave the membrane as H_3O^+ ions. Furthermore, the low magnitude is similar to the results of Okada et al. for the Nafion 117 [25], and suggests that proton migration in the Selemion CMVN occurs in part via the Grotthuss mechanism as reported for the Nafion 117.

With the exception of HCl, there is a definite trend of decreasing water transference coefficients, t_w, as one moves down a periodic table group. The same trend was observed for Li^+, Na^+ and K^+ in Nafion 117 membranes by Okada et al. and Xie et al. [29,39], as well as for the polymeric ion-exchange membranes Neosepta CM2 by Larchet et al. [34], and for

MC-40, MC-41, MA-41 and similar types by Berezina et al. [40]. Berezina and coworkers measured the volume flow as electric current was passed through the membrane, but they seemingly did not correct for the effect of the partial molar volume of the dissolved salt [40]. The reported decreasing water transference coefficient as the external solution salt concentration increases may therefore be overestimated, especially at high concentrations of salt. Without correcting for the partial molar volume of the salt, the value obtained is instead a volumetric permeability, as calculated by Barragán et al. for the Nafion 117 [41]. Larchet and coworkers did correct for the partial molar volume of the salt, but scaled its contribution by the ionic transport number instead of the apparent transport number. This choice may explain the concentration dependency observed by Larchet et al. [34]. We did not observe such a concentration dependency for t_w of the Selemion CMVN with Na^+ and K^+, but a dependency appears to be present for Li^+, cf. Figure 4. Future studies should investigate the consistency of the water transference coefficient obtained from volume flow, streaming potential and electromotive force methods in more detail.

The water transference trend coincides with the measurements of the cation radii of hydration for bulk electrolyte solutions communicated by Nightingale [42]. The relationship is shown in Figure 5, where the results are also compared with the water transference coefficients obtained by Okada et al. and Xie et al. for the Nafion 117 membrane [29,39]. It is striking how the trends observed for the water transference coefficient of Selemion CMVN are also seen in the Nafion 117, but shifted to higher values. The water transference coefficient trend lines are seemingly discontinuous when going from monovalent to divalent cations, but this is mainly due to the water transference coefficient being defined per mole of electrons transferred in the external circuit. Evaluating z^+t_w yields a trend that monotonically increases as r_h increases, as shown in Figure 5b. Obtaining ionic transport numbers and water transference coefficients of many more types of ion-exchange membranes may aid in the development of general models for transport phenomena.

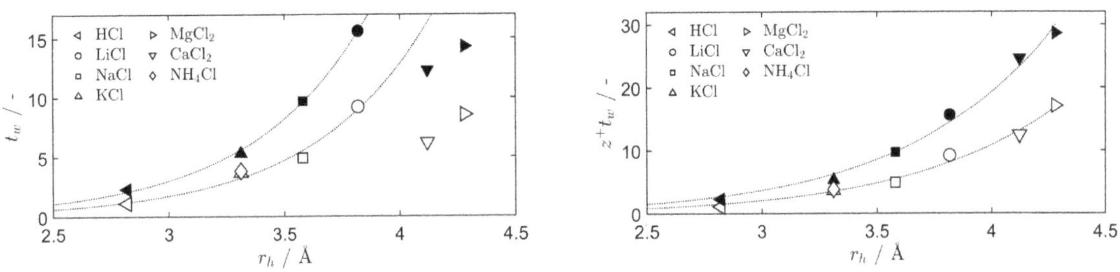

(a) The water transference coefficients. (b) Moles of water transported per cation.

Figure 5. The membrane water transference coefficients, t_w, from the linear region of the curves of Figure 4, compared to the hydrated cation radius in bulk solutions, r_h, communicated by Nightingale [42]. The filled symbols show the values for Nafion 117 [29,39], and the dotted lines show general trends.

The water transference coefficients were correlated with the membrane water uptake, which, for the Nafion 117 equilibrated with the relevant salts, was in the range of 10–25 mol $mol_{SO_3^-}^{-1}$ [26]. The Selemion CMVN has a water uptake of around 0.182 g $g_{polymer}^{-1}$ and an ion-exchange capacity of 1.40 mmol $g_{polymer}^{-1}$ [37], leading to 7 mol $mol_{SO_3^-}^{-1}$. Nafion 117 membranes typically experience swelling in aqueous solutions, while the Selemion CMVN polymer is cross-linked and relatively dense.

Both the water transference coefficient and the ionic mobility of species in the perfluorinated Nafion 117 membrane were investigated by Okada et al. [26]. The magnitude of the water transference coefficient was correlated with the ionic mobility of a species in

Nafion 117 membranes, where cations such as Li^+ showed lower mobility than Na^+ [26]. In the interior of the membrane, larger hydration shells can contribute to the friction a cation experiences during migration. The water transference coefficient, as well as the membrane hydrophobicity, can therefore be important parameters to optimise for the efficiency and selectivity of ion-exchange membranes.

The definition of the apparent transport number, Equation (6), suggests that the water transference coefficient can be found from the apparent transport number if the ionic transport number and the average cation molality are known. The permselectivity analysis further suggests that $t_{M^{z+}} = 1$, but the cation molality can be difficult to estimate because it takes on a mean value in the $\Delta\mu_{MCl_{z+}}$ interval. It is therefore of interest to estimate this mean cation molality and compare it to the arithmetic mean of the external concentrations.

The average cation concentration across the membrane was estimated using the apparent transport numbers together with the ionic transport number and water transference coefficient from the linear region of the permselectivity. Concentrations for some salts are shown in Figure 6, but the same trend was also observed for the remaining salts. For comparison, the Selemion CMVN features a water uptake of around 0.182 g $g_{polymer}^{-1}$, and an ion-exchange capacity of 1.40 mmol $g_{polymer}^{-1}$ [37]. This leads to a molal concentration of charged functional groups of around 7.7 mol kg $^{-1}$. It is evident that the mean concentration in question is not related to the internal concentration in the membrane, but rather a mean of the external states. We observe that the mean value of the cation molality across the membrane phase is different from and smaller than the arithmetic mean of the external concentrations. If the arithmetic mean of the two external concentrations is used for the determination of the water transference coefficient from the apparent transport number, the resulting values may be systematically smaller than the true water transference coefficient of the membrane.

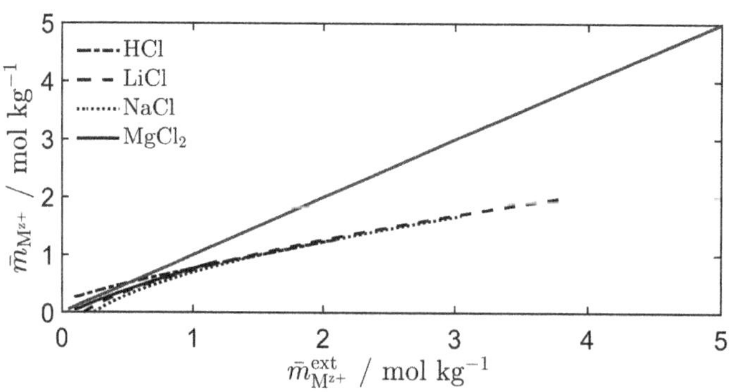

Figure 6. The estimated average cation molal concentration across the membrane, $\bar{m}_{M_{z+}}$, compared to the average external test solution concentration, $\bar{m}^{ext}_{M_{z+}}$. Curves are generated using the apparent transport numbers of Figure 3 together with the ionic transport number and water transference coefficient from the linear region of curves from Figure 4. A solid red line shows the arithmetic mean of the external concentrations.

6. Conclusions

This work builds on previous efforts to characterise ion selectivity and electroosmosis for ion-exchange membranes, in which non-equilibrium thermodynamics is used to describe the observed electric potential across a membrane in a concentration cell. New measurements for a cation-exchange membrane in contact with single-salt solutions of HCl, LiCl, NaCl, $MgCl_2$, $CaCl_2$ and NH_4Cl are presented and analysed.

Evaluating the permselectivity as a function of the ratio of the chemical potentials of water and salt, we find that the Selemion CMVN membrane appears to be perfectly selective to the cation in single-salt solutions, i.e., the transport of Cl^- does not contribute to the electric potential and the net current in the external circuit. Instead, any reduction in permselectivity can be explained by the influence of water transference. For the alkali metals LiCl, NaCl and KCl, we obtain water transference coefficients of $t_w = 9.2$ (± 0.8), 4.9 (± 0.2) and 3.7 (± 0.4), respectively. For the two investigated alkaline earth metals $MgCl_2$ and $CaCl_2$, coefficients of $t_w = 8.5$ (± 0.5) and 6.2 (± 0.6) are observed. For HCl and NH_4Cl, the values are $t_w = 1.1$ (± 0.8) and 3.8 (± 0.5). For LiCl, $MgCl_2$ and $CaCl_2$, the water transference coefficients appear to decrease as the test compartment concentrations are increased past 3.5, 1.3 and 1.4 mol kg^{-1}, respectively. This work suggests that the water transference coefficients remain constant for most salts, but future studies should investigate the concentration dependence of the water transference coefficients of the hygroscopic cations at high concentrations. The magnitude of the water transference coefficients affects the amount of work involved in the transport of ions across the membrane, and it is therefore an important parameter to consider for the design of highly selective and conductive ion-exchange membranes.

Supplementary Materials: The following supporting information can be downloaded at: https://www.mdpi.com/article/10.3390/e27010075/s1, Table S1: Reported measurement and thermodynamic data.

Author Contributions: Conceptualization, S.B.B.S. and O.S.B.; methodology, S.B.B.S.; software, S.B.B.S.; validation, S.B.B.S.; formal analysis, S.B.B.S.; investigation, S.B.B.S., M.H.H. and Z.B.D.; resources, O.S.B.; data curation, S.B.B.S.; writing—original draft preparation, S.B.B.S.; writing—review and editing, S.B.B.S., Z.B.D., M.H.H. and O.S.B.; visualization, S.B.B.S.; supervision, S.B.B.S. and O.S.B.; project administration, O.S.B.; funding acquisition, O.S.B. All authors have read and agreed to the published version of the manuscript.

Funding: S.B.B.S. thanks the Norwegian Research Council for support through the projects KeyTech-NeVe Chino (Project No. 304213), MoreIsLess (Project No. 324077) and 2ND LIFE (Project No. 320760).

Data Availability Statement: Dataset available on request from the authors.

Acknowledgments: The authors wish to thank Børre T. Børresen and Jonas Sottmann for contributing to the supervision of Marte H. Hvamstad during her master's thesis work.

Conflicts of Interest: The authors declare no conflicts of interest.

References

1. Bagastyo, A.Y.; Sinatria, A.Z.; Anggrainy, A.D.; Affandi, K.A.; Kartika, S.W.T.; Nurhayati, E. Resource recovery and utilization of bittern wastewater from salt production: A review of recovery technologies and their potential applications. *Environ. Technol. Rev.* **2021**, *10*, 294–321. [CrossRef]
2. Borges, F.; Roux-de Balmann, H.; Guardani, R. Investigation of the mass transfer processes during the desalination of water containing phenol and sodium chloride by electrodialysis. *J. Membr. Sci.* **2008**, *325*, 130–138. [CrossRef]
3. Zimmermann, P.; Wahl, K.; Tekinalp, Ö.; Solberg, S.B.B.; Deng, L.; Wilhelmsen, Ø.; Burheim, O.S. Selective recovery of silver ions from copper-contaminated effluents using electrodialysis. *Desalination* **2024**, *572*, 117108. [CrossRef]
4. Campione, A.; Gurreri, L.; Ciofalo, M.; Micale, G.; Tamburini, A.; Cipollina, A. Electrodialysis for water desalination: A critical assessment of recent developments on process fundamentals, models and applications. *Desalination* **2018**, *434*, 121–160. [CrossRef]
5. Bazinet, L. Electrodialytic phenomena and their applications in the dairy industry: A review. *Crit. Rev. Food Sci. Nutr.* **2005**, *45*, 307–326. [CrossRef] [PubMed]
6. Singlande, E.; Roux-de Balmann, H.; Lefevbre, X.; Sperandio, M. Improvement of the treatment of salted liquid waste by integrated electrodialysis upstream biological treatment. *Desalination* **2006**, *199*, 64–66. [CrossRef]

7. Galier, S.; Balmann, H.R. Demineralization of glucose solutions by electrodialysis: Influence of the ionic composition on the mass transfer and process performances. *Can. J. Chem. Eng.* **2015**, *93*, 378–385. [CrossRef]
8. Elisseeva, T.; Shaposhnik, V.; Luschik, I. Demineralization and separation of amino acids by electrodialysis with ion-exchange membranes. *Desalination* **2002**, *149*, 405–409. [CrossRef]
9. Shen, J.; Duan, J.; Liu, Y.; Lixin, Y.; Xing, X. Demineralization of glutamine fermentation broth by electrodialysis. *Desalination* **2005**, *172*, 129–135. [CrossRef]
10. Thang, V.H.; Koschuh, W.; Kulbe, K.D.; Novalin, S. Detailed investigation of an electrodialytic process during the separation of lactic acid from a complex mixture. *J. Membr. Sci.* **2005**, *249*, 173–182. [CrossRef]
11. Luis, A.; McClure, D.D.; Lim, K.; Leslie, G.; Coster, H.G.; Barton, G.W.; Kavanagh, J.M. Potential upgrading of bio-refinery streams by electrodialysis. *Desalination* **2017**, *415*, 20–28. [CrossRef]
12. Elmidaoui, A.; Lutin, F.; Chay, L.; Taky, M.; Tahaikt, M.; Hafidi, M.R. Removal of melassigenic ions for beet sugar syrups by electrodialysis using a new anion-exchange membrane. *Desalination* **2002**, *148*, 143–148. [CrossRef]
13. Neytzell-de Wilde, F.G. Demineralization of a molassess distillery waste water. *Desalination* **1987**, *67*, 481–493. [CrossRef]
14. Tedesco, A.T.M.; Cipollina, A.; Micale, G. Towards 1 kW power production in a reverse electrodialysis pilot plant with saline waters and concentrated brines. *J. Membr. Sci.* **2017**, *522*, 226–236. [CrossRef]
15. Mei, Y.; Tang, C.Y. Recent developments and future perspectives of reverse electrodialysis technology: A review. *Desalination* **2018**, *425*, 156–174. [CrossRef]
16. Luo, T.; Abdu, S.; Wessling, M. Selectivity of ion exchange membranes: A review. *J. Membr. Sci.* **2018**, *555*, 429–454. [CrossRef]
17. Zlotorowicz, A.; Strand, R.; Burheim, O.; Wilhelmsen, Ø.; Kjelstrup, S. The permselectivity and water transference number of ion exchange membranes in reverse electrodialysis. *J. Membr. Sci.* **2017**, *523*, 402–408. [CrossRef]
18. Solberg, S.; Zimmermann, P.; Wilhelmsen, Ø.; Lamb, J.J.; Bock, R.; Burheim, O.S. Heat to hydrogen by reverse electrodialysis; using a non-equilibrium thermodynamics model to evaluate hydrogen production concepts utilising waste heat. *Energies* **2022**, *15*, 6011. [CrossRef]
19. Solberg, S.; Zimmermann, P.; Wilhelmsen, Ø.; Bock, R.; Burheim, O.S. Analytical treatment of ion-exchange permselectivity and transport number measurements for high accuracy. *J. Membr. Sci.* **2023**, *685*, 121904. [CrossRef]
20. Solberg, S.B.; Wilhelmsen, Ø.; Burheim, O.S. Transport numbers of ion-exchange membranes in ternary mixtures of KCl, H_2O and ethanol relevant for electrodialysis and desalination processes. *Electrochim. Acta* **2024**, *506*, 145018. [CrossRef]
21. Kjelstrup, S.; Bedeaux, D. *Non-Equilibrium Thermodynamics of Heterogeneous Systems*; World Scientific: Singapore, 2008.
22. De Groot, S.; Mazur, P. *Non-Equilibrium Thermodynamics*; Dover Books on Physics; Dover Publications: Mineola, NY, USA, 1984; ISBN 9780486647418.
23. Jung, T.; Wang, A.A.; Monroe, C.W. Overpotential from cosolvent imbalance in battery electrolytes: Lipf6 in emc:ec. *ACS Omega* **2023**, *8*, 21133–21144. [CrossRef] [PubMed]
24. Okada, T.; Nakamura, N.; Yuasa, M.; Sekine, I. Ion and water transport characteristics in membranes for polymer electrolyte fuel cells containing H^+ and Ca^{2+} cations. *J. Electrochem. Soc.* **1997**, *144*, 2744. [CrossRef]
25. Okada, T.; Møller-Holst, S.; Gorseth, O.; Kjelstrup, S. Transport and equilibrium properties of nafion® membranes with H^+ and Na^+ ions. *J. Electroanal. Chem.* **1998**, *442*, 137–145. [CrossRef]
26. Okada, T.; Satou, H.; Okuno, M.; Yuasa, M. Ion and water transport characteristics of perfluorosulfonated ionomer membranes with H^+ and alkali metal cations. *J. Phys. Chem. B* **2002**, *106*, 1267–1273. [CrossRef]
27. Zawodzinski, T.A.; Derouin, C.; Radzinski, S.; Sherman, R.J.; Smith, V.T.; Springer, T.E.; Gottesfeld, S. Water uptake by and transport through nafion® 117 membranes. *J. Electrochem. Soc.* **1993**, *140*, 1041. [CrossRef]
28. Fuller, T.F.; Newman, J. Experimental determination of the transport number of water in nafion 117 membrane. *J. Electrochem. Soc.* **1992**, *139*, 1332. [CrossRef]
29. Xie, G.; Okada, T. Water transport behavior in nafion 117 membranes. *J. Electrochem. Soc.* **1995**, *142*, 3057. [CrossRef]
30. Ottøy, M.; Førland, T.; Ratkje, S.K.; Møller-Holst, S. Membrane transference numbers from a new emf method. *J. Membr. Sci.* **1992**, *74*, 1–8. [CrossRef]
31. Harned, H.S.; Owen, B.B. *The Physical Chemistry of Electrolytic Solutions*; American Chemical Society Monograph Series, 137; Reinhold Publishing Corporation: New York, NY, USA, 1958; ISBN 9780278917293.
32. Solberg, S.B.; Gómez-Coma, L.; Wilhelmsen, Ø.; Forsberg, K.; Burheim, O.S. Electrodialysis for efficient antisolvent recovery in precipitation of critical metals and lithium-ion battery recycling. *Chem. Eng. J.* **2024**, *486*, 150281. [CrossRef]
33. Berezina, N.P.; Kononenko, N.A.; Dyomina, O.A.; Gnusin, N.P. Characterization of ion-exchange membrane materials: Properties vs structure. *Adv. Colloid Interface Sci.* **2008**, *139*, 3–28. [CrossRef] [PubMed]
34. Larchet, C.; Auclair, B.; Nikonenko, V. Approximate evaluation of water transport number in ion-exchange membranes. *Electrochim. Acta* **2004**, *49*, 1711–1717. [CrossRef]
35. Hamer, W.J.; Wu, Y. Osmotic coefficients and mean activity coefficients of uni-univalent electrolytes in water at 25 °C. *J. Phys. Chem. Ref. Data* **1972**, *1*, 1047–1100. [CrossRef]

36. Pitzer, K.S.; Mayorga, G. Thermodynamics of electrolytes. ii. activity and osmotic coefficients for strong electrolytes with one or both ions univalent. *J. Phys. Chem.* **1973**, *77*, 2300–2308. [CrossRef]
37. Espinoza, C.; Kitto, D.; Kamcev, J. Counter-ion conductivity and selectivity trade-off for commercial ion-exchange membranes at high salinities. *ACS Appl. Polym. Mater.* **2023**, *5*, 10324–10333. [CrossRef]
38. Box, G.; Hunter, J.; Hunter, W. *Statistics for Experimenters: Design, Innovation, and Discovery;* Wiley Series in Probability and Statistics; Wiley: Hoboken, NJ, USA, 2005; ISBN 9780471718130.
39. Okada, T.; Kjelstrup Ratkje, S.; Hanche-Olsen, H. Water transport in cation exchange membranes. *J. Membr. Sci.* **1992**, *66*, 179–192. [CrossRef]
40. Berezina, N.P.; Gnusin, N.; Dyomina, O.; Timofeyev, S. Water electrotransport in membrane systems. Experiment and model description. *J. Membr. Sci.* **1994**, *86*, 207–229. [CrossRef]
41. Barragán, V.M.; Ruiz-Bauzá, C.; Villaluenga, J.P.G.; Seoane, B. Simultaneous electroosmotic and permeation flows through a Nafion membrane: 1. Aqueous electrolyte solutions. *J. Colloid Interface Sci.* **2004**, *277*, 176–183. [CrossRef]
42. Nightingale, E.R.J. Phenomenological theory of ion solvation. effective radii of hydrated ions. *J. Phys. Chem.* **1959**, *63*, 1381–1387. [CrossRef]

entropy

MDPI

Article

Local Equilibrium in Transient Heat Conduction

Kirill Glavatskiy

School of Information and Physical Sciences, The University of Newcastle, Callaghan 2308, NSW, Australia; kirill.glavatskiy@newcastle.edu.au

Abstract: Extended irreversible thermodynamics (EIT) has been widely used to overcome the deficiencies of classical irreversible thermodynamics in describing fast transport phenomena. By employing fluxes as additional independent variables and rejecting local equilibrium hypothesis, EIT may provide a thermodynamically consistent framework for high-frequency and non-local processes. Here, we propose an alternative approach to EIT that shares the same objective but does not reject local equilibrium hypothesis. Using the rates of change of the energy density as the additional independent variable, we illustrate this approach for two typical problems of transient heat conduction: the Cattaneo-type flux model with thermodynamic inertia and the two-temperature model of energy transfer in a phonon–electron system.

Keywords: local equilibrium; irreversible thermodynamics; thermodynamic inertia

1. Introduction

The classical formulation of irreversible thermodynamics (CIT) provides a consistent framework for the local description of a thermodynamic system in which quantities change with position and time [1]. It employs the hypothesis of *local equilibrium*, assuming that macroscopic thermodynamic relations (such as an equation of state or the Gibbs relation) are valid at the meso-scale. The validity of local equilibrium has been demonstrated in various homogeneous [1] and heterogeneous [2–4] systems and even at the nano-scale [5]. However, this question remains open for systems that involve high-frequency phenomena, such as the heating of solids via laser pulses, ultrasound in gases, phonon hydrodynamics, etc., in which local equilibrium appears to be violated [6]. Specifically, in such systems, the thermodynamic quantities at a certain position depend on the properties of the system at the adjacent positions, giving rise to various non-local descriptions [7]. Furthermore, the assumption of local equilibrium in such systems appears to violate the second law of thermodynamics. Such inconsistencies have been resolved by Extended irreversible thermodynamics (EIT), which provides a thermodynamically consistent framework for the description of fast transport phenomena at the micro-and nano-scale [6].

In this paper, we propose an alternative to EIT approach that does not reject local equilibrium hypothesis, while it still provides a thermodynamically consistent description. Following the steps of EIT, we demonstrate this approach for the famous Maxwell–Cattaneo–Vernotte (MCV) equation [8–10] and the two-temperature model of electron–phonon thermal coupling [7].

Let $\mathbf{q}(\mathbf{r}, t)$ be the heat flux and $T(\mathbf{r}, t)$ the temperature in the system of interest, both of which depend on position and time. Using the dot over the symbol is used to indicate the partial time derivative, the MCV-type energy transport equation can be written as

$$\mathbf{q}_C + \tau \dot{\mathbf{q}}_C = -\lambda \nabla T \tag{1}$$

where τ is the relaxation time, λ is the thermal conductivity, and the subscript c refers to the MCV-type behavior. Compared with Fourier's law, $\mathbf{q}_F = -\lambda\nabla T$, Equation (1) contains an additional term, $\tau\dot{\mathbf{q}}$, which accounts for the so-called thermodynamic inertial effect. Such inertia is crucial in fast processes, where the transient dynamics demonstrate a delayed response.

Although Equation (1) accurately describes the dynamics of the aforementioned phenomena, it appears to violate the second law of thermodynamics. The latter states that the local entropy production σ, defined through the entropy balance equation

$$\rho\,\dot{s} = -\nabla\cdot\mathbf{J}_s + \sigma, \tag{2}$$

and represented as the product of the heat flux $\mathbf{q}$ and the thermodynamic force $\mathbf{X}$, must be non-negative. In CIT, $\mathbf{X}_{CIT} = \nabla(1/T)$. If the heat flux is described by Fourier's law, then σ_{CIT} is non-negative, as expected:

$$\sigma_{CIT,F} = \mathbf{q}_F\cdot\mathbf{X}_{CIT} = \mu_{CIT}\,\mathbf{q}_F^2 = \frac{1}{\mu_{CIT}}\mathbf{X}_{CIT}^2 \tag{3}$$

where $\mu_{CIT} \equiv 1/(\lambda T^2)$. However, if the heat flux is described by Equation (1), then the entropy production

$$\sigma_{CIT,C} = \mu_{CIT}\,\mathbf{q}_C^2 - \tau\,\mathbf{q}_C\cdot\dot{\mathbf{q}}_C \tag{4}$$

may become negative, thus violating the second law of thermodynamics.

EIT proposes a modification to local thermodynamic equations by extending the number of independent variables, which control an irreversible evolution of a thermodynamic system [6]. In particular, in the case of heat conduction, EIT postulates that the entropy depends explicitly on the heat flux $\mathbf{q}$, in addition to the classical dependence on the internal energy, u. This, in turn, modifies the Gibbs relation for the entropy, so for the specific (per unit of mass) quantities it reads as

$$ds_{EIT}(u,\mathbf{q}) = \frac{1}{\theta}\,du - \frac{\tau\mu_{EIT}}{\rho}\,\mathbf{q}\cdot d\mathbf{q} \tag{5}$$

where θ is the generalized temperature, which is the non-equilibrium analogue of the equilibrium temperature T, ρ is the material mass density, and $\mu_{EIT} \equiv 1/(\lambda\theta^2)$, where λ is the thermal conductivity. This results in the thermodynamic force having an additional term compared to CIT

$$\mathbf{X}_{EIT} = \nabla\frac{1}{\theta} - \tau\mu_{EIT}\,\dot{\mathbf{q}}_C \tag{6}$$

and hence the entropy production restoring its non-negativity:

$$\sigma_{EIT,C} = \mathbf{q}_C\cdot\mathbf{X}_{EIT} = \mu_{EIT}\,\mathbf{q}_C^2 = \frac{1}{\mu_{EIT}}\mathbf{X}_{EIT}^2 \tag{7}$$

The approach of EIT has been demonstrated to be consistent with various microscopic descriptions in non-equilibrium statistical physics, as well as specific applications of microscopic heat transfer. The dependence of the non-equilibrium entropy s on the flux $\mathbf{q}$ makes EIT a non-local theory.

In this paper, we propose an alternative approach, which provides a thermodynamically consistent description of fast phenomena that, however, is truly *local*. Our approach achieves the same objective as EIT, but keeps the local equilibrium hypothesis intact. We consider two typical problems that are relevant in this context: the MVC-type model (1) itself is derived in Section 2 and the so-called two-temperature model, which arises in the description of energy transfer in an electron–phonon system, is derived in Section 3.

2. Transient Heat Conduction

We assume that the non-equilibrium entropy (per unit of mass) depends on the internal energy (per unit of mass) and the time rate of change of the internal energy [11]:

$$s(\mathbf{r}, t) = s[u(\mathbf{r}, t), \dot{u}(\mathbf{r}, t)] \tag{8}$$

so that its change in some process is governed by the modified Gibbs relation

$$ds(u, \dot{u}) = \beta\, du + \alpha\, d\dot{u} \tag{9}$$

where

$$\begin{aligned} \beta(u, \dot{u}) &\equiv \left.\frac{\partial s}{\partial u}\right|_{\dot{u}} \\ \alpha(u, \dot{u}) &\equiv \left.\frac{\partial s}{\partial \dot{u}}\right|_{u} \end{aligned} \tag{10}$$

Given that $\beta(u, 0) = 1/T$, we may identify $1/\beta$ as the non-equilibrium temperature. The physical meaning of α will become apparent below. It is convenient, however, to write it as $\alpha(u, \dot{u}) = \widetilde{\alpha}[\beta(u, \dot{u}), \dot{u}]$, so that

$$d\alpha = \left.\frac{\partial \widetilde{\alpha}}{\partial \beta}\right|_{\dot{u}} d\beta + \left.\frac{\partial \widetilde{\alpha}}{\partial \dot{u}}\right|_{\beta} d\dot{u} \tag{11}$$

We next assume that $(\partial \widetilde{\alpha}/\partial \dot{u})_\beta = 0$, so that the dependence of α on $\dot{u}$, if any, is only manifested through the corresponding dependence of the non-equilibrium temperature $1/\beta$:

$$\alpha(u, \dot{u}) = \widetilde{\alpha}[\beta(u, \dot{u})] \tag{12}$$

This assumption may be justified by viewing Equation (9) as the first-order approximation in the expansion of s in the powers of $\dot{u}$. Indeed, an explicit dependence of α on $\dot{u}$ would result in s being dependent on higher than linear terms in $\dot{u}$. Thus, Equation (12) implies linear dependence of the non-equilibrium entropy on the time rate of change of the internal energy.

Heat conduction is governed by the energy balance equation

$$\rho\, \dot{u} = -\nabla \cdot \mathbf{q} \tag{13}$$

Taking the time derivative of Equation (13), we obtain

$$\rho\, \ddot{u} = -\nabla \cdot \dot{\mathbf{q}} \tag{14}$$

Substituting Equations (13) and (14) in Equation (9), we obtain the time rate of change of the entropy density

$$\rho\, \dot{s} = -\beta \nabla \cdot \mathbf{q} - \alpha \nabla \cdot \dot{\mathbf{q}} \tag{15}$$

which, after rearrangement, results in

$$\rho\, \dot{s} = -\nabla \cdot (\beta \mathbf{q} + \alpha \dot{\mathbf{q}}) + \left(\mathbf{q} + \dot{\mathbf{q}} \frac{d\widetilde{\alpha}}{d\beta}\right) \cdot \nabla \beta \tag{16}$$

Comparing it with the entropy balance Equation (2), we identify the entropy flux and the entropy production as

$$\begin{aligned} \mathbf{J}_s &= \beta\,\mathbf{q} + \alpha\,\dot{\mathbf{q}} \\ \sigma &= \left(\mathbf{q} + \dot{\mathbf{q}}\,\frac{d\tilde{\alpha}}{d\beta}\right)\cdot\nabla\beta \end{aligned} \tag{17}$$

In order for the entropy production to be non-negative, we must require

$$\mathbf{q} + \dot{\mathbf{q}}\,\frac{d\tilde{\alpha}}{d\beta} = \frac{1}{\mu}\,\nabla\beta \tag{18}$$

where μ is positive. Comparing Equation (18) with MCV Equation (1), we note that these are the same equations. This essentially means that the assumptions (9) and (12) produce MCV Equation (1) in a thermodynamically consistent manner.

We can now identify the coefficients α and μ. Comparing again Equation (18) with (1), we identify $\mu = \beta^2/\lambda$, which thus has the same meaning as μ_{CIT} and μ_{EIT}. Furthermore, we observe that $d\tilde{\alpha}/d\beta = \tau$, and hence

$$\alpha(u, \dot{u}) = \int \tau(\beta)\,d\beta \tag{19}$$

We will further refer to the combination

$$\tilde{\mathbf{q}} \equiv \mathbf{q} + \dot{\mathbf{q}}\,\frac{d\tilde{\alpha}}{d\beta} \tag{20}$$

as the *generalized flux*. Identifying the thermodynamic force as $\mathbf{X} \equiv \nabla\beta$, we obtain for the entropy production

$$\sigma = \tilde{\mathbf{q}}\cdot\mathbf{X} = \mu\,\tilde{\mathbf{q}}^2 = \frac{1}{\mu}\,\mathbf{X}^2 \tag{21}$$

3. Two-Temperature Model

We now consider a heterogeneous system that is made of two components with their own temperatures. A reference example of a such system is heat transport in metals, where one can observe the electron gas at the temperature T_e and the phonon gas (the lattice) with the different temperature T_p [7]. In such a system, heat is transferred in two modes: (i) via spatial heat fluxes $\mathbf{q}_e$ and $\mathbf{q}_p$ of electrons and phonons, respectively, and (ii) via a (scalar) heat flux q_{ep} between the electron and phonon subsystems at the same position.

When the energy densities (per unit of volume) u_e and u_p of the corresponding subsystems are introduced, the energy balance equation for each of the subsystems is

$$\begin{aligned} \dot{u}_e &= -\nabla\cdot\mathbf{q}_e - q_{ep} \\ \dot{u}_p &= -\nabla\cdot\mathbf{q}_p + q_{ep} \end{aligned} \tag{22}$$

We note that the thermodynamic behavior of the electron and phonon subsystems is different, resulting in certain simplifications for Equation (22). We will, however, carry the initial derivation in general terms (keeping the notation of the e and p subsystems), reducing to the particular cases later.

We next generalize Equation (8) so that the entropy density (per unit of volume) s of the system can be represented as a sum of the corresponding densities, s_e and s_p of the subsystems, each depending on the corresponding volumetric internal energy density and its time rate of change:

$$s(\mathbf{r}, t) = s_e[u_e(\mathbf{r}, t), \dot{u}_e(\mathbf{r}, t)] + s_p[u_p(\mathbf{r}, t), \dot{u}_p(\mathbf{r}, t)] \tag{23}$$

The corresponding Gibbs equation for the overall system takes the following form:

$$ds(u_e, u_p, \dot{u}_e, \dot{u}_p) = \beta_e \, du_e + \beta_p \, du_p + \alpha_e \, d\dot{u}_e + \alpha_p \, d\dot{u}_p \tag{24}$$

with β and α defined similarly to Equation (10), and α approximated similarly to Equation (12):

$$\begin{aligned} \alpha_e(u_e, \dot{u}_e) &= \widetilde{\alpha}_e[\beta(u_e, \dot{u}_e)] \\ \alpha_p(u_p, \dot{u}_p) &= \widetilde{\alpha}_p[\beta(u_p, \dot{u}_p)] \end{aligned} \tag{25}$$

Following the same process as in Section 2, we obtain, for the balance of the entropy density, (2), identifying the entropy flux $\mathbf{J}_s$ and the entropy production σ as

$$\begin{aligned} \mathbf{J}_s \equiv\ & \beta_e \, \mathbf{q}_e + \beta_p \, \mathbf{q}_p + \alpha_e \, \dot{\mathbf{q}}_e + \alpha_p \, \dot{\mathbf{q}}_p \\ \sigma \equiv\ & \widetilde{\mathbf{q}}_e \cdot \nabla\beta_e + \widetilde{\mathbf{q}}_p \cdot \nabla\beta_p + q_{ep}(-\beta_e + \beta_p) + \dot{q}_{ep}(-\widetilde{\alpha}_e(\beta_e) + \widetilde{\alpha}_p(\beta_p)) \end{aligned} \tag{26}$$

where the generalized fluxes $\widetilde{\mathbf{q}}_e$ and $\widetilde{\mathbf{q}}_p$ are defined similarly to Equation (20).

To ensure that σ is non-negative, and taking into account that scalar and vectorial fluxes do not couple, we must require that

$$\begin{aligned} \nabla\beta_e &= \mu_{ee} \, \widetilde{\mathbf{q}}_e + \mu_{ep} \, \widetilde{\mathbf{q}}_p \\ \nabla\beta_p &= \mu_{pe} \, \widetilde{\mathbf{q}}_p + \mu_{pp} \, \widetilde{\mathbf{q}}_p \end{aligned} \tag{27}$$

and

$$\begin{aligned} q_{ep} &= g_{00} \left(-\beta_e + \beta_p\right) + g_{01} \left(-\widetilde{\alpha}_e(\beta_e) + \widetilde{\alpha}_p(\beta_p)\right) \\ \dot{q}_{ep} &= g_{10} \left(-\beta_e + \beta_p\right) + g_{11} \left(-\widetilde{\alpha}_e(\beta_e) + \widetilde{\alpha}_p(\beta_p)\right) \end{aligned} \tag{28}$$

with the coefficients μ_{ee}, μ_{ep}, μ_{pe}, μ_{pp} and g_{00}, g_{01}, g_{10}, g_{11} being non-negative. Equations (27) and (28) represent heat transport equations for an isotropic two-temperature system. The identification of the transport coefficients is performed for each particular phenomenon below.

3.1. Electron–Phonon Coupling

The two-temperature model for the electron–phonon energy transfer is [7]

$$\begin{aligned} C_e \, \dot{T}_e &= \nabla \cdot \mathbf{q}_e - G\,(T_e - T_p) \\ C_p \, \dot{T}_p &= G\,(T_e - T_p) \\ \mathbf{q}_e + \tau_F \, \dot{\mathbf{q}}_e &= -\lambda_e \, \nabla T_e \end{aligned} \tag{29}$$

Here, C_e and C_p are the volumetric heat capacities of the electrons and the phonons, respectively, λ_e is the thermal conductivity of the electron gas, and G is the phonon–electron coupling factor describing the rate of energy transfer between electrons and phonons [12], while τ_F is the relaxation time of the electron gas calculated at the Fermi surface [13].

The first two of Equation (29) are the energy balance equations for the electron and phonon subsystems, respectively, i.e., the particular manifestations of Equation (22). We note that the phonon balance equation misses the $\nabla \cdot \mathbf{q}_p$ term, which is due to the fact that spatial equilibrium in the phonon gas is reached much faster than the one in the electron gas. For the same reason, the transport equation for the phonon subsystem is missing, while the transport equation for the electron subsystem (the third equation in Equation (29)) has the form of MCV Equation (1). Comparing Equations (27) and (28) with Equation (29) allows us to identify the new quantities, similarly to how it has been done in Section 2.

We first identify $\beta_e = 1/T_e$ and $\beta_p = 1/T_p$ as the non-equilibrium temperatures of the electron and phonon gas. Furthermore, $(d\tilde{\alpha}_e/d\beta_e) \equiv \tau_F$, and $\alpha_e = \int \tau_F(\beta_e) d\beta_e$.

Similarly, $\alpha_p = \int \tau_p(\beta_p) d\beta_p$. We note, however, that since Equation (29) does not contain the transport equation for the phonon flux, τ_p is undefined. Because the phonon gas reaches equilibrium much faster than the electron gas, one may consider the thermal conductivity of the phonon gas $\lambda_p = \infty$ to be infinite (or, equivalently, the phonon thermal resistivity to be zero, $\mu_{pp} = 0$). This, in turn, means that τ_p may be considered zero; hence, $\alpha_p = 0$, and therefore, the corresponding term $\alpha_p\, \dot{u}_p$ is absent from the Gibbs relation (24).

Next, we observe that the spatial coupling coefficients $\mu_{ep} = \mu_{pe} = 0$. Thus, Equation (27) reduces to a simple equation, $\nabla \beta_e = \mu_{ee}\, \tilde{\mathbf{q}}_e$, which is equivalent to the third of Equation (29) with $\mu_{ee} \equiv 1/(\lambda_e T_e^2)$.

Finally, we observe that the electron–phonon coupling manifests as a single term, $q_{ep} = G\,(T_e - T_p)$, so that $g_{01} = g_{10} = g_{11} = 0$ and $g_{00} \equiv G\, T_e\, T_p$.

3.2. Coupling Between Similar Components

Consider now the two-temperature model for a system that consists of "similar" components. We define such "similarity" as α_e and α_p having the same functional form; i.e., for any β

$$\alpha_e(\beta) = \alpha_p(\beta) = \alpha(\beta) \tag{30}$$

In this case, assuming that β_e and β_p are not too different, we can further approximate

$$\tilde{\alpha}(\beta_e) \approx \tilde{\alpha}(\beta_p) + \frac{d\tilde{\alpha}}{d\beta}\,(\beta_e - \beta_p) \tag{31}$$

Then, the scalar part of the entropy production in Equation (26) becomes

$$\sigma^{(s)} = (-\beta_e + \beta_p)\left(q_{ep} + \frac{d\tilde{\alpha}}{d\beta}\,\dot{q}_{ep}\right) \tag{32}$$

where we have taken into account that, due to Equation (30), $\tilde{\alpha}_e(\beta_p) - \tilde{\alpha}_p(\beta_p) = 0$. Therefore, Equation (28) simplifies to

$$q_{ep} + \frac{d\tilde{\alpha}}{d\beta}\,\dot{q}_{ep} = g\,(-\beta_e + \beta_p) \tag{33}$$

where g is non-negative. With the "equilibrium" notation, it has the form of MCV Equation (1) for the flux q_{ep} between two subsystems:

$$q_{ep} + \tau_{ep}\,\dot{q}_{ep} = G\,(T_e - T_p) \tag{34}$$

with $g \equiv G T_e T_p$.

4. Discussion

We have proposed an alternative approach to EIT, which describes transient heat conduction in a thermodynamically consistent manner.

When the expressions for the entropy production from CIT, EIT, and the proposed approach (Equation (3), Equation (7), and Equation (21), respectively) are compared, it is interesting to note that the entropy production is represented by the same functional bi-linear form of thermodynamic forces and fluxes in all three approaches. The difference, however, lies in the explicit identification of those forces and fluxes. In particular, the heat flux $\mathbf{q}_F$ in CIT is the solution of Fourier's law, and the heat flux $\mathbf{q}_C$ in EIT is the solution of the MCV equation, while the heat flux $\tilde{\mathbf{q}}$ in the transient approach is the generalized flux, defined by Equation (20). Similarly, the thermodynamic force in CIT is $\nabla(1/T)$, and the thermodynamic force in EIT is $\nabla(1/\theta) - \tau\mu\dot{\mathbf{q}}_C$, while the thermodynamic force in the transient approach is $\nabla\beta$. This suggests the universality of the force-flux bi-linearity of the entropy production, while the choice of the approach is dictated by the particular problem of interest.

The proposed transient approach and EIT address the same type of phenomena while providing different solutions. The main difference between the two approaches is the utilization of local equilibrium. EIT is a non-local theory, with the local thermodynamic properties being dependent on the heat flux. In contrast, the proposed transient approach is a local theory, with the local thermodynamic properties being dependent on the time rate of change of the internal energy.

It may appear, however, that such a difference between the proposed transient approach and EIT is merely philosophical. Indeed, the additional independent variable, the heat flux, is measured locally, at the same position as the thermodynamic densities. While this may be true empirically, there exists an important conceptual difference. Thermodynamic density is evaluated at a point that is the center of a small volume element. In contrast, flux is evaluated at a point that is the boundary between two adjacent volume elements. Thus, employing the heat flux as an independent variable means involving two neighboring volume elements, while employing the time rate of change of the internal energy as an independent variable means involving a single volume element. The latter description is more succinct and thus may, in certain circumstances, be preferred.

Another important difference between EIT and the proposed transient approach is the behavior in the stationary state. There, the transient approach reduces to the classical irreversible thermodynamics. In contrast, EIT remains distinct from CIT with the stationary heat flux still being used as the independent thermodynamic variable. This is, in principle, not forbidden; however, it does appear to deviate from the original motivation of EIT to describe fast transient processes. This suggests that EIT may be better suited for low-frequency phenomena in spatially inhomogeneous systems, such as nano-systems [14], while the proposed transient approach is better suited to high-frequency transient processes in homogeneous environments.

Finally, the analysis of the two-temperature model may suggest a simple solution to the problem of non-equilibrium temperature raised in EIT. In particular, in non-equilibrium, different degrees of freedom may have different temperatures. This creates difficulties in the identification of the unique non-equilibrium temperature, θ. The proposed transient approach does not require the existence of the unique non-equilibrium temperature, sustaining its consistency with multiple non-equilibrium temperatures defined in the system.

Funding: This research received no external funding.

Institutional Review Board Statement: Not applicable.

Data Availability Statement: This work does not use or produce any data.

Acknowledgments: The author is grateful to Signe Kjelstrup for her mentorship in non-equilibrium thermodynamics and to Dick Bedeaux for the opportunity to contribute this work.

Conflicts of Interest: The authors declare no conflicts of interest.

References

1. de Groot, S.R.; Mazur, P. *Non-Equilibrium Thermodynamics*; Dover: New York, NY, USA, 1984.
2. Kjelstrup, S.; Bedeaux, D. *Non-Equilibrium Thermodynamics of Heterogeneous Systems*, 2nd ed.; World Scientific: Singapore, 2020.

3. Glavatskiy, K.S.; Bedeaux, D. Numerical solution of the nonequilibrium square-gradient model and verification of local equilibrium for the Gibbs surface in a two-phase binary mixture. *Phys. Rev. E* **2009**, *79*, 031608. [CrossRef] [PubMed]
4. Johannessen, E.; Bedeaux, D. The nonequilibrium van der Waals square gradient model. (II). Local equilibrium of the Gibbs surface. *Physica A* **2003**, *330*, 354. [CrossRef]
5. Bedeaux, D.; Kjelstrup, S.; Schnell, S.K. *Nanothermodynamics: Theory and Applications*; World Scientific: Singapore, 2023.
6. Jou, D.; Casas-Vázquez, J.; Lebon, G. *Extended Irreversible Thermodynamics*, 4rd english ed.; Springer: New York, NY, USA; Dordrecht, The Netherlands; Heidelberg, Germany; London, UK, 2010.
7. Tzou, D.Y. *Macro-to-Microscale Heat Transfer. The Lagging Behaviour*, 2nd ed.; Willey: London, UK, 2015.
8. Maxwell, J.C. On the Dynamical Theory of Gases. Philosophical Transactions of the Royal Society of London. *Philos. Trans. R. Soc. Lond.* **1867**, *157*, 49.
9. Cattaneo, C. Sulla conduzione del calore. Attidel Seminario Matematico e Fisicodella Università di Modena. *Atti Sem. Mat. Fis. Univ. Modena* **1948**, *3*, 83.
10. Vernotte, M.P. Les Paradoxes de la Théorie Continue de L'équation de la Chaleur. *C. R. Hebd. Séances Acad. Sci.* **1958**, *246*, 3154.
11. Glavatskiy, K.S. Local equilibrium and the second law of thermodynamics for irreversible systems with thermodynamic inertia. *J. Chem. Phys.* **2015**, *143*, 164101. [CrossRef]
12. Kaganov, M.I.; Lifshitz, I.M.; Tanatarov, M.V. Relaxation Between Electrons and Crystalline Lattices. *Sov. Phys. JETP* **1975**, *4*, 173–178.
13. Qiu, T.Q.; Tien, C.L. Heat Transfer Mechanisms During Short-Pulse Laser Heating of Metals. *ASME J. Heat Transf.* **1993**, *115*, 835–841. [CrossRef]
14. Sellitto, A.; Cimmelli, V.A.; Jou, D. *Mesoscopic Theories of Heat Transport in Nanosystems*; Springer International Publishing: Cham, Switzerland; Heidelberg, Geramny; New York, NY, USA; Dordrecht, The Netherlands; London, UK, 2016.

entropy

MDPI

Article

Entropy Production in a System of Janus Particles

Andrés Arango-Restrepo *,†, Juan David Torrenegra-Rico † and J. Miguel Rubi †

Condensed Matter Department, Universitat de Barcelona, 08028 Barcelona, Spain;
jdtorrenegrar@unal.edu.co (J.D.T.-R.); mrubi@ub.edu (J.M.R.)
* Correspondence: aarangor@unal.edu.co
† These authors contributed equally to this work.

Abstract: Entropy production is a key descriptor of out-of-equilibrium behavior in active matter systems, providing insights into both single-particle dynamics and emergent collective phenomena. It helps determine transport coefficients and phoretic velocities and serves as a crucial tool for understanding collective phenomena such as structural transitions, regime shifts, clustering, and self-organization. This study investigates the role of entropy production for individual active (catalytic Janus) particles and in systems of active particles interacting with one another and their environment. We employ a multiscale framework to bridge microscopic particle dynamics and macroscopic behavior, offering a thermodynamic perspective on active matter. These findings enhance our understanding of the fundamental principles governing active particle systems and create new opportunities for addressing unresolved questions in non-equilibrium thermodynamics.

Keywords: active particles; energy dissipation; entropy production; janus particles; self-assembly

1. Introduction

Active particles are small entities that can convert the chemical energy of their environment into mechanical motion. Unlike passive particles, which eventually settle into a stable, equilibrium state (called thermodynamic equilibrium), active particles consume energy continuously, keeping them in a constant state of activity. This continuous consumption of energy leads to unique and complex behaviors that are very different from those observed in systems at equilibrium. Examples of active particles include synthetic micromotors, Janus colloids (particles designed with two distinct sides), and biological entities such as bacteria and cells. These systems are studied in various fields because they provide information on how self-organization, transport, and energy conversion occur in both living and artificial systems [1–9].

Thermodynamic principles are essential to understand the behavior of active particles, which interact with their environment by continuously consuming energy and forming out-of-equilibrium structures [10,11]. In systems in equilibrium, energy flows are balanced and entropy production vanishes. In contrast, active particles maintain motion by consuming energy locally, maintaining stable out-of-equilibrium states characterized by persistent dissipation and entropy production [1,12]. For the purposes of this study, we emphasize that the presence of surface gradients, which lead to gradients in surface tension, is the central mechanism driving the system's behavior. These gradients, arising from propulsion mechanisms such as chemical reactions, light absorption, or thermal effects, create effective potentials [13,14] and induce local fluxes in the surrounding medium [15,16]. This dynamic interplay between surface tension gradients, motion, and dissipation highlights the necessity of extending classical thermodynamic frameworks to accurately describe systems far from equilibrium [17].

To rigorously study these behaviors, researchers employ advanced theoretical and experimental tools. Fluctuation theorems play a critical role by describing the statistical properties of entropy production and work in systems driven far from equilibrium [18,19], while stochastic thermodynamics models energy exchanges on the scale of individual particles, tracking heat, work, and entropy along their trajectories [20–23]. For collective behaviors, mesoscopic non-equilibrium thermodynamics (MNET) has emerged as a powerful approach, capturing nonlinearities through Fokker–Planck type kinetic equations that describe the evolution of mesoscopic variables and their associated entropy production [24]. MNET provides a robust framework for studying out-of-equilibrium systems and acts as a bridge between microscopic dynamics and macroscopic behaviors. It extends classical thermodynamics by incorporating mesoscopic variables, allowing the analysis of systems with inherent fluctuations and complex, nonlinear interactions [25]. In particular, MNET captures nonlinear relationships between fluxes and forces, leading to the law of mass action in chemical reactions [26], where the driving force is not affinity but a difference in fugacity [27].

These methodologies provide a framework for analyzing the connection between individual and collective dynamics and fundamental thermodynamic processes. However, our approach emphasizes mesoscopic entropy production rather than stochastic thermodynamics, as our goal is to describe mesoscale collective behavior while linking entropy production to dynamics. Specifically, we aim to (1) use entropy production to characterize different regimes, (2) apply interfacial entropy production to estimate transport coefficients, and (3) investigate entropy production at the particle surface due to chemical reactions while accounting for entropy production in the bulk. Our study explores the utility of entropy production in active particle systems, focusing on catalytic Janus particles. Rather than simply calculating entropy production, we aim to demonstrate its potential as a powerful descriptor, positioning it not merely as a non-equilibrium quantity but also as a fundamental variable for understanding and describing active matter systems.

This paper presents a detailed investigation of the theories and methodologies for analyzing active particles, focusing on the calculation and application of entropy production. By exploring both the dynamics of individual particles and systems of active particles interacting with the medium and each other, we highlight entropy production as a central thermodynamic quantity for understanding non-equilibrium processes in active particle dynamics. Our study examines how entropy production at the mesoscale links microscopic particle behavior with emergent macroscopic phenomena, such as clustering and self-organization. This approach provides new insights into the thermodynamic principles governing active systems, which advances our understanding of non-equilibrium dynamics and offers potential applications in the design of functional materials and the study of biological processes at all scales.

2. Active Particles Dynamics

Entropy production and particle dynamics are intrinsically interconnected. Historically, however, dynamics and thermodynamics have often been treated as distinct disciplines, and the field of active matter is no exception. Despite this separation, models describing the dynamics of active particles are fundamental to understanding the non-equilibrium nature of these systems. In this section, we present a range of models that capture the dynamics of active particles, spanning from the molecular to the macroscopic scale.

Figure 1 illustrates the system and key variables for various modeling approaches, including dissipative molecular dynamics, Langevin/Viscek models, multiscale frameworks, and macroscopic continuum models. Dissipative molecular dynamics focuses on describing the dynamics of individual particles, capturing interfacial processes and gradients near the particle surface [28,29]. Entropy production inherently fluctuates on this scale, and its time average is calculated under thermostated steady-state conditions in open systems [30]. Langevin models are designed to study the emergence of collective behavior, accounting for the positions and orientations of active particles [7,9,31]. In these models, entropy production is typically calculated as a path-averaged quantity [21]. Multiscale models have recently gained popularity due to their ability to couple the discrete nature of particles with continuous representations of temperature, concentration, and hydrodynamic fields [32–35]. Given the interplay of multiple scales in describing active particle dynamics and thermodynamic quantities, a mesoscopic approach to entropy production is often the most appropriate [25]. Continuum models simplify numerical and analytical calculations by assuming a continuous particle distribution and using free energy functionals to describe self-assembling process through Ginzburg–Landau dynamics and kinetic theory [36,37]. Although some microscopic details are lost, these models allow analytical studies to be performed [38].

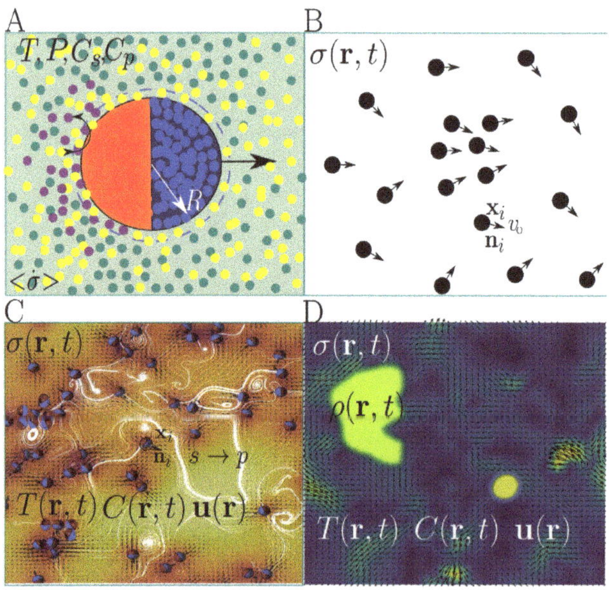

Figure 1. Models for active particle dynamics and thermodynamics. (**A**) Dissipative molecular dynamics: Focuses on the dynamics of individual particles, capturing interfacial concentration gradients, C, and temperature, T, near the particle surface. The entropy production at this scale fluctuates; the time-averaged entropy production, $\langle \sigma \rangle$, is calculated under open, stable, and thermostated conditions. (**B**) Langevin/Vicsek models. Mainly used to track particle positions, $\mathbf{r}_i$, and orientations, $\mathbf{n}_i$, with a fixed active velocity, v_0. The entropy production is calculated as a time-averaged quantity, $\langle \sigma \rangle_t$, over the particle trajectories. (**C**) Multiscale models: Combine the effects of concentration fields, $C(\mathbf{r}, t)$, temperature, $T(\mathbf{r}, t)$, and velocity, $u(\mathbf{r}, t)$, on the dynamics and thermodynamics of active particles. They integrate the continuous nature of thermodynamic fields with the discrete nature of particles, enabling the calculation of the local entropy production, $\sigma(\mathbf{r}, t)$, using mesoscopic approaches. (**D**) Macroscopic/continuous models: Assume a continuous particle distribution, $\rho(\mathbf{r}, t)$. In this framework, entropy production, $\sigma(\mathbf{r}, t)$, is derived using mesoscopic or linear irreversible thermodynamic approach.

Here we use the MNET formalism to compute entropy production and describe the dynamics, because it unites microscopic and macroscopic descriptions and effectively accounts for nonlinearities and dissipation due to surface gradients, in bulk and interactions [39]. However, combining MNET with elements of stochastic thermodynamics can provide additional information on fluctuations at the individual particle level, especially in systems with very small particles.

Non-equilibrium molecular dynamics simulations are widely employed to explore the microscopic behavior of active particles, offering insight into surface phenomena [28,40–42]. In these simulations, the solvent and the active particles are typically represented with Lennard-Jones particles that interact according to the potential

$$u_{ij} = \left[4\varepsilon_{ij}\left(\left(\frac{a}{r}\right)^{12} - \left(\frac{a}{r}\right)^{6} - u(r_c)\right)\right]\Theta(r_c - r) \tag{1}$$

in which ε_{ij} is the interaction strength between particles of fluid (i) and active particles (j), a is the diameter of the particles, r is the distance between particles i and j, Θ is the Heaviside function, and r_c is an interaction cut-off. These simulations enable the direct computation of thermodynamic quantities and transport coefficients for non-interacting active particles, providing insights into entropy production and energy dissipation at the molecular scale.

Langevin dynamics offers stochastic descriptions of the collective motion of particles. Langevin equations account for the interactions among particles, propulsion forces, thermal noise, and dissipation, providing a trajectory-level perspective on work and heat exchange. The typical set of equations is inspired by the Vicsek model [31], which describes the overdamped Langevin dynamics for the position $\mathbf{r}_i$ and orientation $\mathbf{n}_i$ of the ith-particle in a solution in the presence of a substrate s with a concentration C_s

$$\dot{\mathbf{r}}_i = v_0\mathbf{n}_i + \xi_t\nabla C_s(\mathbf{r}_i) + \mathbf{R}_i(t) \tag{2}$$

$$\dot{\mathbf{n}}_i = \xi_r(1 - \mathbf{n}_i\mathbf{n}_i)\nabla C_s(\mathbf{r}_i) + \mathbf{N}_i(t) \times \mathbf{n}_i \tag{3}$$

Here, the random terms arising from thermal fluctuations satisfy the fluctuation–dissipation theorems $\langle \mathbf{R}_i\mathbf{R}_i \rangle = 2D_t\mathbf{I}\delta(t - t')$ and $\langle \mathbf{N}_i\mathbf{N}_i \rangle = 2D_r\mathbf{I}\delta(t - t')$; v_0 is the active velocity, ξ_t is the diffusiophoretic coefficient, ξ_s is the phoretic rotation coefficient, D_t is the particle diffusivity, and D_r is the rotation diffusivity of the particle [7].

The substrate dynamics has been analyzed to account for non-constant active velocities by expressing the self-diffusiophoretic velocity as a function of the reaction rate and substrate concentration [39]. The resulting Langevin dynamics for the particle position and the conservation equation for the substrate are

$$\dot{\mathbf{r}}_i = \xi_s C_s\mathbf{n}_i + \xi_t\nabla C_s(\mathbf{r}_i) + \mathbf{R}_i(t) \tag{4}$$

$$\frac{\partial C_s}{\partial t} = D_s\nabla^2 C_s - \sum_i^{N_p} J_i(\mathbf{r}_i) \tag{5}$$

Here, the diffusiophoretic velocity is $\xi_s C_s$, with ξ_s the self-diffusiophoretic coefficient and $J_i(\mathbf{r}_i)$ the reaction rate at the particle position. As discussed in Ref. [39], fluctuations in diffusive and reactive fluxes are negligible compared to the implicit noise in J_i caused by the thermal fluctuations of $\mathbf{r}_i$, particularly when the particle radius exceeds 100 nm. Consequently, Equation (5) does not include an explicit noise source term, and the entropy production rate does not explicitly depend on noise but implicitly on fluctuating particle position.

On the experimental side, various techniques have been developed to measure entropy production and energy flows in active systems from active particle dynamics. For example, high-resolution particle tracking allows one to extract trajectory-level data, which facilitates the computation of dissipative forces and energy consumption [43,44]. Calorimetric methods are used to directly measure the heat released during propulsion, which provides information on the efficiency of energy conversion [45]. Other approaches include using tracer particles to study the hydrodynamic and chemical fields generated by active particles, providing indirect measurements of the energy fluxes involved [46]. These tools offer a comprehensive view of the thermodynamics governing active particles, bridging the gap between microscopic processes and macroscopic emergent phenomena. In the next sections, we focus on computing entropy production at both particle and active system levels.

3. Single-Particle Entropy Production

Entropy production is a key thermodynamic concept for understanding the motion of active particles, particularly in phoretic mechanisms where surface tension gradients at the particle–fluid interface drive propulsion. Figure 2a illustrates a typical schematic representation of a Janus catalytic particle. In this system, a chemical reaction converts substrate s to product p, releasing heat Q_{rxn}. The reaction generates surface gradients in surface tension γ, caused by gradients in concentration (C_s,C_p) and temperature (T) at the particle–bulk interface. These variations in surface tension induce a slip velocity, $\mathbf{v}_{slip}$, that defines the velocity of the active particle.

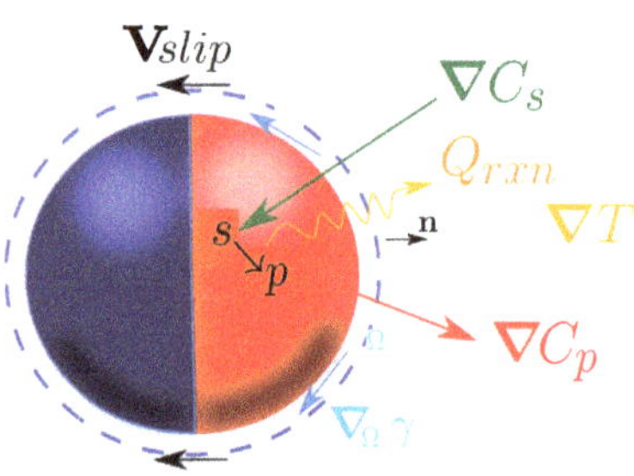

Figure 2. Schematic representation of a catalytic Janus particle, illustrating the reactive interface where the chemical reaction $s \rightarrow p$ occurs. This reaction generates concentration gradients, ∇C_s and ∇C_p, as well as temperature gradients, ∇T, both at the interface and in the surrounding medium. These gradients result in changes in surface tension, $\nabla_\Omega \gamma$, which induce the slip velocity, $\mathbf{v}_{slip}$.

By using Onsager's reciprocal relations [47,48], one can calculate slip velocities in the presence of temperature, concentration, or pressure gradients [11,15,16,49–51] from entropy production, which is derived by combining mass, energy, and momentum balances with the second law of thermodynamics [26]. These gradients induce radial flows and fluid flows, linking chemical reactions with mechanical motion and providing a framework for analyzing energy dissipation. The entropy production rate at the particle surface is given by

$$\sigma^\Omega = -\frac{1}{T}\mathbf{m} \cdot \mathbf{P}^\Omega \cdot (\mathbf{1} - \mathbf{mm}) \cdot \mathbf{v}_{slip} - \frac{1}{T}\mathbf{J}_s^\Omega \cdot \nabla_\Omega \mu_s^\Omega - \frac{1}{T}\mathbf{J}_p^\Omega \cdot \nabla_\Omega \mu_p^\Omega - \frac{1}{T^2}\mathbf{J}_q^\Omega \cdot \nabla_\Omega T^\Omega \quad (6)$$

in which the dissipative forces are the slip velocity, $\mathbf{v}_{slip}$, chemical potential surface gradients of substrate, $\nabla_\Omega \mu_s^\Omega$, and product, $\nabla_\Omega \mu_p^\Omega$, and temperature surface gradient, $\nabla_\Omega T^\Omega$. From now on, using simplified notation, all gradients and variables will be assumed to be interfacial unless otherwise specified. From Equation (6), we can obtain equations for the interfacial pressure, $\mathbf{P}$; $\mathbf{m}$ is the normal vector over the surface, and the equation also includes the surface diffusive fluxes of the substrate, $\mathbf{J}_s$, and product, $\mathbf{J}_p$, as well as the surface heat flux, $\mathbf{J}_q$ [11,50]

$$\mathbf{m}\cdot\mathbf{P}\cdot(\mathbf{1}-\mathbf{mm}) = -\frac{L_{vv}}{T}\mathbf{v}_{slip} - \frac{L_{vs}}{T}\nabla\mu_s - \frac{L_{vp}}{T}\nabla\mu_p - \frac{L_{vq}}{T^2}\nabla T \tag{7}$$

$$\mathbf{J}_s = -\frac{L_{sv}}{T}\mathbf{v}_{slip} - \frac{L_{ss}}{T}\nabla\mu_s - \frac{L_{sp}}{T}\nabla\mu_p - \frac{L_{sq}}{T^2}\nabla T \tag{8}$$

$$\mathbf{J}_p = -\frac{L_{pv}}{T}\mathbf{v}_{slip} - \frac{L_{ps}}{T}\nabla\mu_s - \frac{L_{pp}}{T}\nabla\mu_p - \frac{L_{pq}}{T^2}\nabla T \tag{9}$$

$$\mathbf{J}_q = -\frac{L_{qv}}{T}\mathbf{v}_{slip} - \frac{L_{qs}}{T}\nabla\mu_s - \frac{L_{qp}}{T}\nabla\mu_p - \frac{L_{qq}}{T^2}\nabla T \tag{10}$$

in which the coefficients L_{ij} are the Onsager's matrix elements. From this analysis, and considering the conservation equations for species and energy, the slip velocity can be computed, from which the active particle velocity can be derived [11,50].

Phoretic propulsion mechanisms are based on interfacial asymmetries induced by concentration and temperature gradients. In self-diffusiophoresis, surface reactions generate gradients of reactants and products, resulting in sliding velocities that drive particle motion [52–54]. Similarly, in thermophoresis and electrophoresis, temperature and electrochemical gradients induce directed motion. Classical models, such as the Derjaguin–Anderson model [13,14], describe these processes, while recent advances in chemical–mechanical coupling have improved the quantification of transport coefficients, dissipation rates, and energy conversion efficiencies [55].

From Equation (7), the slip velocity at the particle–bulk interface can be determined, providing a detailed description of the interfacial flow. By integrating this slip velocity times the friction coefficient over the particle's surface, the force over the particle is obtained:

$$\begin{aligned}\mathbf{F} = \xi\int_\Omega \mathbf{v}_{slip}d\Omega &= -\xi\int_\Omega \frac{T}{L_{vv}}\mathbf{m}\cdot\mathbf{P}\cdot(\mathbf{1}-\mathbf{mm})d\Omega \\ &\quad -\xi\int_\Omega \frac{L_{vs}}{L_{vv}}\nabla\mu_s d\Omega - \xi\int_\Omega \frac{L_{vp}}{L_{vv}}\nabla\mu_p d\Omega - \xi\int_\Omega \frac{L_{vq}}{TL_{vv}}\nabla T d\Omega \\ &= \mathbf{F}_H + \mathbf{F}_{ph}\end{aligned} \tag{11}$$

This force is expressed as a combination of contributions from the pressure component (hydrodynamic force $\mathbf{F}_H$) and the temperature and chemical potential/concentration gradients at the surface (phoretic force $\mathbf{F}_{ph}$). This result aligns with the Faxén theorem, which relates the phoretic force on a particle to the integral of the surface gradient of the surface tension [16]. Crucially, this equivalence enables the identification of Onsager coefficients, allowing for the computation of phoretic coefficients that quantitatively connect interfacial processes to the resulting particle motion [56].

Since chemical reactions are the main source of entropy production and directly modify interfacial concentrations and temperatures, they provide key information on the self-phoretic mechanism. This understanding allows not only the calculation of slip velocities but also the determination of self-phoretic coefficients and interfacial transport properties. Assuming symmetry in the azimuthal angle, φ, describing the particle surface, the entropy production at the interface as a function of the polar angle, θ, is [56]

$$\sigma \approx \frac{D_s R_g T_0}{C_0 T R^2}\left(\frac{\partial C_s}{\partial\theta}\right)^2 + \frac{D_s R_g T_0}{C_0 T R^2}\left(\frac{\partial C_p}{\partial\theta}\right)^2 + \frac{\kappa}{T^2R^2}\left(\frac{\partial T}{\partial\theta}\right)^2 + \frac{R_g T_0}{T}k_r(\Delta z)^2 > 0 \tag{12}$$

The first and second terms represent the entropy production contributions from the diffusion of substrate and product, respectively, both proportional to the square of their concentration gradients. The third term captures the entropy production due to conductive heat transport along the interface, depending on the square of the temperature gradient. Finally, the last term quantifies the entropy production associated with chemical reactions, expressed as the square of the fugacity difference, $\Delta z = e^{\mu_s/k_B T} - e^{\mu_p/k_b T}$. This expression uses the formulation of entropy production based on fugacity rather than affinity, providing an alternative framework that extends beyond the classical linear relationship with affinity [27,57]. This approach provides a more general and accurate description of standard chemical reactions. Additionally, the entropy production depends on various thermodynamic properties and transport coefficients, including the universal gas constant, R_g, the initial substrate concentration, C_0, substrate diffusivity at the interface, thermal conductivity at the interface, κ, and the reaction rate constant, k_r.

Figure 3A illustrates a typical profile of the entropy production rate along the interface, plotted as a function of the orientation angle of the particle, θ. Since entropy production is influenced by various transport coefficients, such as the interfacial viscosity, η (which affects the surface diffusion coefficient D_s), the figure depicts several entropy profiles corresponding to different values of η. This variation highlights the sensitivity of entropy production to changes in interfacial transport properties, offering insights into how these coefficients influence the thermodynamic behavior of the system.

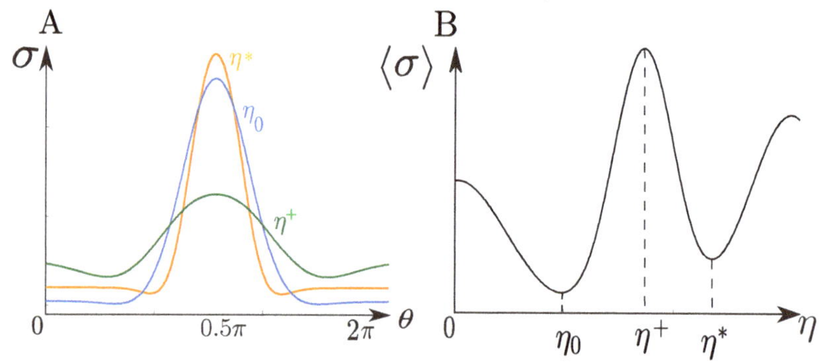

Figure 3. Sketch illustrating the interfacial entropy production for a single catalytic Janus particle. (**A**) Interfacial entropy production rate as a function of the polar angle, θ, describing the particle surface, for different values of the interfacial viscosity. (**B**) Average entropy production rate at the interface as a function of interfacial viscosity, η. Here, η_0 represents the bulk viscosity, η^+ denotes a possible value for the interfacial viscosity, and η^* corresponds to the actual interfacial viscosity.

The average interfacial energy dissipation rate (entropy production rate multiplied by temperature) for a catalytic Janus particle undergoing a first-order exothermic reaction is given by [56]

$$\langle T\sigma \rangle \approx f(U, k_r, D_s)\langle C_s \rangle^2 + g(U, k_r, D_s, U_q, \Delta H_r)\langle C_s \rangle \tag{13}$$

in which f and g are functions of the transport coefficients and thermodynamic properties. The energy dissipation was computed by accounting for substrate and product diffusion, heat conduction, and chemical reactions along the interface [56]. It was observed that dissipation scales linearly and quadratically with the average substrate concentration at the interface.

The entropy production depends on the mass transport coefficient, U, heat transport coefficient, U_q, and the interfacial diffusivity, D_s, quantities that are often challenging to estimate. Notably, the analysis revealed the existence of local maxima in the average energy dissipation as a function of these transport coefficients. This leads to the conclusion that the actual values of these coefficients can be optimized to maximize such non-equilibrium quantities [56] following the maximum entropy production principle [58]. Figure 3B illustrates a schematic profile of the average entropy production at the particle–bulk interface as a function of interfacial viscosity, η. This depiction highlights the potential for estimating the actual value of η by identifying characteristic trends or critical points in the entropy production profile.

4. Active Particles Entropy Production During Self-Organization

The thermodynamic framework for collective behavior in active systems advances our understanding of non-equilibrium principles, with implications for material design and engineering. Insights into clustering, phase transitions, and pattern formation can inform the development of self-assembling materials and microfluidic systems, revealing universal principles applicable in physics, biology, and engineering [59,60]. By framing these behaviors within a thermodynamic context, one can uncover universal principles governing non-equilibrium systems, with applications spanning physics, biology, and engineering.

Self-organization in active matter arises from individual particles' ability to generate local energy flows and respond collectively to interactions, giving rise to phenomena such as swarming, vortex formation, and motility-induced phase separation [17,36,61–65]. On the other hand, traditional tools, such as order parameters, are used to characterize out-of-equilibrium phase transitions. However, these approaches often ignore the irreversible dynamics that drive these transitions and their critical properties. A key bridging concept is entropy production, which has been explored to characterize phase transitions [66,67]. Analyses have been conducted for both continuous and discontinuous phase transitions, extending beyond mean field theories to capture richer nonlocal dynamics, calculating the local entropy production, and providing a more comprehensive understanding of phase behavior in out-of-equilibrium systems [68]. For instance, in clustering and swarming, stochastic thermodynamics have provided tools to quantify entropy production and dissipation [22,69–71] to understand the existence of difference regimes.

Mesoscopic non-equilibrium thermodynamics offers a robust framework to address the nonlinearities in active matter, bridging micro- and macro-scales [25,72]. It proposes a probability density, p, of finding the reactive system composed of active particles at state Γ and time t, which obeys the conservation probability

$$\frac{\partial p}{\partial t} = -\nabla_\Gamma \cdot \mathbf{J} \tag{14}$$

with $\mathbf{J}$ being the probability current in Γ-space. The theory formulates the Gibbs equation, which describes entropy changes in the system:

$$\delta S = -\frac{1}{T}\int \mu(\Gamma,t)\delta p(\Gamma,t)d\Gamma \tag{15}$$

By differentiating the entropy variations with respect to time and using the probability conservation law, we obtain the total entropy production rate in the system:

$$\sigma_{(tot)} = -\frac{1}{T}\int \mathbf{J}(\Gamma,t)\cdot\nabla_\Gamma \mu(\Gamma,t)d\Gamma \tag{16}$$

from which we can infer the probability current

$$\mathbf{J}(\Gamma,t) = -\frac{L(\Gamma)}{k_B T}p(\Gamma,t)\cdot\nabla_\Gamma \mu(\Gamma,t) \tag{17}$$

with $L(\Gamma)$ being an Onsager coefficient and k_B the Boltzmann constant. Substituting the current in the probability conservation law leads to the Fokker–Planck equation:

$$\frac{\partial p}{\partial t} = \nabla_\Gamma \cdot L \nabla_\Gamma p + \frac{Lp}{k_B T} \nabla_\Gamma \Psi \tag{18}$$

where we have used the expression for the chemical potential, $\mu = k_B T \ln p(\Gamma, t) + \Psi(\Gamma)$, with Ψ being a generalized free-energy potential. From the Fokker–Planck equation (Equation (18)), we can obtain the Langevin equations (Equations (2) and (3)) describing the dynamics of the particles. By solving Equations (2) and (3), several structural regimes can be identified, as illustrated in Figure 4A–D. These regimes correspond to different values of the phoretic coefficients (ξ_t,ξ_r) raging between 0 and 22, and -1 and 2, respectively.

In Ref. [39], the MNET formalism was employed to derive Langevin and continuous equations consistent with the second law of thermodynamics, capturing the dynamics of catalytic Janus particles over time. In that framework, the total entropy production rate, $\sigma_{(tot)}$, is composed of two contributions: the interfacial entropy production, σ_i, evaluated at the i-th particle position and mainly given by the chemical reaction, and the bulk entropy production, $\sigma^{(B)}$, which is predominantly governed by substrate diffusion in bulk ($\nabla_{\mathbf{r}} C_s^{(B)}$) and bulk variables ($D_s^{(B)}, T_0^{(B)}, C_0^{(B)}, T^{(B)}$).

$$\sigma_{(tot)} = \sum_{i=1}^{N} \sigma_i + \sigma^{(B)} \approx \frac{R_g T_0}{T} k_r \sum_{i=1}^{N} (\Delta z(\mathbf{r}_i))^2 + \frac{D_s^{(B)} R_g T_0^{(B)}}{C_0^{(B)} T^{(B)}} \| \nabla_{\mathbf{r}} C_s^{(B)} \|^2 \tag{19}$$

When averaged over time, the mesoscopic entropy production provides a measure of dissipation (see Figure 4E). A variation in entropy as a function of the structural parameter, specifically the fraction of aggregated particles, is observed, highlighting the connection between system organization and thermodynamics.

A non-equilibrium free energy for active particles, μ_{AP}, was also derived [39], as a function of the total entropy produced, $\Sigma = \int \int \sigma_{(tot)} d\mathbf{r} dt$, which provides fundamental insights into the thermodynamic feasibility of structure formation and the transition between different structures:

$$N \Delta \mu_{AP} = -\frac{\partial H_p}{\partial \phi_c} + T \frac{\partial S_p}{\partial \phi_c} + T \frac{\partial \Sigma}{\partial \phi_c} \tag{20}$$

in which the free energy depends not only on particles enthalpy, H_p, and entropy, S_p, but also on the total entropy produced, Σ, derivative with respect to the fraction of assembled particles, ϕ_c. This framework serves as an analogue to its equilibrium counterpart, extending free energy principles to systems operating far from equilibrium. Figure 4E is obtained from Equation (19), and it can be analyzed using Equation (20), as demonstrated in Ref. [39], to gain insights into regime transitions. Figure 4E was generated by calculating ϕ and $\sigma_{(tot)}$ for various combinations of ξ_t and ξ_r.

Hydrodynamic effects significantly influence the collective dynamics of active particles, as revealed in recent studies [35]. Analyzing total entropy changes $\Delta S = \Delta_r S + \Sigma$, including reversible and irreversible contributions, as a function of hydrodynamic interaction strength, k_H, offers valuable insights into the most thermodynamically stable regime of active particles in a self-organized state resulting from the condition

$$\frac{\partial \Delta S}{\partial k_H} = 0 \tag{21}$$

This relationship highlights how hydrodynamic effects influence the thermodynamic stability and organization of the system.

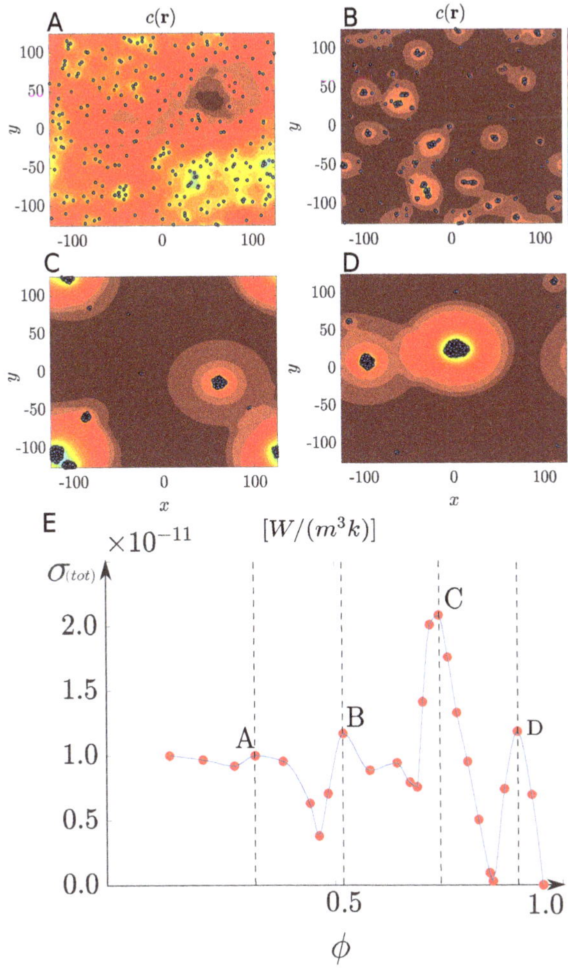

Figure 4. Structural regimes formed by active particles and their non-equilibrium thermodynamic characterization. The representations include both the structural types and the corresponding concentration fields for (**A**) active gas, (**B**) worm-like formations, (**C**) clustering, and (**D**) chemotactic collapse regimes. These results were obtained using Equations (2) and (3), with the $c(\mathbf{r})$ profile obtained in [7], for different dimensionless values of ξ_t and ξ_r. The parameters used were $N = 250$, area fraction 2%, $Pe = 19$, $\xi_t = [0,\ 22]$, $\xi_r = [-1,\ 2]$, particle radius of 1×10^{-6} m, and $D_s = 1 \times 10^{-11}$ m^{2}s^{-1}. To avoid overlapping, a WCA potential was implemented [73]. (**E**) Entropy production computed from Equation (19), after solving particle dynamics, chemical reaction, and substrate diffusion equations, as a function of the structural parameter ϕ, for $T = T_0$, $T^{(B)} = T_0^{(B)} = T_0$, $C_0^{(B)} = 0.1$ M, and $k_r = 0.05$ s^{-1}. The dashed lines indicate the transitions between distinct regimes.

5. Discussions and Conclusions

This work emphasizes entropy production as a fundamental concept to understand not only the dynamics and thermodynamics of active particles but also to calculate phoretic and transport coefficients, as well as to characterize the collective structures that form and the different regimes in which they are encountered.

For single-particle dynamics, entropy production provides a rigorous framework for quantifying interfacial processes, elucidating the role of transport coefficients and the thermodynamic forces driving the motion, such as phoretic velocities. By capturing the interplay between temperature gradients, concentration, and surface properties, it offers a detailed characterization of the irreversible processes governing self-propulsion.

At the collective level, entropy production serves as a unifying metric to analyze regime transitions, structural organization, and the stability of emergent behaviors. From active gas formation to clustering to chemotactic collapse, it provides a consistent thermodynamic perspective on non-equilibrium phenomena. Recent theoretical advances in mesoscopic non-equilibrium thermodynamics bridge the gap between microscopic interactions and macroscopic behavior, enabling models that consistently describe active particle systems at all scales.

The integration of entropy production into the study of active particles reveals the underlying thermodynamic principles governing their behavior, offering a powerful tool for predicting regime transitions and characterizing structural phases. It also provides a basis for quantifying dissipation and energy conversion efficiency in these systems. Challenges remain, particularly in understanding the interplay between entropy production, concentration, temperature, and velocity fields in particle self-organization, as well as in manipulating transport coefficients to control particle dynamics. These open questions suggest interesting avenues for future research aimed at deepening our understanding of active matter.

Author Contributions: All authors have contributed equally. All authors have read and agreed to the published version of the manuscript.

Funding: The authors are grateful for the financial support of MICIU (Spanish Government) under grant No. PID2021-126570NB-I00.

Institutional Review Board Statement: Not applicable.

Data Availability Statement: The original contributions presented in this study are included in the article. Further inquiries can be directed to the corresponding author.

Conflicts of Interest: The authors declare no conflicts of interest.

References

1. Fu, Y.; Yu, H.; Zhang, X.; Malgaretti, P.; Kishore, V.; Wang, W. Microscopic swarms: From active matter physics to biomedical and environmental applications. *Micromachines* **2022**, *13*, 295. [CrossRef] [PubMed]
2. Gompper, G.; Winkler, R.G.; Speck, T.; Solon, A.; Nardini, C.; Peruani, F.; Löwen, H.; Golestanian, R.; Kaupp, U.B.; Alvarez, L.; et al. The 2020 motile active matter roadmap. *J. Phys. Condens. Matter* **2020**, *32*, 193001. [CrossRef] [PubMed]
3. Kurzthaler, C.; Gentile, L.; Stone, H.A. *Out-of-Equilibrium Soft Matter*; The Royal Society of Chemistry: London, UK, 2023. [CrossRef]
4. Liebchen, B.; Mukhopadhyay, A.K. Interactions in active colloids. *J. Phys. Condens. Matter* **2021**, *34*, 083002. [CrossRef] [PubMed]
5. Mendelson, N.H.; Lega, J. A Complex Pattern of Traveling Stripes Is Produced by Swimming Cells of Bacillus subtilis. *J. Bacteriol.* **1998**, *180*, 3285–3294. [CrossRef] [PubMed]
6. Bowick, M.J.; Fakhri, N.; Marchetti, M.C.; Ramaswamy, S. Symmetry, thermodynamics, and topology in active matte. *Phys. Rev. X* **2022**, *12*, 010501.
7. Pohl, O.; Stark, H. Dynamic clustering and chemotactic collapse of self-phoretic active particles. *Phys. Rev. Lett.* **2014**, *112*, 238303. [CrossRef]
8. Toschi, F.; Sega, M. (Eds.) *Flowing Matter*; Soft and Biological Matter; Springer: Cham, Switzerland, 2019. [CrossRef]
9. Fadda, F.; Matoz-Fernandez, D.A.; van Roij, R.; Jabbari-Farouji, S. The interplay between chemo-phoretic interactions and crowding in active colloids. *Soft Matter* **2023**, *19*, 2297–2310. [CrossRef]
10. Gaspard, P.; Kapral, R. Communication: Mechanochemical fluctuation theorem and thermodynamics of self-phoretic motors. *J. Chem. Phys.* **2017**, *147*, 211101. [CrossRef] [PubMed]
11. Gaspard , P.; Kapral, R. Nonequilibrium thermodynamics and boundary conditions for reaction and transport in heterogeneous media. *J. Chem. Phys.* **2018**, *148*, 194114. [CrossRef]
12. Prigogine, I. *Non-Equilibrium Statistical Mechanics*; Courier Dover Publications: New York, NY, USA, 2017.
13. Derjaguin, B.; Sidorenkov, G.; Zubashchenkov, E.; Kiseleva, E. Kinetic phenomena in boundary films of liquids. *Kolloidn. Zh.* **1947**, *9*, 335–347.

14. Anderson, J.L. Colloid Transport by Interfacial Forces. *Annu. Rev. Fluid Mech.* **1989**, *21*, 61–99. [CrossRef]
15. Bedeaux, D.; Albano, A.; Mazur, P. Boundary conditions and non-equilibrium thermodynamics. *Phys. Stat. Mech. Its Appl.* **1976**, *82*, 438–462. [CrossRef]
16. Bafaluy, J.; Pagonabarraga, I.; Rubí, J.; Bedeaux, D. Thermocapillary motion of a drop in a fluid under external gradients. Faxén theorem. *Phys. Stat. Mech. Its Appl.* **1995**, *213*, 277–292. [CrossRef]
17. Takatori , S.C.; Brady, J.F. Towards a thermodynamics of active matter. *Phys. Rev. E* **2015**, *91*, 032117. [CrossRef]
18. Chaudhuri, D. Active Brownian particles: Entropy production and fluctuation response. *Phys. Rev. E* **2014**, *90*, 022131. [CrossRef] [PubMed]
19. Burkholder, E.W.; Brady, J.F. Fluctuation-dissipation in active matter. *J. Chem. Phys.* **2019**, *150*, 184901. [CrossRef]
20. Speck, T.; Jack, R.L. Ideal bulk pressure of active Brownian particles. *Phys. Rev. E* **2016**, *93*, 062605. [CrossRef] [PubMed]
21. Seifert, U. From Stochastic Thermodynamics to Thermodynamic Inference. *Annu. Rev. Condens. Matter Phys.* **2019**, *10*, 171–192. [CrossRef]
22. Fodor , E.; Jack, R.L.; Cates, M.E. Irreversibility and Biased Ensembles in Active Matter: Insights from Stochastic Thermodynamics. *Annu. Rev. Condens. Matter Phys.* **2022**, *13*, 215–238. [CrossRef]
23. Cates, M.E.; Fodor, E.; Markovich, T.; Nardini, C.; Tjhung, E. Stochastic Hydrodynamics of Complex Fluids: Discretisation and Entropy Production. *Entropy* **2022**, *24*, 254. [CrossRef]
24. Arango-Restrepo, A.; Rubi, J.M.; Barragán, D. Understanding Gelation as a Nonequilibrium Self-Assembly Process. *J. Phys. Chem.* **2018**, *122*, 4937–4945. [CrossRef]
25. Reguera, D.; Rubi, J.M.; Vilar, J.M.G. The mesoscopic dynamics of thermodynamic systems. *J. Phys. Chem. B* **2005**, *109*, 21502–21515. [CrossRef] [PubMed]
26. De Groot, S.R.; Mazur, P. *Non-Equilibrium Thermodynamics*; Courier Corporation: Amsterdam, The Netherlands, 2013.
27. Rubi, J.; Bedeaux, D.; Kjelstrup, S.; Pagonabarraga, I. Chemical cycle kinetics: Removing the limitation of linearity of a non-equilibrium thermodynamic description. *Int. J. Thermophys.* **2013**, *34*, 1214–1228. [CrossRef]
28. Olarte-Plata, J.D.; Bresme, F. Orientation of Janus particles under thermal fields: The role of internal mass anisotropy. *J. Chem. Phys.* **2020**, *152*, 204902. [CrossRef] [PubMed]
29. Huang, M.J.; Schofield, J.; Kapral, R. Chemotactic and hydrodynamic effects on collective dynamics of self-diffusiophoretic Janus motors. *New J. Phys.* **2017**, *19*, 125003. [CrossRef]
30. Gilbert, T.; Dorfman, J.R. Entropy production: From open volume-preserving to dissipative systems. *J. Stat. Phys.* **1999**, *96*, 225–269. [CrossRef]
31. Vicsek, T.; Czirók, A.; Ben-Jacob, E.; Cohen, I.; Shochet, O. Novel Type of Phase Transition in a System of Self-Driven Particles. *Phys. Rev. Lett.* **1995**, *75*, 1226–1229. [CrossRef] [PubMed]
32. Saha, S.; Golestanian, R.; Ramaswamy, S. Clusters, asters, and collective oscillations in chemotactic colloids. *Phy. Rev. E* **2014**, *89*, 062316. [CrossRef] [PubMed]
33. Nasouri, B.; Golestanian, R. Exact Phoretic Interaction of Two Chemically Active Particles. *Phys. Rev. Lett.* **2020**, *124*, 168003. [CrossRef] [PubMed]
34. Scagliarini, A.; Pagonabarraga, I. Unravelling the role of phoretic and hydrodynamic interactions in active colloidal suspensions. *Soft Matter* **2020**, *16*, 8893–8903. [CrossRef] [PubMed]
35. Torrenegra-Rico, J.D.; Arango-Restrepo, A.; Rubi, J.M. Self-organization of Janus particles: Impact of hydrodynamic interactions in substrate consumption for structure formation. *J. Chem. Phys.* **2024**, *161*, 224101. [CrossRef]
36. Zhao, H.; Košmrlj, A.; Datta, S.S. Chemotactic Motility-Induced Phase Separation. *Phys. Rev. Lett.* **2023**, *131*, 118301. [CrossRef] [PubMed]
37. Ezhilan, B.; Pahlavan, A.A.; Saintillan, D. Chaotic dynamics and oxygen transport in thin films of aerotactic bacteria. *Phys. Fluids* **2012**, *24*, 091701. [CrossRef]
38. Peng, Z.; Kapral, R. Self-organization of active colloids mediated by chemical interactions. *Soft Matter* **2024**, *20*, 1100–1113. [CrossRef] [PubMed]
39. Torrenegra-Rico, J.D.; Arango-Restrepo, A.; Rubí, J.M. Nonequilibrium thermodynamics of Janus particle self-assembly. *J. Chem. Phys.* **2022**, *157*, 104103. [CrossRef]
40. Bresme, F.; Olarte-Plata, J.D.; Chapman, A.; Albella, P.; Green, C. Thermophoresis and thermal orientation of Janus nanoparticles in thermal fields. *Eur. Phys. J.* **2022**, *45*, 59. [CrossRef] [PubMed]
41. Jiang, M.; Chapman, A.; Olarte-Plata, J.D.; Bresme, F. Controlling local thermal gradients at molecular scales with Janus nanoheaters. *Nanoscale* **2023**, *15*, 10264–10276. [CrossRef]
42. Olarte-Plata, J.D.; Bresme, F. Impact of the Interfacial Kapitza Resistance on Colloidal Thermophoresis. *ACS Omega* **2024**, *9*, 43779–43784. [CrossRef] [PubMed]
43. Cao, Z.; Su, J.; Jiang, H.; Hou, Z. Effective entropy production and thermodynamic uncertainty relation of active Brownian particles. *Phys. Fluids* **2022**, *34*, 053310. [CrossRef]

44. Ro, S.; Guo, B.; Shih, A.; Phan, T.V.; Austin, R.H.; Levine, D.; Chaikin, P.M.; Martiniani, S. Model-Free Measurement of Local Entropy Production and Extractable Work in Active Matter. *Phys. Rev. Lett.* **2022**, *129*, 220601. [CrossRef]
45. Dolai, P.; Maes, C.; Netočný, K. Calorimetry for active systems. *SciPost Phys.* **2023**, *14*, 126. [CrossRef]
46. Paoluzzi, M. Measuring Entropy in Active-Matter Systems. *Physics* **2022**, *15*, 179. [CrossRef]
47. Onsager, L. Reciprocal Relations in Irreversible Processes. II. *Phys. Rev.* **1931**, *38*, 2265–2279. [CrossRef]
48. Dufty, J.W.; Rub, J.M. Generalized Onsager symmetry. *Phys. Rev. A* **1987**, *36*, 222–225. [CrossRef]
49. Paxton, W.F.; Kistler, K.C.; Olmeda, C.C.; Sen, A.; St. Angelo, S.K.; Cao, Y.; Mallouk, T.E.; Lammert, P.E.; Crespi, V.H. Catalytic Nanomotors:Autonomous Movement of Striped Nanorods. *J. Am. Chem. Soc.* **2004**, *126*, 13424–13431. [CrossRef]
50. Gaspard, P.; Kapral, R. The stochastic motion of self-thermophoretic Janus particles. *J. Stat. Mech. Theory Exp.* **2019**, *2019*, 074001. [CrossRef]
51. Gaspard, P.; Kapral, R. Active Matter, Microreversibility, and Thermodynamics. *Research* **2020**, *2020*, 9739231. [CrossRef] [PubMed]
52. Michelin, S.; Lauga, E. Phoretic self-propulsion at finite Péclet numbers. *J. Fluid Mech.* **2014**, *747*, 572–604. [CrossRef]
53. Popescu, M.N.; Uspal, W.E.; Dietrich, S. Self-diffusiophoresis of chemically active colloids. *Eur. Phys. J. Spec. Top.* **2016**, *225*, 2189–2206. [CrossRef]
54. Moran, J.L.; Posner, J.D. Phoretic Self-Propulsion. *Annu. Rev. Fluid Mech.* **2017**, *49*, 511–540. [CrossRef]
55. De Corato, M.; Pagonabarraga, I. Onsager reciprocal relations and chemo-mechanical coupling for chemically active colloids. *J. Chem. Phys.* **2022**, *157*, 084901. [CrossRef] [PubMed]
56. Arango-Restrepo, A.; Rubi, J.M. Interplay of phoresis and self-phoresis in active particles: Transport properties, phoretic, and self-phoretic coefficients. *J. Chem. Phys.* **2024**, *161*, 054906. [CrossRef] [PubMed]
57. Pagonabarraga, I.; Pérez-Madrid, A.; Rubí, J. Fluctuating hydrodynamics approach to chemical reactions. *Phys. Stat. Mech. Its Appl.* **1997**, *237*, 205–219. [CrossRef]
58. Martyushev, L.; Seleznev, V. Maximum entropy production principle in physics, chemistry and biology. *Phys. Rep.* **2006**, *426*, 1–45. [CrossRef]
59. Arango-Restrepo, A.; Barragán, D.; Rubi, J.M. Self-assembling outside equilibrium: Emergence of structures mediated by dissipation. *Phys. Chem. Chem. Phys.* **2019**, *21*, 17475–17493. [CrossRef]
60. Popescu, M.N.; Uspal, W.E.; Bechinger, C.; Fischer, P. Chemotaxis of Active Janus Nanoparticles. *Nano Lett.* **2018**, *18*, 5345–5349. [CrossRef] [PubMed]
61. Klamser, J.U.; Kapfer, S.C.; Krauth, W. Thermodynamic phases in two-dimensional active matter. *Nat. Commun.* **2018**, *9*, 1–8.
62. Cates, M.E.; Tailleur, J. Motility-induced phase separation. *Annu. Rev. Condens. Matter Phys.* **2015**, *6*, 219–244. [CrossRef]
63. Stenhammar, J.; Marenduzzo, D.; Allen, R.J.; Cates, M.E. Phase behaviour of active Brownian particles: The role of dimensionality. *Soft Matter* **2014**, *10*, 1489–1499. [CrossRef] [PubMed]
64. Takatori, S.C.; Brady, J.F. Forces, stresses and the (thermo?) dynamics of active matter. *Curr. Opin. Colloid Interface Sci.* **2016**, *21*, 24–33. [CrossRef]
65. Caprini, L.; Marini Bettolo Marconi, U.; Puglisi, A. Activity induced delocalization and freezing in self-propelled systems. *Sci. Rep.* **2019**, *9*, 1–9. [CrossRef]
66. Zhang, Y.; Barato, A.C. Critical behavior of entropy production and learning rate: Ising model with an oscillating field. *J. Stat. Mech. Theory Exp.* **2016**, *2016*, 113207. [CrossRef]
67. Seara, D.S.; Yadav, V.; Linsmeier, I.; Tabatabai, A.P.; Oakes, P.W.; Tabei, S.M.A.; Banerjee, S.; Murrell, M.P. Entropy production rate is maximized in non-contractile actomyosin. *Nat. Commun.* **2018**, *9*, 4948. [CrossRef]
68. Noa, C.E.F.; Harunari, P.E.; de Oliveira, M.J.; Fiore, C.E. Entropy production as a tool for characterizing nonequilibrium phase transitions. *Phys. Rev. E* **2019**, *100*, 012104. [CrossRef] [PubMed]
69. Shklyaev, O.E.; Shum, H.; Balazs, A.C. Using Chemical Pumps and Motors To Design Flows for Directed Particle Assembly. *Accounts Chem. Res.* **2018**, *51*, 2672–2680. [CrossRef] [PubMed]
70. O'Byrne, J.; Solon, A.; Tailleur, J.; Zhao, Y. An Introduction to Motility-induced Phase Separation. In *Out-of-Equilibrium Soft Matter: Active fluids*; The Royal Society of Chemistry: London, UK, 2023.
71. Mandal, D.; Klymko, K.; DeWeese, M.R. Entropy Production and Fluctuation Theorems for Active Matter. *Phys. Rev. Lett.* **2017**, *119*, 258001. [CrossRef]
72. Arango-Restrepo, A.; Barragán, D.; Rubi, J.M. A Criterion for the Formation of Nonequilibrium Self-Assembled Structures. *J. Phys. Chem.* **2021**, *125*, 1838–1845. [CrossRef]
73. Weeks, J.D.; Chandler, D.; Andersen, H.C. Role of repulsive forces in determining the equilibrium structure of simple liquids. *J. Chem. Phy.* **1971**, *54*, 5237–5247. [CrossRef]

entropy

MDPI

Article

Thermodynamics-like Formalism for Immiscible and Incompressible Two-Phase Flow in Porous Media

Alex Hansen *,† and Santanu Sinha †

PoreLab, Department of Physics, Norwegian University of Science and Technology NTNU, N-7491 Trondheim, Norway; santanu.sinha@ntnu.no

* Correspondence: alex.hansen@ntnu.no

† These authors contributed equally to this work.

Abstract: It is possible to formulate an immiscible and incompressible two-phase flow in porous media in a mathematical framework resembling thermodynamics based on the Jaynes generalization of statistical mechanics. We review this approach and discuss the meaning of the emergent variables that appear, agiture, flow derivative, and flow pressure, which are conjugate to the configurational entropy, the saturation, and the porosity, respectively. We conjecture that the agiture, the temperature-like variable, is directly related to the pressure gradient. This has as a consequence that the configurational entropy, a measure of how the fluids are distributed within the porous media and the accompanying velocity field, and the differential mobility of the fluids are related. We also develop elements of another version of the thermodynamics-like formalism where fractional flow rather than saturation is the control variable, since this is typically the natural control variable in experiments.

Keywords: flow in porous media; immiscible two-phase flow; co-moving velocity

1. Introduction

In spite of the immiscible multi-phase flow in porous media being a field of great industrial importance for more than a century, basic research in this field has not received the attention that it deserves. The main reason for this is that the field is fragmented, being spread over many different disciplines such as biology, geology, geophysics, physics, chemistry, and materials science. Another reason is that it is easy to get the impression that it is not "clean" in the sense that it is a field where theories may be built. Rather, it seems to be a collection of phenomena in need of classification. It is the aim of this paper to convince the readership that this is indeed not the case. There is a physics of flow in porous media [1] where general principles may be formulated. We will, in the following, present the readers with a review of our ongoing attempts at formulating a mathematical framework for immiscible and incompressible two-phase flow in porous media.

In 1907, Buckingham introduced several of the central concepts used today in describing immiscible and incompressible fluid flows in porous media [2,3], but without joining them through equations. This was first performed by Richards in 1931 [4] who, as Buckingham, placed the flow problem in the context of water movement in soils. In 1936, building on the work by Wyckoff and Botset [5], Muscat and Meres [6] introduced the concept of relative permeability, a concept generalizing the notion of "capillary conductance" that Buckingham introduced [2]. With the introduction of the capillary pressure curve by Leverett [7] into the framework established by Muscat and Meres, the generalized Darcy constitutive equations that are generally used today in, e.g., reservoir simulation, were in place. The formulation of the problem in the context of water in soils that Buckingham pioneered, from the work of Richards, forms an alternate formulation of essentially the same ideas.

We review briefly relative permeability theory—the only approach to immiscible two-phase flow in porous media in practical use today—in Section 2. Relative permeability theory is purely phenomenological, and has well-known deficiencies. As a result, there is an ongoing effort to derive effective equations going beyond relative permeability theory based on the physics at the pore scale. This effort is based on *homogenization* which we review briefly in Section 3.

In Section 4, we introduce our approach to immiscible two-phase flow in porous media based on a generalization of statistical mechanics by Jaynes [8,9]. Statistical mechanics in its traditional role is to connect a molecular description of matter to a description at the much larger scales where matter appears continuous, namely thermodynamics. In adopting this idea to two-phase flow in porous media, we find a thermodynamics-like framework on large scales that together with conservation laws provide a closed set of equations that has the equations of relative permeability theory as a special case. As in ordinary thermodynamics, the difference between intensive and extensive variables is central. We explore this in Section 5. The intensive variables appearing in the thermodynamics-like formalism we derive are emergent in that they have no equivalent on the pore scale. This is discussed in Section 6. The temperature-like intensive variable, *agiture*, we conjecture to be proportional to the pressure gradient in Section 6.1. This conjecture has as a consequence in that the *configurational entropy*, which is an example of Shannon's information entropy, is related to the differential mobility of the fluids. The differential mobility is the derivative of the flow velocity with respect to the pressure gradient, and the mobility is the flow velocity divided by the pressure gradient. The *flow derivative* we discuss in Section 6.2 is the conjugate variable to the saturation, which is the fractional volume. We end the section by discussing the conjugate of the porosity, the *flow pressure*, in Section 6.3.

In Section 7, we introduce the *co-moving velocity* which relates the thermodynamic velocities—corresponding to partial molar volumes in ordinary thermodynamics—to the seepage velocities, which are the average pore velocity of each of the immiscible fluids. This concept has the potential to become a very useful concept in practical calculations.

Section 8 contains material that has not been presented before. We start by pointing out a choice made when constructing the Jaynes statistical mechanics in Section 4, and how it could have been made differently. Then we discuss what the ensuing thermodynamics-like formalism would be as a consequence: rather than having saturation as one of the control parameters, we find fractional flow rate being a control parameter. This is useful, as this is the control parameter used in core flooding experiments.

We end by summarizing and concluding in Section 9.

2. Relative Permeability Theory

The core idea of the relative permeability approach may be summarized as follows: We have two immiscible fluids, one more wetting with respect to the porous matrix than the other fluid. We will refer to them as the wetting (w) and non-wetting (n) fluid, respectively. From the perspective of the wetting fluid, the pore space it sees is the total pore space of the porous medium minus the pore space occupied by the non-wetting fluid, and vice versa. The relative reduction of pore space for each fluid implies a reduction in effective permeability for each fluid. We express these statements in the form of two constitutive equations, one for each fluid [1],

$$\vec{v}_w = -\frac{Kk_{rw}(S_w)}{\mu_w\phi S_w}\nabla P_w\,, \tag{1}$$

$$\vec{v}_n = -\frac{Kk_{rn}(S_w)}{\mu_n\phi S_n}\nabla P_n\,. \tag{2}$$

Here, $\vec{v}_w$ and $\vec{v}_n$ are the pore velocities of the wetting and non-wetting fluids, K is the (isotropic) permeability of the porous medium, μ_w and μ_n are the viscosities of each fluid, ϕ is the porosity of the porous medium, P_w and P_n are the fluid pressures, S_w and S_n are the wetting and non-wetting saturations, and $k_{rw}(S_w)$ and $k_{rn}(S_w)$ are the relative permeabilities of the two fluids. It is an essential *assumption* in this theory that these two quantities are functions of the saturations alone. The saturations S_w and S_n are defined as the fraction of pore space occupied by each fluid so that

$$S_w + S_n = 1\,. \tag{3}$$

The difference in pressure between the two fluids is defined as the capillary pressure curve P_c,

$$P_n - P_w = P_c(S_w)\,. \tag{4}$$

It is an assumption that the capillary pressure curve only depends on the saturation S_w. The capillary pressure curve is a particularly difficult quantity, both conceptually and in terms of measurement (see, e.g., [10]). We define the average pore velocity $\vec{v}_p$ as

$$\vec{v}_p = S_w\vec{v}_w + S_n\vec{v}_n\,, \tag{5}$$

i.e., we are using a volume average. This makes sense since we are assuming the fluids to be incompressible,

$$\nabla\cdot[\phi\vec{v}_p] = 0\,. \tag{6}$$

Volume conservation then gives

$$\phi\frac{\partial S_w}{\partial t} + \nabla\cdot[\phi S_w\vec{v}_w] = 0\,, \tag{7}$$

$$\phi\frac{\partial S_n}{\partial t} + \nabla\cdot[\phi S_n\vec{v}_n] = 0\,, \tag{8}$$

where t is time. This set of equations is closed as long as $k_{rw}(S_w)$, $k_{rn}(S_w)$, and $P_c(S_w)$ are provided. In an industrial context, Special Core Analysis (SCAL) forms the work flow for obtaining this information.

3. Homogenization

It is well known that assuming the three constitutive functions depend only on the saturation is generally not correct. The relative permeabilities do depend not only on the saturation, but also how the fluids arrange themselves in the porous medium, and this depends, in turn, on the flow rate. The capillary pressure curve is highly hysteretic, signaling that there are missing variables. However, it has turned out to be difficult to go beyond this almost-90-year-old phenomenological framework. The dominating approach is based on homogenization. This approach may be divided into two branches: that which focuses on momentum transfer, and that which focuses on energy transfer. In the first case, one starts with the hydrodynamic equations on the pore scale [11–16], and in the second case, one sets up a thermodynamic description at the pore scale [17–20]. Then comes the coarse-graining, i.e., spatial averaging. There are several ways of performing this; perhaps that of Whitaker is the most well-known [21,22]. It is based on equating the average of the gradient of a variable associated with pore space to the gradient of the average variable plus an integral over the surface area of the pores. Since the surface area of porous media scales as the volume (i.e., the surface area is *extensive*, the defining property of porous media), this integral does not vanish as one goes up in scale. We split the variable appearing in the surface integral into an average part and a fluctuating part. This leaves us with having the average, and gradients of the average expressed in terms of the fluctuations of the original variable. The last step is to make an independent assumption on how the fluctuating variables is related to the averages—a closure assumption. With this, the equations between the original, pore-scale variables have been turned into equations between spatial averages of these variables.

Whereas the first of these two approaches is based on momentum transfer, the second approach is based on energy transfer, i.e., thermodynamics. One starts with a thermodynamic description on the pore level which is then homogenized. This approach has developed into Thermodynamically Constrained Averaging Theory (TCAT) [23–26].

A recent paper by McClure et al. [27] attempts to derive the relative permeability equations from an energy budget based on thermodynamic considerations and homogenization. The relative permeability equations do appear as a first term in a series expansion. However, it is not shown in [27] that the higher order terms are negligible.

The topology of a geometric object such as a porous medium may be described using the four *Minkowski functionals:* volume, surface area, mean curvature, and the Euler characteristics. The *Hadwiger theorem* states that the Minkowski functionals form a complete basis set for all extensive functions that are invariant with respect to the orientation of the object [28]. The use of this theorem to characterize the free energy of fluids in a porous medium combined with homogenization forms another approach to the scale-up problem [29–32]. Homogenization as seen so far is based on *spatial* averages only. McClure et al. [33,34] emphasize that there is also time which one should average over and that different processes work on different time scales.

An approach circumventing the complexities associated with homogenization is based on classical non-equilibrium thermodynamics [35–40]. By using the extensiveness of the internal energy of the fluids in the porous medium, the Euler theorem for homogeneous functions allows for defining thermodynamic variables such as pressure and chemical potentials on the Darcy scale. Gradients in the intensive variables are introduced, and the machinery of classical non-equilibrium thermodynamics [35,36] is then set in motion. The underlying homogenization is somewhat hidden in this approach, but it is underlying the way a Representative Elementary Volume (REV) is defined and used.

The main practical difficulty with homogenization is the complexity of the ensuing equations and the large number of variables that are necessary. The root of this difficulty is that homogenization can only produce averages over the original variables, and thus cannot produce new types of variables which go hand-in-hand with capturing emergent properties [41].

4. Statistical Mechanics

The essence of these efforts has been to find a description of immiscible and incompressible two-phase flow in porous media at scales where the medium acts as a continuum. Including compressibility would necessitate switching from conserved fluid volume to conserved fluid mass. It would also couple the theory to ordinary thermodynamics through equations of state for the compressible fluids. We wish to avoid these complications at the present stage. This is an example of *upscaling*. The situation is thus that we have a continuum scale phenomenological theory; a relative permeability theory that is simple enough to be used in practical situations, but which is approximate. Attempts at going beyond this theory has, however, not led to practical applications due to the complexity that arises. However, if we look beyond the problem we discuss here, there is an example of successful upscaling that is very familiar: The scale-up from a molecular description to thermodynamics based on statistical mechanics. At molecular scales, the position and momentum of each molecule are the variables that describe a monoatomic gas. This works even if we are dealing with millions of molecules in a molecular dynamics simulation [42]. However, it is rather of the order 10^{23} molecules that are relevant in, e.g., chemistry. A molecular description at such scales is useless. Here, *emergent variables* such as temperature and pressure adequately describe the gas. Statistical mechanics converts the molecular description to thermodynamics which not only contains emergent variables, but also relations between them. These variables and their relations have no meaning within a molecular description.

In the mid-fifties, Jaynes generalized statistical mechanics from being a theory for molecular matter to being based on information theory [8]. This was performed by replacing Boltzmann's statistical interpretation of thermodynamic entropy by Shannon's generalized information entropy [43]. The information entropy is a quantitative measure of what is *not* known about a system. Suppose we have a stochastic process that can produce N possible outcomes. We number the outcomes from 1 to N, assuming a probability p_i for the ith outcome, x_i. A quantitative measure of our ignorance about this process must be independent of how we group events together. For example, if we consider composite events such as $p_{ij} = p_i p_j$, i.e., the probability that an event i is followed by an event j, then the information entropy must remain the same. If we know nothing about the process, its information entropy must be at a maximum, since our ignorance is maximal. In order to utilize this, Shannon generalized to N outcomes the Laplace principle of insufficient reason [44], which states that the optimal choice of probabilities for a stochastic process with *two* possible outcomes is to assign them equal probability. Hence, the optimal choice when there are N outcomes is to assign all probabilities the same value, $p_i = 1/N$. There is only one function that can be constructed from p_i with these properties,

$$S_I = -\sum_{i=1}^{N} p_i \ln p_i \; , \tag{9}$$

which is then the information or Shannon entropy. Jaynes then posed the question: what happens if we *do* know something about the system? Suppose we know the average of the outcomes,

$$\langle x \rangle = \frac{1}{N} \sum_{i=1}^{N} p_i x_i \; . \tag{10}$$

Generalizing the principle of insufficient reason further, Jaynes proposed that the probabilities p_i in this case should be those that maximize the information entropy (9) given a fixed value for $\langle x \rangle$ (Equation (10)).

Jaynes provided a set of circumstances for his generalized statistical mechanics to be applicable. Hansen et al. [9] demonstrated that these circumstances are met for steady-state immiscible two-phase flow in porous media, and hence proceeded to construct such a statistical mechanics. The generalized statistical mechanics will produce a thermodynamics-like framework at large scales. This thermodynamic-like framework is the end goal of this approach.

Suppose we have a cylindrical pore sample as shown in Figure 1. Two immiscible fluids enter at the bottom and leave at the top. The side walls are close off. We assume that the pore structure and chemical composition of the matrix to be statistically uniform throughout the sample. At some distance from the lower edge of the sample, one may assume that the fluids have mixed sufficiently to assume steady-state flow. That is, if we make transversal cuts through the sample, the structure of the porous medium and the distribution of fluids in the pore space to be statistically uniform: Comparing one cut to another, we will not be able to determine which is closest to the inlet. Figure 1 shows one such cut. Within this cut, we pick out a *Representative Elementary Area* (REA), which is large enough to represent the statistics of the sample and small enough to have the flow fluctuate. There is an volumetric flow rate Q_p passing through the REA which may be split into a volumetric flow rate of the wetting fluid Q_w and the volumetric flow rate of the non-wetting fluid Q_n, i.e.,

$$Q_p = Q_w + Q_n \,. \tag{11}$$

The area of the REA is A. An area $A_p < A$ cuts through pore space. This area may be divided into the area cutting through the pore space filled with wetting fluid, A_w and the area cutting though the pore space filled with the non-wetting fluid, A_n, and we have that

$$A_p = A_w + A_n \,. \tag{12}$$

At a given time for a given REA, the fluid configuration, characterized by the velocity field and where the fluids are, is X. The quantities that describe the REA depend on X, so that we have $Q_p(X)$, $Q_w(X)$, $Q_n(X) = Q_p(X) - Q_w(X)$, $A_p(X)$, $A_w(X)$, and $A_n(X) = A_p(X) - A_w(X)$. Of course, $A_p(X)$ is not dependent on the velocity field, nor the distribution of the two fluids, but only on the shape of pore space. Let us denote $p(X)$ the probability density to find fluid and pore space configuration X. We define a *configurational entropy*,

$$S = -\int dX\, p(X)\, \ln p(X) \,. \tag{13}$$

The probability density $p(X)$ is *continuous*, so that we are dealing with differential entropy [45] in Equation (13). Another way of dealing with the transition from discrete to continuous systems is that of Jaynes [46].

We assume that we know the averages (in time and position of the REA) of the variables just described, i.e.,

$$Q_u = \int dX\, p(X)\, Q_u(X) \,, \tag{14}$$

$$A_w = \int dX\, p(X)\, A_w(X) \,, \tag{15}$$

$$A_p = \int dX\, p(X)\, A_p(X) \,. \tag{16}$$

The variable Q_u is related to the volumetric flow rate Q_p, but we will defer its definition until Equation (30).

Following the Jaynes maximum entropy principle, we maximize the entropy (13) given the constraints (14) to (16), finding that

$$p(X;\lambda_u,\lambda_w,\lambda_p) = \frac{1}{Z(\lambda_u,\lambda_w,\lambda_p)} \exp\left[-\lambda_u Q_u(X) - \lambda_w A_w(X) - \lambda_p A_p(X)\right] \,, \tag{17}$$

where

$$Z(\lambda_u,\lambda_w,\lambda_p) = \int dX\, \exp\left[-\lambda_u Q_u(X) - \lambda_w A_w(X) - \lambda_p A_p(X)\right] \tag{18}$$

is the normalization factor, also known as the partition function. Three new *emergent variables* have appeared, λ_u, λ_w, and λ_p. We write the partition function as

$$Z(\lambda_u,\lambda_w,\lambda_p) = \exp\left[-\lambda_u Q_z(\lambda_u,\lambda_w,\lambda_p)\right] \,. \tag{19}$$

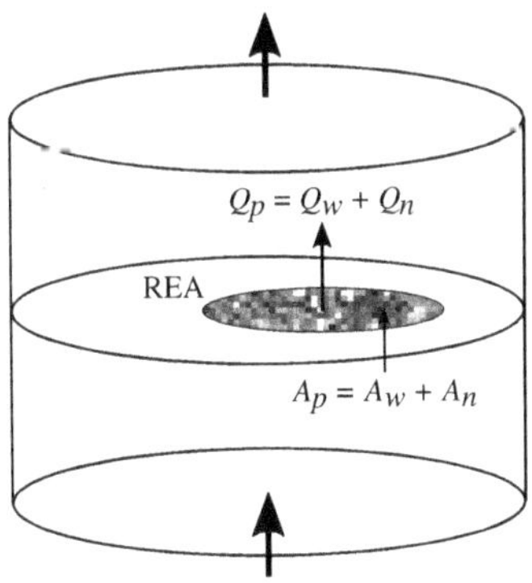

Figure 1. We define a plane that cuts through the cylindrical pore sample orthogonally to the average flow direction. We define Representative Elementary Area (REA) within the plane. There is a wetting fluid flow rate Q_w and a non-wetting fluid flow rate Q_n passing through the REA. The total flow rate is Q_p. The wetting fluid covers an area A_w of the REA and the non-wetting fluid an area A_n. The total pore area of the REA is A_p.

The partition function is in some sense the "end product" of a statistical mechanics approach to a problem. It is the generating function for the probability distribution for microscopic configurations. With it, all macroscopic averages may be calculated.

In ordinary statistical mechanics, a Hamiltonian, i.e., a microscopic energy function is necessary input for calculating the partition function. In the present problem we have three pore-level functions, $Q_u(X)$, $A_w(X)$ and $A_p(X)$. The problem in ordinary statistical mechanics, is that even for seemingly simple problems such as the two-dimensional Ising model for magnetic systems, calculating the partition function may be extremely difficult (cfr. Onsager's calculation of the partition function for this system [47]). The problem is surely no less difficult in the porous media context. Perhaps, it is even worse in that we do not have good models for the three microscopic functions in contrast to ordinary statistical mechanics where model Hamiltonians are legio.

A byproduct of the statistical mechanics approach is the emergence of a set of thermodynamics-like relations between the REA-scale variables. This framework will be in place whether it is possible to calculate the partition function or not. We will now develop this thermodynamics-like framework for immiscible two-phase flow in porous media.

We start by calculating the entropy S by inserting Equation (17) in Equation (13) and using Equation (19). We obtain

$$S(Q_z, A_w, A_p) = -\lambda_u Q_z(\lambda_u, \lambda_w, \lambda_p) + \lambda_u Q_u + \lambda_w A_w + \lambda_p A_p \,. \tag{20}$$

We rewrite this equation as

$$Q_z = Q_u - \frac{1}{\lambda_u}\, S + \frac{\lambda_w}{\lambda_u}\, A_w + \frac{\lambda_p}{\lambda_u}\, A_p \,. \tag{21}$$

It is convenient to define the new variables

$$\theta = +\frac{1}{\lambda_u}\,, \tag{22}$$

$$\mu = -\frac{\lambda_w}{\lambda_u}\,, \tag{23}$$

$$\pi = -\frac{\lambda_p}{\lambda_u}\,. \tag{24}$$

We may then write Equation (21) as

$$Q_u(S, A_w, A_p) = Q_z(\theta, \mu, \pi) + \theta S + \mu\, A_w + \pi\, A_p\,, \tag{25}$$

where we have used that this is a Legendre transform (assuming convexity of the involved functions), since

$$S = -\left(\frac{\partial Q_z}{\partial \theta}\right)_{\mu,\pi}, \tag{26}$$

$$A_w = -\left(\frac{\partial Q_z}{\partial \mu}\right)_{\theta,\pi}, \tag{27}$$

$$A_p = -\left(\frac{\partial Q_z}{\partial \pi}\right)_{\theta,\mu}. \tag{28}$$

We have finally arrived at defining the volumetric flow rate Q_p in Equation (11),

$$Q_p(\theta, A_w, A_p) = Q_z(\theta, \mu, \pi) + \mu A_w + \pi A_p\,, \tag{29}$$

and the relation between Q_u and Q_p is then

$$Q_p(\theta, A_w, A_p) = Q_u(S, A_w, A_p) - S\theta\,. \tag{30}$$

We see that θ (Equation (22)) plays the same role as temperature in ordinary thermodynamics. Hansen et al. [9] named it *agiture*, which stands for *agitation temperature*. They also named μ (Equation (27)) the *flow derivative*. It is analogous to *chemical potential* in ordinary thermodynamics. Lastly, π, defined in Equation (28), they called the *flow pressure*. It has no equivalent in ordinary thermodynamics. We note that the unit of the agiture is flow rate, whereas the unit of the flow derivative and the flow pressure is velocity.

5. Extensive and Intensive Variables

The variables Q_u, Q_z, Q_p, A_w, A_p, and S are extensive in the area of the REA: double its area A, and all these variable double. On the other hand, the variables θ, μ, and π are intensive. They do not change when A is changed.

We rescale Q_p, A_w, and A_p by A, defining

$$\phi = \frac{A_p}{A}, \tag{31}$$

$$S_w\phi = \frac{A_w}{A} = \frac{A_w}{A_p}\phi\,, \tag{32}$$

$$v_d = v_p\phi = \frac{Q_p}{A} = \frac{Q_p}{A_p}\phi\,, \tag{33}$$

$$s = \frac{S}{A}, \tag{34}$$

where ϕ is the porosity, S_w the wetting saturation, v_d the Darcy velocity, v_p the pore or seeping velocity, and s the entropy density.

In order to keep track of what happens to the variables in, e.g., $Q_p(\theta, A_w, A_p, A)$, we express the extensivity through the scaling relation

$$\lambda Q_p(\theta, A_w, A_p, A) = Q_p(\theta, \lambda A_w, \lambda A_p, \lambda A)\,. \tag{35}$$

We set the scale factor $\lambda = 1/A$, finding

$$v_d(\theta, S_w, \phi) = v_p(\theta, S_w, \phi)\phi = \frac{1}{A} Q_p(\theta, A_w, A_p, A) = Q_p(\theta, S_w\phi, \phi, 1) \,. \tag{36}$$

Using the same arguments, we find that

$$s = s(\theta, S_w, \phi) \,. \tag{37}$$

6. Emergent Variables

The three variables θ, μ, and π emerge as a result of the scale-up process. They are the conjugate variables to the entropy S, the wetting area A_w, and the pore area A_p, respectively. How are we to interpret them?

We follow Hansen et al. [9], and consider a cylindrical porous medium (see Figure 2). We imagine a plane running through the cylinder axis parallel to the average flow direction. On one side of the plane, the porous medium has one set of properties; on the other side of the plane, it has another set of properties. The difference may be in chemical composition of the matrix, or it may consist of different porosity or topological structure. We name the two halves "A" and "B", respectively. We assume the two halves to be statistically homogeneous along the axis.

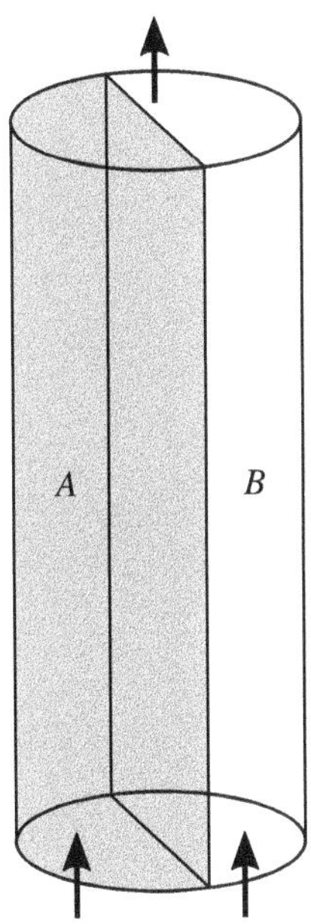

Figure 2. A cylindrical porous media sample consisting of two halves, A and B, having different properties with respect to the two immiscible fluids injected at the bottom edge. The two halves are in direct contact and fluids may pass unrestricted between the two halves.

We now assume that two immiscible and incompressible fluids are simultaneously pushed through the porous cylinder. Away from the edge where the two fluids are injected, the flow is in a steady state. This means that the flow statistics is invariant along the axis. However, it does not mean that the fluid–fluid interfaces do not move. At sufficient flow rates, they do and as a result, fluid clusters will merge and break up.

As the statistical distributions describing the flow is invariant along the flow axis, the configurational entropy S is invariant along this axis. As shown by Hansen et al. [9], this means that the agiture in sectors A and B, θ^A and θ^B, are equal,

$$\theta^A = \theta^B \,. \tag{38}$$

In ordinary thermodynamics, the rule is that the conjugate of a conserved quantity is constant in a heterogeneous system at equilibrium. This is a generalization of the argument for the temperature being the same everywhere in a system at equilibrium as the entropy is conserved and the temperature is its conjugate. This reasoning may be repeated for the flow problem.

Neither A_w nor A_p are conserved along the flow axis at the pore scale. This would imply that $\mu^A \neq \mu^B$ and $\pi^A \neq \pi^B$. This remains true at the continuum scale for A_p. However, A_w is conserved at the continuum scale; see Equation (7). This is an emergent conservation law. We therefore conjecture that

$$\mu^A = \mu^B \,. \tag{39}$$

6.1. Agiture

The agiture θ plays the role of a temperature in the thermodynamics-like formalism that we are developing. But, can we express it in terms of more familiar variables?

We note first that in the formalism we have developed in Sections 4 and 5, there is one variable that is missing: the pressure P. Intuitively, one would expect the agiture to be related to the pressure gradient ∇P: a higher pressure gradient should indicate a higher agiture.

Figure 3 shows a cylindrical porous medium consisting of two halves A and B. They have different properties such as porosity, chemical composition of the matrix, or the pore spaces have different topology. At the bottom one third of the porous cylinder, the boundary between the two halves is impenetrable. This is indicated by the dark section of the boundary. Above, the two halves are in direct contact. We inject the immiscible fluids into each of the halves A and B at the bottom. We may perform this by injecting at different volumetric flow rates $Q_p^A = Q_w^A + Q_n^A$ and $Q_p^B = Q_w^B + Q_n^B$. This gives rise to flow derivatives μ_A and μ_B, and agitures θ_A and θ_B in the two halves far enough from the inlet for the flow to be in a steady state. The pressure gradient will then point along the axis of the cylinder, the z-direction, and we will in the following use the notation $P' = \partial P/\partial z$ for the gradient. We find a pressure gradient P'_A and P'_B in A and B, respectively. Further into the cylindrical sample above the impenetrable wall, the flow will adjust to a new steady state characterized by equal agiture θ and flow derivative μ, and in both halves according to Equations (38) and (39). Likewise the pressure gradient P' must be equal in the two halves since, there cannot be any net flow in either direction across the boundary between the two halves, apart from local fluctuations.

Since the variables, θ and μ on one side, and P' on the other side, form alternate descriptions, we must assume that they are related. The simplest relation between them we may write down is

$$P' = -c_\theta \theta + c_\mu \mu \,. \tag{40}$$

The reason for the minus sign in front of the θ term is due to the flow direction is the opposite of the pressure gradient direction. We note that the unit of c_θ is viscosity, and that of c_μ is viscosity times area.

We note that Equation (36) defines the seepage velocity $v_p(\theta, S_w, \phi)$ where S_w rather than the flow derivative μ is a control variable. At the same time, the seepage velocity depends on the local pressure gradient, $v_p(P', S_w, \phi)$. Hence, we conjecture that $c_\mu = 0$, so that

$$P' = -c_\theta \theta \, . \tag{41}$$

We note that $Q_p = Q_p(\theta, A_w, A_p)$, defined in Equation (29), may be written as

$$Q_p = Q_p(P', A_w, A_p) = A_p v_p(P', S_w, \phi) \, , \tag{42}$$

where we have also used Equation (36). We see that this is a constitutive equation relating seepage velocity to local pressure gradient, saturation, and porosity.

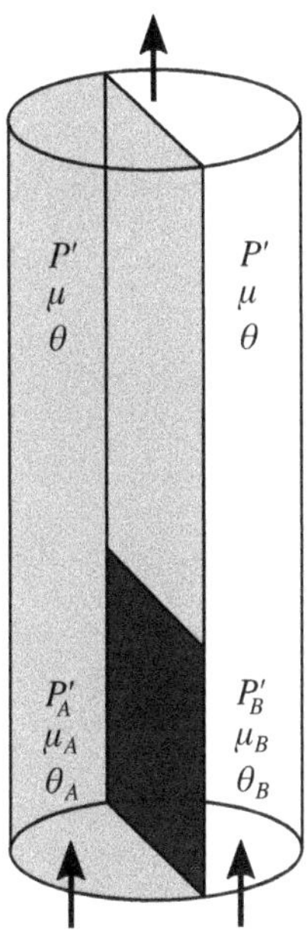

Figure 3. A cylindrical porous media sample consisting of two halves, A and B, having different properties with respect to the two immiscible fluids injected at the bottom edge. There is a impenetrable wall separating the two halves at the bottom third of the boundary. This is indicated by the dark section. Above, the two halves are in direct contact, as in Figure 2. A mixture of the two fluids are injected into the two halves at the lower edge. The pressure gradient is P'_A on the A side and P'_B on the B side up to the upper edge of the impenetrable wall. Likewise the flow derivatives are μ_A and μ_B and the agitures are θ_A and θ_B. Higher up, where the two halves communicate, the pressure gradients, the flow derivatives, and the agitures become pairwise equal.

To be concrete, we return for a moment to Section 2 and the constitutive equations forming the core of relative permeability theory, (1) and (2), which give

$$v_p(P', S_w, \phi) = -\frac{K}{\phi}\left[\frac{k_{rw}(S_w)}{\mu_w} + \frac{k_{rn}(S_w)}{\mu_n}\right] P' \, . \tag{43}$$

The agiture is the conjugate variable to the entropy. We may express it as

$$\theta = \left(\frac{\partial Q_u}{\partial S}\right)_{A_w, A_p} \, , \tag{44}$$

The entropy is given by

$$S = -\left(\frac{\partial Q_p}{\partial \theta}\right)_{A_w,A_p} = c_\theta \left(\frac{\partial Q_p}{\partial P'}\right)_{A_w,A_p}, \tag{45}$$

or, in terms of the entropy density,

$$s = c_\theta \left(\frac{\partial v_p}{\partial P'}\right)_{S_w,\phi}. \tag{46}$$

We note that the *differential mobility* is defined as

$$m(P', S_w, \phi) = -\left(\frac{\partial v_p}{\partial P'}\right)_{S_w,\phi}. \tag{47}$$

Using the relative permeability constitutive Equation (43), we find the mobility in this case to be

$$m(S_w, \phi) = \frac{K}{\phi}\left[\frac{k_{rw}(S_w)}{\mu_w} + \frac{k_{rn}(S_w)}{\mu_n}\right]. \tag{48}$$

Mobility and differential mobility are equal in this case, as the velocity is linear in the pressure gradient.

Comparing Equations (46) and (47), we see that

$$s = -c_\theta m. \tag{49}$$

Since we are dealing with *differential entropy*, it is not a problem that it is negative. It is surprising that the differential mobility and the entropy should be related.

Equation (41) relates the agiture θ and the pressure gradient P' and the way we have argued is by demonstrating that the pressure gradient behaves as one would expect an agiture (=agitation temperature) should do. However, one could test Equation (41) directly by measuring the left hand side and compare it to the right hand side. This is difficult, as it would entail measuring the configurational entropy and then using Equation (44), and measuring entropy is also difficult in ordinary thermodynamics.

6.2. Flow Derivative

Hansen et al. [48] defined the two *thermodynamic velocities* as

$$\hat{v}_w = \left(\frac{\partial Q_p}{\partial A_w}\right)_{\theta,A_n}, \tag{50}$$

$$\hat{v}_n = \left(\frac{\partial Q_p}{\partial A_n}\right)_{\theta,A_w}, \tag{51}$$

where the control variables are θ, A_w, and A_n, defined in (12), making A_p a dependent variable. Changing the variables $(\theta, A_w, A_n) \to (\theta, S_w, A_p)$, these two equations may be written as

$$\hat{v}_w = v_p + S_n\left(\frac{\partial v_p}{\partial S_w}\right)_{\theta,\phi}, \tag{52}$$

$$\hat{v}_n = v_p - S_w\left(\frac{\partial v_p}{\partial S_w}\right)_{\theta,\phi}. \tag{53}$$

The flow derivative μ, which is the conjugate of the wetting area A_w, and thereby also the saturation S_w, is given by

$$\mu = -\left(\frac{\partial Q_p}{\partial A_w}\right)_{\theta,A_p} = \left(\frac{\partial v_p}{\partial S_w}\right)_{\theta,\phi} . \tag{54}$$

Hence, Equation (53) may be recognized as a Legendre transformation substituting $S_w \to \mu$,

$$\hat{v}_n(\theta,\mu,\phi) = v_p(\theta,S_w,\phi) - S_w(\theta,\mu,\phi)\mu . \tag{55}$$

In other words, the non-wetting thermodynamic velocity is the Legendre transformation of the average seepage velocity with respect to the saturation.

6.3. Flow Pressure

The flow pressure π is the conjugate of the pore area A_p, and thereby the porosity ϕ,

$$\pi = -\left(\frac{\partial Q_p}{\partial A_p}\right)_{\theta,A_w} = \left(\frac{\partial v_p}{\partial \phi}\right)_{\theta,S_w} . \tag{56}$$

There are no conservation laws associated with this variable.

We expect the flow pressure π to be of less practical use, as we expect that the flow rate Q_p to be proportional to the pore area A_p. This makes the Legendre transform between A_p and π break down due to lack of convexity. The flow pressure will nevertheless play a role, e.g., when porosity gradients are present.

7. Seepage Velocities and the Co-Moving Velocity

The thermodynamic velocities defined in Equations (50) and (51) are not the seepage velocities, which we define as

$$v_w = \frac{Q_w}{A_w} , \tag{57}$$

$$v_n = \frac{Q_n}{A_n} . \tag{58}$$

These are the velocities that measured in the laboratory, whereas the thermodynamic velocities are not. We rewrite Equation (11) as

$$Q_p = Q_w + Q_n = v_w A_w + v_n A_n . \tag{59}$$

We express the saturations in terms of the variables A_w and A_n,

$$S_w = \frac{A_w}{A_w + A_n} , \tag{60}$$

and

$$S_n = \frac{A_n}{A_w + A_n} , \tag{61}$$

leading to Equation (3). We may then write Equation (59) as

$$v_p = v_w S_w + v_n S_n , \tag{62}$$

where v_p, the average velocity, is defined in Equation (36).

We now turn to the thermodynamic velocities, Equations (50) and (51). The average flow rate Q_p obeys the Euler scaling relation,

$$\lambda Q_p(P', A_w, A_n) = Q_p(P', \lambda A_w, \lambda A_n) . \tag{63}$$

We then have from the Euler theorem for homogeneous functions,

$$Q_p = \left(\frac{\partial Q_p}{\partial A_w}\right)_{A_n,\phi,P'} A_w + \left(\frac{\partial Q_p}{\partial A_n}\right)_{A_w,\phi,P'} A_n = \hat{v}_w A_w + \hat{v}_n A_n . \tag{64}$$

Dividing this equation by A_p gives

$$v_p = \hat{v}_w S_w + \hat{v}_n S_n . \tag{65}$$

Comparing Equations (62) and (65),

$$v_p = v_w S_w + v_n S_n = \hat{v}_w S_w + \hat{v}_n S_n . \tag{66}$$

This equation does not imply that $v_w = \hat{v}_w$ and $v_n = \hat{v}_n$. Rather, the most general relation between the thermodynamic and seeping velocities are

$$v_w = \hat{v}_w - S_n v_m , \tag{67}$$

$$v_n = \hat{v}_n + S_w v_m , \tag{68}$$

where v_m is the *co-moving velocity* [48–50]. We may combine these two equations with Equations (52) and (53) to find

$$v_w = v_p + S_n\left[\left(\frac{\partial v_p}{\partial S_w}\right)_{\phi,P'} - v_m\right] , \tag{69}$$

$$v_n = v_p - S_w\left[\left(\frac{\partial v_p}{\partial S_w}\right)_{\phi,P'} - v_m\right] . \tag{70}$$

By taking the derivative of Equation (66) with respect to S_w, and using Equations (67) and (68), we find

$$v_m = \left(\frac{\partial v_p}{\partial S_w}\right)_{\phi,P'} - v_w + v_n = S_w\left(\frac{\partial v_w}{\partial S_w}\right)_{\phi,P'} + S_n\left(\frac{\partial v_n}{\partial S_w}\right)_{\phi,P'} . \tag{71}$$

We now return to Equation (55), which may now be expressed in terms of the seepage velocity v_n, rather than the thermodynamic velocity $\hat{v}_n$,

$$v_n(\theta,\mu,\phi) - S_w(\theta,\mu,\phi) v_m(\theta,\mu,\phi) = v_p(\theta,S_w,\phi) - S_w(\theta,\mu,\phi)\mu , \tag{72}$$

signifying that (θ,μ,ϕ) are the natural variables for the co-moving velocity. Measurements of the co-moving velocity suggest that it has the simple functional form [51–53],

$$v_m(\theta,\mu,\phi) = a(\theta,\phi) + b(\theta,\phi)\mu , \tag{73}$$

based on analysis of relative permeability data, on dynamic pore network modeling, and on lattice Boltzmann simulations on reconstructed sandstones.

It has recently been realized that the co-moving velocity has an equivalent in the thermodynamics of two-fluid mixtures [54], where a *co-molar volume* ties the partial molar volumes of a binary mixture together with the Voronoi volumes of each of the fluids in the mixture. When plotting the co-molar volume against the variable corresponding to μ, it is *nearly* linear, but not quite.

We note that the thermodynamic velocities $(\hat{v}_w, \hat{v}_n)$ can be found by knowing v_p alone; see Equations (52) and (53). The seepage velocities, on the other hand, requires additional knowledge, i.e., v_p and v_m, which are given by Equations (69) and (70). They provide a two-way mapping,

$$\begin{pmatrix} v_w \\ v_n \end{pmatrix} \leftrightarrow \begin{pmatrix} v_p \\ v_m \end{pmatrix} . \tag{74}$$

8. Control Variables

In Section 4, we made a choice when focusing on the averages Q_u, A_w, and A_p in Equations (14) to (16), leading to a thermodynamics-like formalism based upon the extensive variables Q_p, A_w, and A_p, and their intensive conjugates θ $(= -c_\theta P')$, μ, and π. In core flooding, these are not the natural control variables. Rather, one controls the flow rates Q_w and Q_n into the core, making A_w (i.e., the saturation) and the pressure gradient the dependent variables. It is possible to build a Jaynes statistical mechanics based on Q_u, Q_w, and A_p, rather than Q_u, A_w, and A_p. We will do this elsewhere. However, we can still develop the ensuing thermodynamics-like formalism as long as we work only with the extensive variables. We will do so in the following.

Controlling the Fractional Flow Rate

Our starting point is again the REA shown in Figure 1, and Equations (11) and (12). We now assume that our control variables are Q_w and Q_n together with A_p, the REA pore space. The dependent variable is the pressure gradient P'. They are related through the constitutive equation

$$P' = P'(Q_w, Q_n, A_p) . \tag{75}$$

We may rescale the extensive variables in this expression, $A_p \to \lambda A_p$, $Q_w \to \lambda Q_w$, and $Q_n \to \lambda Q_n$. In practice, changing A_p means changing the sample. Hence, this is a theoretical exercise. Nevertheless, mathematics works. This leaves P' unchanged, i.e.,

$$P'(\lambda Q_w, \lambda Q_n, \lambda A_p) = P'(Q_w, Q_n, A_p) . \tag{76}$$

It is, however, convenient to rewrite Equation (75) as

$$A_p = A_p(P', Q_w, Q_n) , \tag{77}$$

making it implicit with respect to P'. The scaling relation (76) then takes the form

$$\lambda A_p(P', Q_w, Q_n) = A_p(P', \lambda Q_w, \lambda Q_n) . \tag{78}$$

We then use the Euler theorem for homogeneous functions,

$$A_p = \left(\frac{\partial A_p}{\partial Q_w}\right)_{Q_n,\phi,P'} Q_w + \left(\frac{\partial A_p}{\partial Q_n}\right)_{Q_w,\phi,P'} Q_n = \hat{p}_w Q_w + \hat{p}_n Q_n , \tag{79}$$

where we have defined the *thermodynamic paces*,

$$\hat{p}_w = \left(\frac{\partial A_p}{\partial Q_w}\right)_{Q_n,\phi,P'} , \tag{80}$$

and

$$\hat{p}_n = \left(\frac{\partial A_p}{\partial Q_n}\right)_{Q_w,\phi,P'} . \tag{81}$$

Pace is inverse velocity. We also define the *physical paces*,

$$p_p = \frac{A_p}{Q_p} = \frac{1}{v_p} , \tag{82}$$

$$p_w = \frac{A_w}{Q_w} = \frac{1}{v_w} , \tag{83}$$

and

$$p_n = \frac{A_n}{Q_n} = \frac{1}{v_n} . \tag{84}$$

This allows us to write Equation (12) as

$$A_p = p_w Q_w + p_n Q_n . \tag{85}$$

Combining this equation with Equation (79) gives

$$A_p = p_w Q_w + p_n Q_n = \hat{p}_w Q_w + \hat{p}_n Q_n . \tag{86}$$

We now introduce the fractional flow rates,

$$F_w = \frac{Q_w}{Q_w + Q_n} , \tag{87}$$

and

$$F_n = \frac{Q_n}{Q_w + Q_n} , \tag{88}$$

so that

$$F_w + F_n = 1 . \tag{89}$$

We may then write Equation (86) as

$$p_p = p_w F_w + p_n F_n = \hat{p}_w F_w + \hat{p}_n F_n , \tag{90}$$

where p_p, the average pace, is defined in Equation (82).

We now make a coordinate transformation,

$$(P', Q_w, Q_n) \to (P', Q_p, F_w) , \tag{91}$$

so that we have

$$A_p(P', Q_p, F_w) = A_p(P', Q_w, Q_n) . \tag{92}$$

We use extensivity,

$$\lambda A_p(P', Q_p, F_w) = A_p(P', \lambda Q_p, F_w) , \tag{93}$$

leading to

$$A_p(P', Q_p, F_w) = Q_p A_p(P', 1, F_w) = Q_p p_p(P', F_w) , \tag{94}$$

when setting $\lambda = 1/Q_p$ and defining $p_p(P', F_w) = A_p(P', 1, F_w)$.

We express the thermodynamic paces Equations (80) and (81) in terms of (P', Q_p, F_w), finding

$$\hat{p}_w = p_p + F_n \left(\frac{\partial p_p}{\partial F_w}\right)_{\phi,P'} , \tag{95}$$

and

$$\hat{p}_n = p_p - F_w \left(\frac{\partial p_p}{\partial F_w} \right)_{\phi,P'} . \tag{96}$$

We may define a *co-moving pace function*, p_m, by the expressions

$$\hat{p}_w = p_w + F_n p_m , \tag{97}$$

and

$$\hat{p}_n = p_n - F_w p_m , \tag{98}$$

where p_m is the most general function to fulfill Equation (90). Combining these two equations with Equations (96) and (97) gives

$$p_w = p_p + F_n \left[\left(\frac{\partial p_p}{\partial F_w} \right)_{\phi,P'} - p_m \right] , \tag{99}$$

$$p_n = p_p - F_w \left[\left(\frac{\partial p_p}{\partial F_w} \right)_{\phi,P'} - p_m \right] , \tag{100}$$

By taking the derivative of Equation (90) with respect to F_w, and using Equations (97) and (98), we find

$$p_m = \left(\frac{\partial p_p}{\partial F_w} \right)_{\phi,P'} - p_w + p_n = F_w \left(\frac{\partial p_w}{\partial F_w} \right)_{\phi,P'} + F_n \left(\frac{\partial p_n}{\partial F_w} \right)_{\phi,P'} . \tag{101}$$

We may at this point ask whether the co-moving pace p_m is a linear function in the variable

$$\nu = \left(\frac{\partial p_p}{\partial F_w} \right)_{\phi,P'} \tag{102}$$

in the same way that the co-moving velocity v_m is a linear function of $\mu = (\partial v_p / \partial S_w)_{\phi,P'}$; see Equation (73). The answer is no. This may, for example, be seen by testing the relative permeability pair proposed by Picchi and Battiato [15],

$$k_{rw} = k_{rw}^0 S_w^2 , \tag{103}$$

$$k_{rn} = k_{rn}^0 S_n^2 \left[1 + 2 \frac{\alpha \mu_n}{\mu_w} \frac{S_w}{S_n} \right] , \tag{104}$$

where $0 \leq \alpha \leq 1$ is a parameter. This pair leads to the co-moving velocity

$$v_m = \alpha k_{rn}^0 + \frac{\mu}{2} , \tag{105}$$

in units of $v_0 = -KP'/\mu_w \phi$. The corresponding co-moving pace is a complex curve when plotting p_m vs. ν; see Figure 4.

We note that we used the following relations between fractional flow F_w and saturation S_w in calculating p_m above,

$$F_w = \frac{Q_w}{Q_p} = \frac{A_p S_w v_w}{A_p v_p} = S_w \frac{v_w(P', S_w, \phi)}{v_p(P', S_w, \phi)} . \tag{106}$$

Likewise, we have

$$S_w = \frac{A_w}{A_p} = \frac{Q_p F_w p_w}{Q_p p_p} = F_w \frac{p_w(P', F_w, \phi)}{p_p(P', F_w, \phi)} . \tag{107}$$

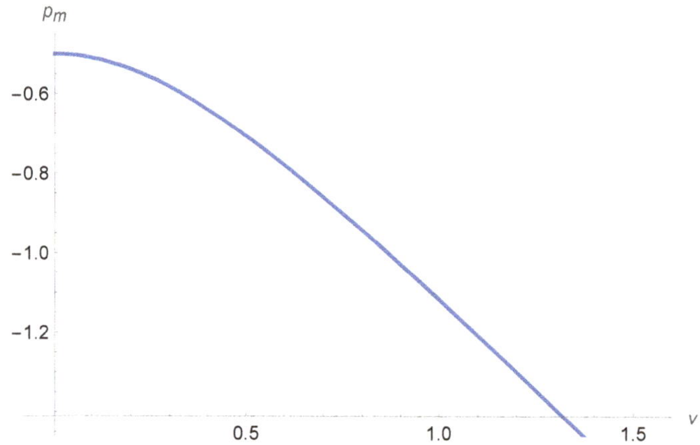

Figure 4. The co-moving pace p_m against v for the Picchi and Battiato pair of relative permeabilities [15]; Equations (106) and (107) with $\mu_w = \mu_n$, $k^0_{rw} = k^0_{rn}/2$, and $\alpha = 1/2$.

9. Discussion and Conclusions

The aim of this review has been to present the central ideas of the thermodynamics-like formalism for immiscible two-phase flow in porous media that was first introduced by Hansen et al. [48] in the context of steady-state flow. However, the ultimate goal of this approach is to be able to handle flow that is not in a steady state. This means handling gradients in the conjugate variables to configurational entropy, saturation and porosity: agiture θ, flow derivative μ, and flow pressure π, turning the equilibrium thermodynamics-like formalism into a non-equilibrium thermodynamics-like formalism.

We noted in Section 4 that actually calculating the partition function (Equation (19)) may be impossible in the present context. The consequence of this is that we are unable to derive directly the constitutive equation $Q_p(P', A_w, A_p)$ from the pore-level physics. It must be found by other means, and then used as input in the thermodynamics-like formalism.

We have conjectured here that the agiture, the temperature-like emergent variable conjugate to the configurational entropy, is proportional to the pressure gradient. A surprising consequence of this conjecture is that the differential mobility of the fluids is directly related to the entropy: higher differential mobility means lower configurational entropy.

The question of what are the control variables and the dependent variables is central, and much of the thermodynamics-like formalism concerns how to switch between the extensive variables Q_p, A_w, and A_p, and their intensive conjugates, θ, μ, and π. However, there was already a choice made when the three extensive variables were picked. We could have made other choices. In this review, we discuss, for the first time, another choice: Q_p, Q_w and A_p. Without developing a Jaynes-type statistical mechanics based on this choice, we nevertheless explore aspects of the ensuing thermodynamics-like formalism. We find a set of equations that parallel those found in [48], but where the fractional flow rate F_w rather than the saturation S_w is a control variable. Since this is the natural variable in core flooding experiments, not the saturation, such a formalism is necessary.

Author Contributions: Conceptualization, A.H.; Validation, S.S.; Writing—original draft, A.H.; Writing—review & editing, S.S. All authors have read and agreed to the published version of the manuscript.

Funding: This work was partly supported by the Research Council of Norway through its Centers of Excellence funding scheme, project number 262644. We furthermore acknowledge funding from the European Research Council (Grant Agreement 101141323 AGIPORE).

Data Availability Statement: The original contributions presented in this study are included in the article. Further inquiries can be directed to the corresponding author.

Acknowledgments: We thank Steffen Berg for valuable discussions and comments.

Conflicts of Interest: The authors declare no conflicts of interest.

References

1. Feder, J.; Flekkøy, E.G.; Hansen, A. *Physics of Flow in Porous Media*; Cambridge University Press: Cambridge, UK, 2022.
2. Buckingham, E. *Studies on the Movement of Soil Moisture*; Bureau of Soils Bulletin no. 38; U. S. Department of Agriculture: Washington, DC, USA, 1907.
3. Narasimhan, T.N. Central Ideas of Buckingham (1907): A Century Later. *Vadose Zone J.* **2007**, *6*, 687. [CrossRef]
4. Richards, L.A. Capillary conduction of liquids through porous mediums. *Physics* **1931**, *1*, 318. [CrossRef]
5. Wyckoff, R.D.; Botset, H.G. The flow of gas-liquid mixtures through unconsolidated sands. *Physics* **1936**, *7*, 325. [CrossRef]
6. Muscat, M.; Meres, M.W. The flow of heterogeneous fluids through porous media. *Physics* **1936**, *7*, 346. [CrossRef]
7. Leverett, M.C. Capillary behavior in porous sands. *Trans. AIMME* **1940**, *12*, 152.
8. Jaynes, E.T. Information theory of statistical mechanics. *Phys. Rev.* **1957**, *106*, 620. [CrossRef]
9. Hansen, A.; Flekkøy, E.G.; Sinha, S.; Slotte, P.A. A statistical mechanics framework for immiscible and incompressible two-phase flow in porous media. *Adv. Water Res.* **2023**, *171*, 104336. [CrossRef]
10. Moura, M.; Fiorentino, E.-A.; Måløy, K.J.; Schäfer, G.; Toussaint, R. Impact of sample geometry on the measurement of pressure-saturation curves: Experiments and simulations. *Water Res. Res.* **2015**, *51*, 8900. [CrossRef]
11. Auriault, J.L.; Sanchez-Palencia, E. Remarques sur la loi de Darcy pour les écoulements biphasiques en milieu poreux. *J. Theor. Appl. Mech.* **1986**, 141–156.
12. Auriault, J.L. Nonsaturated deformable porous media: Quasistatics. *Transp. Porous Media* **1987**, *2*, 45. [CrossRef]
13. Auriault, J.L.; Lebaigue, O.; Bonnet, E. Dynamics of two immiscible fluids flowing through deformable porous media. *Transp. Porous Media* **1989**, *4*, 105. [CrossRef]
14. Picchi, D.; Battiato, I. The impact of pore-scale flow regimes on upscaling of immiscible two-phase flow in porous media. *Water Res. Res.* **2018**, *54*, 6683. [CrossRef]
15. Picchi, D.; Battiato, I. Relative permeability scaling from pore-scale flow regimes. *Water Res. Res.* **2019**, *55*, 3215. [CrossRef]
16. Lasseux, D.; Valdés-Parada, F.J. A macroscopic model for immiscible two-phase flow in porous media. *J. Fluid Mech.* **2022**, *944*, A43. [CrossRef]
17. Hassanizadeh, S.M.; Gray, W.G. General conservation equations for multi-phase systems: 1. Averaging procedure. *Adv. Water Res.* **1979**, *2*, 131. [CrossRef]
18. Hassanizadeh, S.M.; Gray, W.G. Mechanics and thermodynamics of multiphase flow in porous media including interfase boundaries. *Adv. Water Res.* **1990**, *13*, 169. [CrossRef]
19. Hassanizadeh, S.M.; Gray, W.G. Towards an improved description of the physics of two-phase flow. *Adv. Water Res.* **1993**, *16*, 53. [CrossRef]
20. Hassanizadeh, S.M.; Gray, W.G. Thermodynamic basis of capillary pressure in porous media. *Water Resour. Res.* **1993**, *29*, 3389. [CrossRef]
21. Whitaker, S. Flow in porous media I: A theoretical derivation of Darcy's law. *Trans. Porous Media* **1986**, *1*, 3. [CrossRef]
22. Whitaker, S. Flow in porous media II: The governing equations for immiscible, two-phase flow. *Trans. Porous Media* **1986**, *1*, 105. [CrossRef]
23. Gray, W.G.; Miller, C.T. Thermodynamically constrained averaging theory approach for modeling flow and transport phenomena in porous medium systems: 1. Motivation and overview. *Adv. Water Res.* **2005**, *28*, 161. [CrossRef]
24. Gray, W.G.; Miller, C.T. Thermodynamically constrained averaging theory approach for modeling flow and transport phenomena in porous medium systems: 2. Foundation. *Adv. Water Res.* **2005**, *28*, 181. [CrossRef]
25. Niessner, J.; Berg, S.; Hassanizadeh, S.M. Comparison of two-phase Darcy's law with a thermodynamically consistent approach, *Transp. Porous Media* **2011**, *88*, 133. [CrossRef]
26. Gray, W.G.; Miller, C.T. *Introduction to the Thermodynamically Constrained Averaging Theory for Porous Medium Systems*; Springer: Berlin/Heidelberg, Germany, 2014.
27. McClure, J.E.; Fan, M.; Berg, S.; Armstrong, R.T.; Berg, C.F.; Li, Z.; Ramstad, T. Relative permeability as a stationary process: Energy fluctuations in immiscible displacement. *Phys. Fluids* **2022**, *34*, 092011. [CrossRef]
28. Mecke, K.R. Additivity, Convexity, and Beyond: Applications of Minkowski Functionals in Statistical Physics. *Springer Lect. Notes Phys.* **2000**, *554*, 111.
29. McClure, J.E.; Armstrong, R.T.; Berrill, M.A.; Schlüter, S.; Berg, S.; Gray, W.G.; Miller, C.T. Geometric state function for two-fluid flow in porous media. *Phys. Rev. Fluids* **2018**, *3*, 084306. [CrossRef]

30. Khanamiri, H.H.; Berg, C.F.; Slotte, P.A.; Sclüter, S.; Torsæter, O. Description of free energy for immiscible two-fluid flow in porous media by integral geometry and thermodynamics. *Water Res. Res.* **2018**, *54*, 9045. [CrossRef]
31. Armstrong, R.T.; McClure, J.E.; Robins, V.; Liu, Z.; Arns, C.H.; Schlüter, S.; Berg, S. Porous media characterization using Minkowski functionals: Theories, applications and future directions. *Transp. Porous Media* **2019**, *130*, 305. [CrossRef]
32. McClure, J.E.; Armstrong, R.T.; Berg, S. Geometric evolution as a source of discontinuous behavior in soft condensed matter. *arXiv* **2019**, arXiv:1906.04073.
33. McClure, J.E.; Armstrong, R.T.; Berg, S. Thermodynamics of fluctuations based on time-and-space averages. *Phys. Rev. E* **2021**, *104*, 035106. [CrossRef]
34. McClure, J.E.; Berg, S.; Armstrong, R.T. Capillary fluctuations and energy dynamics for flow in porous media. *Phys. Fluids* **2021**, *33*, 083323. [CrossRef]
35. de Groot, S.R.; Mazur, P. *Non-Equilibrium Thermodynamics*; Dover: New York, NY, USA, 1984.
36. Kjelstrup, S.; Bedeaux, D.; Johannesen, E.; Gross, J. *Non-Equilibrium Thermodynamics for Engineers*; World Scientific: Singapore, 2017.
37. Kjelstrup, S.; Bedeaux, D.; Hansen, A.; Hafskjold, B.; Galteland, O. Non-isothermal transport of multi-phase fluids in porous media. The entropy production. *Front. Phys.* **2018**, *6*, 126. [CrossRef]
38. Kjelstrup, S.; Bedeaux, D.; Hansen, A.; Hafskjold, B.; Galteland, O. Non-isothermal transport of multi-phase fluids in porous media: Constitutive equations. *Front. Phys.* **2019**, *6*, 150. [CrossRef]
39. Winkler, M.; Gjennestad, M.A.; Bedeaux, D.; Kjelstrup, S.; Cabriolu, R.; Hansen, A. Onsager-symmetry obeyed in athermal mesoscopic systems: Two-phase flow in porous media. *Front. Phys.* **2020**, *8*, 60. [CrossRef]
40. Bedeaux, D.; Kjelstrup, S. Fluctuation-dissipiation theorems for multiphase flow in porous media. *Entropy* **2022**, *24*, 46. [CrossRef]
41. Anderson, P.W. More is Different. *Science* **1972**, *177*, 393. [CrossRef] [PubMed]
42. Frenkel, D.; Smit, B. *Understanding Molecular Simulation*, 3rd ed.; Academic Press: London, UK, 2023.
43. Shannon, C.E. A mathematical theory of communication. *Bell Syst. Tech. J.* **1948**, *27*, 379. [CrossRef]
44. Laplace, P.-S. *A Philosophical Essay on Probabilities*; Dover: Mineola, NY, USA, 2003.
45. Michaelowicz, J.V.; Nichols, J.M.; Bucholtz, F. *Handbook of Differential Entropy*; Routhledge: New York, NY, USA, 2014.
46. Jaynes, E.T. Information theory and statistical mechanics. In *Brandeis University Summer Institute Lectures in Theoretical Physics*; Ford, K.W., Ed.; W. A. Benjamin: New York, NY, USA, 1963.
47. Onsager, L. Crystal Statistics. I. A Two-Dimensional Model with an Order-Disorder Transition. *Phys. Rev.* **1944**, *65*, 117. [CrossRef]
48. Hansen, A.; Sinha, S.; Bedeaux, D.; Kjelstrup, S.; Gjennestad, M.A.; Vassvik, M. Relations between seepage velocities in immiscible, incompressible two-phase flow in porous media. *Transp. Porous Media* **2018**, *125*, 565. [CrossRef]
49. Roy, S.; Sinha, S.; Hansen, A. Flow-area relations in immiscible two-phase flow in porous media. *Front. Phys.* **2020**, *8*, 4. [CrossRef]
50. Pedersen, H.; Hansen, A. Parameterizations of immiscible two-phase flow in porous media. *Front. Phys.* **2023**, *11*, 1127345. [CrossRef]
51. Roy, S.; Pedersen, H.; Sinha, S.; Hansen, A. The co-moving velocity in immiscible two-phase flow in porous media. *Transp. Porous Media* **2022**, *143*, 69. [CrossRef]
52. Alzubaidi, A.; McClure, J.E.; Pedersen, H.; Hansen, A.; Berg, C.F.; Mostaghimi, P.; Armstrong, R.T. The impact of wettability on the co-moving velocity of two-fluid flow in porous media. *Transp. Porous Media* **2024**, *151*, 1967. [CrossRef]
53. Hansen, A. Linearity of the co-moving velocity. *Transp. Porous Media* **2024**, *151*, 2477. [CrossRef]
54. Olsen, K.P.; Hafskjold, B.; Hansen, A.; Lervik, A. A new thermodynamic function for binary mixtures: The co-molar volume. *Proc. Natl. Acad. Sci. USA* **2025**, *submitted*.

entropy

MDPI

Article

Thermodynamic Properties of Hydrogen Adsorbed on Graphite Surfaces at Temperatures Above 100 K: A Molecular Dynamics and Classical Density Functional Theory Study

Vegard G. Jervell, Morten Hammer, Øivind Wilhelmsen * and Thuat T. Trinh *

Porelab, Department of Chemistry, Norwegian University of Science and Technology, Høgskoleringen 5, NO-7491 Trondheim, Norway; vegard.g.jervell@ntnu.no (V.G.J.); morten.hammer@ntnu.no (M.H.)
* Correspondence: oivind.wilhelmsen@ntnu.no (Ø.W.); thuat.trinh@ntnu.no (T.T.T.)

Abstract: Improved technological solutions for the transport and storage of hydrogen are crucial for the widespread adoption of hydrogen as a clean energy carrier. Graphite-based materials have been identified as potential candidates due to their high surface area and ability to adsorb hydrogen molecules. In this study, we investigate the adsorption and thermodynamic properties of hydrogen adsorbed on a graphite surface using molecular dynamics (MD) simulation and classical density functional theory (cDFT). We demonstrate how to use the MD parameters for graphite to derive an effective wall potential for hydrogen–graphite interactions that can be used in the cDFT calculations. The methodology results in good agreement between cDFT and MD, with the enthalpy and entropy of adsorption differing by 3.5% and 7%, respectively. We determine the enthalpy and entropy of adsorption at 298 K to be in the ranges of $-6.37\,\mathrm{kJ\,mol^{-1}}$ to $-6.16\,\mathrm{kJ\,mol^{-1}}$ and $-75.42\,\mathrm{J\,mol^{-1}\,K^{-1}}$ to $-79.95\,\mathrm{J\,mol^{-1}\,K^{-1}}$, respectively. We find that the adsorbed hydrogen has a $12.4\,\mathrm{J\,mol^{-1}\,K^{-1}}$ to $11.4\,\mathrm{J\,mol^{-1}\,K^{-1}}$ lower heat capacity than the bulk hydrogen in the temperature range from 150 K to 400 K. This suggests that the adsorbed molecules are bound to adsorption sites that arrest nearly all the translational degrees of freedom.

Keywords: thermodynamics; hydrogen; graphite; Mie potential; classical density functional theory; Langmuir equation; adsorption

1. Introduction

To reduce carbon dioxide emissions and mitigate global warming, it has been shown that hydrogen (H_2) must be adopted as a key energy carrier and reduction agent [1–5]. Hydrogen adsorption on graphite is a frequently discussed topic in materials science and energy storage due to its potential applications in hydrogen storage systems, fuel cells, catalysis, and sensing technologies. Graphite, a form of carbon characterized by its layered structure, exhibits unique properties that make it an attractive material for hydrogen adsorption applications [6–9]. Each graphite layer consists of hexagonally arranged carbon atoms bonded together via strong covalent forces within the plane, while weaker van der Waals interactions exist between adjacent layers. These unique structural features contribute to a high surface area, which can potentially be of relevance in hydrogen storage technologies.

A key property of materials used in storage technologies is the amount of fluid adsorbed per surface area, referred to as the adsorption. Several models for the adsorption have been presented in the literature [10–19]. The Langmuir model and its extensions are among the most popular. The simplicity of the Langmuir model makes it appealing, and it has been shown to be accurate and reliable within its range of validity. In order to apply the Langmuir model, two parameters must be determined: the adsorption capacity of the surface with respect to the adsorbent and the adsorption equilibrium constant. In this work, we will determine the adsorption capacity and adsorption equilibrium constant, and evaluate the range of validity of the Langmuir model for the graphite–hydrogen system at temperatures above 100 K.

In addition to the adsorption, thermodynamic properties such as the enthalpy, entropy, and heat capacity of adsorption processes are of high relevance for developing hydrogen storage systems [20–22]. The enthalpy change during adsorption and desorption is strongly associated with the energy demand. If the enthalpy of adsorption is too large, this may prevent efficient use of the material [23]. A thorough investigation of the thermodynamic properties of the adsorbed phase is therefore necessary, both to determine the feasibility of the technology, and to guide the design of more efficient materials for hydrogen storage applications.

Within the Gibbs formulation of interfacial thermodynamics, excess variables are used to describe vapor–liquid interfaces and adsorbed phases [24,25]. The results from molecular dynamics (MD) simulations or other tools such as classical density functional theory [26–31] (cDFT) can be analyzed within this framework to obtain useful properties such as adsorption isotherms and energies. In cDFT, the Helmholtz energy is given as a functional of the inhomogeneous density field. Helmholtz energy functionals capable of accurately predicting properties such as surface tensions, the curvature dependence of the surface tension, adsorption isotherms, and interfacial resistances of real fluids have been developed based on statistical associating fluid theory [31–37].

In this work, we will demonstrate how cDFT can be utilized in conjunction with MD, both to calculate the properties of the hydrogen–graphite interface, and as a framework to ensure thermodynamically consistent analysis of the results obtained from MD simulations. In this study, we present an investigation of hydrogen adsorption on graphite above 100 K, using Mie (generalized Lennard-Jones) potentials to describe hydrogen–hydrogen and hydrogen–carbon interactions. The results from MD and cDFT both agree with the Langmuir adsorption equation, which allows us to extract an enthalpy adsorption that is in excellent agreement with experimental results.

2. Theory

In the following, the core theories used in this work will be introduced. In Section 2.1, cDFT will be described. Next, we show in Section 2.2 how the carbon–hydrogen interaction potential and graphite crystal structure from MD simulations can be incorporated into cDFT calculations. Finally, in Section 2.3, we show how the Gibbs dividing surface should be defined in order to analyze the results with the Langmuir adsorption equation. The procedure for determining the location of the dividing surface in MD simulations is detailed in Section 3.2.

2.1. Classical Density Functional Theory

Classical density functional theory (cDFT) is a framework to study inhomogeneous systems [27,29]. The grand potential of a single-component system is expressed as a functional of the density field,

$$\Omega[\rho] = \int k_B T\phi(\mathbf{r}) + \rho(\mathbf{r})(V_{ex}(\mathbf{r}) - \mu)\mathrm{d}\mathbf{r}, \tag{1}$$

where ϕ is the reduced Helmholtz energy density at position r, ρ is the density, V_{ex} is an external potential, μ is the chemical potential, and k_B is Boltzmann's constant. Following the variational principle, the equilibrium density field may be found by solving the equation

$$\frac{\delta\Omega[\rho]}{\delta\rho(\mathrm{r})} = 0. \tag{2}$$

In this work, we approximate hydrogen as monoatomic Mie particles and use a Helmholtz energy functional that is consistent with the SAFT-VR Mie equation of state [38,39]. The reduced Helmholtz energy density is then

$$\phi = \phi^{id} + \phi^{hs} + \phi^{disp}, \tag{3}$$

where superscripts *id*, *hs*, and *disp* are the ideal gas, hard sphere, and dispersion contributions to the Helmholtz energy density, respectively. The ideal gas contribution is $\phi^{id}(\mathrm{r}) = k_B T\rho(\mathrm{r})[\ln(\rho(\mathrm{r})\Lambda^3) - 1]$, where $\Lambda \equiv \sqrt{\frac{2\pi\hbar^2}{mk_BT}}$ is the de Broglie wavelength. The hard sphere contribution corresponds to the Helmholtz energy density of a fluid consisting of hard spheres with diameter

$$d_{BH} = \int_0^{\sigma}\left[1 - \exp\left(-\frac{u(r)}{k_BT}\right)\right]\mathrm{d}r, \tag{4}$$

where $u(r)$ is the hydrogen interaction potential and σ is the potential size parameter. This contribution is computed using the White Bear fundamental measure theory [30]. For the dispersion term, ϕ^{disp}, we adopt the contribution presented by Sauer and Gross [26]. For further reading, the reader is referred to Refs. [28,40,41]. The cDFT implementation used in this work is available open source on GitHub in the `SurfPack` repository under the ThermoTools project [42].

2.2. Solid–Fluid Interaction

In the following, we will elaborate on how information about the fluid–solid interactions and the crystal structure of the solid can be used to develop an effective external potential, also referred to as a wall potential, that can be used in cDFT [26,43].

The external potential felt by a hydrogen molecule at position r due to interactions with the carbon atoms in the graphite is $u_w(r) = \sum_j u(r_j)$, where r_j is the distance to the jth carbon atom. This sum may be treated by Fourier analysis in order to incorporate the full symmetry of the crystal structure, as shown by Steele [43]. However, it has been found that a simple yet accurate approximation to the full Fourier analysis is

$$u_w(z) = u_{w,0}(z) + u_{w,\infty}(z), \tag{5}$$

where z is the dimension normal to the surface and

$$\begin{aligned}\frac{u_{w,0}(z)}{C\epsilon} &= 2\pi\rho_p\int_z^{\infty} u(r)r\mathrm{d}r \\ &= 2\pi\rho_p\left[\frac{\sigma^2}{\lambda_r - 2}\left(\frac{\sigma}{z}\right)^{\lambda_r-2} - \frac{\sigma^2}{\lambda_a - 2}\left(\frac{\sigma}{z}\right)^{\lambda_a-2}\right]\end{aligned} \tag{6}$$

is the potential resulting from interactions with the outermost crystal plane. The contribution $u_{w,\infty}(z)$ is the result of interactions with all remaining crystal planes, and it is found to be well represented by

$$u_{w,\infty}(z) = -\frac{2\pi\rho_p C\epsilon\sigma^3}{(\lambda_a - 2)(\lambda_a - 3)\Delta}\left(\frac{\sigma}{z + d\Delta}\right)^{\lambda_a-3}. \tag{7}$$

Here, ρ_p is the planar density of the graphene layers and Δ is the graphite layer spacing. Note that the naive approach of obtaining $u_{w,\infty}$ by integrating the attractive part of the atomic interaction potential over the bulk of the solid leads to the result that $d = 1$; however, Steele finds that the wall potential u_w better represents the result of the full Fourier analysis when setting $d = 0.61$, and we follow this recommendation. As such, the wall potential obtained from Equations (5)–(7) should be interpreted as an approximation to the full Fourier analysis shown by Steele [43]. Furthermore, the exponents of the potential given by Equations (5)–(7) are not necessarily those of the widely used 10-4-3 potential [26,44], which are recovered only if the carbon–hydrogen interactions are assumed to be governed by Lennard-Jones potentials (12-6).

2.3. Adsorption

The adsorption of hydrogen on a graphite surface may be expressed by the equilibrium

$$H_2(g) \leftrightharpoons H_2(ads), \tag{8}$$

where (g) denotes the gas phase and (ads) denotes the adsorbed gas on the solid surface. The chemical potential of a gas adsorbed to a surface is

$$\mu^{ads} = \mu^{ads,0} + RT \ln a^{ads} = \mu^{ads,0} + RT \ln \gamma^{ads} \frac{\Gamma}{\Gamma_\infty}, \tag{9}$$

where $\mu^{ads,0}$ is the standard chemical potential of the adsorbed phase; a^{ads} is the activity of the adsorbed phase; Γ is the surface excess concentration in particles per unit area; and γ^{ads} denotes the activity coefficient. The standard state is chosen as a hypothetical ideal state with $\gamma^s = 1$ at layer saturation ($\Gamma = \Gamma_\infty$). The equilibrium constant of Equation (8) is given by

$$\ln K_L = -\frac{\Delta_{ads} G}{RT}, \tag{10}$$

where $\Delta_{ads} G$ is the Gibbs energy of adsorption. By combining the equations above with the assumptions of ideal gas and $\gamma^{ads} = 1$, we obtain the well-known Langmuir adsorption model [8,19,45],

$$\theta \equiv \frac{\Gamma}{\Gamma_\infty} = \frac{K_L \frac{P}{P^\circ}}{1 + K_L \frac{P}{P^\circ}}, \tag{11}$$

where P is the pressure and P° is the standard state pressure. To determine the Langmuir constant, K_L, and the adsorption capacity, Γ_∞, the adsorption must be measured or predicted. In this work, we use both MD and cDFT to determine the adsorption at varying temperatures and pressures.

Following Gibbs [24], the adsorption Γ is defined as $\Gamma \equiv \frac{1}{A}(N - \rho_b V_f)$, where A is the surface area, N is the total number of particles in the system, ρ_b is the bulk density of the fluid, and V_f is the fluid volume. In order to compute the adsorption, a choice of the Gibbs dividing surface, which makes the fluid volume well defined, must be made [21,27]. The computed adsorption depends on this (non-unique) choice of dividing surface. The Langmuir equation requires a positive adsorption. When aiming to compute the Langmuir constant and adsorption capacity, the choice of dividing surface should thus be consistent with the constraint that $\Gamma \geq 0$.

At infinite dilution, cDFT gives an exact solution for the density profile in the presence of an external field V_{ex},

$$\lim_{\rho_b \to 0} \frac{\rho(z)}{\rho_b} = \exp\left(-\frac{V_{ex}(z)}{k_B T}\right) \tag{12}$$

where ρ_b is the fluid bulk density and k_B is Boltzmann's constant. Denoting the position of the Gibbs dividing surface as α, with the fluid phase located at $z \geq \alpha$, we have

$$A\Gamma = \int_{-\infty}^{\infty} \rho(z)\mathrm{d}z - \int_{\alpha}^{\infty} \rho_b \mathrm{d}z, \tag{13}$$

and inserting the exact solution for the density profile at infinite dilution yields

$$A \lim_{\rho_b \to 0} \frac{\Gamma}{\rho_b} = \int_{-\infty}^{\alpha} \exp\left(-\frac{V_{ex}(z)}{k_B T}\right)\mathrm{d}z + \int_{\alpha}^{\infty} \left[\exp\left(-\frac{V_{ex}(z)}{k_B T}\right) - 1\right]\mathrm{d}z. \tag{14}$$

The left-hand integral is always positive, while the right-hand integral may change sign depending on the temperature and choice of dividing surface. Denoting the location of the potential root as σ_w, such that $V_{ex}(z \geq \sigma_w) \leq 0$, we note that the right-hand integral can be guaranteed to be positive for any choice of $\alpha \geq \sigma_w$. Further, we note from Equation (13) that the adsorption can be made arbitrarily large by selecting a large value of α. We thus select $\alpha = \sigma_w$ as the Gibbs dividing surface when computing the adsorption, and remark that a similar choice of dividing surface was used in the work by Bråten et al. [25]. The relation between the density profile, adsorption, and dividing surface is illustrated in Figure 1. In Section 3.2, we show how the location of this dividing surface is determined in MD simulations.

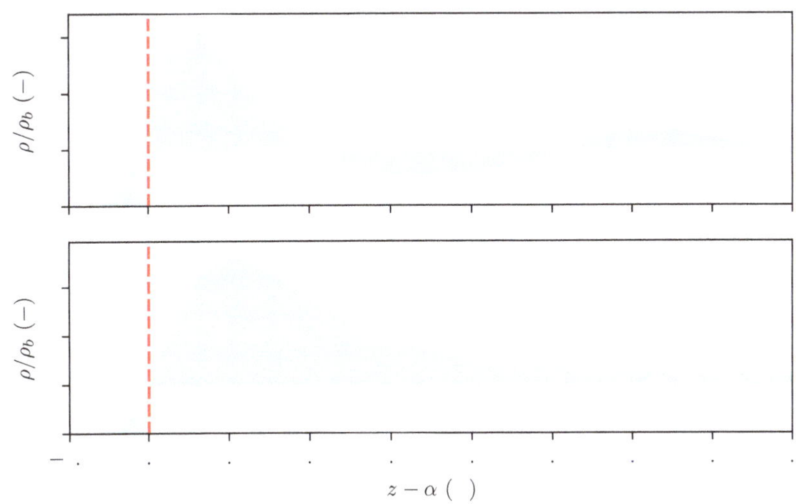

Figure 1. The density profile of a fluid adsorbed on a solid wall at high (top) and low (bottom) bulk fluid densities computed using cDFT. The red dashed line indicates the Gibbs dividing surface and the shaded region shows the ratio $(\rho - \rho_b)/\rho_b$, where ρ is the local density and ρ_b is the bulk density.

3. Methods

3.1. Molecular Dynamics Simulations

We carried out classical MD simulations using the LAMMPS package [46] to study hydrogen adsorption on a defect-free graphite crystal. The graphite structure consisted of 32,000 carbon atoms and 10 stacked graphene sheets, with each sheet oriented such that its surface was perpendicular to the z-direction. The simulated graphite surface measured 98×85 Å^2. The system included two layers of graphite separated by a pore width (W) of 50 Å, as illustrated in Figure 2. Periodic boundary conditions were applied in all directions. The simulations were performed across a temperature range of 100 K to

400 K. A representative snapshot of the gas mixture in equilibrium with the graphite is shown in Figure 2.

The MD simulations employed a time step of 1 fs. The initial configuration involved randomly distributing H_2 gas molecules over the graphite surface. To ensure system stability, we conducted 2 ns of simulations in the canonical ensemble (NVT) using the Nose–Hoover thermostat [47], which effectively maintained a constant temperature. Once thermal equilibrium was achieved, we performed another 2 ns run under microcanonical ensemble (NVE) conditions to investigate adsorption and transport properties. The average temperatures and pressures in the NVE simulation were consistent with expected values. In total, 4 million MD steps were executed, which is sufficiently long to obtain reliable statistics and consistent trajectories.

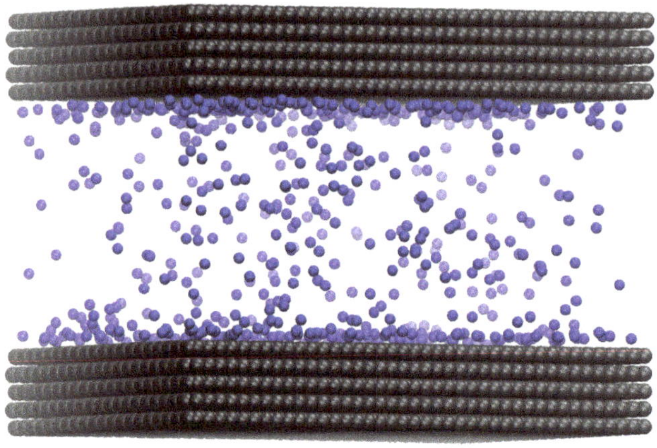

Figure 2. Snapshot of simulated system with $^{N}H_2$ = 700.

We employed a flexible graphite model [20,48,49] to represent the solid phase. The carbon–carbon bond stretching in graphite was described using the harmonic potential

$$V_S(r_{ij}) = \frac{1}{2}k_S(r_{ij} - r_0)^2 \tag{15}$$

where k_S is the force constant. The graphite bending was calculated with the harmonic potential

$$V_B\left(\theta_{ijk}\right) = \frac{1}{2}k_B\left(\theta_{ijk} - \theta_0\right)^2 \tag{16}$$

where θ_{ijk} is the angle between atoms i, j, and k. k_B is the force constant. The torsion energy in graphite was calculated as

$$V_T(\omega) = \frac{1}{2}k_T[1 - \cos(2\omega)]^2 \tag{17}$$

where ω is the torsion angle and k_T is the force constant. Interactions between carbon atoms in separate graphene sheets, as well as hydrogen–hydrogen and hydrogen–carbon interactions, were represented by a Mie potential

$$\frac{u_{ij}(r_{ij})}{C\epsilon_{ij}} = \left(\frac{\sigma_{ij}}{\mathrm{r}_{ij}}\right)^{\gamma_{r,ij}} - \left(\frac{\sigma_{ij}}{\mathrm{r}_{ij}}\right)^{\gamma_{a,ij}}, \quad C = \frac{\gamma_{r,ij}}{\gamma_{r,ij} - \gamma_{a,ij}}\left(\frac{\gamma_{r,ij}}{\gamma_{a,ij}}\right)^{\frac{\gamma_{a,ij}}{\gamma_{a,ij}-\gamma_{r,ij}}} \tag{18}$$

where γ_r and γ_a are the repulsive and attractive exponents of the Mie potential, r_{ij} is the distance, ϵ_{ij} is the Mie well depth, and σ_{ij} is the characteristic length scale corresponding to the distance where the Mie interparticle potential is zero. This potential has previously been shown to capture the equilibrium properties of hydrogen vapor and liquid at higher temperatures [50]. At lower temperatures, quantum corrections need to be incorporated.

The mixing rule for the Mie potential between hydrogen and the carbon of graphite was calculated as

$$\lambda_{k,ij} - 3 = \sqrt{(\lambda_{k,ii} - 3)(\lambda_{k,ij} - 3)}, \quad k = \text{a, r} \tag{19}$$

$$\sigma_{ij} = (1 - l_{ij})\frac{\sigma_{ii} + \sigma_{ij}}{2} \tag{20}$$

$$\epsilon_{ij} = (1 - k_{ij})\frac{\sqrt{\sigma_{ij}^3 \sigma_{jj}^3}}{\sigma_{ij}^3}\sqrt{\epsilon_{ii}\epsilon_{jj}} \tag{21}$$

where k_{ij} and l_{ij} are adjustable parameters that can be fitted to experimental data on binary mixtures. In this work, k_{ij} and l_{ij} are set to zero. All parameters for the force fields are listed in Table 1.

Table 1. The potential parameters for flexible graphite [20,48,49] and hydrogen gas [50]. The lattice spacing $\Delta = 3.35$ Å and bulk density $\rho_b = 0.114$ Å^{-3} were used for the graphite Steele potential derived from the carbon–hydrogen Mie potential.

Species	σ **(Å)**	ϵ **(K)**	γ_r	γ_a
H_2	3.26	17.93	8	6
Carbon	3.33	26.0	12	6
Graphite	Force constant (kJ mol^{-1})		Equilibrium position	
Bond stretching	4393.9 Å^{-2}		1.42 Å	
Angle bending	418.2		$2\pi/3$	
Torsion	26.15		—	

3.2. *Locating the Dividing Surface*

When computing the adsorption, a non-unique choice of the Gibbs dividing surface must be made. As discussed in Section 2.3, we have chosen to use the root of the wall potential V_{ex} to define the location of the dividing surface, α. Since the computed adsorption depends on the choice of dividing surface, it is key that the choice of dividing surface is consistently applied to both the cDFT and the MD results in order for the results to be comparable.

When using cDFT to compute the adsorption, the external potential V_{ex} is explicitly imposed, and solving $V_{ex}(\alpha) = 0$ to determine α poses no extra challenge. When analyzing the MD results, however, the situation is more complex.

Because the graphite phase is treated explicitly in MD, it is subject to thermal expansion. Additionally, the graphite planes are not a perfectly flat planes, but are allowed to flex and vibrate. In order to determine the effective potential seen by the hydrogen molecules in MD, we utilize the exact cDFT result from Equation (12), and define

$$V_{eff}^{\circ}(z;T) = \lim_{\rho_b \to 0} V_{eff}(z;T,\rho_b), \quad V_{eff}(z;T,\rho_b) = -k_B T \ln \frac{\rho(z)}{\rho_b}, \tag{22}$$

such that the location of the dividing surface is given by the solution to $V_{eff}^{\circ}(\alpha;T) = 0$. Denoting the solution to $V_{eff}(z;T,\rho_b) = 0$ as $\sigma_{w,eff}$, we can find the location of the dividing surface at a given temperature by determining $\sigma_{w,eff}$ at several bulk densities and extrapolating to zero density.

3.3. Adsorption Parameters

Using adsorptions obtained from MD or cDFT, we determined the parameters of the Langmuir equation by use of linear regression. For this purpose, the Langmuir equation was linearized as

$$\frac{1}{\Gamma} = \frac{P^{\circ}}{\Gamma_{\infty} K_L P} + \frac{1}{\Gamma_{\infty}}. \tag{23}$$

Note that the adsorption capacity, Γ_{∞}, is assumed to be independent of temperature [8,13,19], while the Langmuir constant K_L is a strong function of temperature (viz. Equation (25)). Thus, in order to obtain consistent values from the regression, all isotherms must be fitted simultaneously in such a way that a single value is obtained for Γ_{∞}, while one value of K_L is obtained for each isotherm.

With the Langmuir constant determined across a range of temperatures, the adsorption enthalpy and entropy may be determined from the temperature dependence as

$$\ln K_L = -\frac{\Delta_{\text{ads}} H^{\circ}}{RT} + \frac{\Delta_{\text{ads}} S^{\circ}}{R}. \tag{24}$$

If the heat capacity of the adsorbed phase and the fluid phase are equal, $\Delta_{\text{ads}} H$ and $\Delta_{\text{ads}} S$ do not vary with temperature. On the other hand, if the difference in heat capacity between the two phases, $\Delta_{\text{ads}} C_p$, is included and assumed constant, the temperature dependence of the Langmuir constant becomes

$$\ln K_L = -\frac{1}{RT}\left[\Delta_{\text{ads}} H^{\circ}(T_0) + \Delta_{\text{ads}} C_p (T - T_0)\right] + \frac{1}{R}\left[\Delta_{\text{ads}} S^{\circ}(T_0) + \Delta_{\text{ads}} C_p \ln \frac{T}{T_0}\right]. \tag{25}$$

4. Results and Discussion

In the following, we will compare the results from the MD simulations to cDFT calculations for the adsorption of hydrogen on a graphite surface. The outermost graphene layer is placed at the $z = 0$ coordinate, with the fluid phase extending in the positive direction.

4.1. Location of the Dividing Surface

The location of the dividing surface in the MD simulations is determined by finding the position at which $\rho(\sigma_{w,eff}) = \rho_b$ and extrapolating to zero density, as described in Section 3.2. As shown in Figure 3, $\sigma_{w,eff}$ exhibits a clear linear dependence on density, such that a linear extrapolation to zero density is well justified.

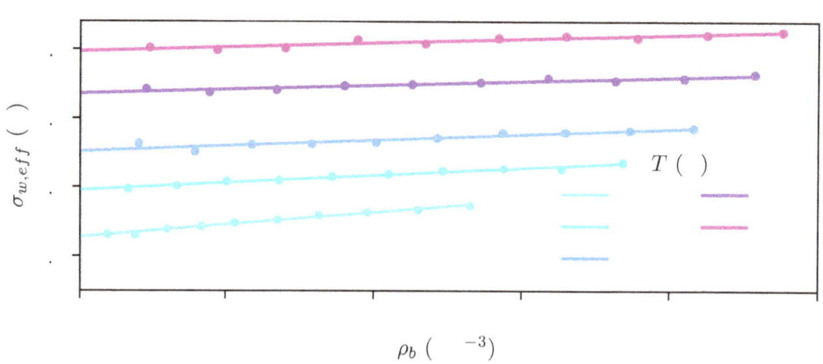

Figure 3. Location at which $\rho(\sigma_{w,eff}) = \rho_b$ (dots) from the MD simulations, and extrapolation to zero density (lines) for all temperatures investigated. The root of the external potential applied in the cDFT is located 2.735 Å from the graphite surface.

It is evident from Figure 3 that the dividing surface moves outwards towards the fluid bulk phase as the temperature increases. The thermal expansion coefficient of graphite is on the order of $10^{-6}\,\mathrm{K}^{-1}$ [51], which is far too small to fully explain the observed shift of the dividing surface. We attribute the shift to a combination of thermal expansion of the graphite and thermal agitation causing the outermost layer to become more "diffuse" at higher temperatures.

We would like to emphasize the importance of choosing a consistent dividing surface when computing adsorption isotherms. If a dividing surface is placed at a fixed position relative to the graphite surface, shifting of the density profile as a result of graphite expansion would erroneously be interpreted as a decrease in the adsorption. This would in turn lead to an incorrect estimate of the temperature dependency of the adsorption, and invalidate the results for the enthalpy, entropy, and heat capacity of adsorption.

The root of the potential used in the cDFT calculations is found to be 2.735 Å from the graphite surface, which is significantly closer than the root of the effective potential from the MD simulations, which ranges from 3.03 Å at 100 K to 3.30 Å at 400 K. Since the Steele potential is derived for a static crystal [43], the cDFT potential should be interpreted as the zero-temperature limit of the effective potential, where the solid structure is static. Correcting this potential to account for thermal agitation of the crystal structure may be possible using a procedure analogous to the Feynman–Hibbs correction that accounts for quantum swelling [50,52–54].

4.2. Adsorption

To investigate the adsorption of H_2 on graphite, we first compare the calculated adsorption profile from the MD simulations to the cDFT calculations. As shown in Figure 4, the H_2 molecules form a single adsorption layer, which is different from the CO_2 adsorption observed as two layers in previous studies [21,22]. This suggests that the Langmuir equation is suitable for describing the adsorption of hydrogen on graphite at the temperatures considered in this work [8,19].

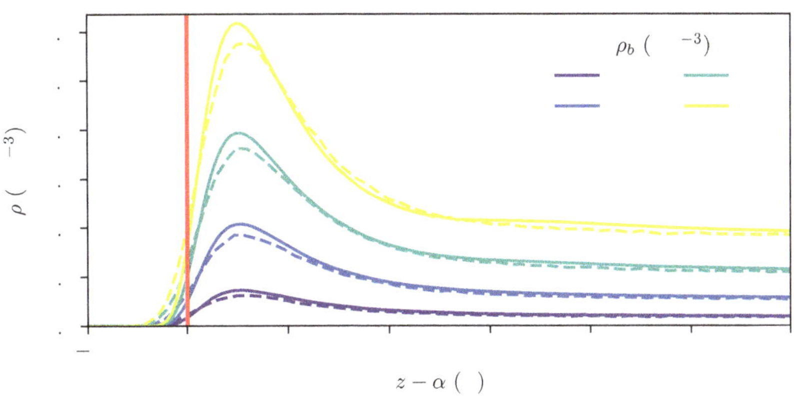

Figure 4. Computed density profile from MD (dashed lines) and cDFT (solid lines) at 300 K for several bulk densities. The density profiles have been shifted to place the dividing surface (red line) at the origin for both MD and cDFT calculations.

Even though the location of the dividing surface relative to the outermost graphene layer is significantly different in the MD and cDFT calculations, as illustrated in Figure 3, the density profiles relative to the dividing surface are in excellent agreement, as shown in Figure 4. This underscores the importance of rigorously defining a dividing surface, and appropriately applying the chosen definition to both the MD and cDFT results if they are to be comparable.

At higher bulk densities, the MD results show a slightly higher hydrogen density on the negative side of the dividing surface and a slightly lower peak density than that predicted by cDFT, as seen for the $\rho_b = 4.6\,\mathrm{nm}^{-3}$ density profile in Figure 4. This observation is consistent with the hypothesis that thermal agitation causes the outermost graphene layer to become more diffuse, as it has been observed that the increase in the potential root caused by such "smearing" is accompanied by a softening of the potential and a reduction in the potential well depth [50,52].

Adsorptions and the corresponding bulk pressures computed from the MD and cDFT density profiles at temperatures ranging from 100 K to 400 K are shown in Figure 5. The bulk pressures in the cDFT calculations were computed using the SAFT-VR Mie equation of state (EoS) implemented in the open-source library `ThermoPack` [38,39,55,56], which is consistent with the Helmholtz energy functional employed in the cDFT.

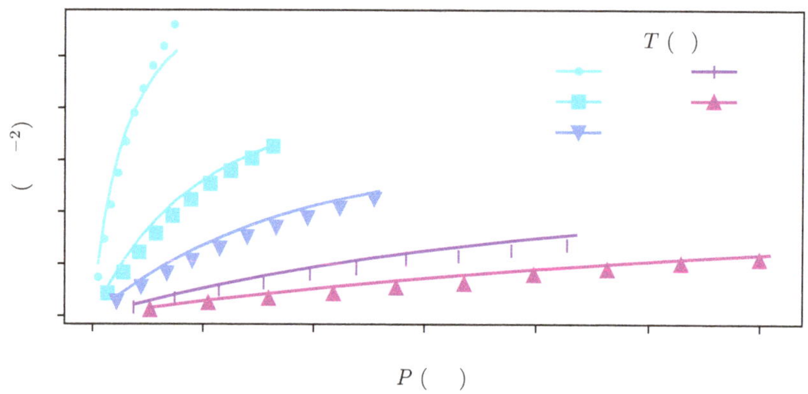

Figure 5. Adsorption isotherms of H_2 on graphite obtained from MD simulations (marks) and cDFT calculations (lines) at various temperatures ranging from 100 K to 400 K.

We find that the adsorptions predicted by cDFT deviate from those obtained from MD by up to 38 %, with systematic over-prediction at temperatures $T \geq 150\,\mathrm{K}$. At the 100 K isotherm, there is an under-prediction of −10 % at the highest investigated pressure of ≈ 37 bar, and an over-prediction of 35 % at the lowest pressure of ≈ 2.5 bar. Deviations in the computed pressure are about an order of magnitude smaller, with the pressure obtained from the EoS deviating by 1–2 % from the simulations.

Despite these apparently large relative deviations, we observe that there is excellent agreement between the cDFT calculations and MD simulations regarding the temperature dependency of the adsorption, which is the key to extracting reliable information about the adsorption enthalpy, entropy, and heat capacity. Furthermore, we observe that the largest relative deviations in adsorption between cDFT and MD are found at lower pressures, where the absolute adsorption itself is very small.

The adsorption capacity of graphite was determined by linear regression, as detailed in Section 3.3, with the regression results shown in Table 2. The Langmuir constant $K_L(T)$ and adsorption capacity Γ_∞ were fitted simultaneously to the isotherms from 150 K to 400 K, giving coefficients of determination $R^2 = 0.99$ and $R^2 = 0.97$ for the MD and cDFT computations, respectively. Being consistent with the assumptions made when deriving the Langmuir equation, the adsorption capacity should be treated as a constant across all temperatures. Following this procedure, the MD simulations and cDFT calculations yield adsorption capacities of $10.9\,\mu\mathrm{mol\,m}^{-2}$ and $7.1\,\mu\mathrm{mol\,m}^{-2}$, respectively. This is in reasonable agreement with previously reported simulation results from Simon et al. [20], which report an adsorption capacity of 18.5 µmol m^{-2}, and experimental studies [57–60] have yielded saturation adsorption capacities ranging from 8.93 µmol m^{-2} to 56.2 µmol m^{-2}.

Table 2. The adsorption capacity of graphite for hydrogen as determined by linear regression of the Langmuir equation to the adsorption isotherms computed from cDFT and MD simulations. Results are shown for regressions to individual isotherms, as well as simultaneous regression of all isotherms from 150 K to 400 K.

	Isotherm	**150 K–400 K**	**100 K**	**150 K**	**200 K**	**300 K**	**400 K**
Γ_∞ (µmol m^{-2})	cDFT	7.10	12.04	9.99	8.32	6.36	5.36
	MD	10.91	18.10	13.40	13.48	10.80	7.97
$-\ln K_L$	cDFT	–	–	4.54	5.50	6.59	7.28
	MD	–	–	3.75	4.72	5.86	6.56

Despite the good agreement with previously reported adsorption capacities, regressions to the individual isotherms show that the predicted adsorption capacity increases rapidly when going from 150 K to 100 K. Closer inspection of the density profiles from the MD and cDFT reveals the emergence of a second adsorption layer, which becomes especially prominent at 100 K. The Langmuir equation is based on the assumption that adsorption only occurs in a single monolayer, and the assumption that the adsorption capacity is constant breaks down when this is not the case. For this reason, the 100 K isotherm was omitted from the regression used to obtain the Langmuir constants used for further analysis in Section 4.3. Our results thus indicate that care should be taken when applying the Langmuir equation to the hydrogen–graphite system at low temperatures, especially when calculating the adsorption capacity.

4.3. Thermodynamic Properties of Adsorbed Hydrogen

Understanding the thermodynamics of adsorbed hydrogen is crucial for optimizing materials and systems for hydrogen storage applications. In addition to Langmuir constants, the enthalpy, entropy, and change in the heat capacity of adsorption are important parameters, as explained in Section 3.3. These were determined by linear regression, with the regression coefficients presented in Table 3 and the fitted curves shown in Figure 6.

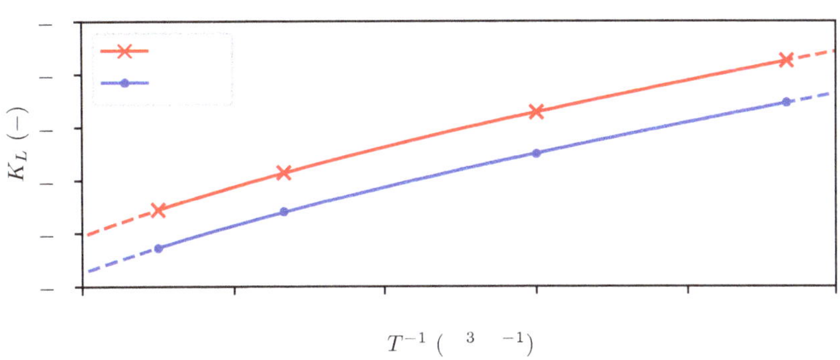

Figure 6. The Langmuir constant computed using MD (dots) and cDFT (crosses), and the resulting linear regressions (lines).

The computed Langmuir constants shown in Figure 6 display a distinct non-linearity. This is a clear indicator that the difference in heat capacity between the adsorbed phase and fluid phase is non-negligible, and justifies the use of Equation (25) rather than the simpler

Equation (24) in the regression. Since there is a physical justification of Equation (25), neglecting $\Delta_{\text{ads}}C_p$ will likely lead to increased uncertainty in the estimate of the adsorption enthalpy.

Table 3 shows excellent agreement between the results from MD simulations and cDFT: the enthalpy and entropy of adsorption at 298 K obtained from the two methods deviate by only 3.5% and 7%, respectively. The deviation of 9% between the two methods in the computed change in heat capacity is somewhat larger. Previously reported values for the enthalpy of adsorption of hydrogen on graphite range from $-5\,\text{kJ}\,\text{mol}^{-1}$ to $-7\,\text{kJ}\,\text{mol}^{-1}$ [9,20], in excellent agreement with the results obtained in this work.

Table 3. The enthalpy, entropy, and change in heat capacity of adsorption for hydrogen on graphite, computed using MD simulation results and cDFT. The enthalpy and entropy of adsorption are reported at 298 K.

Method	$\Delta_{\text{ads}}H^\circ$ (298 K) (kJ mol^{-1})	$\Delta_{\text{ads}}S^\circ$ (298 K) (J mol^{-1} K^{-1})	$\Delta_{\text{ads}}C_p$ (J mol^{-1} K^{-1})
cDFT	−6.373	−69.95	−12.43
MD	−6.162	−75.42	−11.38

An interesting observation is that both cDFT and MD yield a negative change in heat capacity for the adsorption equilibrium (Equation (8)). This means that the adsorbed phase has a lower heat capacity than the bulk fluid. We hypothesize that this is because molecules in the adsorbed layer have fewer translational degrees of freedom than particles in the bulk fluid. If all translational degrees of freedom of the molecules are arrested because they are bound to adsorption sites, $\Delta_{\text{ads}}C_p = -\frac{3R}{2} \approx -12.48\,\text{J}\,\text{mol}^{-1}\,\text{K}^{-1}$, which agrees well with the values obtained from MD and cDFT.

In order to investigate the impact of including the difference in heat capacity between the adsorbed phase and bulk fluid, a separate regression was conducted under the assumption that $\Delta_{\text{ads}}C_p = 0$ (i.e., using Equation (24)). This yielded a significantly poorer fit to the computed Langmuir constants, and an enthalpy and entropy of adsorption that deviated significantly from previously reported values [9,20]. This further supports the hypothesis that the difference in heat capacity is non-negligible, and underscores the importance of including the heat capacity difference as a variable when regressing the enthalpy and entropy of reaction to equilibrium constant data over a large temperature range.

5. Conclusions and Outlook

In this work, we have used molecular dynamics (MD) and classical density functional theory (cDFT) to study the adsorption of hydrogen on graphite at temperatures ranging from 100 K to 400 K, and bulk fluid pressures ranging from 2.5 bar to 300 bar. It has been demonstrated how an effective wall potential can be constructed from the MD parameters for graphite in order to use cDFT calculations to study the same system. Furthermore, it has been shown that a consistent definition of the dividing surface must be used to compare the adsorptions from the two methodologies. Using a fixed position relative to the graphite as the dividing surface in the MD simulations will lead to incorrect calculation of the adsorption enthalpy.

Overall, there is good agreement between results from MD simulations and cDFT. The largest discrepancies were seen for the absolute adsorptions, with relative differences of up to 38 % between the two methods. However, this is likely due to the small absolute values of the adsorption. We find that both the temperature and pressure dependency of the adsorption obtained from the two methods are in excellent agreement at temperatures above 150 K.

The saturation capacity of graphite was determined to be 7.1 µmol m^{-2} from cDFT, and 10.9 µmol m^{-2} from MD. The cDFT and MD results show that the adsorption capacity increases significantly at lower temperatures, suggesting that care should be exercised when applying the Langmuir equation at temperatures below 100 K. The computed density profiles at 100 K from cDFT and MD showed the emergence of a second adsorption layer, meaning that the assumptions behind the Langmuir equation begin to break down in these conditions.

The standard enthalpy of adsorption at 298 K was determined to be in the range −6.37 kJ mol^{-1} to −6.16 kJ mol^{-1}, with an agreement between cDFT and MD within 3.5%. The standard entropy of adsorption at 298 K was determined to be between −75.4 J mol^{-1} K^{-1} and −70.0 J mol^{-1} K^{-1}, with the results obtained from cDFT and MD differing by 7 %. We found that the difference in heat capacity of the adsorbed hydrogen and bulk hydrogen is to a very good approximation constant from 150 K to 400 K, with bulk hydrogen having a heat capacity between 11.38 J mol^{-1} K^{-1} and 12.43 J mol^{-1} K^{-1} higher than the adsorbed hydrogen. The discrepancy between values obtained from cDFT and MD was largest for the difference in heat capacity, with a difference of 9 % between the values obtained using the two methods.

When comparing the location of the Gibbs dividing surface relative to the outermost graphene plane obtained from MD simulations to that in the cDFT calculations, we found that the graphite bulk appears to expand outwards as temperature increases, and that this effect is far larger than what can be attributed to thermal expansion. We also found that the effective wall potential experienced by hydrogen in the MD simulations appears to become softer at higher temperatures. We hypothesize that this effect is due to thermal agitation causing the outermost graphene layer to "smear out" in space, and note the possibility of including this effect in cDFT calculations as an intriguing possibility for future investigations.

To improve the accuracy of the results and better understand the behavior of hydrogen on graphite, further work is needed to refine the potential that describes graphite–hydrogen interactions. Furthermore, quantum effects need to be incorporated to extend the analysis to cryogenic temperatures.

Author Contributions: Conceptualization, Ø.W.; Methodology, V.G.J., M.H. and Ø.W.; Software, M.H. and V.G.J.; Investigation, T.T.T.; Writing—original draft, V.G.J. and T.T.T.; Writing—review & editing, V.G.J., M.H., Ø.W. and T.T.T.; Supervision, Ø.W. and T.T.T. All authors have read and agreed to the published version of the manuscript.

Funding: This research was funded by the Research Council of Norway (RCN), the Center of Excellence Funding Scheme, Project No. 262644, PoreLab, and the European Research Council (ERC) under the European Union's Horizon Europe research and innovation program (grant agreement No. 101115669).

Institutional Review Board Statement: Not applicable.

Data Availability Statement: The data presented in this study are available on request from the corresponding author.

Conflicts of Interest: The authors declare no conflicts of interest.

References

1. Dillon, A.C.; Heben, M.J. Hydrogen storage using carbon adsorbents: Past, present and future. *Appl. Phys. A Mater. Sci. Process.* **2001**, *72*, 133–142. [CrossRef]
2. Patchkovskii, S.; Tse, J.S.; Yurchenko, S.N.; Zhechkov, L.; Heine, T. Graphene nanostructures as tunable storage media for molecular hydrogen. *Proc. Natl. Acad. Sci. USA* **2005**, *102*, 10439–10444. [CrossRef] [PubMed]
3. Schlapbach, L.; Züttel, A. Hydrogen-storage materials for mobile applications. *Nature* **2001**, *414*, 353–358. [CrossRef]

4. Züttel, A. Hydrogen storage methods. *Naturwissenschaften* **2004**, *91*, 157–172. [CrossRef] [PubMed]
5. Steele, B.C.H.; Heinzel, A. Materials for fuel-cell technologies. *Nature* **2001**, *414*, 345–352. [CrossRef]
6. Balog, R.; Jørgensen, B.; Nilsson, L.; Andersen, M.; Rienks, E.; Bianchi, M.; Fanetti, M.; Lægsgaard, E.; Baraldi, A.; Lizzit, S.; et al. Bandgap opening in graphene induced by patterned hydrogen adsorption. *Nat. Mater.* **2010**, *9*, 315–319. [CrossRef]
7. Popov, V. Carbon nanotubes: Properties and application. *Mater. Sci. Eng. R Rep.* **2004**, *43*, 61–102. [CrossRef]
8. Kanashenko, S.; Gorodetsky, A.; Chernikov, V.; Markin, A.; Zakharov, A.; Doyle, B.; Wampler, W. Hydrogen adsorption on and solubility in graphites. *J. Nucl. Mater.* **1996**, *233-237*, 1207–1212. [CrossRef]
9. Mattera, L.; Rosatelli, F.; Salvo, C.; Tommasini, F.; Valbusa, U.; Vidali, G. Selective adsorption of $^1H_2$ and $^2H_2$ on the (0001) graphite surface. *Surf. Sci.* **1980**, *93*, 515–525. [CrossRef]
10. Gao, Y.; Li, Y.; Zhang, L.; Huang, H.; Hu, J.; Shah, S.M.; Su, X. Adsorption and removal of tetracycline antibiotics from aqueous solution by graphene oxide. *J. Colloid Interface Sci.* **2012**, *368*, 540–546. [CrossRef]
11. Mills, G.; Jónsson, H.; Schenter, G.K. Reversible work transition state theory: Application to dissociative adsorption of hydrogen. *Surf. Sci.* **1995**, *324*, 305–337. [CrossRef]
12. Yu, J.G.; Yu, L.Y.; Yang, H.; Liu, Q.; Chen, X.H.; Jiang, X.Y.; Chen, X.Q.; Jiao, F.P. Graphene nanosheets as novel adsorbents in adsorption, preconcentration and removal of gases, organic compounds and metal ions. *Sci. Total Environ.* **2015**, *502*, 70–79. [CrossRef]
13. Ye, Z.; Chen, D.; Pan, Z.; Zhang, G.; Xia, Y.; Ding, X. An improved Langmuir model for evaluating methane adsorption capacity in shale under various pressures and temperatures. *J. Nat. Gas Sci. Eng.* **2016**, *31*, 658–680. [CrossRef]
14. Kapoor, A.; Ritter, J.; Yang, R.T. An extended Langmuir model for adsorption of gas mixtures on heterogeneous surfaces. *Langmuir* **1990**, *6*, 660–664. [CrossRef]
15. Kurniawan, A.; Sutiono, H.; Indraswati, N.; Ismadji, S. Removal of basic dyes in binary system by adsorption using rarasaponin–bentonite: Revisited of extended Langmuir model. *Chem. Eng. J.* **2012**, *189-190*, 264–274. [CrossRef]
16. Zhang, P.; Wang, L. Extended Langmuir equation for correlating multilayer adsorption equilibrium data. *Sep. Purif. Technol.* **2010**, *70*, 367–371. [CrossRef]
17. Ayawei, N.; Ebelegi, A.N.; Wankasi, D. Modelling and Interpretation of Adsorption Isotherms. *J. Chem.* **2017**, *2017*, 3039817. [CrossRef]
18. Yao, C. Extended and improved Langmuir equation for correlating adsorption equilibrium data. *Sep. Purif. Technol.* **2000**, *19*, 237–242. [CrossRef]
19. Langmuir, I. The adsorption of gases on plane surfaces of glass, mica and platinum. *J. Am. Chem. Soc.* **1918**, *40*, 1361–1403. [CrossRef]
20. Simon, J.M.; Haas, O.E.; Kjelstrup, S. Adsorption and Desorption of H_2 on Graphite by Molecular Dynamics Simulations. *J. Phys. Chem. C* **2010**, *114*, 10212–10220. [CrossRef]
21. Trinh, T.; Bedeaux, D.; Simon, J.M.; Kjelstrup, S. Thermodynamic characterization of two layers of CO_2 on a graphite surface. *Chem. Phys. Lett.* **2014**, *612*, 214–218. [CrossRef]
22. Trinh, T.; Bedeaux, D.; Simon, J.M.; Kjelstrup, S. Calculation of the chemical potential and the activity coefficient of two layers of CO_2 adsorbed on a graphite surface. *Phys. Chem. Chem. Phys.* **2015**, *17*, 1226–1233. [CrossRef] [PubMed]
23. Kojima, Y. Hydrogen storage materials for hydrogen and energy carriers. *Int. J. Hydrogen Energy* **2019**, *44*, 18179–18192. [CrossRef]
24. Gibbs, J.W. *On the Equilibrium of Heterogeneous Substances*; The Academy New Haven : New Haven, CT, USA, 1874.
25. Bråten, V.; Zhang, D.T.; Hammer, M.; Aasen, A.; Schnell, S.K.; Wilhelmsen, Ø. Equation of state for confined fluids. *J. Chem. Phys.* **2022**, *156*, 244504 . [CrossRef] [PubMed]
26. Sauer, E.; Gross, J. Classical Density Functional Theory for Liquid–Fluid Interfaces and Confined Systems: A Functional for the Perturbed-Chain Polar Statistical Associating Fluid Theory Equation of State. *Ind. Eng. Chem. Res.* **2017**, *56*, 4119–4135. [CrossRef]
27. Rehner, P. Interfacial Properties Using Classical Density Functional Theory: Curved Interfaces and Surfactants. Ph.D. Thesis, University of Stuttgart, Stuttgart, Germany, 2021.
28. Stierle, R.; Sauer, E.; Eller, J.; Theiss, M.; Rehner, P.; Ackermann, P.; Gross, J. Guide to efficient solution of PC-SAFT classical Density Functional Theory in various Coordinate Systems using fast Fourier and similar Transforms. *Fluid Phase Equilibria* **2020**, *502*, 112306. [CrossRef]
29. Evans, R.; Oettel, M.; Roth, R.; Kahl, G. New developments in classical density functional theory. *J. Phys. Condens. Matter* **2016**, *28*, 240401. [CrossRef] [PubMed]
30. Tarazona, P. Free-energy density functional for hard spheres. *Phys. Rev. A* **1985**, *31*, 2672–2679. [CrossRef]
31. Aasen, A.; Jervell, V.G.; Hammer, M.; Strøm, B.A.; Skarsvåg, H.L.; Wilhelmsen, Ø. Bulk and interfacial thermodynamics of ammonia, water and their mixtures. *Fluid Phase Equilibria* **2024**, *584*, 114125. [CrossRef]
32. Gross, J. A density functional theory for vapor-liquid interfaces using the PCP-SAFT equation of state. *J. Chem. Phys.* **2009**, *131*, 204705. [CrossRef]

33. Rehner, P.; Gross, J. Multiobjective Optimization of PCP-SAFT Parameters for Water and Alcohols Using Surface Tension Data. *J. Chem. Eng. Data* **2020**, *65*, 5698–5707. [CrossRef]
34. Sauer, E.; Gross, J. Prediction of Adsorption Isotherms and Selectivities: Comparison between Classical Density Functional Theory Based on the Perturbed-Chain Statistical Associating Fluid Theory Equation of State and Ideal Adsorbed Solution Theory. *Langmuir* **2019**, *35*, 11690–11701. [CrossRef]
35. Aasen, A.; Blokhuis, E.M.; Wilhelmsen, Ø. Tolman lengths and rigidity constants of multicomponent fluids: Fundamental theory and numerical examples. *J. Chem. Phys.* **2018**, *148*, 204702. [CrossRef] [PubMed]
36. Wilhelmsen, Ø.; Trinh, T.T.; Kjelstrup, S.; van Erp, T.S.; Bedeaux, D. Heat and mass transfer across interfaces in complex nanogeometries. *Phys. Rev. Lett.* **2015**, *114*, 065901. [CrossRef] [PubMed]
37. Rehner, P.; Aasen, A.; Wilhelmsen, Ø. Tolman lengths and rigidity constants from free-energy functionals—General expressions and comparison of theories. *J. Chem. Phys.* **2019**, *151*, 244710. [CrossRef] [PubMed]
38. Lafitte, T.; Apostolakou, A.; Avendaño, C.; Galindo, A.; Adjiman, C.S.; Müller, E.A.; Jackson, G. Accurate statistical associating fluid theory for chain molecules formed from Mie segments. *J. Chem. Phys.* **2013**, *139*, 154504. [CrossRef]
39. Hammer, M.; Aasen, A.; Ervik, Å.; Wilhelmsen, Ø. Choice of reference, the influence of non-additivity and challenges in thermodynamic perturbation theory for mixtures. *J. Chem. Phys.* **2020**, *152*, 134106. [CrossRef]
40. Roth, R. Introduction to Density Functional Theory of Classical Systems: Theory and Applications; Lecture Notes, 2006. Available online: https://bytebucket.org/knepley/cdft-git/wiki/papers/Lecture_Notes_on_DFT__Roland_Roth.pdf (accessed on 6 February 2025).
41. Hammer, M.; Bauer, G.; Stierle, R.; Gross, J.; Wilhelmsen, Ø. Classical density functional theory for interfacial properties of hydrogen, helium, deuterium, neon, and their mixtures. *J. Chem. Phys.* **2023**, *158*, 104107. [CrossRef]
42. NTNU Department of Chemistry. ThermoTools: SurfPack Open Source Surface- and Interfacial Properties. 2024. Available online: https://github.com/thermotools/SurfPack (accessed on 6 February 2025).
43. Steele, W.A. The physical interaction of gases with crystalline solids: I. Gas-solid energies and properties of isolated adsorbed atoms. *Surf. Sci.* **1973**, *36*, 317–352. [CrossRef]
44. Albesa, A.G.; Llanos, J.L.; Vicente, J. Comparative Study of Methane Adsorption on Graphite. *Langmuir* **2008**, *24*, 3836–3840. [CrossRef]
45. Chang, C.K.; Tun, H.; Chen, C.C. An activity-based formulation for Langmuir adsorption isotherm. *Adsorption* **2020**, *26*, 375–386. [CrossRef]
46. Plimpton, S. Fast parallel algorithms for short-range molecular dynamics. *J. Comput. Phys.* **1995**, *117*, 1–19. [CrossRef]
47. Hoover, W.G. Canonical dynamics: Equilibrium phase-space distributions. *Phys. Rev. A* **1985**, *31*, 1695–1697. [CrossRef]
48. Trinh, T.T.; Vlugt, T.J.; Hägg, M.B.; Bedeaux, D.; Kjelstrup, S. Simulation of Pore Width and Pore Charge Effects on Selectivities of CO_2 vs. H_2 from a Syngas-like Mixture in Carbon Mesopores. *Energy Procedia* **2015**, *64*, 150–159. [CrossRef]
49. Mayo, S.L.; Olafson, B.D.; Goddard, W.A. DREIDING: A generic force field for molecular simulations. *J. Phys. Chem.* **1990**, *94*, 8897–8909. [CrossRef]
50. Aasen, A.; Hammer, M.; Ervik, Å.; Müller, E.A.; Wilhelmsen, Ø. Equation of state and force fields for Feynman–Hibbs-corrected Mie fluids. I. Application to pure helium, neon, hydrogen, and deuterium. *J. Chem. Phys.* **2019**, *151*, 064508. [CrossRef]
51. Morgan, W. Thermal expansion coefficients of graphite crystals. *Carbon* **1972**, *10*, 73–79. [CrossRef]
52. Aasen, A.; Hammer, M.; Müller, E.A.; Wilhelmsen, Ø. Equation of state and force fields for Feynman–Hibbs-corrected Mie fluids. II. Application to mixtures of helium, neon, hydrogen, and deuterium. *J. Chem. Phys.* **2020**, *152*, 074507. [CrossRef]
53. Aasen, A.; Hammer, M.; Lasala, S.; Jaubert, J.N.; Wilhelmsen, Ø. Accurate quantum-corrected cubic equations of state for helium, neon, hydrogen, deuterium and their mixtures. *Fluid Phase Equilibria* **2020**, *524*, 112790. [CrossRef]
54. van Westen, T.; Bauer, G.; Gross, J. Corresponding-states framework for classical and quantum fluids—Beyond Feynman–Hibbs. *J. Chem. Phys.* **2025**, *162*, 031101. [CrossRef]
55. Wilhelmsen, Ø.; Aasen, A.; Skaugen, G.; Aursand, P.; Austegard, A.; Aursand, E.; Gjennestad, M.A.; Lund, H.; Linga, G.; Hammer, M. Thermodynamic Modeling with Equations of State: Present Challenges with Established Methods. *Ind. Eng. Chem. Res.* **2017**, *56*, 3503–3515. [CrossRef]
56. SINTEF Energy Research/NTNU Department of Chemistry. ThermoTools: ThermoPack Open Source Thermodynamics Library. 2024. Available online: https://github.com/thermotools/thermopack (accessed on 6 February 2025).
57. Benard, P.; Chahine, R. Determination of the adsorption isotherms of hydrogen on activated carbons above the critical temperature of the adsorbate over wide temperature and pressure ranges. *Langmuir* **2001**, *17*, 1950–1955. [CrossRef]
58. Hirscher, M.; Panella, B. Nanostructures with high surface area for hydrogen storage. *J. Alloys Compd.* **2005**, *404*, 399–401. [CrossRef]

59. Panella, B.; Hirscher, M.; Roth, S. Hydrogen adsorption in different carbon nanostructures. *Carbon* **2005**, *43*, 2209–2214. [CrossRef]
60. Zhou, L.; Zhou, Y.; Sun, Y. Studies on the mechanism and capacity of hydrogen uptake by physisorption-based materials. *Int. J. Hydrogen Energy* **2006**, *31*, 259–264. [CrossRef]

entropy

MDPI

Article

Alkali Halide Aqueous Solutions Under Pressure: A Non-Equilibrium Molecular Dynamics Investigation of Thermal Transport and Thermodiffusion

Guansen Zhao and Fernando Bresme *

Department of Chemistry, Molecular Sciences Research Hub Imperial College, London W12 0BZ, UK; g.zhao21@imperial.ac.uk
* Correspondence: f.bresme@imperial.ac.uk

Abstract: Thermal gradients induce thermodiffusion in aqueous solutions, a non-equilibrium effect arising from the coupling of thermal and mass fluxes. While thermal transport processes have garnered significant attention under standard conditions, thermal transport at high pressures and temperatures, typical of the Earth's crust, has escaped scrutiny. Non-equilibrium thermodynamics theory and non-equilibrium molecular dynamics simulations provide an excellent means to quantify thermal transport under extreme conditions and establish a connection between the behaviour of the solutions and their microscopic structure. Here, we investigate the thermal conductivity and thermal diffusion of NaCl and LiCl solutions in the GPa pressure regime, targeting temperatures between 300 K and 1000 K at 1 molal concentration. We employ non-equilibrium molecular dynamics simulations along with the Madrid-2019 and TIP4P/2005 force fields. The thermal conductivity of the solutions increases significantly with pressure, and following the behaviour observed at standard pressure, the thermal conductivity is lower than that of pure water. The reduction in thermal conductivity is significant in the GPa pressure regime, $\sim$3% for 1 molal NaCl and LiCl solutions. We demonstrate that under GPa pressure conditions, the solutions feature thermophobic behaviour, with ions migrating towards colder regions. The pronounced impact of pressure is more evident in LiCl solutions, which display a thermophilic to thermophobic "transition" at pressures above 0.25 GPa. We discuss a correlation between the solution's thermophobicity and the disruption of the water hydrogen bond structure at high pressure, where the water structure resembles that observed in simple liquids.

Keywords: thermodoffusion; alkali halide solutions; GPa pressure; thermal conductivity; water

1. Introduction

Electrolyte solutions experience high pressures of approximately 0.1 to 1 GPa in the Earth's mantle [1,2] and in the deep oceanic trenches or extraterrestrial oceans beneath thick ice crusts [3]. Previous studies have explained how compression affects the rates of physicochemical [4] and biological [5] processes. Experimental [6,7] and computational [8,9] investigations have demonstrated that high pressure in the GPa range considerably modifies the hydrogen bonding network of liquid water, resulting in a structure reminiscent of that observed in simple liquids. The alteration of the hydrogen-bond network influences the thermal transport properties of water, affecting the energy transfer efficiency between molecules [10,11].

The Ludwig-Soret effect, discovered in the late 19th century, is a non-equilibrium coupling effect between heat and mass fluxes [12,13]. It has been demonstrated that applying a thermal gradient to an aqueous electrolyte solution induces a concentration gradient. This coupling effect between heat and mass fluxes can be elucidated at the macroscopic scale using linear non-equilibrium thermodynamics theory and the corresponding phenomenological equations, which quantify the extent of mass separation in response to the thermal gradient [14]. The tendency of solutes to migrate towards either hot (thermophilic) or cold (thermophobic) regions in the thermal field is believed to influence, for example, the thermophoretic response of colloidal suspensions [15,16]. Furthermore, the Soret-Ludwig phenomenon provides a theoretical foundation for the design of analytical devices to monitor the thermophoretic motion of biomolecules [17,18].

The Soret coefficient $s_T = D_T/D$ is defined by the ratio of the thermal diffusion coefficient (D_T) and the inter-diffusion coefficient (D), and it is often employed to quantify the magnitude of the thermodiffusion effect. The sign change and inversion point of the Soret coefficient, where $s_T = 0$, have been reported in experiments involving liquid mixtures and electrolyte aqueous solutions. At the inversion point, the Soret coefficient changes from thermophilic ($s_T < 0$) to thermophobic ($s_T > 0$) as the temperature increases [15,19]. Molecular dynamics simulations of alkali halide solutions reproduced the experimental behaviour using various forcefields [10,20–22]. Moreover, minima in the Soret coefficient have been reported experimentally and in simulations for different electrolyte solutions [23–26]. The simulations highlighted a correlation between the thermodiffusion response and the minimum in the Soret coefficient. Furthermore, recent studies of electrolyte solutions at standard pressure conditions demonstrated a correlation between thermophobicity and the disruption of the oxygen-oxygen structural characteristics of water [21]. This is an aspect we aim to address in this work, since the structure of water is significantly modified at high pressure, and we expect, based on previous observations, an enhancement of the thermophobic response. The predominance of thermophobic response is relevant to the behaviour of solutions in, for example, oceans or the Earth's mantle since heat-mass coupling would lead to a preferential accumulation of salt in colder regions.

The impact of pressure on ionic solutions has garnered significant interest, as high pressure and temperature conditions can be found in the Earth's crust. Several studies have focused on the structure of ion solutions, electric conductivity, solution dielectric permittivity, and the properties of pure water at pressures in the GPa range [27–35]. Metamorphic fluids contain water as one major component, and understanding the properties of water, such as mass transport within the Earth, is of particular interest. There is evidence that the properties of these fluids differ considerably from those of fluids under standard conditions [36,37]. The properties of high-pressure fluids are also of interest in other bodies within our solar system, such as Europa, where oceans have been proposed with pressures reaching the 0.1 GPa range [3].

In this work, we investigate the impact of pressure on the thermal transport properties of NaCl and LiCl solutions. NaCl is a significant component of sea salt, and we have chosen a concentration of 1 m close to that found in the ocean. The lithium ion is relevant in energy applications, and it exhibits stronger thermophilicity than other alkali cations, such as Na^+ [20,23,24,38]. Also, alkali chlorides are major components of deep crustal fluids [37]. We report both Soret coefficients and thermal conductivities for these salts, analyze the water structure and hydration shells of ions at high pressure and discuss their impact on thermal transport and thermal diffusion. These thermophysical properties of water and its solutions at high pressure might be of interest to inform thermodynamic models of the upper mantle [36] since water plays a significant role in the heat and mass transport inside the Earth [37].

We employ the TIP4P/2005 water [39] and Madrid-2019 forcefields [40] to simulate water molecules and ions, respectively. TIP4P/2005 is one of the most accurate non-polarizable rigid water models, widely used to study supercooled water and ice polymorphs [41–43]. The Madrid-2019 forcefield is designed to conduct simulations with TIP4P/2005 water, utilising scaled charges to replicate the thermophysical and structural properties of electrolyte solutions [40]. A recent investigation demonstrated that the Madrid-2019 model accurately reproduces the dependence of thermal conductivity and thermal diffusion of NaCl and LiCl under standard pressure conditions [21].

2. Materials and Methods

To compute the thermal conductivity and Soret coefficient of the electrolyte solutions, we employed boundary-driven non-equilibrium molecular dynamics (NEMD) simulations [44]. All NEMD simulations were performed using LAMMPS (vs. 3 Mar 2020) [45] with a 1 fs timestep. This approach involves thermostatting the solutions in specific regions, located at the centre and edges of the simulation box, to predefined hot (T_H) and cold (T_C) temperatures (see Figure 1). The thermostat regions had a thickness of 8 Å , and the temperature of the molecules and ions were reset every timestep using the Langevin thermostat [46,47] with a coupling constant of 500 fs. The total momentum of the simulation box was reset every timestep. The electrostatic interactions and long-range dispersion interactions were calculated using the particle-particle particle-mesh (PPPM) method [48] using a force accuracy of 5×10^{-5} kcal/(mol Å) for the coulombic interactions, and 10^{-4} and 2×10^{-4} kcal/(mol Å) for the real and k-space contributions of the dispersion interactions, respectively. In setting the thermal gradients, we followed the approach discussed in reference [43], where it was shown that thermal gradients similar to those employed here result in a linear response.

We used cuboid simulation boxes with a size ratio $\{L_x : L_y : L_z\} = \{1 : 1 : 3\}$, and $L_x \approx 40$ Å. The heat flux was set up along the z direction. A typical simulation box contained 6663 TIP4P/2005 water molecules [39] and 120 ion pairs corresponding to a 1 mol/kg concentration. We used the Madrid-2019 forcefield [40] to describe the ion-ion and ion-water interactions. The system was pre-equilibrated for 5 ns using the isothermal-isobaric (NPT) ensemble at the target pressure and a temperature equal to the average of T_{hot} and T_{cold}. Then, the average density during the NPT simulation was calculated, and the simulation cell vectors were set to the average values obtained from the NPT simulation. The thermostats were applied, and the NEMD simulations were performed at constant volume conditions. Following equilibration over 10 ns, the system reached the stationary state, and we calculated the local temperature and concentration profiles, averaging over at least 50 ns production run and 10 replicas, providing an accumulated sampling time of 0.5 µs. The thermal conductivity and the Soret coefficient were obtained from the analysis of the temperature and concentration profiles.

The thermal conductivity, $\lambda(z)$, was determined using Fourier's law by calculating the local thermal gradient $\nabla T(z)$, and the local temperature profile in the z direction,

$$\lambda(z) = -\frac{J_q}{\nabla T(z)}, \quad (1)$$

where J_q denotes the heat flux in the direction of the thermal gradient. The heat flux was calculated using the continuity equation:

$$J_q = \frac{\dot{Q}}{2A}. \quad (2)$$

Here, $\dot{Q}$ represents the heat rate; A is the simulation box's cross-sectional area perpendicular to the heat flux direction, and the factor "2" accounts for the two heat fluxes within the simulation cell. The heat rate was computed from the energy exchanged at the hot and cold thermostats, which should be equal and of opposite sign (see Figure 2). Our method provides excellent energy conservation, with the energy exchanged at the cold and hot thermostats being the same within computational accuracy.

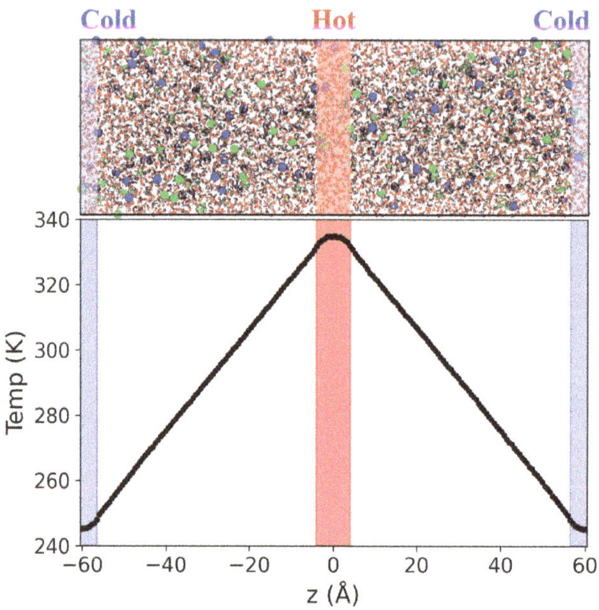

Figure 1. (**Top**) Snapshot of the simulations cell employed in the NEMD simulation of 1 molal LiCl aqueous solution. The red, white, blue and green spheres represent the oxygen, hydrogen atoms, and Li^+ and Cl^- ions, respectively. The thermostatting regions are highlighted in red (hot, center of the box) and blue (cold, box edges). (**Bottom**) Temperature profile obtained with our NEMD simulation method.

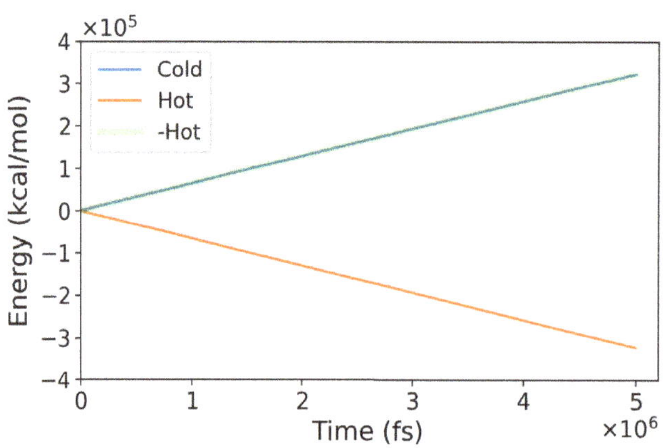

Figure 2. Energy exchanged at the hot and cold thermostats during a representative NEMD run. The blue and orange lines indicate the energy exchange in the cold and hot regions. The negative of the energy exchanged at the hot thermostats has been plotted in green for an easier comparison and test of energy conservation.

For the Soret coefficient investigation, we considered the linear non-equilibrium formalism. The entropy production (σ) of a binary system based on the non-equilibrium thermodynamics [14,49]:

$$\sigma = -\frac{1}{T^2}\mathbf{J}_q \cdot \nabla T - \frac{1}{T}\sum_{i=1}^{2}\mathbf{J}_i \cdot \nabla_T \mu_i, \tag{3}$$

where $\mathbf{J}_q$ and $\mathbf{J}_i$ are the heat and mass fluxes, respectively; $\nabla_T \mu_i$ represents the chemical potential gradient of the ith component at constant temperature. In Equation (3), we have ignored charge fluxes since we did not consider external electrostatic fields in the simulations reported below.

Considering $\sum_i \mathbf{J}_i = 0$, we can write the phenomenological linear equations [14,21]:

$$\mathbf{J}_q = -L_{qq}\frac{\nabla T}{T^2} - \frac{L_{q1}}{T}\nabla_T(\mu_1 - \mu_2), \tag{4}$$

$$\mathbf{J}_1 = -L_{1q}\frac{\nabla T}{T^2} - \frac{L_{11}}{T}\nabla_T(\mu_1 - \mu_2), \tag{5}$$

where $L_{\alpha\beta}$ are the Onsager coupling coefficients. When the system reaches the stationary state $J_1 = 0$, and Equation (5) can be rewritten as,

$$\frac{\nabla_T(\mu_1 - \mu_2)}{\nabla T} = -\frac{1}{T}\frac{L_{1q}}{L_{11}} = s_T\frac{d\mu_1}{d\ln w_1}, \tag{6}$$

where w_1 is the solute weight fraction and s_T is the Soret coefficient. Then the Soret coefficient can be obtained [14],

$$s_T = -\frac{1}{T}\left(\frac{\nabla \ln(w_1/w_2)}{\nabla \ln T}\right)_{J_1=0} = -\frac{1}{x_1 x_2}\left(\frac{\nabla x_1}{\nabla T}\right)_{J_1=0} = -\frac{1}{b}\left(\frac{db}{dT}\right)_{J_1=0}. \tag{7}$$

In the equation above, x_1 and x_2 represent the molar fractions of solute and water, respectively; b is the solute molality. The derivation of the right hand side of Equation (7) can be found in ref. [21].

We also conducted equilibrium NPT simulations with Gromacs (vs. March 2021 [50]) to investigate the structural properties and the speed of sound of the solutions. The equilibrium simulations included 926 water molecules and 17 ion pairs, and the production runs were performed for at least 60 ns with a 1 fs timestep. Using the NPT simulation trajectories, we calculated the radial distribution functions (RDF) and the ion-water coordination numbers. The code TRAVIS [51,52] was employed to compute the spatial 3D density distribution functions (SDF) [53] of hydrogen and oxygen atoms in water. Additionally, we performed the mean square displacement analysis to calculate the diffusion coefficient of ions using the Einstein relation [54].

The fluctuation properties including isobaric thermal expansion coefficient (α) and isothermal compressibility (k_T) were computed employing the fluctuation relations [55] $\langle \delta V \delta H \rangle_{NPT} = k_B T^2 V \alpha$, and $\langle \delta V^2 \rangle_{NPT} = V k_B T k_T$, respectively. In the above equations, V is the volume; H is the enthalpy, k_B is the Boltzmann constant; and $\delta\mathcal{P} = \mathcal{P} - \mathcal{P}_{ens}$ where "ens" represents the ensemble average of property $\mathcal{P}$. The speed of sound was obtained from:

$$c_s = \sqrt{\frac{C_P}{C_V}\frac{1}{\rho k_T}}, \tag{8}$$

where ρ is the mass density; C_P and C_V denote the isobaric and isochoric heat capacities, respectively.

3. Results

To evaluate the performance of the forcefield under high pressure, we computed the equation of state of water for the 290 K isotherm. The agreement between the simulated and experimental results up to 0.8 GPa is excellent (see Figure A1 in the Appendix A). Moreover, the equations of states obtained from NEMD simulations at standard conditions (see Table A2) agree with previous reports for the same force field using small samples and equilibrium conditions (see refs. [41,43]). It is important to note that TIP4P/2005 exhibits a shift in temperature with respect to the experimental coexistence diagram, and the liquid phase remains stable at pressures exceeding 0.8 GPa.

Figure 3a shows the thermal conductivities of 1 m NaCl and LiCl solutions as a function of pressure for the 290 K isotherm. Our previous simulations demonstrated that the thermal conductivity converges within 10–30 ns at temperatures above 220 K for TIP4P/2005 water at various isobars, ranging from 1 to 1200 bar [43]. We followed this approach to obtain running averages during the production simulation phase.

The thermal conductivity of LiCl and NaCl solutions increases with pressure. We do not find a significant difference in the thermal conductivity of these solutions. This result agrees with previous studies of these solutions at near-standard conditions [21]. In that study, it was shown that the thermal conductivity of alkali halide solutions features a stronger dependence on the anion mass and a weaker dependence on the cation type. The thermal conductivity of the electrolyte solutions is lower than that of pure water (see Figure 3). We observe a 1–3% reduction in conductivity for 1 m NaCl and LiCl solutions, and the reduction increases with rising pressure. The reduction in the thermal conductivity in solutions relative to pure water is consistent with experimental and computer simulation studies at standard conditions [21,56]. Previous investigations [21] have demonstrated the mechanism behind the decrease in thermal conductivity upon adding salts: a decrease in the thermal conductivity correlates with an increase in the molar volume of the solution. However, the speed of sound varies depending on the type of salt. The reduction in thermal conductivity is also correlated with a disruption of the hydrogen bond structure of water, which is stronger at higher salt concentrations.

Advancing the discussion below, we note that water's characteristic tetrahedral structure is considerably disrupted at high pressures, >0.1 GPa. Therefore, the high thermal conductivity observed in the high-pressure range is unrelated to water's hydrogen bonding network.

Our Soret coefficients are shown in Figure 3b. The Soret coefficients of both chloride solutions increase with pressure up to 0.5 GPa, reaching a plateau at higher pressures, where the Soret coefficient becomes nearly independent of pressure. At these elevated pressures, water loses the typical tetrahedral order (see Figure 4, and the radial distributions resemble those observed in simple liquids. The analysis of the spatial distribution function of water (see Figure 5) also supports this notion.

We find that the solutions become more thermophobic under pressure. This effect is particularly evident in the Li^+ salt, whose Soret coefficient exhibits a sign change at $P{\sim}0.186$ GPa. At this pressure, the solution transitions from thermophilic to thermophobic. This transition correlates with the systematic loss of tetrahedral order, as indicated in our radial distribution functions (see Figure 4). The impact of the pressure is also reflected in the self-diffusion coefficient, which we calculated from equilibrium simulations (see Table A3) and the results obtained at standard conditions are consistent with the experimental data [57]. The self-diffusion coefficients of both cations and anions remain nearly constant at pressures below 0.5 Gpa, but exhibit a sharp decreasing trend upon compression, decreasing to approximately half of their initial value from 0.5 to 2 GPa.

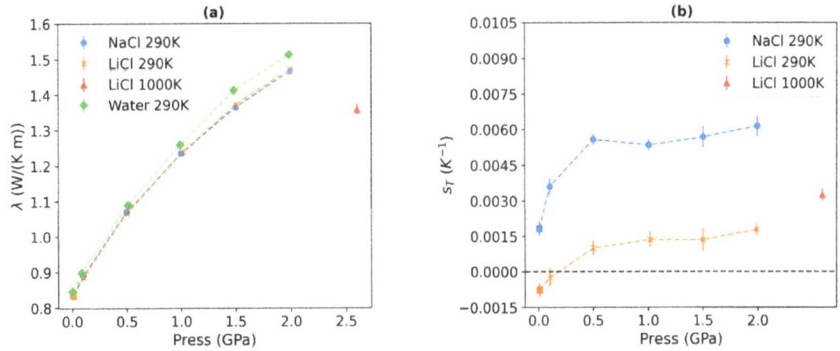

Figure 3. (**a**) Thermal conductivities and (**b**) Soret coefficients at 290 K temperature, 10^{-4}, 0.1, 0.5, 1, 1.5, and 2 GPa pressures for 1 m NaCl and LiCl solutions. The square represents the thermal conductivity at 298 K and 31 bar reported in Ref. [21]. The green points in (**a**) represent our thermal conductivity data for pure water at 290 K temperature. The red triangles represent the thermal conductivity and Soret coefficient of 1 m LiCl solutions at 1000 K and 2.6 GPa pressure. The numerical data are compiled in Table A2.

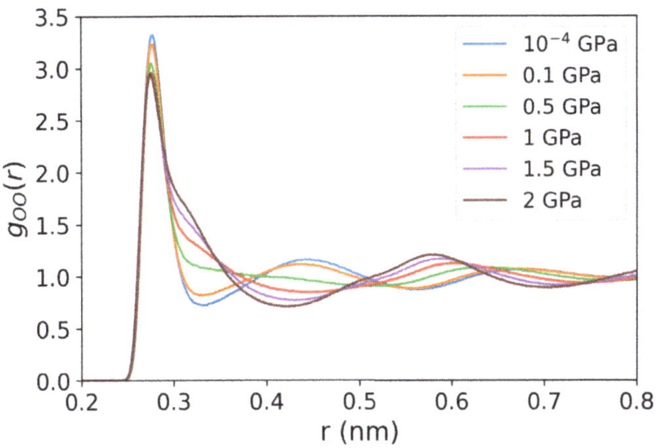

Figure 4. Oxygen-oxygen radial distribution functions of pure water at 290 K and various pressures (10^{-4}, 0.1, 0.5, 1, 1.5, and 2 GPa).

Recently, Zhang et al. investigated pure water under extreme conditions using neural network (NN) potentials, targeting pressures up to 22 GPa and temperatures in the 1000–2000 K range [11]. These conditions are found in the Earth's mantle. The authors concluded that the thermal conductivity under these extreme conditions exhibits slight dependence on temperature while increasing with pressure [11]. Understanding the properties of ionic solutions at high pressure (approximately 1 GPa) and temperature (approximately 1000 K) is crucial for modelling deep fluids [37]. Therefore, we conducted additional NEMD simulations of 1 m LiCl solutions at high temperatures (1000 K) and extreme pressures (2.6 GPa) (see Figure 3). The density predicted with the TIP4P/2005 model for the 1 m LiCl solution (1.17 ± 0.003 g/cm^3) is comparable to the NN potential predictions for pure water (1.20 g/cm^3) [11]. Our thermal conductivity for the LiCl solution is 1.36 ± 0.02 W/(K m), a value similar to that reported under the same thermodynamic conditions for pure water modelled with the NN potential at 1.14 ± 0.22. The Soret coefficient under extreme conditions is more thermophobic than that of the LiCl solutions at 290 K in the 0.1–2 GPa range. This observation supports our perspective that disrupting the characteristic hydrogen bond structure of water (through increased temperature and pressure) enhances the thermophobicity of the solutions. Our results demonstrate that thermodiffusion is a significant coupling process under conditions encountered in the Earth's mantle.

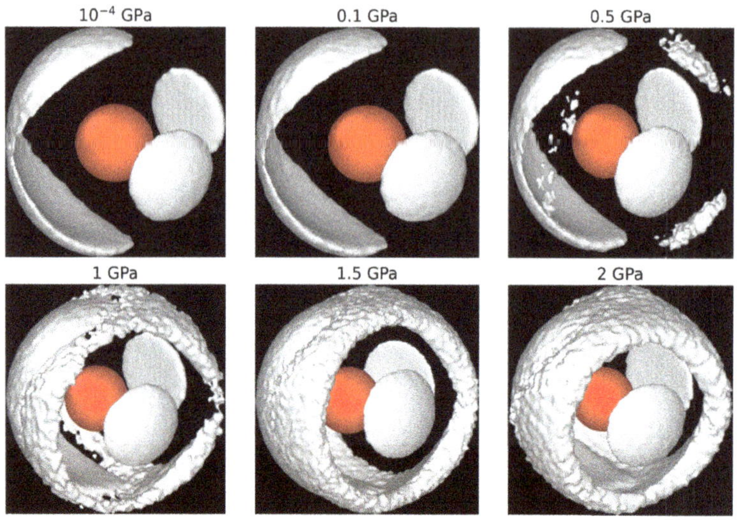

(**a**) NaCl

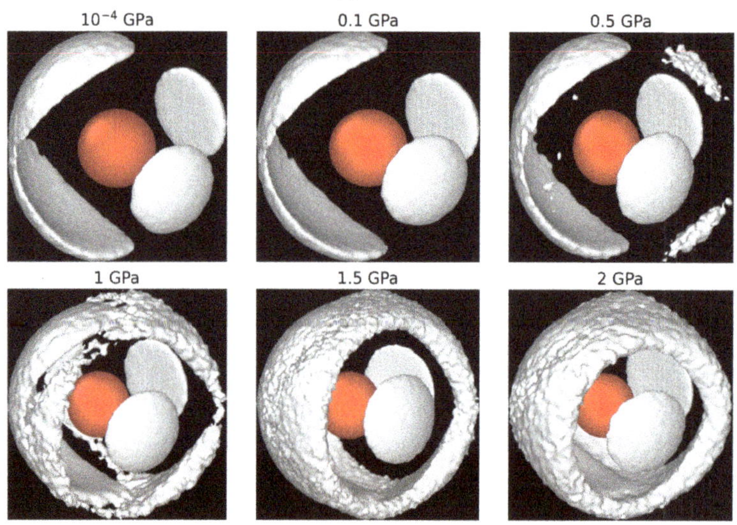

(**b**) LiCl

Figure 5. Spatial distribution functions $g_{OO}(r, \Omega)$ for the 1m NaCl (**a**) and LiCl (**b**) solutions at 290 K and 10^{-4}, 0.1, 0.5, 1, 1.5, and 2 GPa pressures. The central oxygen, colored in red, is shown for reference. Image generated with OVITO [58].

We now discuss the structural changes the solutions undergo upon increasing the pressure. We show in Figure 4 the oxygen-oxygen RDFs for pure water at 290 K temperature and pressures from 10^{-4} to 2 GPa. A considerable disruption of the water structure is observed at pressures higher than 0.5 GPa, evinced by the shift of the second intermolecular peak around $r = 0.45$ nm to longer inter-molecular distances at 0.5 GPa pressure and above, the second-highest peak is located at $\sim 2\, r_{max}$, where r_{max} denotes the location of the main maximum in the radial distribution function, consistent with the structure for a simple liquid. The O-O RDFs of NaCl and LiCl solutions show similar behavior, including the peak positions and values in response to rising pressures (see Figure A2 in the Appendix A).

The SDFs of the aqueous solutions in Figure 5 further support the observations reported above. A significant loss of local tetrahedral order in water is observed at 0.5 GPa, and the tetrahedral order completely disappears at higher pressures. This observation applies to both chloride solutions, LiCl and NaCl. The liquid structures at 1.5 and 2 GPa pressures closely resemble a simple liquid, agreeing with the findings of pure water at extreme conditions in Ref. [9], and the SDFs reported in analyses of X-ray scattering of aqueous solutions at high pressure [33].

We computed the difference between the O-O RDFs in the electrolyte solutions and pure water at standard pressure to investigate how ion-water interactions vary at elevated pressures (see Figure 6). The notable change at around the second peak (0.4–0.5 nm) as pressure goes up is associated with the loss of tetrahedral structure, and the disruption increases with rising pressure. At low pressures (10^{-4} and 0.1 GPa), LiCl has a small impact on the O-O RDF, mainly around the region near the first coordination shell, whereas NaCl induces more significant changes at distances 0.35–0.5 nm. As pressure increases, the disruption of the O-O structure in the 0.35–0.5 nm range rises significantly, At 0.5 GPa and above, the water hydrogen bond orientational structure disappears (see Figure 6), and the observed changes are independent of the type of cation, Li^+ or Na^+. Overall, we confirm a correlation between thermophobicity and the loss of tetrahedral structure, particularly in the 0.35–0.5 nm range. This observation is consistent with the conclusions in reference [21] for chloride solutions at standard pressure.

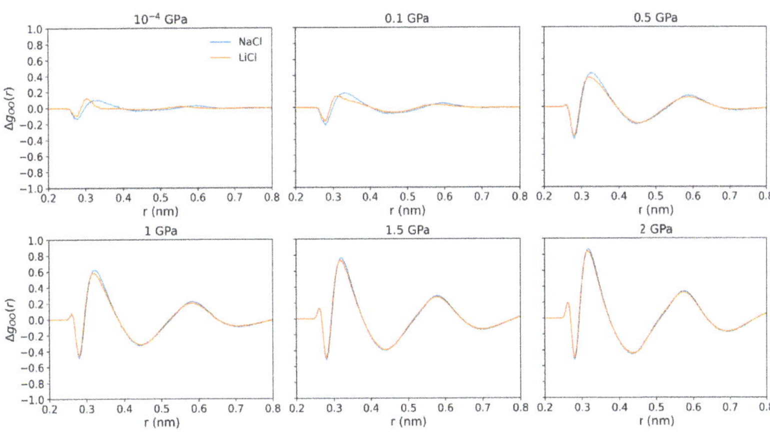

Figure 6. Perturbation of the oxygen-oxygen radial distribution function in 1 m NaCl and LiCl solutions at 290 K and pressures of 10^{-4}, 0.1, 0.5, 1, 1.5, and 2 GPa. The different RDFs are presented using the O-O RDFs of water at standard pressure, i.e., $\Delta g_{OO}(r) = g_{OO,solution}(r) - g_{OO,purewater\ (10^{-4}\ GPa)}(r)$.

We show the ion-oxygen RDFs in Figure 7. The coordination numbers obtained by integrating the radial distribution function up to its first minimum are reported in Table A1. The pressure has a minor impact on the cation-water structure, especially for Li^+, where the Li-O coordination number equals 4 for all the isobars we studied. In contrast, the Cl-O RDFs are significantly influenced by compression. The second Cl-O peak decays from 0.1 GPa onwards and disappears above 1.5 GPa. These changes in the RDF indicate a systematic loss of a well-defined solvation structure, and the Cl-O coordination number increases significantly up to ~20 at ~2 GPa. This large increase in chloride coordination number (and shift of the main Cl-O peak to longer distances) agrees with the data reported in previous X-ray diffraction measurements, where an increase from 6 to 16 was reported at 1.7 GPa pressure and 300 K [33]. Similarly, the small impact of pressure on the cation-coordination number observed here is consistent with those X-ray diffraction measurements. The disruption of the hydration shell around the anions at these high pressures is linked to the drastic alteration in the hydrogen bonding network. Significant effects in the anion hydration were also reported in x-ray absorption studies of RbBr at 2.8 GPa pressure, along with a concomitant loss of orientation order in the hydration shells [59].

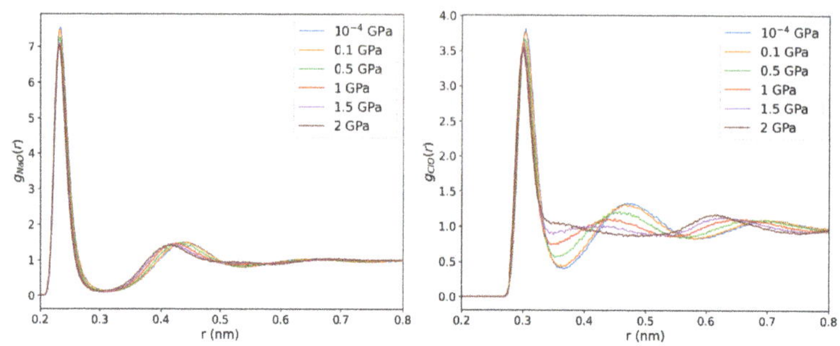

(**a**) NaCl

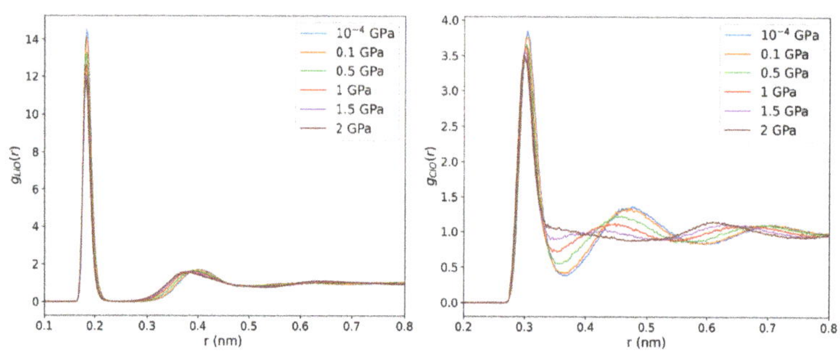

(**b**) LiCl

Figure 7. Oxygen-ion radial distribution functions for 1m NaCl (**a**) and LiCl (**b**) solutions at 290 K and 10^{-4}, 0.1, 0.5, 1, 1.5, and 2 GPa pressures. The left and right panels show the oxygen-cation and oxygen-anion RDFs, respectively.

We performed additional analyses of the ion-ion correlations to quantify the ion association. Figure A3 in the Appendix A shows the cation-anion RDFs and corresponding running coordination numbers at 2 GPa for the NaCl and LiCl solutions. The radial distances for the first minimum in Na-Cl and Li-Cl RDFs are ∼0.28 and 0.33 nm, respectively, consistent with the positions found in previous simulation studies [20,27,28]. Sakuma and Ichiki simulated NaCl solutions in the 670–2000 K temperature range and up to 2 GPa pressure, and they found that the cation-anion coordination numbers tend to decrease with increasing pressure and decreasing temperature [27,28]. At 290 K and 2 GPa, we do not find evidence of significant ion pair formation, particularly for the LiCl solutions (see Figure A3 in the Appendix A).

Finally, we explore the dependence of the response functions—isobaric thermal expansion and isothermal compressibility on pressure. Figure 8 displays these properties for pure water and the aqueous solutions. Moreover, we have included experimental data for the isobaric thermal expansion and isothermal compressibility of water. These properties were extracted from the equation of state reported in Ref. [60]. Figure 8 shows small differences between the data for pure water and the aqueous solutions. For both the simulated and experimental isobaric thermal expansion coefficient (Figure 8a), all the liquids exhibit a local maximum at around 0.5 GPa. This pressure corresponds to the same range where we observe a significant disruption of the tetrahedral orientational order (see Figure 5). As the pressure increases, the thermal expansion coefficient difference between water and ionic solutions diminishes. The simulated thermal expansion coefficients are 15% higher than the experimental data on average. This deviation is similar to the ~9% overestimation observed at 298 K and 1 bar with TIP4P/2005 [39].

Our simulation results align closely with the experimental isothermal compressibility and speed of sound data, confirming the accuracy of the TIP4P/2005 model. The isothermal compressibility (see Figure 8b) decreases in the whole pressure range, and the speed of sound (see Figure 8c) increases with rising pressures. The thermal conductivity is correlated with the speed of sound via the Bridgmann equation [61]. While this equation is not always accurate, it contains the key ingredients to understand qualitatively the dependence of thermal conductivity on the speed of sound. The Bridgman equation has inspired a recent formulation where the speed of sound appears explicitly as a key variable [11]. Moreover, the lack of dependence of the speed of sound on the cation type is consistent with the similar thermal conductivities of NaCl and LiCl (see Figure 3a).

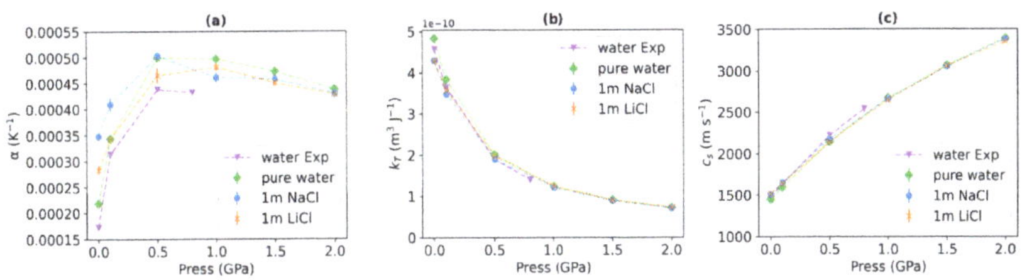

Figure 8. (**a**) Isobaric thermal expansion coefficient, (**b**) isothermal compressibility, and (**c**) speed of sound of pure water and 1m NaCl, and 1m LiCl solutions at 290 K. The data correspond to the following pressures: 10^{-4}, 0.1, 0.5, 1, 1.5, and 2 GPa. The purple triangles represent the experimental data from Ref. [60] at 10^{-4}, 0.1, 0.5, and 0.8 GPa pressures and 290 K.

4. Conclusions

We have examined, using non-equilibrium molecular dynamics and state-of-the-art rigid empirical force fields, the thermal transport properties and thermodiffusion of two alkali halide solutions, 1 m NaCl and 1 m LiCl, under GPa pressures. These pressure conditions are significant for crustal fluids.

The thermal conductivity of the solutions does not change substantially with the type of cation. In line with experiments and previous simulations, adding LiCl and NaCl diminishes the thermal conductivity of water, with the reduction being more pronounced (approximately 3%) at GPa pressures. The thermal conductivity varies considerably with pressure, increasing nearly twofold compared to pure water at standard conditions. This increase correlates with the substantial reduction of the solution compressibility as pressure rises, along with the accompanying increase in the speed of sound. This correlation is clearly observed when comparing the data in Figures 3a and 8. Therefore, the solutions and pure water at GPa pressure are efficient heat conductors.

The Soret coefficients of LiCl and NaCl increase with pressure, reaching a plateau above 0.5 GPa. The thermophobicity of the solutions is closely linked to the loss of tetrahedral order and hydrogen bonding in water. At high pressure (approximately 0.5 GPa), the water structure resembles a simple liquid. The hydration structure of Li^+ does not significantly change with pressure, and the tetrahedral water coordination of Li^+ is maintained up to 20 GPa. This result indicates that the thermophobicity of the alkali halide solutions observed at GPa pressures is associated with the loss of the hydrogen bond structure in water, highlighting the role of hydrogen bonds in determining the transition from thermophilic to thermophobic states. This correlation aligns with previous conclusions regarding thermodiffusion processes under standard conditions [16]. Besides, This type of behavior has also been observed in concentrated aqueous solutions [21], and it is noted here when we apply high pressures, where water loses its orientational order (see Figure 8). This led us to conclude that thermophilicity is favored in situations with weak hydrogen bonding, while strong hydrogen bonding leads to thermophilicity, and ions are excluded from the colder regions in that case. Based on this, we expect that thermal gradients will lead to salt migration towards cooler regions in crustal fluids.

The simulations reported in this work were performed with rigid water models. Bond and angular degrees of freedom might be relevant in extreme conditions occurring in the earth's mantle, and these effects might be included. Previous simulations of the TIP4P/2005 flexible model reported slightly higher thermal conductivities than the rigid model at 300 K and standard pressure [62]. Machine learning potentials and the introduction of quantum corrections [63] might also provide a focus for future investigations.

Author Contributions: Conceptualization, F.B.; Methodology, G.Z. and F.B.; Validation, G.Z. and F.B.; Formal analysis, G.Z. and F.B.; Investigation, G.Z. and F.B.; Resources, F.B.; Writing—original draft, G.Z. and F.B.; Writing—review & editing, G.Z. and F.B.; Visualization, G.Z.; Supervision, F.B. All authors have read and agreed to the published version of the manuscript.

Funding: We acknowledge the ICL RCS High-Performance Computing facility and the UK Materials and Molecular Modelling Hub for computational resources, partially funded by the EPSRC (Grant Nos. EP/P020194/1 and EP/T022213/1.

Data Availability Statement: The original contributions presented in this study are included in the article/Appendix A. Further inquiries can be directed to the corresponding author.

Acknowledgments: We are pleased to dedicate this article to Prof. Signe Kjelstrup, who has made major and pioneering contributions to the field of non-equilibrium thermodynamics.

Conflicts of Interest: The authors declare no conflicts of interest.

Abbreviations

The following abbreviations are used in this manuscript:

NEMD	Non-equilibrium Molecular Dynamics
PPPM	Particle-Particle Particle-Mesh
NPT	Isothermal-isobaric
RDF	Radial Distribution Function
SDF	Spatial Distribution Function
Symbols	
A	Cross-sectional area of simulation box
b	Solute molality
c_s	Speed of sound
C_P	Isobaric heat capacity
C_V	Isochoric heat capacity
D	Inter-diffusion coefficient
D_T	Thermal diffusion coefficient
g	Ratio of the local particle density in the radial distribution function
H	Enthalpy
J_i	Mass flux
J_q	Heat flux

k_B	Boltzmann constant
k_T	Isothermal compressibility
L_x, L_y, L_z	Simulation box length in x, y, z dimensions
r_{max}	Location of the main maximum in the radial distribution function
s_T	Soret coefficient
T_H, T_C	Temperature of the hot and cold thermostatting regions
V	Volume
w	Weight fraction
x	Molar fraction
α	Isobaric thermal expansion coefficient
λ	Thermal conductivity
$\nabla\mu$	Chemical potential gradient
∇T	Thermal gradient
ρ	Density
σ	Entropy production
$\dot{Q}$	Heat rate

Appendix A

Table A1. Coordination numbers between the ions and oxygen in water, and the radial distance for the calculation of the coordination numbers: 1 m NaCl (**Top**) and LiCl (**Bottom**) solutions at 290 K and 10^{-4}, 0.1, 0.5, 1, 1.5, and 2 GPa pressures.

NaCl Press (GPa)	n_{Na-OW}	$r_{Na-OW}(nm)$	n_{Cl-OW}	$r_{Cl-OW}(nm)$
10^{-4}	5.55	0.32	5.94	0.37
0.1	5.64	0.32	5.98	0.36
0.5	5.84	0.31	6.05	0.35
1	5.95	0.31	6.20	0.35
1.5	6.03	0.31	23.03	0.52
2	6.06	0.30	19.87	0.49
LiCl Press (GPa)	n_{Li-OW}	$r_{Li-OW}(nm)$	n_{Cl-OW}	$r_{Cl-OW}(nm)$
10^{-4}	4.00	0.27	5.81	0.36
0.1	4.00	0.26	5.99	0.37
0.5	4.00	0.25	6.15	0.36
1	4.00	0.26	6.55	0.35
1.5	4.00	0.24	22.66	0.51
2	4.00	0.24	18.38	0.47

Table A2. Results of the NEMD simulations. The pressures are in GPa, temperature in K, density (ρ) in g/cm^3, thermal conductivity (λ) in $W/(K\,m)$ and Soret coefficient (s_T) in the unit of K^{-1}. $\delta\lambda$ and δs_T are the standard deviations calculated from the replicas.

1 M NaCl						
Press	Temp	ρ	λ	$\delta\lambda$	s_T	δs_T
0.10	290.1	1.08	0.889	0.001	3.60×10^{-4}	3.52×10^{-4}
0.50	290.2	1.19	1.069	0.002	5.57×10^{-3}	2.20×10^{-4}
1.00	290.1	1.28	1.235	0.005	5.35×10^{-3}	2.44×10^{-4}
1.50	290.2	1.34	1.363	0.004	5.67×10^{-3}	4.60×10^{-4}
1.99	290.4	1.39	1.466	0.003	6.13×10^{-3}	4.32×10^{-4}
1 M LiCl						
Press	Temp	ρ	λ	$\delta\lambda$	s_T	δs_T
0.10	290.2	1.06	0.884	0.002	-2.55×10^{-4}	3.80×10^{-4}
0.50	290.1	1.17	1.066	0.002	9.82×10^{-4}	3.10×10^{-4}

Table A2. *Cont.*

1.01	290.1	1.26	1.239	0.005	1.35×10^{-3}	3.54×10^{-4}
1.50	290.0	1.33	1.371	0.006	1.35×10^{-3}	4.96×10^{-4}
1.99	290.2	1.37	1.472	0.003	1.78×10^{-3}	2.55×10^{-4}
2.60	999.9	1.17	1.357	0.002	3.24×10^{-3}	2.60×10^{-4}
			Pure water			
Press	Temp	ρ	λ	$\delta\lambda$		
1.70×10^{-3}	289.9	1.00	0.845	0.002		
0.87	289.6	1.04	0.897	0.002		
0.51	289.9	1.16	1.089	0.002		
0.99	290.0	1.25	1.258	0.006		
1.48	289.2	1.31	1.412	0.001		
1.98	289.9	1.36	1.512	0.003		

Table A3. Self diffusion coefficients of ions ($Na^+(D_{Na})$, $Li^+(D_{Li})$, and $Cl^-(D_{Cl})$) and corresponding standard errors in the unit of 10^{-10} m^2/s: 1 molal NaCl (**Top**) and LiCl (**Bottom**) solutions at 290 K and pressures of 10^{-4}, 0.1, 0.5, 1, 1.5, and 2 GPa.

NaCl Press (GPa)	D_{Na}	δD_{Na}	D_{Cl}	δD_{Cl}
10^{-4}	5.56	0.26	9.64	0.10
0.1	5.29	0.46	9.63	0.16
0.5	5.77	0.09	9.10	0.09
1	5.65	0.41	8.51	0.12
1.5	3.41	0.43	5.70	0.28
2	2.82	0.39	5.04	0.38
LiCl Press (GPa)	D_{Li}	δD_{Li}	D_{Cl}	δD_{Cl}
10^{-4}	5.86	0.25	11.93	0.49
0.1	5.84	0.47	10.31	0.27
0.5	5.75	0.17	10.37	0.69
1	4.09	0.19	7.08	0.14
1.5	3.39	0.55	6.10	0.57
2	1.86	0.51	5.19	0.28

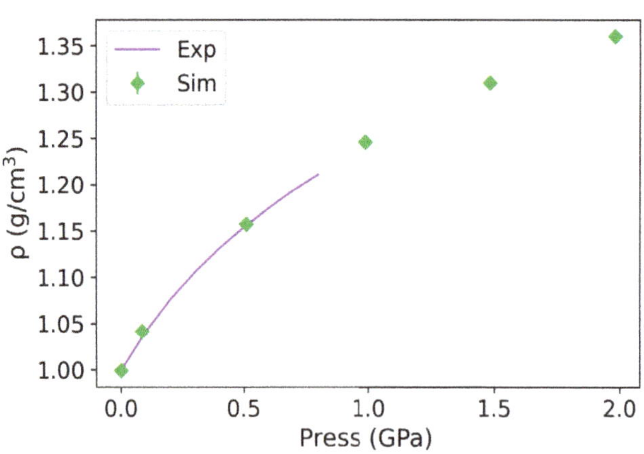

Figure A1. Equation of state from equilibrium NPT simulations (green points) at 10^{-4}, 0.1, 0.5, 1, 1.5, and 2 GPa pressures and 290 K temperature. The simulation results are compared with experimental data (purple curve) from 10^{-4} to 0.8 GPa [60].

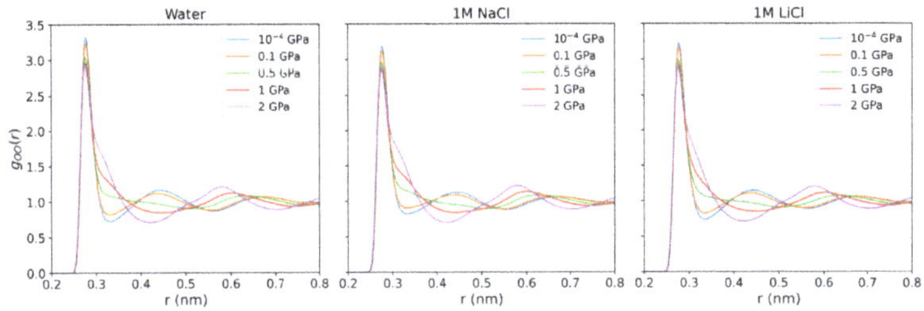

Figure A2. Oxygen-oxygen radial distribution functions for pure water, 1 m NaCl, and LiCl solutions at 290 K and various pressures (10^{-4}, 0.1, 0.5, 1, and 2 GPa).

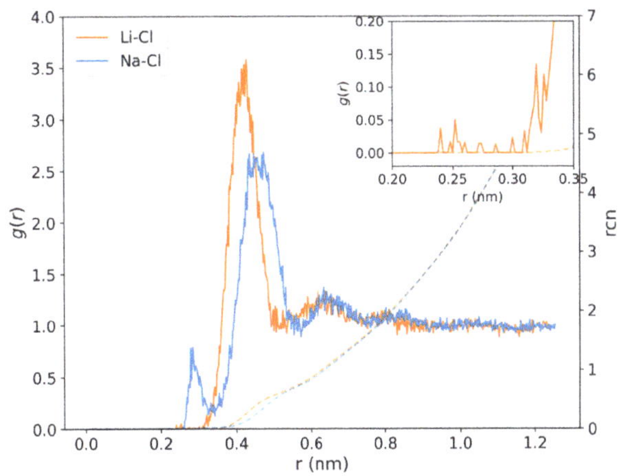

Figure A3. Sodium-chloride (blue curve) and lithium-chloride (orange curve) radial distribution functions and corresponding running coordination numbers (dashed line) for NaCl and LiCl solutions at 290 K temperature and 2 GPa pressure. The inset represents the zoomed-in region for the radial distance between 0.2 and 0.35 nm, providing a closer view of the first minimum in the radial distribution functions for Li-Cl.

References

1. Meade, C.; Jeanloz, R. Deep-Focus Earthquakes and Recycling of Water into the Earth's Mantle. *Science* **1991**, *252*, 68–72. [CrossRef]
2. Bina, C.R.; Navrotsky, A. Possible presence of high-pressure ice in cold subducting slabs. *Nature* **2000**, *408*, 844–847. [CrossRef] [PubMed]
3. Kargel, J.S.; Kaye, J.Z.; Head, J.W., III; Marion, G.M.; Sassen, R.; Crowley, J.K.; Ballesteros, O.P.; Grant, S.A.; Hogenboom, D.L. Europa's Crust and Ocean: Origin, Composition, and the Prospects for Life. *Icarus* **2000**, *148*, 226–265. [CrossRef]
4. Byrne, R.H.; Laurie, S.H. Influence of pressure on chemical equilibria in aqueous systems—With particular reference to seawater (Technical Report). *Pure Appl. Chem.* **1999**, *71*, 871–890. [CrossRef]
5. Barciszewski, J.; Jurczak, J.; Porowski, S.; Specht, T.; Erdmann, V.A. The role of water structure in conformational changes of nucleic acids in ambient and high-pressure conditions. *Eur. J. Biochem.* **1999**, *260*, 293–307. [CrossRef] [PubMed]
6. Katayama, Y.; Hattori, T.; Saitoh, H.; Ikeda, T.; Aoki, K.; Fukui, H.; Funakoshi, K. Structure of liquid water under high pressure up to 17 GPa. *Phys. Rev. B Condens. Matter Mater. Phys.* **2010**, *81*, 014109. [CrossRef]
7. Strässle, T.; Saitta, A.M.; Le Godec, Y.; Hamel, G.; Klotz, S.; Loveday, J.S.; Nelmes, R.J. Structure of Dense Liquid Water by Neutron Scattering to 6.5 GPa and 670 K. *Phys. Rev. Lett.* **2006**, *96*, 067801. [CrossRef] [PubMed]
8. Starr, F.W.; Sciortino, F.; Stanley, H.E. Dynamics of simulated water under pressure. *Phys. Rev. E Stat. Phys. Plasmas Fluids Relat. Interdiscip. Top.* **1999**, *60*, 6757. [CrossRef]
9. Bresme, F.; Römer, F. Heat transport in liquid water at extreme pressures: A non equilibrium molecular dynamics study. *J. Mol. Liq.* **2013**, *185*, 1–7. [CrossRef]

10. Römer, F.; Wang, Z.; Wiegand, S.; Bresme, F. Alkali halide solutions under thermal gradients: Soret coefficients and heat transfer mechanisms. *J. Phys. Chem. B* **2013**, *117*, 8209–8222. [CrossRef] [PubMed]
11. Zhang, C.; Puligheddu, M.; Zhang, L.; Car, R.; Galli, G. Thermal Conductivity of Water at Extreme Conditions. *J. Phys. Chem. B* **2023**, *127*, 7011–7017. [CrossRef]
12. Ludwig, C. Diffusion zwischen ungleich erwärmten Orten gleich zusammengesetzter Lösung. *Sitzungsber. Bayer. Akad. Wiss. Math. Naturwiss. Kl.* **1856**, *20*, 539.
13. Soret, C. Sur l' état d' équilibre que prend au point de vue de sa concentration une dissolution saline primitivement homogéne dont deux parties sont portèes á des tempèratures diffèrentes. *Arch. Sci. Phys. Nat.* **1879**, *2*, 48–61.
14. de Groot, S.; Mazur, P. *Non-Equilibrium Thermodynamics*; Dover Publications: New York, NY, USA, 1984.
15. Wiegand, S. Thermal diffusion in liquid mixtures and polymer solutions. *J. Condens. Matter Phys.* **2004**, *16*, R357. [CrossRef]
16. Niether, D.; Wiegand, S. Thermophoresis of biological and biocompatible compounds in aqueous solution. *J. Condens. Matter Phys.* **2019**, *31*, 503003. [CrossRef]
17. Wienken, C.J.; Baaske, P.; Rothbauer, U.; Braun, D.; Duhr, S. Protein-binding assays in biological liquids using microscale thermophoresis. *Nat. Commun.* **2010**, *1*, 100. [CrossRef] [PubMed]
18. Mast, C.B.; Braun, D. Thermal trap for DNA replication. *Phys. Rev. Lett.* **2010**, *104*, 188102. [CrossRef] [PubMed]
19. Wittko, G.; Köhler, W. On the temperature dependence of thermal diffusion of liquid mixtures. *EPL* **2007**, *78*, 46007. [CrossRef]
20. Lecce, S.D.; Albrecht, T.; Bresme, F. The role of ion-water interactions in determining the Soret coefficient of LiCl aqueous solutions. *Phys. Chem. Chem. Phys.* **2017**, *19*, 9575–9583. [CrossRef]
21. Bresme, F.; Vasey, E. Thermal transport of alkali halide aqueous solutions: A non-equilibrium molecular dynamics investigation. *Mol. Phys.* **2024**, *122*, e2388302. [CrossRef]
22. Hutchinson, A.J.; Torres, J.F.; Corry, B. Modeling thermodiffusion in aqueous sodium chloride solutions—Which water model is best? *J. Chem. Phys.* **2022**, *156*, 164503. [CrossRef] [PubMed]
23. Colombani, J.; Bert, J.; Dupuy-Philon, J. Thermal diffusion in (LiCl, RH_2O). *J. Chem. Phys.* **1999**, *110*, 8622–8627. [CrossRef]
24. Lecce, S.D.; Albrecht, T.; Bresme, F. A computational approach to calculate the heat of transport of aqueous solutions. *Sci. Rep.* **2017**, *7*, srep44833. [CrossRef] [PubMed]
25. Lecce, S.D.; Bresme, F. Soret coefficients and thermal conductivities of alkali halide aqueous solutions via non-equilibrium molecular dynamics simulations. *Mol. Simul.* **2019**, *45*, 351–357. [CrossRef]
26. Mohanakumar, S.; Kriegs, H.; Briels, W.J.; Wiegand, S. Overlapping hydration shells in salt solutions causing non-monotonic Soret coefficients with varying concentration. *Phys. Chem. Chem. Phys.* **2022**, *24*, 27380–27387. [CrossRef]
27. Sakuma, H.; Ichiki, M. Density and isothermal compressibility of supercritical H2O-NaCl fluid: Molecular dynamics study from 673 to 2000 K, 0.2 to 2 GPa, and 0 to 22 wt % NaCl concentrations. *Geofluids* **2016**, *16*, 89–102. [CrossRef]
28. Sakuma, H.; Ichiki, M. Electrical conductivity of NaCl-H_2O fluid in the crust. *J. Geophys. Res. Solid Earth* **2016**, *121*, 577–594. [CrossRef]
29. Rozsa, V.; Galli, G. Solvation of simple ions in water at extreme conditions. *J. Chem. Phys.* **2021**, *154*, 144501. [CrossRef]
30. Pan, D.; Spanu, L.; Harrison, B.; Sverjensky, D.A.; Galli, G. Dielectric properties of water under extreme conditions and transport of carbonates in the deep Earth. *Proc. Natl. Acad. Sci. USA* **2013**, *110*, 6646–6650. [CrossRef] [PubMed]
31. Sherman, D.M.; Collings, M.D. Ion association in concentrated NaCl brines from ambient to supercritical conditions: Results from classical molecular dynamics simulations. *Geochem. Trans.* **2002**, *3*, 102–107. [CrossRef] [PubMed]
32. Brodholt, J.P. Molecular dynamics simulations of aqueous NaCl solutions at high pressures and temperatures. *Chem. Geol.* **1998**, *151*, 11–19. [CrossRef]
33. Yamaguchi, T.; Fukuyama, N.; Yoshida, K.; Katayama, Y. Ion Solvation and Water Structure in an Aqueous Sodium Chloride Solution in the Gigapascal Pressure Range. *J. Phys. Chem. Lett.* **2021**, *12*, 250–256. [CrossRef] [PubMed]
34. Bove, L.E.; Pietrucci, F.; Saitta, A.M.; Klotz, S.; Teixeira, J. On the link between polyamorphism and liquid-liquid transition: The case of salty water. *J. Chem. Phys.* **2019**, *151*, 044503. [CrossRef] [PubMed]
35. Mishima, O. Melting of the Precipitated Ice IV in LiCl Aqueous Solution and Polyamorphism of Water. *J. Phys. Chem. B* **2011**, *115*, 14064–14067. [CrossRef]
36. Galvez, M.E.; Manning, C.E.; Connolly, J.A.; Rumble, D. The solubility of rocks in metamorphic fluids: A model for rock-dominated conditions to upper mantle pressure and temperature. *Earth Planet. Sci. Lett.* **2015**, *430*, 486–498. [CrossRef]
37. Manning, C.E. Fluids of the Lower Crust: Deep Is Different. *Annu. Rev. Earth Planet. Sci.* **2018**, *46*, 67–97. [CrossRef]
38. Lee, N.; Mohanakumar, S.; Briels, W.J.; Wiegand, S. Non-monotonic Soret coefficients of aqueous LiCl solutions with varying concentrations. *Phys. Chem. Chem. Phys.* **2024**, *26*, 7830–7836. [CrossRef] [PubMed]
39. Abascal, J.L.F.; Vega, C. A general purpose model for the condensed phases of water: TIP4P/2005. *J. Chem. Phys.* **2005**, *123*, 234505. [CrossRef] [PubMed]
40. Zeron, I.M.; Abascal, J.L.F.; Vega, C. A Force Field of Li^+, Na^+, K^+, Mg^{2+}, Ca^{2+}, Cl^-, and $SO_4{}^{2-}$ in Aqueous Solution Based on the TIP4P/2005 Water Model and Scaled Charges for the Ions. *J. Chem. Phys.* **2019**, *151*, 134504. [CrossRef]

41. Vega, C.; Abascal, J.L.F. Simulating water with rigid non-polarizable models: A general perspective. *Phys. Chem. Chem. Phys.* **2011**, *13*, 19663–19688. [CrossRef] [PubMed]
42. Vega, C.; Abascal, J.L.; Conde, M.; Aragones, J. What ice can teach us about water interactions: A critical comparison of the performance of different water models. *Faraday Discuss.* **2009**, *141*, 251–276. [CrossRef] [PubMed]
43. Zhao, G.; Bresme, F. Thermal Transport and Thermal Polarization of Water in the Supercooled Regime. *J. Phys. Chem. Lett.* **2024**, *15*, 9774–9779. [CrossRef] [PubMed]
44. Bresme, F.; Lervik, A.; Armstrong, J. Non-equilibrium Molecular Dynamics. In *Experimental Thermodynamics Volume X: Non-Equilibrium Thermodynamics with Applications*; The Royal Society of Chemistry: London, UK, 2015. [CrossRef]
45. Plimpton, S. Fast parallel algorithms for short-range molecular dynamics. *J. Comp. Phys.* **1995**, *117*, 1–19. [CrossRef]
46. Schneider, T.; Stoll, E. Molecular-dynamics study of a three-dimensional one-component model for distortive phase transitions. *Phys. Rev. B* **1978**, *17*, 1302–1322. [CrossRef]
47. Dünweg, B.; Paul, W. Brownian dynamics simulations without Gaussian random numbers. *Int. J. Mod. Phys. C* **1991**, *2*, 817–827. [CrossRef]
48. Isele-Holder, R.E.; Mitchell, W.; Hammond, J.R.; Kohlmeyer, A.; Ismail, A.E. Reconsidering Dispersion Potentials: Reduced Cutoffs in Mesh-Based Ewald Solvers Can Be Faster Than Truncation. *J. Chem. Theory Comput.* **2013**, *9*, 5412–5420. [CrossRef] [PubMed]
49. Kjelstrup, S.; Bedeaux, D. *Non-Equilibrium Thermodynamics of Heterogeneous Systems*; WORLD SCIENTIFIC: Singapore, 2008. [CrossRef]
50. Bekker, H.; Berendsen, H.; Dijkstra, E.; Achterop, S.; Drunen, R.; van der Spoel, D.; Sijbers, A.; Keegstra, H.; Reitsma, B.; Renardus, M. Gromacs: A parallel computer for molecular dynamics simulations. *Phys. Comput.* **1993**, *92*, 252–256.
51. Brehm, M.; Thomas, M.; Gehrke, S.; Kirchner, B. TRAVIS—A free analyzer for trajectories from molecular simulation. *J. Chem. Phys.* **2020**, *152*, 164105. [CrossRef] [PubMed]
52. Brehm, M.; Kirchner, B. TRAVIS—A Free Analyzer and Visualizer for Monte Carlo and Molecular Dynamics Trajectories. *Chem. Inf. Model.* **2011**, *51*, 2007–2023. [CrossRef] [PubMed]
53. Kusalik, P.G.; Svishchev, I.M. The spatial structure in liquid water. *Science* **1994**, *265*, 1219–1221. [CrossRef]
54. Hess, B.; Kutzner, C.; Spoel, D.V.D.; Lindahl, E. GROMACS 4: Algorithms for highly efficient, load-balanced, and scalable molecular simulation. *J. Chem. Theory Comput.* **2008**, *4*, 435–447. [CrossRef]
55. Allen, M.P.; Tildesley, D.J. *Computer Simulation of Liquids*; Oxford University Press: London, UK, 1987.
56. Ramires, M.L.; de Castro, C.A.; Fareleira, J.M.; Wakeham, W.A. Thermal Conductivity of Aqueous Sodium Chloride Solutions. *J. Chem. Eng. Data* **1994**, *39*, 186–190. [CrossRef]
57. Zhang, J.; Clennell, M.B.; Dewhurst, D.N. Transport Properties of NaCl in Aqueous Solution and Hydrogen Solubility in Brine. *J. Phys. Chem. B* **2023**, *127*, 8900–8915. [CrossRef] [PubMed]
58. Stukowski, A. Visualization and analysis of atomistic simulation data with OVITO—The Open Visualization Tool. *Model. Simul. Mater. Sci. Eng.* **2009**, *18*, 015012. [CrossRef]
59. Filipponi, A.; De Panfilis, S.; Oliva, C.; Ricci, M.A.; D'Angelo, P.; Bowron, D.T. Ion Hydration under Pressure. *Phys. Rev. Lett.* **2003**, *91*, 165505. [CrossRef] [PubMed]
60. Wagner, W.; Pruß, A. The IAPWS formulation 1995 for the thermodynamic properties of ordinary water substance for general and scientific use. *J. Phys. Chem. Ref. Data* **2002**, *31*, 387–535. [CrossRef]
61. Bridgman, P.W. The Thermal Conductivity of Liquids. *Proc. Natl. Acad. Sci. USA* **1923**, *9*, 341–345. [CrossRef] [PubMed]
62. Chapman, A.; Bresme, F. Polarisation of water under thermal fields: The effect of the molecular dipole and quadrupole moments. *Phys. Chem. Chem. Phys.* **2022**, *24*, 14924–14936. [CrossRef] [PubMed]
63. Xu, K.; Hao, Y.; Liang, T.; Ying, P.; Xu, J.; Wu, J.; Fan, Z. Accurate prediction of heat conductivity of water by a neuroevolution potential. *J. Chem. Phys.* **2023**, *158*, 204114. [CrossRef]

entropy

Article

Adsorption Kinetics Model of Hydrogen on Graphite

Jean-Marc Simon * and Guilherme Carneiro Queiroz da Silva

Laboratoire Interdisciplinaire Carnot de Bourgogne, UMR-6303 CNRS-Université Bourgogne Europe (UBE), 9 Av. A. Savary, 21000 Dijon, France; gcarneiroq@gmail.com

* Correspondence: jmsimon@ube.fr

Abstract: A new kinetic equation for the adsorption and desorption of H_2 on graphite is derived based on the adsorption and desorption equilibrium rates obtained from the molecular dynamics. These rates are proportional to the activity in the gas and the adsorbed phase and thus do not obey Langmuir kinetics. The new equation offers a new route for understanding experimental results. It is used to simulate the kinetics under different thermodynamic conditions, both isothermal and non-isothermal. The characteristic times of adsorption and desorption are in good agreement with the data from the literature. The relation between the kinetics and the mass flow equation is discussed within the framework of the non-equilibrium thermodynamics of heterogeneous systems. Finally, expressions for the transport coefficients are proposed for both the transfer of mass and the coupling between the mass and heat fluxes.

Keywords: adsorption kinetics model; hydrogen on graphite; transport equation; transport coefficients

1. Introduction

Hydrogen-based technologies are an alternative for reducing the production of greenhouse gases like CO_2. The current research aims to optimise the handling of this resource under safe and economically viable conditions [1–3]. Its production through electrolysis as a solution for storing energy and the recent discovery of geosourced hydrogen [4–6] have generated growing interest in its recovery and valorisation. However, a better understanding of the physical chemistry of hydrogen is essential to addressing challenges related to its production, storage, transport, and final use.

The traditional methods of storage, like mechanical compression at high pressure or liquefaction at cryogenic temperatures, though effective, must overcome challenges related to safety, corrosion, and energy costs. New solutions are exploring the interaction between hydrogen and different types of materials [1–3]. It can simply be adsorbed onto adsorbent materials such as nanostructured carbon materials, Metal– and Covalent Organic Frameworks (MOFs and COFs), polymers, or zeolites. These materials are generally characterised by their safe and good reversibility in use (adsorption/desorption) and can offer high storage capacities. For the development of safe and efficient conditions of storage, it is important to consider not only the thermodynamics of adsorption but its kinetics as well.

In order to investigate the interplay between the thermodynamics and dynamics in physical adsorption processes, Simon et al., in a previous paper [7], studied a simple system using molecular dynamics: a gas phase of di-hydrogen in contact with a graphite surface. Equilibrium states were simulated at temperatures ranging from 70 K to 390 K using different amounts of hydrogen molecules. The population of adsorbed molecules

 https://doi.org/10.3390/e27030229

was recorded, and the thermodynamic isotherms were computed. The results showed good agreement with Langmuir's adsorption isotherm:

$$\theta_{eq} = \frac{Ka^g}{1 + Ka^g}, \text{ with } a^g = p/p^0. \tag{1}$$

where a^g, p, p^0, and θ_{eq} are the activity of the gas phase and its pressure, normal pressure (1 bar), and the coverage under equilibrium conditions. The definition of θ in the equations above follows Langmuir's original definition, which is the fractional coverage of a monolayer. In other words, this is the ratio between the surface concentration, c^s, and the saturated surface concentration, c^{sat}. K is Langmuir's constant. Langmuir's model is based on non-interacting and non-mobile adsorbed molecules, while the simulations clearly evidence that the hydrogen molecules are mobile and interact between themselves. An interesting discussion on the interpretation of the ideality or non-ideality of the adsorbed phase in terms of the interaction between the adsorbed molecules is proposed in [8] based on the results [7]. The qualitative results we present here do not change with the choice of one interpretation over another; however, quantitative differences may appear.

Following [7], the adsorbed phase shows non-ideal behaviour that falls outside of Henry's regime (at low coverage), which is revealed in the expression of the chemical potential:

$$\mu^{ads} = \mu^\star + RT \ln a^{ads}, \text{ with } a^{ads} = \frac{\theta}{1-\theta} = \gamma\theta. \tag{2}$$

Here, T and R are the temperature (K) and the perfect gas constant, and $\mu^\star$ is the standard state in Henry's law. a^{ads} is the chemical activity of the adsorbed phase, and $\gamma = 1/(1-\theta)$ is the activity coefficient. Outside of Henry's regime ($\theta \approx 0$), γ deviates from one, which is a sign of non-ideal thermodynamic behaviour.

The condition of equilibrium demands that the chemical potential of the adsorbed phase μ^{ads} and that of the gas phase μ^g are equal. Using μ^0, the standard state of the gas phase, the chemical potential of the gas phase is

$$\mu^g = \mu^0 + RT \ln a^g, \tag{3}$$

In the same molecular dynamics study, Simon et al. [7] computed the unidirectional rates of adsorption and desorption under equilibrium conditions. They proposed new kinetic equations for the adsorption/desorption of mobile hydrogen molecules that did not agree with the non-mobile model of Langmuir kinetics. The rates, J_a (adsorption) and J_d (desorption), are shown to be proportional to the thermodynamic activity of the gas, a^g, and of the adsorbed phase, a^{ads}:

$$J_a = k_a a^g = k_a p/p^0, \tag{4}$$

$$J_d = k_d a^{ads} = k_d \frac{\theta}{1-\theta}. \tag{5}$$

The coefficients k_a and k_d are the adsorption and desorption rate constants, respectively. Using Equations (4) and 5 with $J_a = J_d$, the equilibrium conditions, we obtain the adsorption Langmuir equation:

$$k_a \frac{p}{p^0} = k_d \frac{\theta_{eq}}{1-\theta_{eq}}. \tag{6}$$

Compared with Equation (1), the Langmuir constant is given by the ratio $k_a/k_d = K$.

For the adsorption of hydrogen on graphite, it was reported [7] that both the rate constant and the saturated surface concentration follow an Arrhenius law:

$$\begin{aligned} c^{sat} &= 1.35 \times 10^{-5} \exp\left(\frac{180}{RT}\right), &(7)\\ k_a &= 3.5 \times 10^{3} \exp\left(\frac{1500}{RT}\right), &(8)\\ k_d &= 1.8 \times 10^{7} \exp\left(\frac{-4400}{RT}\right). &(9) \end{aligned}$$

The units of the activation energy and of the constant are, respectively, J/mol and mol/m^{2}s, while c^{sat} is given in mol/m^2.

The equations are consistent with the thermodynamics of a Langmuir process (isothermal adsorption) but, as mentioned above, not with Langmuir kinetics:

$$\begin{aligned} J_a^L &= k_a^L p(1-\theta), &(10)\\ J_d^L &= k_d^L \theta. &(11) \end{aligned}$$

Here, J_a^L, J_d^L, k_a^L, and k_d^L are the Langmuir adsorption and desorption fluxes and rate constants.

It is worth noticing that for a low coverage limit, in Henry's regime with $1-\theta=1$, both the Langmuir and the proposed kinetic models lead to the same expressions for the fluxes:

$$\begin{aligned} J_a^H &= k_a^H p, &(12)\\ J_d^H &= k_d^H \theta, &(13) \end{aligned}$$

where J_a^H, J_d^H, k_a^H, and k_d^H are Henry's adsorption and desorption fluxes and rate constants, respectively. So, when $1-\theta \approx 1$, we have the following:

$$\begin{aligned} k_a^H &= k_a = k_a^L, &(14)\\ k_d^H &= k_d = k_d^L \theta. &(15) \end{aligned}$$

One particular feature of the proposed model is that it takes into account the non-ideality of the adsorbed phase. This is facilitated by including the dependence of the desorption rate on the activity coefficient (γ). The new expressions therefore apply to non-ideal mobile adsorbed phases.

The aim of this work is to develop a set of equations that are able to describe the kinetics of the adsorption/desorption of H_2 on graphite under isothermal and non-isothermal conditions. This paper is divided as follows: First, theoretical considerations comparing the proposed model to the classical Langmuir expressions are given. Then, the link between the new model and transport equations are discussed. In the last part, the numerical results obtained using the new equations are also shown.

2. Theoretical Considerations

2.1. Isothermal Adsorption Kinetics

Common to all reaction schemes is the general expression for the net rate of adsorption, J_{net}:

$$J_{net} = c^{sat}\frac{d\theta(t)}{dt} = J_a - J_d, \tag{16}$$

where t is the time. In this expression, the positive flux is defined as the mass transfer from the gas to the adsorbed phase. Considering the expression for J_a and J_d, Equations (4), (5), and (10)–(13), the net rate is therefore a function of $p(t)$ and $\theta(t)$, which are functions of time. For simplicity, the equations are written without explicit temporal dependence.

Relevant adsorption experiments are often conducted under isobaric conditions. Solving the differential equations at a constant p results in analytical expressions for the adsorption rate curves. Non-isobaric conditions lead to more complex solutions that can be obtained through numerical integration. This latter type will not be considered here.

At this point, it is necessary to discuss whether the expressions of J_a and J_d are valid outside of equilibrium conditions. These expressions assume microscopic reversibility and thus hold close to equilibrium. Far from equilibrium, the two models, Langmuir and non-Langmuir, can be considered as a first approximation. Nonetheless, the results should be regarded with caution. Conditions far from equilibrium will be used to illustrate the behaviour predicted by the mathematical expressions obtained for the mass flux according to the new model.

The flux equations defined in [7] involve one additional assumption that is based on consideration of the Langmuir kinetics. From the solid to the gas phase, three zones are considered. The first is the adsorbed phase, where the molecules are in contact with the solid surface. The second is a surface buffer region a few angstroms thick, where the molecules do not behave in the same way as they do in the bulk/gas phase. In other words, their behaviour is still driven by interactions with the surface. Where H_2 is close to the graphite surface, the adsorbate molecules are more densely packed due to the attractive forces between molecules and both the solid and already adsorbed H_2. The last region is the gas phase, where the effects of the interface are negligible. Note that the gas and solid phases are not in direct contact in this framework. This segmentation was used in ref. [7] to compute the fluxes. J_a was calculated as the part of the flux that reached the adsorbed surface area originating from the gas phase. J_d was the part of the flux that reached the gas phase and came from the adsorbed phase. Thus, these two fluxes are not located at the same plane positions. The definition of the net flux, Equation (16), is the transport of mass across this buffer surface region, and the net flux is, on average, assumed to be constant through it. It is necessary to stress that this buffer zone is not explicitly analysed, but its role in the adsorption kinetics is taken into account through the values of the rate constants in the kinetics, that is, k_a and k_d. A discussion on the link between the Langmuir and non-Langmuir kinetics and the role of correlations across an equivalent of the surface buffer is proposed in [8].

Below, we present equations for the temporal evolution of the coverage, θ. This set of equations is the result of integrating the expressions present in the kinetic models just discussed (Equations (4), (5), (10), and (11)). Henry's regime is also present to give a limited ideal case as a reference for model comparison.

2.1.1. Net Rate—Non-Langmuir Kinetics

Equation (16) is integrated for non-ideal surface kinetics using Equations (4) and (5). Introducing the expressions of the rates, Equations (4) and (5), into Equation (16) results in the following:

$$c^{sat}\frac{d\theta}{dt} = k_a\frac{p}{p^0} - k_d\frac{\theta}{1-\theta}, \tag{17}$$

or

$$c^{sat}(1-\theta)\frac{d(1-\theta)}{dt} = -(1-\theta)\left(k_a\frac{p}{p^0} + k_d\right) + k_d, \tag{18}$$

which can be written as

$$c^{sat}\left(\frac{X}{k_d - AX}\right)dX = dt, \tag{19}$$

with $X = (1-\theta)$ and $A = \left(k_a\frac{p}{p^0} + k_d\right)$.

We integrate this expression from the initial coverage at time zero, θ_i ($X_i = X(t=0)$), into the coverage at time t and obtain the following:

$$\frac{c^{sat}}{A}\left[X_i - X + \frac{k_d}{A}\ln\left(\frac{k_d - AX_i}{k_d - AX}\right)\right] = t, \tag{20}$$

or

$$\frac{c^{sat}}{A}\left[\theta - \theta_i + \frac{k_d}{A}\ln\left(\frac{\theta_i A - k_a \frac{p}{p^0}}{\theta A - k_a \frac{p}{p^0}}\right)\right] = t, \tag{21}$$

where A is a constant under isobaric conditions. A can simply be obtained from the equilibrium isotherm, Equation (6):

$$A = k_a \frac{p}{p^0} + k_d = k_d\left(\frac{1}{1-\theta_{eq}}\right). \tag{22}$$

By introducing the expression for A into Equation (21), we can write t as a function of θ:

$$\frac{c^{sat}(1-\theta_{eq})^2}{k_d}\left[\frac{\theta - \theta_i}{1-\theta_{eq}} + \ln\left(\frac{\theta_i - \theta_{eq}}{\theta - \theta_{eq}}\right)\right] = t. \tag{23}$$

Equation (23) has a well-defined solution for θ as a function of t, even if we are close to saturation (full coverage). When θ_{eq} is close to unity, the first term in the square brackets dominates at a low value of t, making θ proportional to t. The logarithm term dominates in most cases, when θ approaches equilibrium or a high value of t.

Full coverage at $\theta_{eq} = 1$ can be understood as the case of an infinitely large pressure and an infinitely fast adsorption rate, with $t = 0$. This limit cannot be reached for Equation (23) or for Langmuir kinetics; see Equation (27) below. Desorption towards θ_{eq}=0 (zero pressure) is, on the other hand, well described by Equation (23).

The expressions for the characteristic times for adsorption, t_{ads}, and desorption, t_{des}, are the same, as expected for a process which obeys microscopic reversibility. They can be studied analytically from the long-term logarithmic behaviour of Equation (23):

$$t_{ads} = t_{des} = \frac{c^{sat}(1-\theta_{eq})^2}{k_d} = \frac{c^{sat}}{k_d\left(\gamma_{\theta_{eq}}\right)^2}. \tag{24}$$

The characteristic time is a function of θ_{eq} and not of θ_i. This will lead to asymmetry in the time dependence of adsorption vs. desorption. However, Equation (23) involves linear dependence in addition to logarithmic behaviour, so it may be useful to define another characteristic time τ corresponding to a progress of $100(1-1/\mathrm{e}) \approx 63\%$, as for pure logarithmic behaviour (see Langmuir's kinetics and Henry's regime):

$$\tau = \frac{c^{sat}(1-\theta_{eq})^2}{k_d}\left[1 + \frac{(1-1/\mathrm{e})(\theta_{eq} - \theta_i)}{(1-\theta_{eq})}\right]. \tag{25}$$

It is worth noting that τ is a function of both the final coverage and the initial coverage.

2.1.2. Net Rate—Langmuir Kinetics

For comparison, the expressions obtained from Langmuir kinetics (Equations (10) and (11)) are also given. By inserting the expressions of the rates, Equations (10) and (11), into Equation (16), we obtain the following:

$$c^{sat}\frac{d\theta}{dt} = k_a^L \frac{p}{p^0}(1-\theta) - k_d^L \theta. \tag{26}$$

Integrating this equation with time leads to the evolution of the surface coverage.

$$\theta(t) = \theta_{eq} - \left(\theta_{eq} - \theta_i\right) \exp\left[-\frac{k_d^L t}{c^{sat}\left(1 - \theta_{eq}\right)}\right], \qquad t_{ads}^L = t_{des}^L = \frac{c^{sat}\left(1 - \theta_{eq}\right)}{k_d^L}, \tag{27}$$

where t^L is the characteristic time of the adsorption/desorption kinetics from Langmuir expressions. Under the specific equilibrium conditions and using the flux desorption rate expressions, Equations (5) and (11), it follows that $k_d^L = k_d/(1 - \theta_{eq})$, and the expressions for the characteristic times for Langmuir and non-Langmuir kinetics are then identical $t^L = t_{ads}$, but they differ from τ.

2.1.3. Net Rate—Henry's Regime

Under Henry's regime, the interaction between the adsorbed particles is neglected, and the activity coefficient $\gamma = 1 - \theta$ is approximated to 1. Using Equations (12), (13), (16), and (15), we obtain the following:

$$c^{sat} \frac{d\theta}{dt} = k_a \frac{p}{p^0} - k_d \theta. \tag{28}$$

Through integration, we obtain the following:

$$\theta(t) = \theta_{eq} - \left(\theta_{eq} - \theta_i\right) \exp\left(-\frac{k_d t}{c^{sat}}\right), \qquad t_{ads} = t_{des} = \frac{c^{sat}}{k_d}, \tag{29}$$

2.2. *The Transport Equation*

Close to equilibrium, in the linear regime, the transport of heat and mass obeys the phenomenological equations described in the book by S.R. de Groot and P. Mazur [9] on homogeneous systems and in S. Kjelstrup and D. Bedeaux's book [10] on heterogeneous systems. Based on the expression of the entropy production, the fluxes are written as a product of the transport coefficients and thermodynamic forces. In S. Kjelstrup and D. Bedeaux's monograph [10], the transport equations are alternatively given in terms of resistivity coefficients. When we consider a surface, the thermodynamic driving forces are scalar properties and are expressed as the difference in the thermodynamic properties across the surface. Because of the heterogenous nature of the surface, the fluxes are different across the surface, except under stationary conditions. It is then useful to express the transport equations, i.e., the fluxes and transport coefficients, taking two different types of references: the gas side and the adsorbed phase side of the surface buffer, J^g and J^{ads}, respectively. The two references are equivalent and can be used interchangeably depending on the practical criteria. For non-isothermal conditions, the mass flux expressions are as follows:

$$J^g = -L_{\mu q}^g \left(\frac{1}{T^g} - \frac{1}{T^{ads}}\right) + L_{\mu\mu}^g \left(\frac{\mu^g(T^g) - \mu^{ads}(T^g)}{T^g}\right), \tag{30}$$

$$J^{ads} = -L_{\mu q}^{ads} \left(\frac{1}{T^g} - \frac{1}{T^{ads}}\right) + L_{\mu\mu}^{ads} \left(\frac{\mu^g(T^{ads}) - \mu^{ads}(T^{ads})}{T^{ads}}\right), \tag{31}$$

where $L_{\mu\mu}$ and $L_{\mu q}$ are, respectively, the mass and the cross-coupling (heat and mass) transport coefficients. Clearly, J^g and J^{ads} are different except under a stationary state, when the mass flux is constant across the whole surface. In our model, as mentioned above, the net flux, Equation (16), is assumed to be constant, and then $J^g = J^{ads} = J_{net}$. In the following, the mass transfer equation is derived based on Equation (16) for the net flux and using the non-Langmuir kinetics for both isothermal and non-isothermal conditions.

2.2.1. Isothermal Conditions

Under a constant pressure, the activity of the gas phase is constant. At equilibrium $J_{net} = 0$, $k_a a^g = k_d a^{ads}_{eq}$. This is used to defined two sets of transport equations, like for Equations (30) and (31). From Equations (4), (5), and (16), we obtain the following:

$$J_{net} = k_d\left(a^{ads}_{eq} - a^{ads}\right) = k_a\left(a^g - a^{g\phi}\right), \tag{32}$$

$$J_{net} = k_d \exp\left(\frac{-\mu^\star + \mu^{ads}_{eq}}{RT}\right)\left[1 - \exp\left(\frac{\mu^{ads} - \mu^{ads}_{eq}}{RT}\right)\right], \tag{33}$$

$$J_{net} = k_a \exp\left(\frac{-\mu^0 + \mu^g}{RT}\right)\left[1 - \exp\left(\frac{\mu^{g\phi} - \mu^g}{RT}\right)\right], \tag{34}$$

The superscript ϕ for the gas indicates a state of equilibrium with the adsorbed phase $\mu^{ads} = \mu^{g\phi}$. The related activity, $a^{g\phi}$, is simply given by the Langmuir equation, Equation (1). Close to equilibrium, $\mu^{ads}_{eq} \approx \mu^{ads}$ or $\mu^g \approx \mu^{g\phi}$. The exponential function can then be expanded into a series, and the above transport equations can be written as the difference in the chemical potential across the surface, in agreement with [10]:

$$J_{net} = \frac{k_d}{R} a^{ads}_{eq}\left(\frac{\mu^{ads}_{eq} - \mu^{ads}}{T}\right), \tag{35}$$

$$J_{net} = \frac{k_a}{R} a^g\left(\frac{\mu^g - \mu^{g\phi}}{T}\right). \tag{36}$$

The two expressions, Equations (35) and (36), are equivalent but with the reference in the adsorbed phase and in the gas, respectively. Through comparison with Equations (30) and (31), under isothermal conditions, we obtain the following directly:

$$L^{ads}_{\mu\mu} = L^g_{\mu\mu} = k_a a^g/R = k_d a^{ads}_{eq}/R. \tag{37}$$

It is interesting to note in the equation above how the mass transfer coefficient is written in terms of a product involving the rate constant and the activity. In other words, it is described by the kinetic and thermodynamic aspects of the process.

Far from equilibrium, Equations (33) and (34) have to be used without approximations. Thus, the expression of $L_{\mu\mu}$ shows dependence on the difference in the chemical potential between the actual value and the value in equilibrium. From Equations (33) and (34), we find the following:

$$L^{ads,ext}_{\mu\mu} = L^{ads}_{\mu\mu}\left[\frac{1 - \exp(-X)}{X}\right], \text{ with } X = \frac{\mu^{ads}_{eq} - \mu^{ads}}{RT}, \tag{38}$$

$$L^{g,ext}_{\mu\mu} = L^g_{\mu\mu}\left[\frac{1 - \exp(-Z)}{Z}\right], \text{ with } Z = \frac{\mu^g - \mu^{g\phi}}{RT}. \tag{39}$$

where the superscript *ext* indicates the properties extending beyond close equilibrium. Obviously, $Z = X$ and $L^{ads,ext}_{\mu\mu} = L^{g,ext}_{\mu\mu}$ since $L^{ads}_{\mu\mu} = L^g_{\mu\mu}$. It is worth noting that the first corrective term for $L^{ads}_{\mu\mu}$, using Taylor's expansion, is proportional to X: $L^{g,ext}_{\mu\mu} \approx L^g_{\mu\mu}(1 + X/2)$.

2.2.2. Non-Isothermal Conditions

In this part, the temperature of both phases, gas, T^g, and adsorbed, T^{ads}, differs but is kept constant during the adsorption process. The final state is a stationary non-equilibrium situation characterised by constant non-zero heat flux and zero mass flux. At zero mas flux,

$J_{net} = 0$, $k_a(T^g)a^g(T^g) = k_d(T^{ads})a^{ads}_{stat}(T^{ads})$, and an equivalent of the Langmuir adsorption isotherm for non-isothermal conditions can be written using Equations (8) and (9):

$$\theta_{stat}(T^{ads}) = \frac{K_{NI}p(T^g)/p^0}{1+K_{NI}p(T^g)/p^0}, \quad (40)$$

$$K_{NI}(T^g, T^{ads}) = k_a(T^g)/k_d(T^{ads}) = 1.94\times 10^{-4}\exp\left(\frac{1500}{R\,T^g}+\frac{4400}{R\,T^{ads}}\right), \quad (41)$$

where K_{NI} is the non-isothermal Langmuir constant. In this expression, the subscript *stat* is used to identify the final stationary state.

For clarity in the following part, the temperature dependence of the kinetic parameters and thermodynamic properties will be omitted. We first introduce $a^{g\phi}_{stat}$, which refers to the activity of the gas phase in the stationary state with the actual adsorbed phase at a temperature of T^{ads} using Equation (41). This is used to defined two sets of transport equations, like for Equations (30) and (31). From Equations (4), (5), and (16),

$$J_{net} = k_d\left(a^{ads}_{stat} - a^{ads}\right) = k_a\left(a^g - a^{g\phi}_{stat}\right), \quad (42)$$

$$J_{net} = k_d\exp\left(\frac{-\mu^\star + \mu^{ads}_{stat}}{RT^{ads}}\right)\left[1-\exp\left(\frac{\mu^{ads}-\mu^{ads}_{stat}}{RT^{ads}}\right)\right], \quad (43)$$

$$J_{net} = k_a\exp\left(\frac{-\mu^0+\mu^g}{RT^g}\right)\left[1-\exp\left(\frac{\mu^{g\phi}-\mu^g}{RT^g}\right)\right]. \quad (44)$$

Approaching the limit of the stationary state, the net mass flux can be written in a similar way to that for the isothermal case:

$$J_{net} = \frac{k_d}{R}a^{ads}_{stat}\left(\frac{\mu^{ads}_{stat}-\mu^{ads}}{T^{ads}}\right), \quad (45)$$

$$J_{net} = \frac{k_a}{R}a^g\left(\frac{\mu^g-\mu^{g\phi}}{T^g}\right). \quad (46)$$

This expression is very similar to the expression for the isothermal case, Equations (35) and (36). However, it does not make a clear distinction between the contribution of the temperature and the difference in the chemical potential, like in Equations (30) and (31). This can be achieved for the adsorbed phase reference by rewriting Equation (42):

$$J_{net} = k_d(T^g)a^{ads}_{eq}(T^g) - k_d(T^{ads})a^{ads}(T^{ads}), \quad (47)$$

$$= k_d(T^{ads})\exp\left[\frac{-4400}{R}\left(\frac{1}{T^g}-\frac{1}{T^{ads}}\right)\right]a^{ads}_{eq}(T^g) - k_d(T^{ads})a^{ads}(T^{ads}), \quad (48)$$

$$= k_d\exp\left(\frac{-\mu^\star+\mu^{ads}_{eq}}{RT^{ads}}\right)\left\{\exp\left[\frac{-4400}{R}\left(\frac{1}{T^g}-\frac{1}{T^{ads}}\right)\right]-\exp\left(\frac{\mu^{ads}-\mu^{ads}_{eq}}{RT^{ads}}\right)\right\}. \quad (49)$$

In Equation (49), all of the chemical potentials are given for a temperature of T^{ads}, and we use the equilibrium condition $a^{ads}_{eq}(T^g) = a^{ads}_{eq}(T^{ads})$:

$$a^{ads}_{eq}(T) = \frac{\theta_{eq}}{1-\theta_{eq}} = \exp\left(\frac{-\mu^\star(T)+\mu^{ads}_{eq}(T)}{RT}\right). \quad (50)$$

From Equation (42), the net flux can alternatively be written taking the gas phase reference:

$$
\begin{aligned}
J_{net} &= k_a(T^g)a^g(T^g) - k_a(T^{ads})a^{g\phi}(T^{ads}), & (51)\\
&= k_a(T^g)\left\{a^g(T^g) - \exp\left[\frac{-1500}{R}\left(\frac{1}{T^g} - \frac{1}{T^{ads}}\right)\right]a^{g\phi}(T^{ads})\right\}, & (52)\\
&= k_a \exp\left(\frac{-\mu^0 + \mu^{g\phi}}{RT^g}\right)\left\{\exp\left(\frac{\mu^g - \mu^{g\phi}}{RT^g}\right) - \exp\left[\frac{-1500}{R}\left(\frac{1}{T^g} - \frac{1}{T^{ads}}\right)\right]\right\} & (53)
\end{aligned}
$$

where we again used $a^{g\phi}(T^{ads}) = a^{g\phi}(T^g)$.

Approaching the stationary state and imposing the concept of T^g and T^{ads} being close, Equations (49) and (53) can be expressed:

$$
\begin{aligned}
J_{net} &= k_d a_{eq}^{ads}\left[\frac{-4400}{R}\left(\frac{1}{T^g} - \frac{1}{T^{ads}}\right) + \left(\frac{\mu_{eq}^{ads} - \mu^{ads}}{RT^{ads}}\right)\right], & (54)\\
J_{net} &= k_a a^{g\phi}\left[\frac{1500}{R}\left(\frac{1}{T^g} - \frac{1}{T^{ads}}\right) + \left(\frac{\mu^g - \mu^{g\phi}}{RT^g}\right)\right]. & (55)
\end{aligned}
$$

Through comparison with Equations (30) and (31), we obtain the following:

$$
\begin{aligned}
L_{\mu q}^{ads} &= k_d a_{eq}^{ads}\frac{4400}{R} = 4400 L_{\mu\mu}^{ads}, & (56)\\
L_{\mu q}^{g} &= -k_a a^{g\phi}\frac{1500}{R} = -1500 L_{\mu\mu}^{g}. & (57)
\end{aligned}
$$

$L_{\mu\mu}^{ads}$ and $L_{\mu\mu}^{g}$ are expressed the same way as they are for the isothermal case but they are not equal since the temperature difference is not null.

Another transport quantity that quantifies the coupling between the heat and mass fluxes is the heat of transfer [9,10], $q*$. This gives the heat transported with the mass flux at a constant temperature. Bearing in mind the aforementioned transport coefficients, it is defined according to the equations below:

$$
\begin{aligned}
q^{*ads} &= \left(\frac{J_q^{ads}}{J_{net}}\right)_{(T^{ads}=T^g)} = \frac{L_{\mu q}^{ads}}{L_{\mu\mu}^{ads}} = 4400\ \mathrm{J/mol}, & (58)\\
q^{*g} &= \left(\frac{J_q^{g}}{J_{net}}\right)_{(T^{ads}=T^g)} = \frac{L_{\mu q}^{g}}{L_{\mu\mu}^{g}} = -1500\ \mathrm{J/mol}. & (59)
\end{aligned}
$$

Here, J_q^{ads} and J_q^g are the measurable heat fluxes on the adsorbed side and on the gas side of the surface buffer. q^{*ads} and q^{*g} are constant in our non-Langmuir kinetics model and independent of temperature. Similar behaviour has been observed for the direct computation of q^* from molecular dynamics simulations of the adsorption of n-butane onto zeolite silicalite [11]; it is interesting to note that this is not the case for a liquid–vapour interface [10]. As expected [10], the difference $q^{*g} - q^{*ads}$ gives the enthalpy of adsorption, being $\Delta_{ads}H = -5900\ \mathrm{J/mol}$ here.

3. Results and Discussion

3.1. Isothermal Kinetics of Adsorption and Desorption

Equation (23) was used to investigate the isothermal kinetics of adsorption and desorption. We used a pressure of 50 bar and temperatures of 77 K and 293 K. These are the same conditions as those reported by Burress et al. [12]. Through applying Langmuir's isotherm equation, Equation (1), the equilibrium coverages are 0.990 and 0.0987 for 77 and 293 K, respectively.

Figure 1 reports the result at 293 K. The coverage, θ, and the logarithmic function of $|\theta_{eq} - \theta|$ are plotted as a function of time according to Equation (23) for adsorption and desorption. Logarithmic behaviour is observed. The characteristic time is computed from Equation (25), and the results are 4.3 and 4.6 ps for adsorption and desorption, respectively. These results agree well with the characteristic time of 4.9 ps found using the Henry's regime expression, Equation (29). This is expected since the range of coverage is small, lower than 0.1, where Henry's regime is supposed to apply. They are also in the same order of magnitude as those in Burress et al. [12] (16 ps for adsorption and 2.1 ps for desorption), although these authors used Langmuir dynamics.

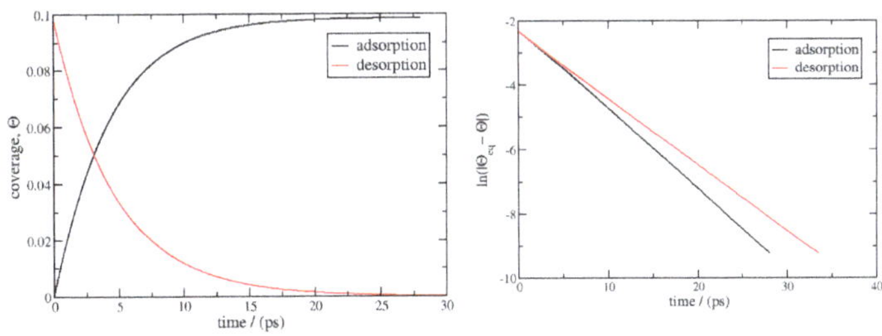

(a) (b)

Figure 1. Adsorption and desorption kinetics using Equation (23); the temperature is 293 K, with the pressure between 0 and 50 bar at a temperature of 293 K. The equilibrium coverages are 0 and 0.0987, respectively. In the left figure (**a**), the coverage θ is plotted as a function of time, t; in the right figure (**b**), $ln(|\theta_{eq} - \theta|)$ is plotted as a function of time.

At 77 K, the equilibrium coverage corresponding to 50 bar is $\theta_{eq} = 0.990$, close to saturation. The kinetics of desorption/adsorption from 0 to θ_{eq} are calculated using Equation (23). The results are given in Figures 2 and 3. The adsorption curve is linear below a coverage of 0.95 and tends towards logarithmic behaviour beyond this point. For desorption, the logarithmic behaviour is spread over a wider coverage range; it is observed below a coverage of 0.2. We obtain a characteristic desorption time of 360 ps using Equation (25), which is much longer than the value for adsorption, at about 6.2 ps. This difference can be explained by the asymmetry in the values of the activity of the gas and adsorbed phases, which results in an asymmetric net flux, as displayed for 77 K in Figure 4. During adsorption, the net flux is at the maximum for most of the coverage, below 0.9, while it is nearly null for desorption in the same range. The desorption kinetics are then much less efficient than the adsorption kinetics. Burress et al. [12], for the same conditions, gave characteristic times about two times higher, at 15 ps for adsorption and 530 ps for desorption. Nonetheless, the model presents the same qualitative features, as the estimated times are in the same order of magnitude and the strong asymmetry between the adsorption and desorption kinetics is reproduced.

The two previous simulations were carried out under Henry's regime and near saturation. The results are reported in Figure 5 for intermediate coverage. At 293 K, the adsorption and desorption from 50 to 450 bar were studied. From Langmuir's isotherm equation, Equation (1), the equilibrium coverages are 0.0987 and 0.496. Although this range is rather wide, about 0.4, the logarithmic domain appears rapidly after 3 ps. The characteristic time τ is 1.9 ps for adsorption and 2.9 ps for desorption, and these values are smaller than that for Henry's regime (4.9 ps) presented above.

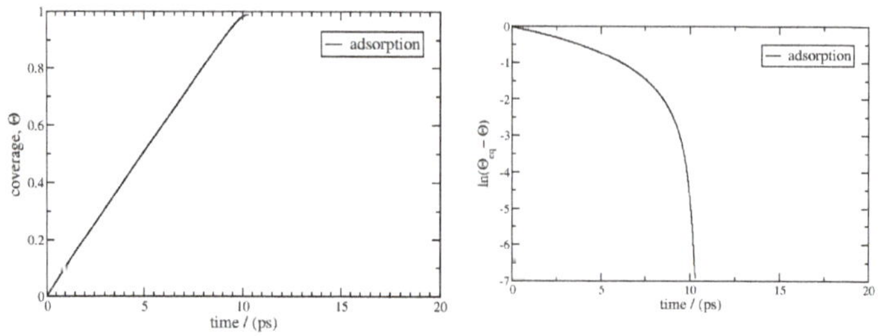

(a) (b)

Figure 2. Adsorption kinetics using Equation (23); the temperature is 77 K, and the initial coverage and the coverage at equilibrium are 0 and 0.990, respectively, at a pressure of 50 bar. In the left figure (**a**), the coverage θ is plotted as a function of time, t; in the right figure (**b**), $ln(|\theta - \theta_{eq}|)$ is plotted as a function of time. The kinetics do not follow logarithmic behaviour in the coverage range of the plot. The characteristic time is 6.2 ps.

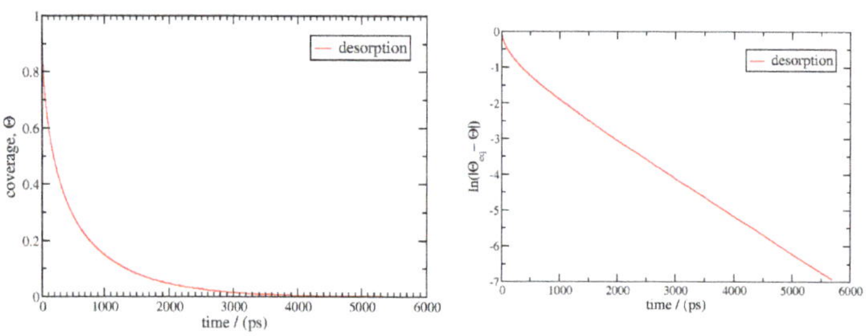

(a) (b)

Figure 3. Desorption kinetics using Equation (23); the temperature is 77 K, and the initial coverage and the coverage at equilibrium are 0.990 and 0, respectively. In the left figure (**a**), the coverage θ is plotted as a function of time, t; in the right figure (**b**), $ln(|\theta - \theta_{eq}|)$ is plotted as a function of time. After 50 ps, the kinetics follow logarithmic behaviour, and the characteristic time scale for desorption is 360 ps.

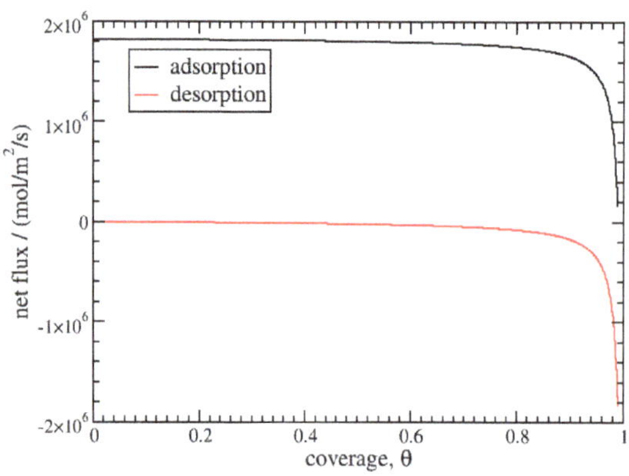

Figure 4. Evolution of the net flux for desorption and adsorption as a function of the coverage at 77 K between a coverage of 0 and 0.990.

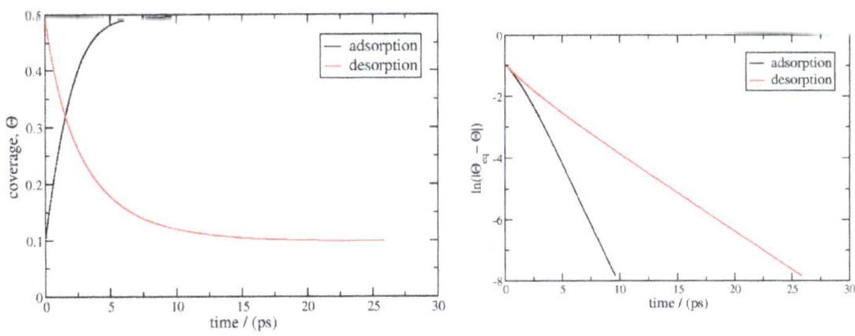

(**a**) (**b**)

Figure 5. Adsorption and desorption kinetics using Equation (23); the pressure ranges between 50 and 450 bar at a temperature of 293 K. The equilibrium coverages are 0.0987 and 0.496, respectively. In the left figure (**a**), the coverage θ is plotted as a function of time, t; in the right figure (**b**), $ln(|\theta_{eq} - \theta|)$ is plotted as a function of time. The characteristic time scales for adsorption and desorption are 1.9 ps and 2.9 ps.

In order to quantify the adsorption kinetics close to equilibrium, the transport coefficients $L_{\mu\mu}$ were computed for four different temperatures using Equation (37). The results are plotted in Figure 6 for 77 K, 273 K, 293 K, and 313 K. The values of $L_{\mu\mu}$ show an increase as a function of both the coverage and pressure. For constant coverage, the values of $L_{\mu\mu}$ increase with temperature, but the trend is the reverse at a constant pressure. These trends are in agreement with those observed in non-equilibrium molecular dynamics studies of the adsorption of n-butane onto zeolite silicalite [11]. A similar dependence of $L_{\mu\mu}$ (or inversely that for resistance) on pressure and temperature has also been observed at liquid–vapour interfaces both in molecular simulations and using theoretical models [10,13–17].

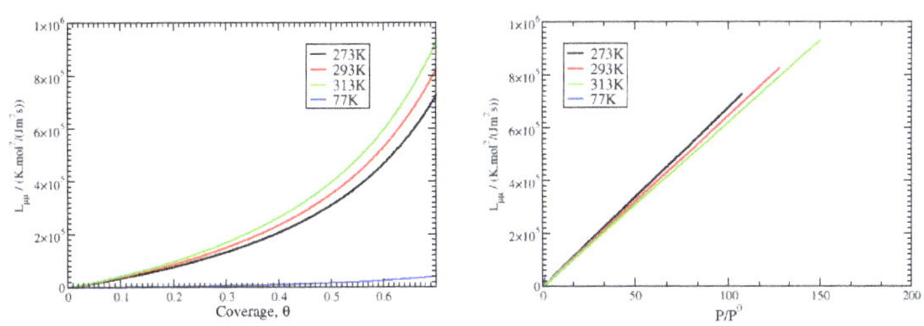

(**a**) (**b**)

Figure 6. Mass transport coefficient $L_{\mu\mu}$ as a function of the coverage (**a**) and pressure (**b**) at 77 K, 273 K, 293 K, and 313 K.

Beyond close equilibrium, the transport coefficients, $L_{\mu\mu}^{ext}$, are computed using Equation (38). In Figure 7, the values for adsorption between 0 and 50 and 50 and 450 bar and for desorption between 450 and 50 bar at 293 K are reported. They are compared with the reference $L_{\mu\mu}$ values. As expected, $L_{\mu\mu}$ and $L_{\mu\mu}^{ext}$ agree close to equilibrium, but they differ significantly outside of this condition. With the same equilibrium values, the desorption between 450 and 50 bar and the adsorption between 0 and 50 bar show continuous linear evolution of the values of $L_{\mu\mu}^{ext}$ around equilibrium. The slope of $L_{\mu\mu}$ with coverage is steeper than that of $L_{\mu\mu}^{ext}$, which indicates closer values to those for equilibrium for the latter. This is important since in general, it is common to consider constant transfer coefficients during the kinetics. From the results, this can be considered to be a good approximation close to the equilibrium state. The asymmetry between the adsorption and desorption kinetics that was mentioned above is also evidenced by comparing the values of $L_{\mu\mu}^{ext}$ during the adsorption and desorption kinetics from 50 to 450 bar.

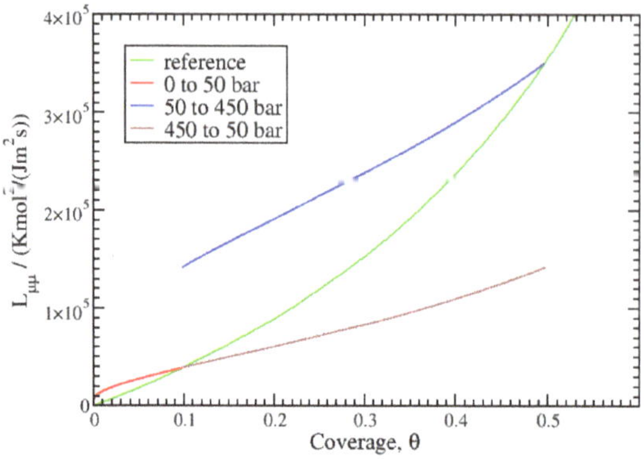

Figure 7. Comparison between the mass transport coefficients $L_{\mu\mu}$ obtained close to equilibrium at 293 K and $L_{\mu\mu}^{ext}$ during adsorption or desorption calculated using Equation (38).

3.2. *Non-Isothermal Kinetics of Adsorption*

So far, Equation (23) has been derived to account for the evolution of θ under isothermal conditions. It is also possible to derive this kind of kinetic equation for non-isothermal conditions with both phases constant but different temperatures and the gas phase at a constant pressure. The final expression is the same as that for Equation (23) but uses the non-isothermal Langmuir constant K_{NI} in the derivation—see Equation (41)—instead of the equilibrium Langmuir constant. The non-isothermal kinetics of adsorption are presented in Figure 8. A 450-bar-pressure gas phase is maintained at 293 K, and different surface temperatures are considered: 273 K, 283 K, 293 K, 303 K, and 313 K. As described in Section 2.2.2, the final stationary state is outside of equilibrium and characterised by zero net mass flux and non-zero heat flux. The adsorption kinetics are very similar to those in isothermal conditions, and as expected, the final coverage decreases as the surface temperature increases.

In the stationary state, the coverage can be obtained directly from Equation (41) under isothermal and non-isothermal conditions. The results are compared in Figure 9 for surface temperatures ranging from 73 to 503 K and using a constant $T^g = 293$ K for the non-isothermal case. The curves exhibits a similar inverse S shape, with a crossing point at the temperature of 293 K. Below this temperature, the non-isothermal coverage is lower than the equilibrium value, and above this point, it is higher; this is easily understandable according to Le Chatelier's principle. The difference between these two curves does not exceed 0.03 in coverage, which, except when approaching zero coverage, can be considered negligible in most cases or at least a secondary effect. This is also the case for the kinetics. It is important to note that because the adsorption is exothermic, strong non-isothermal effects can occur and delay the kinetics of adsorption/desorption. For example, in reference [18], Inzoli et al. have evidenced, using molecular dynamics, that the long-term adsorption kinetics of n-butane on silicalite are governed by the heat of the transfer between the zeolite and the gas phase. This is clearly due to the exothermic effect of adsorption.

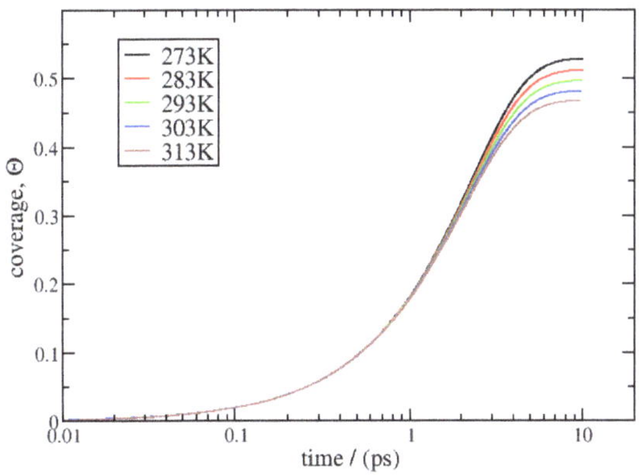

Figure 8. Non-isothermal adsorption kinetics calculated from Equation (23) using an initial bare surface and a constant gas pressure of 450 bar at 293 K with surface temperatures of 273 K, 283 K, 293 K, 303 K, and 313 K. The final state is outside of equilibrium.

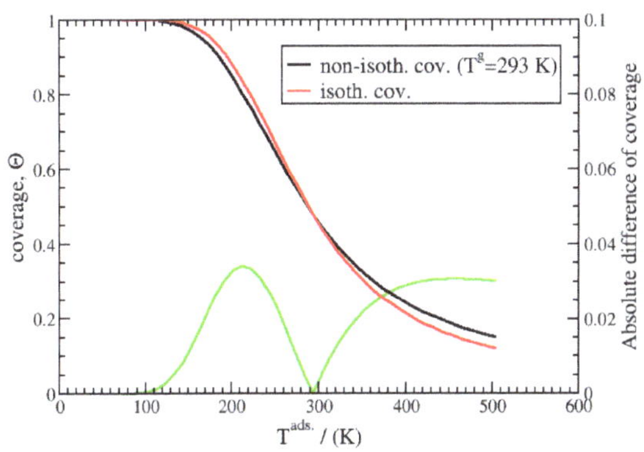

Figure 9. Comparison between stationary coverage under isothermal and non-isothermal conditions using Equation (41) and a constant gas pressure of 50 bar at 293 K for non-isothermal situations. The absolute difference between the two curves for each temperature is given in green.

4. Conclusions

The adsorption and desorption kinetics of H_2 on graphite are studied based on the non-Langmuir adsorption and desorption equilibrium rates obtained in a previous article [7]. New analytical expressions are derived to model the kinetics both under isothermal and non-isothermal conditions. The results show good agreement with the data given by Burress et al. [12]. In particular, the asymmetry in time between the adsorption and the desorption is pointed out, which is due to the asymmetry in the thermodynamic forces between adsorption and desorption.

A link with the transport equations for the non-equilibrium thermodynamics of heterogeneous systems is proposed, and expressions for the transport coefficients $L_{\mu\mu}$ and $L_{\mu q}$ and the heat of transport are also given. They are written as a product of the dynamical and thermodynamical properties. The results are in qualitative agreement with previous studies on the transport of mass across interfaces from molecular dynamics simulations. The values of $L_{\mu\mu}$ increase with the coverage and temperature.

These equations offer new additional tools that can be used for a better analysis and understanding of the kinetics of adsorption obtained both in experiments and molecular simulations.

Author Contributions: Conceptualisation: J.-M.S. Methodology: J.-M.S. Software: J.-M.S. Validation: J.-M.S. and G.C.Q.d.S. Formal analysis: J.-M.S. Writing—original draft preparation: J.-M.S. and G.C.Q.d.S. All authors have read and agreed to the published version of the manuscript.

Funding: This research received no external funding.

Data Availability Statement: Data are the results of the equations provided within the article.

Acknowledgments: G.C.Q.d.S. thanks the CEA for postdoc funding.

Conflicts of Interest: The authors declare no conflicts of interest.

List of Symbols

The following symbols are used in this manuscript; the units are given in parentheses:

Greek symbols

γ	the activity factor of the adsorbed phase;
μ^g, μ^0	the chemical potential of the gas phase and the standard phase (J/mol), where the subscript *eq* indicates the properties under equilibrium;
μ^{ads}, $\mu^\star$	the chemical potential of the adsorbed phase and the reference phase (J/mol), where the subscript *eq* indicates the properties under equilibrium;
τ	the characteristic times for adsorption/desorption under non-Langmuir kinetics (s);
θ, θ_{eq}, θ_i	coverage ($\theta = c^s/c^{sat}$), the equilibrium value and the initial value;
ϕ	this is used as a superscript for μ^g and a^g to indicate that the properties take the value in equilibrium with the actual adsorbed phase and given by the Langmuir equation.

Latin symbols

a^g	gas activity;
a^{ads}	activity of the adsorbed phase;
c^s, c^{sat}	monolayer surface concentration and saturated equivalent (mol/m^2);
eq	the subscript *eq* indicates the properties under equilibrium;
ext	this superscript on $L_{\mu q}$ and $L_{\mu\mu}$ indicates the properties extending beyond close equilibrium;
J^a, J^d	unidirectional adsorption/desorption flux (mol/s/m^2);
J^{ads}	net flux at the limit of the adsorbed phase (mol/s/m^2);
J^g	net flux at the limit of the gas phase (mol/s/m^2);
J^{net}	net flux (mol/s/m^2) ($J^{net} = J^a - J^d$);
k_a, k_d	the unidirectional adsorption/desorption rate constant (mol/s/m^2), where the superscript L refers to the Langmuir kinetics;
K	the Langmuir constant;
K_{NI}	the non-isothermal Langmuir constant;
$L_{\mu q}$	the surface cross-coupling heat and mass transport coefficient (K.mol/s/m^2), where the superscripts *ads* and *g* refer to the limit of the adsorbed and gas phases, respectively;
$L_{\mu\mu}$	the surface mass transport coefficient (K.mol^2/(J.m^2.s), where the superscripts *ads* and *g* refer to the limit of the adsorbed and gas phases, respectively;
p	gas pressure (Pa);
p^0	standard pressure (10^5 Pa);
$R = 8.314$J/K/mol	a perfect gas constant;
$stat$	the subscript *stat* on θ, μ, or a refers to the properties in the stationary state under non-isothermal conditions;
t	time (s);
t_{ads}, t_{des}	adsorption/desorption characteristic time based on the exponential part of the kinetics (s), where the superscript L refers to the Langmuir kinetics;
T	temperature (K);
T^g, T^{ads}	temperatures at the limit of the gas/adsorbed phase (K).

References

1. Pareek, A.; Dom, R.; Gupta, J.; Chandran, J.; Adepu, V.; Borse, P.H. Insights into renewable hydrogen energy: Recent advances and prospects. *Mater. Sci. Energy Technol.* **2020**, *3*, 319–327. [CrossRef]
2. Le, T.T.; Sharma, P.; Bora, B.J.; Tran, V.D.; Truong, T.H.; Le, H.C.; Nguyen, P.Q.P. Fueling the future: A comprehensive review of hydrogen energy systems and their challenges. *Int. J. Hydrogen Energy* **2024**, *54*, 791–816. [CrossRef]
3. Hassan, I.; Ramadan, H.S.; Saleh, M.A.; Hissel, D. Hydrogen storage technologies for stationary and mobile applications: Review, analysis and perspectives. *Renew. Sustain. Energy Rev.* **2021**, *149*, 111311. [CrossRef]
4. Zgonnik, V. The occurrence and geoscience of natural hydrogen: A comprehensive review. *Earth-Sci. Rev.* **2020**, *203*, 103140. [CrossRef]
5. Al-Shafi, M.; Massarweh, O.; Abushaikha, A.S.; Bicer, Y. A review on underground gas storage systems: Natural gas, hydrogen and carbon sequestration. *Energy Rep.* **2023**, *9*, 6251–6266. [CrossRef]
6. Epelle, E.I.; Obande, W.; Udourioh, G.A.; Afolabi, I.C.; Desongu, K.S.; Orivri, U.; Gunes, B.; Okolie, J.A. Perspectives and prospects of underground hydrogen storage and natural hydrogen. *Sustain. Energy Fuels* **2022**, *6*, 3324–3343. [CrossRef]
7. Simon, J.M.; Haas, O.E.; Kjelstrup, S. Adsorption and desorption of H2 on graphite by molecular dynamics simulations. *J. Phys. Chem. C* **2010**, *114*, 10212–10220. [CrossRef]
8. van Erp, T.S.; Trinh, T.; Kjelstrup, S.; Glavatskiy, K.S. On the relation between the Langmuir and thermodynamic flux equations. *Front. Phys.* **2014**, *1*, 36. [CrossRef]
9. de Groot, S.R.; Mazur, P. *Non-Equilibrium Thermodynamics*; Dover Publications, Dover: New York, NY, USA, 1984.
10. Kjelstrup, S.; Bedeaux, D. *Non-Equilibrium Thermodynamics of Heterogeneous Systems*; Series on Advances in Statistical Mechanics; World Scientific Publishing Co. Pte, Ltd.: Singapore, 2008; Volume 16.
11. Inzoli, I.; Kjelstrup, S.; Bedeaux, D.; Simon, J.M. Transport coefficients of n-butane into and through the surface of silicalite-1 from non-equilibrium molecular dynamics study. *Microporous Mesoporous Mater.* **2009**, *125*, 112–125. [CrossRef]
12. Burress, J.W.; Kraus, M.; Beckner, M.; Cepel, R.; Suppes, G.; Wexler, C.; Pfeifer, P. Hydrogen storage in engineered carbon nanospaces. *Nanotechnology* **2009**, *20*, 204026. [CrossRef] [PubMed]
13. Inzoli, I.; Kjelstrup, S.; Bedeaux, D.; Simon, J. Transfer coefficients for the liquid–vapor interface of a two-component mixture. *Chem. Eng. Sci.* **2011**, *66*, 4533–4548. [CrossRef]
14. Xu, J.; Kjelstrup, S.; Bedeaux, D.; Røsjorde, A.; Rekvig, L. Verification of Onsager's reciprocal relations for evaporation and condensation using non-equilibrium molecular dynamics. *J. Colloid Interface Sci.* **2006**, *299*, 452–463. [CrossRef]
15. Røsjorde, A.; Kjelstrup, S.; Bedeaux, D.; Hafskjold, B. Nonequilibrium molecular dynamics simulations of steady-state heat and mass transport in condensation. II. Transfer coefficients. *J. Colloid Interface Sci.* **2001**, *240*, 355–364. [CrossRef] [PubMed]
16. Johannessen, E.; Bedeaux, D. Integral relations for the heat and mass transfer resistivities of the liquid–vapor interface. *Phys. A Stat. Mech. Its Appl.* **2006**, *370*, 258–274. [CrossRef]
17. Glavatskiy, K. Resistances for heat and mass transfer through a liquid—Vapor interface in a binary mixture. *J. Chem. Phys.* **2010**, *133*, 234501. [CrossRef] [PubMed]
18. Inzoli, I.; Simon, J.M.; Kjelstrup, S.; Bedeaux, D. Thermal effects during adsorption of n-butane on a silicalite-1 membrane: A non-equilibrium molecular dynamics study. *J. Colloid Interface Sci.* **2007**, *313*, 563–573. [CrossRef] [PubMed]

entropy

MDPI

Article

Kinetic Description of Viral Capsid Self-Assembly Using Mesoscopic Non-Equilibrium Thermodynamics

Jason Peña [1,*], Leonardo Dagdug [1] and David Reguera [2,3]

[1] Physics Department, Universidad Autónoma Metropolitana-Iztapalapa, Mexico City 09340, Mexico; dll@xanum.uam.mx
[2] Departament de Física de la Matèria Condensada, Universitat de Barcelona, Martí i Franquès 1, 08028 Barcelona, Spain; dreguera@ub.edu
[3] Universitat de Barcelona Institute of Complex Systems (UBICS), Martí i Franquès 1, 08028 Barcelona, Spain
* Correspondence: jason.pena@xanum.uam.mx

Abstract: The self-assembly mechanisms of various complex biological structures, including viral capsids and carboxysomes, have been theoretically studied through numerous kinetic models. However, most of these models focus on the equilibrium aspects of a simplified kinetic description in terms of a single reaction coordinate, typically the number of proteins in a growing aggregate, which is often insufficient to describe the size and shape of the resulting structure. In this article, we use mesoscopic non-equilibrium thermodynamics (MNET) to derive the equations governing the non-equilibrium kinetics of viral capsid formation. The resulting kinetic equation is a Fokker–Planck equation, which considers viral capsid self-assembly as a diffusive process in the space of the relevant reaction coordinates. We discuss in detail the case of the self-assembly of a spherical (icosahedral) capsid with a fixed radius, which corresponds to a single degree of freedom, and indicate how to extend this approach to the self-assembly of spherical capsids that exhibit radial fluctuations, as well as to tubular structures and systems with higher degrees of freedom. Finally, we indicate how these equations can be solved in terms of the equivalent Langevin equations and be used to determine the rate of formation and size distribution of closed capsids, opening the door to the better understanding and control of the self-assembly process.

Keywords: non-equilibrium thermodynamics; entropy production; kinetic equation; Fokker–Planck equation; self-assembly; viral capsid; nucleation; nucleation rate; physical virology

1. Introduction

Viruses have become widespread throughout the world. These tiny cellular-dependent agents are found to be integral components of essentially all ecosystems on the Earth [1,2]. They are believed to invade all life forms in the sea [3]; to assist microbes that contribute to global nutrient cycles [4,5]; and to exist in highly hostile environments [6,7], including terrestrial hot springs [8], alkaline lakes [9], and the hyper-saline Dead Sea [10]. These findings have led to the consideration of viruses as the most prolific biological entities [11–13], attracting the attention of researchers from various scientific fields to investigate their origins, evolution, and physicochemical properties.

In particular, for almost a century or beyond, viruses and virus-like particles have intrigued physicists due to their striking structures and physical properties [14–18], mechanisms of self-assembly [19–22], and strategies for the packaging and delivery of nucleic acids [23–26], among other aspects. Similarly, keen interest has arisen in the self-assembly of other spherical and tubular biological and synthetic shells [27–31], such as carboxysomes [32,33]. However, despite the increasing interest in the self-assembly of viruses and the numerous remarkable and insightful studies on this topic (see, for example, refs. [34–59]), many of the underlying physical mechanisms involved in the viral replication cycle remain poorly understood.

In essence, viruses are composed of a single or multiple chains of nucleic acids wrapped in a protein shell, also known as the capsid [60,61]. Their genomes encode a limited set of instructions that induce the synthesis of one or a few types of capsid proteins (CPs), which constitute the capsid. Additionally, viral genomes compel the host cell to produce copies of the virus's nucleic acids, which will be enclosed within the capsid, thereby enabling the new viral particles to hijack other susceptible cells [60].

The economy and efficiency of viral capsid self-assembly (VCSA) are fascinating and exceptional. Under appropriate conditions, a collection of CPs can aggregate to form a capsid with a structure identical to or very similar to that of native viruses, even in vitro and in the absence of genetic material [62–70]. Furthermore, numerous experimental studies show that viral capsids and other nanostructures exhibit polymorphism [50,62,66,71–75], which means that, under certain conditions, the resulting assembled structures manifest differences in size, shape, and symmetry. Polymorphism can sometimes be inferred from the system's energy surface, which may reveal multiple minima indicating the potential observation of different structural forms [76]. Scientists have successfully reproduced key features of this phenomenon using theoretical and computational approaches [50,76–86]. These include the formation of icosahedral capsids and other geometrical shapes [80], the necessity of well-defined metastable intermediates for self-assembly [82], the distribution of malformed capsids [85], and even control over the shapes of capsids using DNA origami nanostructures as binding and assembly platforms [86]. Nonetheless, there are limitations in the current understanding of the kinetic pathways involved in this process, their role in polymorphism, and the ongoing debate on whether viral capsid self-assembly and polymorphism, in some viruses, are driven by kinetics or thermodynamics [36,38,50,66,87]. More importantly, despite the usefulness of certain kinetic models, often based on the proposal and numerical solution of a large set of reaction equations, these models have significant drawbacks that severely limit their overall predictive capacity [88]. Such limitations may bias or underestimate the role of kinetics in the self-assembly process, as well as its influence on the size distribution of native-like assemblies and aberrant structures or their relative populations observed experimentally [89].

Classical nucleation theory of virus capsids (CNTVC) [87] stands out as an alternative approach that, by relying on the physical modeling of the free energy of capsid formation, overcomes some of these limitations [88] and provides a robust framework for both assembly and disassembly modeling [90]. In the standard description of viral assembly provided by CNTVC, the radius of a growing capsid is fixed and only the number of CPs, or more generally of capsid building blocks (CBBs), which can be clusters of CPs, is used as a single relevant reaction coordinate. However, the radius and aspect ratio of a growing shell are fluctuating variables that might lead to the formation of a final closed structure that is very different from the energetically optimal assembly. In addition, viral self-assembly is fundamentally an out-of-equilibrium process. Hence, the kinetic aspects are very important and require the formulation of proper equations to describe them. This highlights the necessity of a formalism that facilitates the study of the non-equilibrium kinetics of viral capsid formation.

In this context, mesoscopic non-equilibrium thermodynamics (MNET) stands out as the ideal framework to rigorously formulate kinetic equations to describe out-of-equilibrium processes at the nanoscale, where fluctuations are unavoidable.

In this article, we combine CNTVC and MNET to derive the equations governing the non-equilibrium kinetics of viral capsid formation. For this purpose, we first review some of the main features of VCSA, suggesting that nucleation is the underlying mechanism driving this phenomenon. Then, we summarize the fundamental ideas behind CNTVC and briefly discuss the resulting free energy landscape for the formation of spherical (icosahedral) and tubular (elongated) capsids. Thereafter, we provide a comprehensive overview of the general framework of MNET, where entropy production is key to describe the time evolution of an out-of-equilibrium system. Subsequently, we derive a general Fokker–Planck equation describing the kinetics of viral self-assembly with arbitrary degrees of freedom. We then describe in detail the case of the self-assembly of a spherical capsid with a fixed radius, which corresponds to a single degree of freedom, and indicate how to extend this approach to capsids with different architectures and energetic contributions and higher degrees of freedom. Following this, we explore the determination of the size distribution of closed capsids and the computation of steady-state nucleation rates as further applications of this alternative theoretical framework. Finally, we provide our conclusions and a future outlook on the application of MNET to describe VCSA.

2. Kinetic Aspects of Viral Capsid Self-Assembly

In the context of viral capsid formation, the study of the kinetics aims to describe how and how quickly closed capsids self-assemble. To develop a robust kinetic theory that successfully reproduces, at least qualitatively, the general aspects of VCSA, it is necessary to understand the actual behavior of this process. Numerous studies on capsid formation reveal two fundamental characteristics of its kinetics. Initially, there is a delay period during which no capsids are produced, even under conditions that promote self-assembly. Subsequently, the production of capsids increases rapidly, following a sigmoidal curve [77,87,91,92]. The disassembly of the capsid occurs under different physicochemical conditions than those for assembly, leading to a hysteresis phenomenon [93,94]. In other words, once capsids are formed and closed, they do not dissolve easily, making viral capsids resistant to hazardous conditions. Based on these characteristics, several kinetic models have been proposed to describe the self-assembly of viral capsids. They have been used to describe the assembly of viruses such as hepatitis B [66] and C [95], human papillomavirus (HPV) [73], and brome mosaic virus (BMV) [96], as well as to study the influence of nucleic acids and lipid bilayers on the assembly [97]. These contributions have shed light on the underlying mechanisms used by CBBs to assemble complex structures.

The motivation to develop an alternative theoretical framework for the kinetics of VCSA arises from the strong (and somewhat artificial) approximations and assumptions used in some kinetic models. For example, modeling the kinetics of viral capsids with a large set of master equations requires significant computational resources and accurate knowledge of all attachment and detachment rates of CBBs as input [88], but these are typically unknown and must therefore be inferred. Additionally, this approach relies on experimental data and highly specific input parameters for each modeled capsid. This complexity makes the theoretical framework difficult to generalize to situations where the radii of spherical and tubular capsids are not fixed [98], as well as to other macromolecules exhibiting self-assembly.

The kinetic theory that arises from the continuum description of VCSA given by CNTVC stands out as an alternative approach. By relying on the physical modeling of the free energy of capsid formation, this theory is able to overcome some of these limitations and provide closed solutions for (i) the critical size of the capsid, (ii) the nucleation barrier, and (iii) the steady-state nucleation rate [87]. Nonetheless, the standard kinetic description of CNTVC focuses on the formation of capsids with a fixed geometry and radius, using a single variable—typically, the number of CBBs in the growing capsid—as a reaction coordinate. This description may be valid as a first approximation, but, during the assembly of a virus, there might be fluctuations in the radius or shape of the growing shell, which can impact dramatically the final outcome of the process. The generalization of a master equation approach to incorporate these fluctuations involves a substantial amount of mathematics, and it becomes increasingly complex when modeling VCSA with multiple degrees of freedom or two or more types of CBBs [98].

In light of these limitations, MNET can provide kinetic equations to model out-of-equilibrium processes, thereby establishing a systematic approach to analyzing the non-equilibrium properties of the system based on the knowledge of its behavior at equilibrium. This aspect positions MNET as a promising theoretical framework for the description of the out-of-equilibrium characteristics of the self-assembly of not only tubular and spherical viral capsids but also carboxysomes and synthetically assembled structures.

Mesoscopic non-equilibrium thermodynamics relies on knowledge of the equilibrium free energy landscape of the out-of-equilibrium process that one aims to describe. Therefore, the energetic considerations involved in the context of viral capsid formation are discussed briefly in the next section.

3. Continuum Description of the Free Energy Landscape of Viral Capsid Self-Assembly

The primary hypothesis of the continuum description of VCSA is that nucleation is the underlying mechanism for the aggregation of CBBs into closed protein shells [87]. Nucleation is an activated process that considers self-assembly as the overcoming of a free energy barrier to capsid formation. In the original formulation of CNTVC, the self-assembly of a spherical capsid is modeled as the growth of a partial shell that can increase or decrease in area due to the attachment or detachment of individual CBBs. An intermediate aggregate is represented as a spherical cap with fixed radius R_T equal to the final radius of the closed capsid. The growth of this spherical shell is promoted by the energy gain $\Delta\mu$ due to the aggregation of CBBs and hindered by an energy penalty due to the missing contacts at the edge of the growing shell (see Figure 1). The resulting free energy of the self-assembly of empty spherical capsids, ΔG, is given by [87]

$$\Delta G = n\Delta\mu + \tau\, l(n), \tag{1}$$

where n is the instantaneous number of CBBs in a partial capsid, τ is the line tension (analogous to the surface tension of a liquid droplet in homogeneous condensation), and $l(n)$ is the length of the rim. Taking into account the geometry of a spherical cap, the previous expression can be rewritten as [45,87]

$$\Delta G(n) = n\Delta\mu + \lambda\sqrt{n(n_T - n)}, \tag{2}$$

in which n_T is the total number of CBBs in the closed capsid, and $\lambda = 4\pi\tau R_T/n_T$ is an edge energy constant related to the line tension, τ. The competition between these opposite contributions leads to the existence of a nucleation barrier, which has to be surmounted in order to assemble a fully formed capsid.

In Figure 1, we show a schematic representation of a partial capsid having n CBBs (top left corner) and a representative energy landscape for $\Delta\mu = -0.5\,k_BT$ and $\lambda \approx 2.15\,k_BT$, where three important stages of VCSA are highlighted. According to Equation (2), capsid formation starts at $(n = 0, \Delta G = 0)$ (yellow point), representing the situation where all CBBs are free in the solution. As the CBBs start to aggregate due to the thermal fluctuations of the bath, the free energy of the incipient shell increases, indicating that partial capsids with a small number of constituents are unstable assemblies that will tend to dissolve. Only when having a cluster with a critical size n^* does the system reach an unstable state (cyan point) where any perturbation can cause the partial capsid to dissolve or continue assembling until it closes. If the nucleation barrier, ΔG^*, is surmounted, the aggregate will tend to progressively grow, ultimately forming a closed spherical capsid of size n_T (red point).

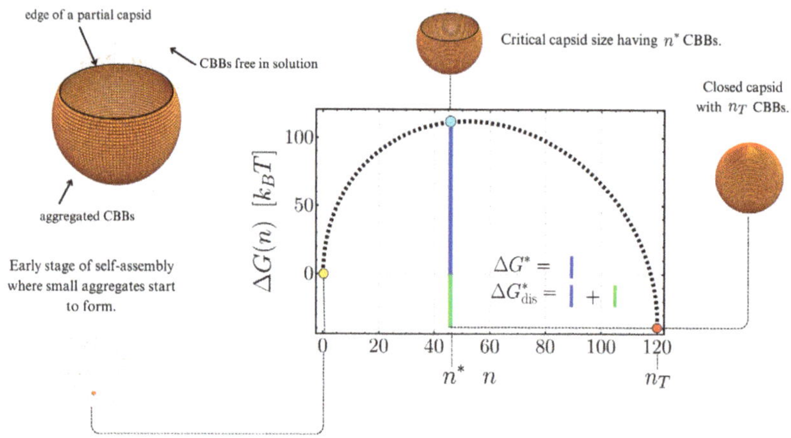

Figure 1. Schematic representation of a partial capsid having n CBBs (top left corner) and the free energy landscape generated by $\Delta G(n)$ in Equation (2) as a function of the number of CBBs, n. The self-assembly of viral capsids is described from left to right, while the disassembly of closed capsids occurs from right to left. The nucleation barrier towards assembly, ΔG^*, differs from the nucleation barrier driving the disassembly of closed capsids, ΔG^*_{dis}, reflecting the system's hysteresis.

This continuum description of VCSA qualitatively aligns with theoretical, experimental, and computational findings regarding VCSA [87], including the existence of a critical capsid size, the lack of a significant population of intermediate capsids, and the presence of hysteresis in the system. The latter inference arises from the fact that, for $\Delta\mu < 0$, $\Delta G^* \neq \Delta G^*_{\text{dis}}$, with ΔG^*_{dis} being the energetic barrier to be surmounted for disassembly, as shown in Figure 1.

Extensions of Classical Nucleation Theory of Viral Capsids Incorporating Elastic Contributions

The standard classical nucleation theory (CNT) description of viral assembly, as shown in Equation (2), only considers the binding energy gain and line tension costs involved in the formation of spherical capsids with a predetermined radius. This picture neglects an important ingredient: the elastic energy arising from the bending and stretching associated with the growth of a curved shell. Incorporating these contributions leads to the following free energy of formation [99–101]:

$$\Delta G = n\Delta\mu + \tau\, l(n) + G_{\text{ela}}, \tag{3}$$

where $G_{\text{ela}} = G_{\text{str}} + G_{\text{bend}}$ is the elastic energy associated with both stretching, G_{str}, and bending, G_{bend}, energies introduced by the curvature of the capsid. The mathematical expressions for the stretching and bending contributions within G_{ela} depend on the specific geometry of the growing aggregate but are generally modeled using thin shell theory (TST) [102–104] and the Helfrich model [105], respectively.

Several studies have explored the elastic properties of viral capsids and complex macromolecules along this line [42,102,103,106–113]. The incorporation of these terms not only modifies the values of the nucleation barrier to surmount but also might alter the radius of the shell and even the preferred geometry of the growing capsid. The competition between the line tension, the difference in chemical potential, and the relative importance of bending and stretching contributions, quantified by the Föppl–von Kármán (FvK) number, γ_{FvK}, can result in the formation of spherical caps but also ribbons and cylindrical patches [99].

At the limit where bending dominates and stretching can be neglected, the free energy of formation of spherical capsids becomes [100,101]

$$\Delta G(n,R) = \lambda\, n_T \left\{ n_r \left[f_r \left(\frac{1}{R_r} - 1 \right)^2 - \Gamma \right] + \sqrt{n_r \left(1 - \frac{n_r}{R_r^2} \right)} \right\}. \tag{4}$$

The derivation of the latter equation is described in detail in Ref. [101] and involves the use of scaled variables and parameters. Specifically, in this formulation, $n_r \equiv n/n_T$ and $R_r \equiv R/R_T$, where R is the instantaneous radius of the capsid, whereas $\Gamma \equiv -\Delta\mu/\lambda$ and $f_r \equiv 2\pi\kappa/\lambda n_T$ are the scaled supersaturation and the relative bending constant, respectively. The new parameters introduced in this extension of CNTVC include κ, which is the bending modulus of the capsid, and n_T and R_T, representing the target number of CBBs and the preferred curvature radius of a capsid, both dictated by the preferred angle of interaction between individual CBBs [101]. In the latter equation, ΔG, depends on both n and R as the model allows changes in the curvature radius of a partial capsid as it progressively grows.

At the limit dominated by stretching energy, i.e., $\gamma_{\mathrm{FvK}} >> 1$, only tubes are formed as these structures can freely grow without any stretching energy penalty. If the line tension is also large, the cylindrical tubes will form from curved square or circular patches, which are the shells minimizing the contour length. In this case, the free energy of formation is found to be [99,114]

$$\Delta G(n,R) = n\Delta\mu + \tau\, l(A) + \frac{1}{2}\frac{\kappa A}{R_T^2}\left[\left(\frac{1}{R_r} - 1 \right)^2 + 1 \right], \tag{5}$$

where $A = na_1$ represents the area of the patch composed of n CBBs, each with nominal area a_1. This free energy landscape for tube formation also depends on two relevant variables: the radius of the growing tube and the number of CBBs. However, if the line tension is small, these tubes will form, in general, from rectangular (or elliptical) curved patches described by three variables: the number of CBBs, n; the radius of the growing tube, R; and the aspect ratio of the patch, f. In terms of these parameters, the free energy of formation of a tube from of a rectangular patch can be written as [114]

$$\Delta G_{\mathrm{rect}}(n,R,f) = n\Delta\mu + \Lambda \frac{(1+f)}{\sqrt{f}} \sqrt{n} + \frac{1}{2}\tilde{\kappa} n \left[\left(\frac{1}{R_r} - 1 \right)^2 + 1 \right], \tag{6}$$

where $\Lambda \equiv 2\tau\sqrt{a_1}$ is a line energy constant and $\tilde{\kappa} \equiv \kappa a_1/R_T^2$ is a bending stiffness constant, which are redefined for convenience. The last term in Equation (6) represents the bending energy cost and has been obtained using the continuum elasticity model introduced by Castelnovo [113], which is a generalization of Helfrich's model [105] for systems with non-zero spontaneous curvature.

All free energies presented above represent extensions of CNTVC that explore different aspects of the self-assembly of spherical (icosahedral) and cylindrical (tubular) assemblies. More specifically, they provide valuable insights for the study of the equilibrium aspects of the self-assembly process. In particular, at equilibrium, the probability density of finding a partial shell with a given size is given by the standard statistical mechanics expression [88,98,115]

$$P_{\text{eq}} \sim \exp\left(\frac{-\Delta G}{k_B T}\right), \tag{7}$$

which, naturally, is a Boltzmann factor. This equation is a key component that will allow us to derive the entropy production of the system and, subsequently, the equations describing the non-equilibrium kinetics of VCSA.

The presence of additional degrees of freedom in the free energy landscape (as described in Equations (4)–(6)) has important consequences for the energetics of capsid formation and the definition of the optimal size and radius of the resulting spherical or tubular structure. These energetic considerations have been analyzed in Refs. [100,101,114]. However, since self-assembly is a dynamic process, these additional degrees of freedom do not only impact the energetics of the capsid. The rate at which a shell changes its radius or aspect ratio can cause the pathway of assembly to deviate from the optimal one (i.e., the minimum free energy path), yielding a distribution of sizes that may be quite different from the expected optimal aggregate. Thus, it is very important to describe the multidimensional kinetics of viral assembly when other degrees of freedom become relevant. This is accomplished using MNET.

In the next section, we briefly review the general framework of MNET [116]. Then, we particularize it to describe the behavior and evolution of the out-of-equilibrium mechanisms driving VCSA.

4. General Framework for Mesoscopic Non-Equilibrium Thermodynamics

Entropy as the Key to the Time Evolution of an Out-of-Equilibrium System

MNET is a theoretical framework that extends non-equilibrium thermodynamics to systems that exist between the microscopic and macroscopic scales, where fluctuations are significant and non-linear effects dominate [116]. Unlike traditional approaches, which are limited to macroscopic systems and linear responses, MNET incorporates a probabilistic interpretation of densities, conservation laws, and entropy principles to describe a broader range of non-equilibrium phenomena. In this context, MNET extends traditional thermodynamic concepts to systematically analyze the stochastic behavior of small systems. Readers interested in the foundations of MNET are referred to [116].

In a nutshell, the main asset of MNET lies in acknowledging that entropy production is key to describing the evolution of a system out of equilibrium. To truly understand this core component and its implications in kinetics, consider a system with i degrees of freedom, $\{\gamma_i\} = \{\gamma_1, \gamma_2, \cdots, \gamma_i\}$, representing any relevant set of coordinates or mesoscopic degrees of freedom that define the state of the system. For example, $i = 1$ and $\gamma_1 = n$ for the standard one-dimensional description of CNTVC (see Equation (2)) or $i = 2$ and $\{\gamma_i\} = \{n, R\}$ if we are characterizing the self-assembly kinetics of a protein shell whose number of constituents and curvature radius vary with thermal fluctuations (as in Equation (4) or Equation (5)). Given that we are examining a statistical system, we must work with probabilities. The probability density in terms of finding the system in the state $\{\gamma, \gamma + \delta\gamma\}$ at time t is $P(\gamma, t)$. The entropy of the system in terms of this probability is expressed by the Gibbs entropy postulate, namely [117]

$$S = S_{eq} - k_B \int P(\gamma, t) \ln\left[\frac{P(\gamma, t)}{P_{\text{eq}(\gamma)}}\right] d\gamma, \tag{8}$$

where S_{eq} is the entropy of the system when the degrees of freedom $\{\gamma\}$ are at equilibrium. The second term of the latter equation considers entropy contributions that arise when the $\{\gamma\}$ variables are out of equilibrium, which, as a consequence, produce a deviation in the probability density $P(\gamma,t)$ from its equilibrium value, $P_{\text{eq}}(\gamma)$. The equilibrium probability density is typically constructed to fulfill the following condition [118]:

$$P_{\text{eq}} \sim \exp\left[\frac{-\Delta W(\gamma)}{k_B T}\right]. \tag{9}$$

This is a relation in which $\Delta W(\gamma)$ is the reversible work or availability required to achieve the particular state defined by γ; k_B is Boltzmann's constant; and T is the temperature. Variations in Equation (8) lead to

$$\Delta S = -k_B \int \Delta P(\gamma,t) \ln\left[\frac{P(\gamma,t)}{P_{\text{eq}}(\gamma)}\right] \mathrm{d}\gamma. \tag{10}$$

Since the probability is a conserved normalized quantity, the probability density in the $\gamma-$space is governed by the continuity equation, allowing us to write

$$\frac{\partial P(\gamma,t)}{\partial t} = -\frac{\partial J(\gamma,t)}{\partial \gamma}, \tag{11}$$

with $J(\gamma,t)$ being the density flux, a function that will be specified later. By taking the time derivative of Equation (10) and using Equation (11), we obtain

$$\frac{\mathrm{d}S}{\mathrm{d}t} = k_B \left\{ \ln\left[\frac{P(\gamma,t)}{P_{\text{eq}}(\gamma)}\right] J(\gamma,t) \right\}_{\text{eval}} - k_B \int J(\gamma,t) \frac{\partial}{\partial \gamma}\left\{ \ln\left[\frac{P(\gamma,t)}{P_{\text{eq}(\gamma)}}\right] \right\} \mathrm{d}\gamma, \tag{12}$$

after integrating by parts. An alternative representation of the latter equation reads

$$\frac{\mathrm{d}S}{\mathrm{d}t} = -\int \frac{\partial}{\partial \gamma} J_s \, \mathrm{d}\gamma + \sigma, \tag{13}$$

where we have defined the entropy flux and entropy production as

$$J_s \equiv k_B \, J(\gamma,t) \ln\left[\frac{P(\gamma,t)}{P_{\text{eq}}(\gamma)}\right], \tag{14}$$

and

$$\sigma \equiv -k_B \int J(\gamma,t) \frac{\partial}{\partial \gamma}\left\{ \ln\left[\frac{P(\gamma,t)}{P_{\text{eq}}(\gamma)}\right] \right\} \mathrm{d}\gamma, \tag{15}$$

respectively. As in standard non-equilibrium thermodynamics (NET) [117,119], the entropy production is the product of a flux multiplied by a thermodynamic force. In this case, the thermodynamic force, in terms of the mesoscopic variables, is identified as the gradient of the factor $\ln\left[P(\gamma,t)/P_{\text{eq}(\gamma)}\right]$. Following the tenets of NET, we assume a linear dependence between fluxes and forces, which leads us to

$$J(\gamma,t) = -k_B T \, \mathcal{L}(\gamma, P(\gamma)) \frac{\partial}{\partial \gamma}\left\{ \ln\left[\frac{P(\gamma,t)}{P(\gamma)}\right] \right\}, \tag{16}$$

in which $\mathcal{L}(\gamma, P(\gamma))$ is an Osanger coefficient that may depend, in general, on the probability density of finding the system at a certain state γ, $P(\gamma)$.

By substituting Equation (16) into Equation (11) and defining a generalized diffusion coefficient $D \equiv [k_B \mathcal{L}(\gamma, P(\gamma))]/P(\gamma,t)$, we find that

$$\frac{\partial P(\gamma,t)}{\partial t} = -\frac{\partial}{\partial \gamma}\left\{-k_B\,\mathcal{L}(\gamma,P(\gamma))\frac{\partial}{\partial \gamma}\ln\left[\frac{P(\gamma,t)}{P_{\text{eq}}(\gamma)}\right]\right\}$$
$$= \frac{\partial}{\partial \gamma}\left[D\,P_{\text{eq}}(\gamma)\frac{\partial}{\partial \gamma}\frac{P(\gamma,t)}{P_{\text{eq}}(\gamma)}\right], \tag{17}$$

which is a diffusion equation. From this relation, we can obtain the equations governing the non-equilibrium kinetics of a system from its behavior at equilibrium. Furthermore, by implementing the equilibrium statistical description of the process into the kinetic non-equilibrium mathematical framework, i.e., Equation (9) into Equation (17), we obtain

$$\frac{\partial P}{\partial t} = \frac{\partial}{\partial \gamma}\left[D\frac{\partial}{\partial \gamma}P(\gamma,t) + \frac{DP(\gamma,t)}{k_B T}\frac{\partial}{\partial \gamma}\Delta W(\gamma)\right], \tag{18}$$

which is the Fokker–Planck equation describing the evolution of the probability density $P(\gamma,t)$, i.e., a kinetic equation describing how $P(\gamma,t)$ changes in time and in the γ phase space.

Under the standard conditions for assembly, consisting of a constant pressure and temperature, the proper thermodynamic potential is the Gibbs free energy, namely $\Delta W \equiv \Delta G = \Delta H - T\Delta S$, where ΔH and ΔS are the changes in enthalpy and entropy, respectively. The resulting equation is a Fokker–Planck equation for a system diffusing over the free energy landscape defined by ΔG. In the context of VCSA, the landscape features an unstable or saddle point state corresponding to the nucleation barrier that must be surmounted to self-assemble a tubular or spherical virus. Implementing the proper free energy of formation for spherical (icosahedral) capsids, i.e., Equation (2) or Equation (4), or for tubular (elongated) capsids, given by Equation (5) or Equation (6), in the general kinetic equation, i.e., Equation (18), leads to specific kinetic equations that can be used to compare the outcomes and rates of formation of different capsid architectures.

Given the structure of Equation (18), MNET considers the evolution of a system in time to be a generalized diffusion process in the space of the mesoscopic variables $\{\gamma\}$ over an energy landscape generated by the equilibrium free energy, ΔG. The MNET framework used to model out-of-equilibrium kinetics has been successfully applied to describe many non-equilibrium phenomena, including diffusion in the presence of entropic barriers [120], activated processes [121–123], molecular motors [124], and the translocation of stiff chains [125].

In the following section, we particularize this general framework to describe the kinetics of VCSA in spherical and tubular viruses.

5. Kinetics of Viral Capsid Self-Assembly from MNET

Once the theoretical framework is outlined, the application to the description of the non-equilibrium aspects of the self-assembly of spherical and tubular shells is straightforward. As an illustrative example, in this section, we discuss the case of the self-assembly of empty spherical capsids with one significant mesoscopic variable, whose free energy is given by Equation (2). Additionally, we extend these ideas to a system with two degrees of freedom; finally, we provide guidance on how to work with more complex systems.

5.1. The Case of One Degree of Freedom: The Assembly of a Spherical Capsid with a Fixed Radius

In the continuum description of the self-assembly of spherical capsids of fixed radius, whose free energy of formation is given by Equation (2), there is a single mesoscopic variable: the instantaneous number of CBBs n. Consequently, particularizing the Fokker–Planck Equation (18) to this system, the corresponding kinetic equation is

$$\frac{\partial P(n,t)}{\partial t} = \frac{\partial}{\partial n}\left[D_n \frac{\partial P(n,t)}{\partial n} + \frac{D_n}{k_B T}\frac{\partial \Delta G}{\partial n} P(n,t)\right], \tag{19}$$

where the diffusion coefficient D_n is related to the rate of attachment of CBBs to a partial capsid, and $\Delta G = n\Delta\mu + \lambda\sqrt{n(n_T - n)}$. The numerical solution of Equation (19) with proper initial and boundary conditions provides valuable insights into the time evolution of the distribution of cluster sizes and the rate of formation of closed capsids for different values of $\Delta\mu$ and λ.

Furthermore, the Fokker–Planck equation, Equation (19), is equivalent to a Langevin equation of the form

$$\frac{\partial n}{\partial t} = -\frac{D_n(n)}{k_B T}\frac{\partial \Delta G}{\partial n} + \frac{\partial D_n(n)}{\partial n} + \zeta_n, \tag{20}$$

where ζ_n represents white Gaussian noise in size, satisfying $\langle \zeta_n(t) \rangle = 0$, $\langle \zeta_n(t)\zeta_n(t') \rangle = 2D_n(n)\delta(t-t')$, and the second term on the right-hand side represents the noise-induced drift term arising in the general situation where the diffusion coefficient $D_n(n)$, which is ultimately the attachment rate, depends on the size of the shell. Equation (20) can be numerically solved using a simple stochastic Euler algorithm [126]. This perspective allows VCSA to be interpreted as a diffusive process along the free energy landscape, where the system explores different physical states. In this framework, the driving force for self-assembly is given by $F = -\partial\Delta G/\partial n$, while Gaussian noise represents thermal fluctuations that facilitate the overcoming of the nucleation barrier and the exploration of energetically unfavorable states. This analogy holds for each dimension in a system with i degrees of freedom.

Finally, the kinetic equation given by Equation (19) provides an alternative method of deriving the rate of capsid formation at a steady state. Starting from Equation (17), Equation (19) leads to a flux of the form

$$J_{\rm nr} = -\left[D_n\, P_{\rm eq}(n) \frac{\partial}{\partial n}\frac{P(n,t)}{P_{\rm eq}(n)}\right], \tag{21}$$

a relation that may be rewritten as

$$\frac{J_{nr}(n)}{D_n\, P_{\rm eq}(n)} = -\frac{\partial}{\partial n}\frac{P(n,t)}{P_{\rm eq}(n)}. \tag{22}$$

By integrating by parts, we find that

$$\int_{n_0}^{n_f} \frac{J_{nr}(n)}{D_n\, P_{\rm eq}(n)} = -\left.\frac{P(n)}{P_{\rm eq}(n)}\right|_{n_0}^{n_f}. \tag{23}$$

At the steady state, J_{nr} is independent of the number of CBBs. If this is the case, the latter equation simplifies to

$$J_{nr}\int_{n_0}^{n_f} \frac{{\rm d}n}{D_n\, P_{\rm eq}(n)} = -\left.\frac{P(n)}{P_{\rm eq}(n)}\right|_{n_0}^{n_f}. \tag{24}$$

Now, the theoretical framework of VCSA relies on the use of concentrations instead of probability distributions; to this end, it is important to note that

$$C(n,t) = N\, P(n,t), \tag{25}$$

with N being the total number of clusters in the system, assumed to be constant. Thus,

$$J_{\text{nr}} \int_{n_0}^{n_f} \frac{\mathrm{d}n}{D_n\, C_{\text{eq}}(n)} = -\frac{C(n)}{C_{\text{eq}}}\bigg|_{n_0}^{n_f}, \tag{26}$$

where $C(n)$ is the instantaneous concentration of aggregates of size n; $C_{\text{eq}}(n)$ is the concentration of these aggregates at equilibrium; $n_0 = 0$ represents a cluster of size 0, namely free CBBs in solution; and $n_f = n_T$, where n_T represents the aggregation number. Then,

$$J_{\text{nr}} \int_{0}^{n_T} \frac{\mathrm{d}n}{D_n\, C_{\text{eq}}(n)} = \frac{C(n=0)}{C_{\text{eq}}(n=0)} - \frac{C(n_T)}{C_{\text{eq}}(n_T)}. \tag{27}$$

The boundary conditions for the self-assembly of viral capsids are a fixed concentration of CBBs equal to the equilibrium concentration and a null concentration of fully formed capsids, which mathematically are translated into $C(n = 0)/C_{\text{eq}(n=0)} \approx 1$ and $C(n_T)/C_{\text{eq}(n_T)} \approx 0$, respectively. Thus,

$$J_{\text{nr}} = \left[\int_{0}^{n_T} \frac{\mathrm{d}n}{D_n\, C_{\text{eq}}(n)}\right]^{-1}. \tag{28}$$

Equation (7) can be used to express the relationship between the equilibrium concentration of aggregates of size n, $C_{\text{eq}(n)}$ in terms of the free energy of self-assembly, $\Delta G(n)$ [98,115]. The resulting equation is $C_{\text{eq}}(n) = C_s \exp\left[-\beta \Delta G(n)\right]$, in which C_s is the concentration of CBBs at the reference state of supersaturation. Therefore,

$$J_{\text{nr}} = C_s \left\{\int_{0}^{n_T} \frac{\exp[\beta \Delta G(n)]}{D_n}\right\}. \tag{29}$$

Finally, by using the steepest descent approach within the context of Becker and Döring kinetics [87,127,128], we ultimately find that

$$J = C_s D_n^* Z \exp(-\beta \Delta G^*), \tag{30}$$

where

$$Z \equiv \sqrt{\frac{W^*}{2\pi k_B T}}, \quad \text{and} \quad W^* = -\left(\frac{\partial^2 \Delta G(n)}{\partial n^2}\right)_{n=n^*}. \tag{31}$$

The latter equation is identical to Equation (15) of Ref. [87]. Thus, MNET offers a direct approach to deducing the nucleation rate of an out-of-equilibrium system.

5.2. Two Degrees of Freedom

Let us now consider the self-assembly of an empty spherical capsid in the bending-dominated case, for which the free energy of formation is given by Equation (4). The relevant mesoscopic variables are $\{n, R\}$ as the number of CBBs and the curvature radius of the spherical cap vary during capsid formation. The out-of-equilibrium kinetic equation describing this system is

$$\begin{aligned} \frac{\partial P(n,R,t)}{\partial t} =& \frac{\partial}{\partial n}\left[D_n \frac{\partial P(n,R,t)}{\partial n} + \frac{D_n}{k_B T}\frac{\partial \Delta G}{\partial n} P(n,R,t)\right] \\ &+ \frac{\partial}{\partial R}\left[D_R \frac{\partial P(n,R,t)}{\partial R} + \frac{D_R}{k_B T}\frac{\partial \Delta G}{\partial R} P(n,R,t)\right], \end{aligned} \tag{32}$$

where we have considered that there are no cross-effects. The latter equation is equivalent to the following set of Langevin equations:

$$\frac{\partial n}{\partial t} = -\frac{D_n}{k_B T}\frac{\partial \Delta G}{\partial n} + \frac{\partial D_n}{\partial n} + \zeta_n, \tag{33}$$

$$\frac{\partial R}{\partial t} = -\frac{D_R}{k_B T}\frac{\partial \Delta G}{\partial R} + \frac{\partial D_R}{\partial R} + \zeta_R, \tag{34}$$

where ζ_n and ζ_R represent white Gaussian noise in size and radius, respectively, satisfying $\langle \zeta_i(t) \rangle = 0$, $\langle \zeta_i(t)\zeta_j(t') \rangle = 2D_i\delta_{ij}\delta(t-t')$. The second term on the right-hand side of the latter equations represents the noise-induced drift term considering the general situation where the diffusion coefficients D_n and D_R may depend on both the size and the radius of the growing shell. As in the one-dimensional case, D_n and D_R are related to the rate of subunit attachment and the rate of radial fluctuation, respectively. The kinetic equation describing the formation of cylindrical assemblies from a squared patch, or any system with two mesoscopic degrees of freedom, is identical to Equation (32) and leads to the same set of Langevin equations that can be solved with a simple stochastic Euler algorithm. The appropriate free energy of self-assembly for tubular assemblies in the limit of high line tension is given by Equation (5).

5.3. Three Degrees of Freedom: The Assembly of Tubes with Arbitrary Line Tension

For the general case of an arbitrary value of the line tension, the aspect ratio f of the growing tubular shell also becomes a fluctuating variable. In this case, the free energy landscape of formation of a cylindrical assembly is given by Equation (6), and the corresponding equation that describes the kinetics of the assembly becomes

$$\begin{aligned}\frac{\partial P(n,R,f,t)}{\partial t} =& \frac{\partial}{\partial n}\left[D_n \frac{\partial P(n,R,f,t)}{\partial n} + \frac{D_n}{k_B T}\frac{\partial \Delta G}{\partial n} P(n,R,f,t)\right] \\ &+ \frac{\partial}{\partial R}\left[D_R \frac{\partial P(n,R,f,t)}{\partial R} + \frac{D_R}{k_B T}\frac{\partial \Delta G}{\partial R} P(n,R,f,t)\right] \\ &+ \frac{\partial}{\partial f}\left[D_f \frac{\partial P(n,R,f,t)}{\partial f} + \frac{D_f}{k_B T}\frac{\partial \Delta G}{\partial f} P(n,R,f,t)\right],\end{aligned} \tag{35}$$

which is equivalent to the following set of Langevin equations:

$$\frac{\partial n}{\partial t} = -\frac{D_n}{k_B T}\frac{\partial \Delta G}{\partial n} + \frac{\partial D_n}{\partial n} + \zeta_n, \tag{36}$$

$$\frac{\partial R}{\partial t} = -\frac{D_R}{k_B T}\frac{\partial \Delta G}{\partial R} + \frac{\partial D_R}{\partial R} + \zeta_R, \tag{37}$$

$$\frac{\partial f}{\partial t} = -\frac{D_f}{k_B T}\frac{\partial \Delta G}{\partial f} + \frac{\partial D_f}{\partial f} + \zeta_f, \tag{38}$$

where ζ_n, ζ_R, and ζ_f represent white Gaussian noise in size, radius, and aspect ratio, respectively, satisfying $\langle \zeta_i(t) \rangle = 0$, $\langle \zeta_i(t)\zeta_j(t') \rangle = 2D_i\delta_{ij}\delta(t-t')$.

In all of these cases, the numerical solution of either the Fokker–Planck equation or the equivalent set of Langevin equations provides the probability density of finding the system in a specific state at time t. When evaluated at an appropriate time, it reveals the size distribution of intermediate aggregates and fully formed capsids. This facilitates the tuning of experimental and computational parameters, such as the concentration of CBBs and the binding energy between them, to explore the possibility of achieving a very efficient assembly of structures of the desired size or to induce the formation of incorrect sizes as a potential strategy to hamper viral replication. These solutions also provide insights into the preferred self-assembly pathways, the stability of intermediate aggregates, and the polydispersity and distribution of capsid sizes formed under given physicochemical conditions. Additionally, the out-of-equilibrium kinetic equation enables the direct determination of the transient and steady-state nucleation rates. These features highlight the advantages, potential applications, and universality of the MNET approach in the context of the self-assembly of viral capsids and other biological structures.

6. Conclusions

We have seen how an appropriate evaluation of entropy production is the key to describing the time evolution of an out-of-equilibrium system. MNET uses this fundamental principle to derive kinetic equations for non-equilibrium systems at small scales where fluctuations are unavoidable. In this work, we have particularized this formalism to describe VCSA. The application of MNET enables the derivation of a universal kinetic framework for VCSA based on the knowledge of the free energy of capsid formation provided by the CNTVC. This approach leads to a general Fokker–Planck equation considering capsid formation as a diffusive process in the space of mesoscopic reaction coordinates. The system's free energy dictates the form of the general kinetic equation, Equation (18), ultimately specifying the corresponding kinetic equations for different capsid architectures. This theoretical framework offers key advantages. It allows for the calculation of the size distribution of fully formed capsids, which, due to the out-of-equilibrium kinetics, may differ significantly from the energetically most stable capsid or the one obtained following the minimum energy path. It also provides a time-dependent description of the distribution of intermediates and the rate at which viral capsids form. Furthermore, this technique offers opportunities to explain interesting phenomena in the theoretical modeling of VCSA, including the formation of capsids of suboptimal size [45,100,101] and the hypothesis that CBBs may have optimal bending angles distinct from those observed in native viruses [101]. This framework could deepen our understanding of VCSA and enable innovative applications, such as antiviral strategies based on kinetic considerations or by inducing misassembly and the design of virus-like particles with efficient and controlled assembly for bio- and nanotechnological purposes. These ideas can also be extended to describe the kinetics of formation of other nanoscale structures.

Author Contributions: Conceptualization, J.P., L.D. and D.R.; Formal analysis, J.P., L.D. and D.R.; Investigation, J.P., L.D. and D.R.; Methodology, J.P., L.D. and D.R.; Writing—original draft, J.P., L.D. and D.R.; Writing—review and editing, J.P., L.D. and D.R. All authors have read and agreed to the published version of the manuscript.

Funding: We acknowledge fruitful discussions with C.I. Mendoza. J.P. acknowledges funding from the Metropolitan Autonomous University in Mexico for a research stay at the University of Barcelona, Spain. D.R. acknowledges funding from the Spanish government through Grant Nos. PID2021-126570NB-I00 (MINECO/FEDER, UE) and RED2022-134654-T.

Institutional Review Board Statement: Not applicable.

Data Availability Statement: No new data were created or analyzed in this study. The original contributions presented in the study are included in the article.

Acknowledgments: This paper is dedicated to Signe Kjelstrup on the occasion of her 75th birthday. It has been a pleasure to know her for a significant fraction of these years. Her kindness, scholarship, and enthusiasm have been an inspiration to us and a driving force for the development of non-equilibrium thermodynamics.

Conflicts of Interest: The authors declare no conflict of interest.

References

1. Zhang, Y.-Z.; Shi, M.; Holmes, E.C. Using Metagenomics to Characterize an Expanding Virosphere. *Cell* **2018**, *172*, 1168–1172. [CrossRef] [PubMed]
2. French, R.K.; Holmes, E.C. An Ecosystems Perspective on Virus Evolution and Emergence. *Trends Microbiol.* **2020**, *28*, 165–175. [CrossRef] [PubMed]
3. Suttle, C.A. Viruses in the sea. *Nature* **2005**, *437*, 356–361. [CrossRef] [PubMed]
4. Pearce, D.A.; Wilson, W.H. Viruses in Antarctic ecosystems. *Antarct. Sci.* **2003**, *15*, 319–331. [CrossRef]
5. Lopez-Simon, J.; Vila-Nistal, M.; Rosenova, A.; De Corte, D.; Baltar, F.; Martinez-Garcia, M. Viruses under the Antarctic Ice Shelf are active and potentially involved in global nutrient cycles. *Nat. Commun.* **2023**, *14*, 8295. [CrossRef]
6. Le Romancer, M.; Gaillard, M.; Geslin, C.; Prieur, D. Viruses in extreme environments. *Rev. Environ. Sci. Biotechnol.* **2007**, *6*, 17–31. [CrossRef]
7. Gil, J.F.; Mesa, V.; Estrada-Ortiz, N.; Lopez-Obando, M.; Gómez, A.; Plácido, J. Viruses in Extreme Environments, Current Overview, and Biotechnological Potential. *Viruses* **2021**, *13*, 81. [CrossRef]
8. Rachel, R.; Bettstetter, M.; Hedlund, B.P.; Häring, M.; Kessler, A.; Stetter, K.O.; Prangishvili, D. Remarkable morphological diversity of viruses and virus-like particles in hot terrestrial environments. *Arch. Virol.* **2002**, *147*, 2419–2429. [CrossRef]
9. Jiang, S.; Steward, G.; Jellison, R.; Chu, W.; Choi, S. Abundance, Distribution, and Diversity of Viruses in Alkaline, Hypersaline Mono Lake, California. *Microb. Ecol.* **2004**, *47*, 9–17. [CrossRef]
10. Oren, A.; Bratbak, G.; Heldal, M. Occurrence of virus-like particles in the Dead Sea. *Extremophiles* **1997**, *1*, 143–149. [CrossRef]
11. Breitbart, M.; Rohwer, F. Here a virus, there a virus, everywhere the same virus? *Trends Microbiol.* **2005**, *13*, 278–284. [CrossRef] [PubMed]
12. Edwards, R.A.; Rohwer, F. Viral metagenomics. *Nat. Rev. Microbiol.* **2005**, *3*, 504–510. [CrossRef] [PubMed]
13. Cobian Guemes, A.G.; Youle, M.; Cantu, V.A.; Felts, B.; Nulton, J.; Rohwer, F. Viruses as Winners in the Game of Life. *Annu. Rev. Virol.* **2016**, *3*, 197–214. [CrossRef]
14. Stanley, W.M. Properties of Viruses. *Medicine* **1939**, *18*, 4. Available online: https://journals.lww.com/md-journal/fulltext/1939/12000/properties_of_viruses.1.aspx (accessed on 20 December 2024). [CrossRef]
15. Pollard, E.C. The Physics of Viruses. *Sci. Am.* **1954**, *191*, 62–71. [CrossRef]
16. Stockley, P.; Twarock, R. (Eds.) *Emerging Topics in Physical Virology*; Imperial College Press: London, UK, 2010.
17. Khaykelson, D.; Raviv, U. Studying Viruses Using Solution X-ray Scattering. *Biophys. Rev.* **2020**, *12*, 41–48. [CrossRef]
18. Mateu, M.G. (Ed.) *Structure and Physics of Viruses: An Integrated Textbook*; Springer: Dordrecht, The Netherlands, 2024.
19. Kushner, D.J. Self-assembly of biological structures. *Bacteriol. Rev.* **1969**, *33*, 302–345. [CrossRef]
20. Bancroft, J.B. The Self-Assembly of Spherical Plant Viruses. In *Advances in Virus Research*; Smith, K.M., Lauffer, M.A., Bang, F.B., Eds.; Academic Press: New York, NY, 1970; Volume 16, pp. 99–134. [CrossRef]
21. Rong, J.; Niu, Z.; Lee, L.A.; Wang, Q. Self-assembly of viral particles. *Curr. Opin. Colloid Interface Sci.* **2011**, *16*, 441–450. [CrossRef]
22. Panahandeh, S.; Li, S.; Dragnea, B.; Zandi, R. Virus Assembly Pathways Inside a Host Cell. *ACS Nano* **2022**, *16*, 317–327. [CrossRef]
23. Sun, S.; Rao, V.B.; Rossmann, M.G. Genome packaging in viruses. *Curr. Opin. Struct. Biol.* **2010**, *20*, 114–120. [CrossRef]
24. Rossmann, M.G.; Rao, V.B. (Eds.) *Viral Molecular Machines*; Springer: Boston, MA, USA, 2012.
25. Garmann, R.F.; Goldfain, A.M.; Tanimoto, C.R.; Beren, C.E.; Vasquez, F.F.; Villarreal, D.A.; Knobler, C.M.; Gelbart, W.M.; Manoharan, V.N. Single-Particle Studies of the Effects of RNA–Protein Interactions on the Self-Assembly of RNA Virus Particles. *Proc. Natl. Acad. Sci. USA* **2022**, *119*, e2206292119. [CrossRef] [PubMed]
26. Iglesias, S.M.; Li, F.; Briani, F.; Cingolani, G. Viral Genome Delivery Across Bacterial Cell Surfaces. *Annu. Rev. Microbiol.* **2024**, *78*, 125–145. [CrossRef] [PubMed]
27. Perham, R.N.; Phillips, D.C.; Radda, G.K. Self-assembly of biological macromolecules. *Philos. Trans. R. Soc. B Biol. Sci.* **1975**, *272*, 123–136. [CrossRef]
28. Whitesides, G.M.; Grzybowski, B. Self-Assembly at All Scales. *Science* **2002**, *295*, 2418–2421. [CrossRef]
29. Fang H.; Tyukodi, B.; Rogers, W.B.; Hagan, M.F. Polymorphic self-assembly of helical tubules is kinetically controlled. *Soft Matter* **2022**, *18*, 6716–6728. [CrossRef]
30. Tyukodi, B.; Mohajerani, F.; Hall, D.M.; Grason, G.M.; Hagan, M.F. Thermodynamic Size Control in Curvature-Frustrated Tubules: Self-Limitation with Open Boundaries. *ACS Nano* **2022**, *16*, 9077–9085. [CrossRef]
31. Pinto, D.E.P.; Šulc, P.; Sciortino, F.; Russo, J. Design strategies for the self-assembly of polyhedral shells. *Proc. Natl. Acad. Sci. USA* **2023**, *120*, e2219458120. [CrossRef]
32. Mohajerani, F.; Sayer, E.; Neil, C.; Inlow, K.; Hagan, M.F. Mechanisms of Scaffold-Mediated Microcompartment Assembly and Size Control. *ACS Nano* **2021**, *15*, 4197–4212. [CrossRef]

33. Li, Y.; Kennedy, N.W.; Li, S.; Mills, C.E.; Tullman-Ercek, D.; Olvera de la Cruz, M. Computational and Experimental Approaches to Controlling Bacterial Microcompartment Assembly. *ACS Cent. Sci.* **2021**, *7*, 658–670. [CrossRef]
34. Caspar, D.L.D.; Klug, A. Physical principles in the construction of regular viruses. *Cold Spring Harb. Symp. Quant. Biol.* **1962**, *27*, 1–24. [CrossRef]
35. Zlotnick, A. To build a virus capsid: An equilibrium model of the self-assembly of polyhedral protein complexes. *J. Mol. Biol.* **1994**, *241*, 59–67. [CrossRef] [PubMed]
36. Bruinsma, R.F.; Gelbart, W.M.; Reguera, D.; Rudnick, J.; Zandi, R. Viral self-assembly as a thermodynamic process. *Phys. Rev. Lett.* **2003**, *90*, 248101. [CrossRef] [PubMed]
37. Zlotnick, A.; Stray, S.J. How does your virus grow? Understanding and interfering with virus assembly. *Trends Biotechnol.* **2003**, *21*, 536–542. [CrossRef] [PubMed]
38. Zandi, R.; Reguera, D.; Bruinsma, R.F.; Gelbart, W.M.; Rudnick, J. Origin of icosahedral symmetry in viruses. *Proc. Natl. Acad. Sci. USA* **2004**, *101*, 15556–15560. [CrossRef]
39. Zandi, R.; Reguera, D. Mechanical properties of viral capsids. *Phys. Rev. E* **2005**, *72*, 021917. [CrossRef]
40. Twarock, R. Mathematical virology: A novel approach to the structure and assembly of viruses. *Philos. Trans. R. Soc. A Math. Phys. Eng. Sci.* **2006**, *364*, 3357–3373. [CrossRef]
41. Roos, W.H.; Ivanovska, I.L.; Evilevitch, A.; Wuite, G.J.L. Viral capsids: Mechanical characteristics, genome packaging and delivery mechanisms. *Cell. Mol. Life Sci.* **2007**, *64*, 1484–1497. [CrossRef]
42. Stephanidis, B.; Adichtchev, S.; Gouet, P.; McPherson, A.; Mermet, A. Elastic properties of viruses. *Biophys. J.* **2007**, *93*, 1354–1359. [CrossRef]
43. Angelescu, D.G.; Linse, P. Viruses as supramolecular self-assemblies: Modelling of capsid formation and genome packaging. *Soft Matter* **2008**, *4*, 1981–1990. [CrossRef]
44. Elrad, O.M.; Hagan, M.F. Encapsulation of a polymer by an icosahedral virus. *Phys. Biol.* **2010**, *7*, 045003. [CrossRef]
45. Luque, A.; Reguera, D.; Morozov, A.; Rudnick, J.; Bruinsma, R. Physics of shell assembly: Line tension, hole implosion, and closure catastrophe. *J. Chem. Phys.* **2012**, *136*, 184507. [CrossRef] [PubMed]
46. Baschek, J.E.; Klein, H.C.R.; Schwarz, U.S. Stochastic Dynamics of Virus Capsid Formation: Direct Versus Hierarchical Self-Assembly. *BMC Biophys.* **2012**, *5*, 22. [CrossRef] [PubMed]
47. Boettcher, M.A.; Klein, H.C.R.; Schwarz, U.S. Role of Dynamic Capsomere Supply for Viral Capsid Self-Assembly. *Phys. Biol.* **2015**, *12*, 016014. [CrossRef]
48. Garmann, R.F.; Comas-Garcia, M.; Knobler, C.M.; Gelbart, W.M. Physical principles in the self-assembly of a simple spherical virus. *Acc. Chem. Res.* **2016**, *49*, 48–55. [CrossRef]
49. Hagan, M.; Zandi, R. Recent advances in coarse-grained modeling of virus assembly. *Curr. Opin. Virol.* **2016**, *18*, 36–43. [CrossRef]
50. Moerman, P.; van der Schoot, P.; Kegel, W. Kinetics versus Thermodynamics in Virus Capsid Polymorphism. *J. Phys. Chem. B* **2016**, *120*, 26, 6003–6009. [CrossRef]
51. Chaudhary, A.; Yadav, R.D. A review on virus protein self assembly. *J. Nanopart. Res.* **2019**, *21*, 254. [CrossRef]
52. Marchetti, M.; Kamsma, D.; Cazares Vargas, E.; Hernandez García, A.; van der Schoot, P.; de Vries, R.; Wuite, G.J.L.; Roos, W.H. Real-Time Assembly of Viruslike Nucleocapsids Elucidated at the Single-Particle Level. *Nano Lett.* **2019**, *19*, 5746–5753. [CrossRef] [PubMed]
53. Garmann, R.F.; Goldfain, A.M.; Manoharan, V.N. Measurements of the self-assembly kinetics of individual viral capsids around their RNA genome. *Proc. Natl. Acad. Sci. USA* **2019**, *116*, 22485–22490. [CrossRef]
54. Zandi, R.; Dragnea, B.; Travesset, A.; Podgornik, R. On virus growth and form. *Phys. Rep.* **2020**, *847*, 1–102. [CrossRef]
55. Buzón, P.; Maity, S.; Roos, W.H. Physical virology: From virus self-assembly to particle mechanics. *WIREs Nanomed. Nanobiotechnol.* **2020**, *12*, e1613. [CrossRef] [PubMed]
56. van Rosmalen, M.G.M.; Kamsma, D.; Biebricher, A.S.; Li, C.; Zlotnick, A.; Roos, W.H.; Wuite, G.J.L. Revealing in Real-Time a Multistep Assembly Mechanism for SV40 Virus-Like Particles. *Sci. Adv.* **2020**, *6*, eaaz1639. [CrossRef] [PubMed]
57. Bruinsma, R.F.; Wuite, G.J.L.; Roos, W.H. Physics of Viral Dynamics. *Nat. Rev. Phys.* **2021**, *3*, 76–91. [CrossRef]
58. Kra, K.; Li, S.; Gargowitsch, L.; Degrouard, J.; Pérez, J.; Zandi, R.; Bressanelli, S.; Tresset, G. Energetics and Kinetic Assembly Pathways of Hepatitis B Virus Capsids in the Presence of Antivirals. *ACS Nano* **2023**, *17*, 12723–12733. [CrossRef]
59. Li, S.; Tresset, G.; Zandi, R. Switchable Conformation in Protein Subunits: Unveiling Assembly Dynamics of Icosahedral Viruses. *arXiv* **2024**, arXiv:2409.00226. [CrossRef]
60. Lostroh, P. *Molecular and Cellular Biology of Viruses*, 1st ed.; Garland Science: New York, NY, USA, 2019.
61. Flint, J.; Racaniello, V.R.; Rall, G.F.; Hatziioannou, T.; Skalka, A.M. *Principles of Virology*, 5th ed.; ASM Press: Washington, DC, USA, 2020; ISBN 978-1-683-67358-3.
62. Bancroft, J.B.; Bracker, C.E.; Wagner, G.W. Structures derived from cowpea chlorotic mottle and brome mosaic virus protein. *Virology* **1969**, *38*, 324–335. [CrossRef]

63. Phillips, B.A. In vitro assembly of poliovirus: II. Evidence for the self-assembly of 14 S particles into empty capsids. *Virology* **1971**, *44*, 307–316. [CrossRef]
64. Xie, Z.; Hendrix, R.W. Assembly in vitro of bacteriophage HK97 proheads. *J. Mol. Biol.* **1995**, *253*, 74–85. [CrossRef]
65. Gross, I.; Hohenberg, H.; Kräusslich, H.G. In vitro assembly properties of purified bacterially expressed capsid proteins of human immunodeficiency virus. *Eur. J. Biochem.* **1997**, *249*, 592–600. [CrossRef]
66. Ceres, P.; Zlotnick, A. Weak Protein-Protein Interactions Are Sufficient to Drive Assembly of Hepatitis B Virus Capsids. *Biochemistry* **2002**, *41*, 11525–11531. [CrossRef]
67. Kegel, W.; van der Schoot, P. Competing hydrophobic and screened-coulomb interactions in hepatitis B virus capsid assembly. *Biophys. J.* **2004**, *86*, 3905–3913. [CrossRef] [PubMed]
68. Li, T.-C.; Takeda, N.; Miyamura, T.; Matsuura, Y.; Wang, J.C.Y.; Engvall, H.; Hammar, L.; Xing, L.; Cheng, R.H. Essential Elements of the Capsid Protein for Self-Assembly into Empty Virus-like Particles of Hepatitis E Virus. *J. Virol.* **2005**, *79*, 12999–13006. [CrossRef] [PubMed]
69. Mukherjee, S.; Thorsteinsson, M.V.; Johnston, L.B.; DePhillips, P.A.; Zlotnick, A. A quantitative description of in vitro assembly of human papillomavirus 16 virus-like particles. *J. Mol. Biol.* **2008**, *381*, 229–237. [CrossRef]
70. Shoemaker, G.K.; van Duijn, E.; Crawford, S.E.; Uetrecht, C.; Baclayon, M.; Roos, W.H.; Wuite, G.J.L.; Estes, M.K.; Prasad, B.V.V.; Heck, A.J.R. Norwalk Virus Assembly and Stability Monitored by Mass Spectrometry. *Mol. Cell. Proteom.* **2010**, *9*, 1742–1751. [CrossRef]
71. Salunke, D.M.; Caspar, D.L.D.; Garcea, R.L. Polymorphism in the assembly of polyomavirus capsid protein VP1. *Biophys. J.* **1989**, *56*, 887–900. [CrossRef] [PubMed]
72. Kanesashi, S.; Ishizu, K.; Kawano, M.; Han, S.; Tomita, S.; Watanabe, H.; Kataoka, K.; Handa, H. Simian virus 40 VP1 capsid protein forms polymorphic assemblies in vitro. *J. Gen. Virol.* **2003**, *84*, 1899–1905. [CrossRef]
73. Casini, G.L.; Graham, D.; Heine, D.; Garcea, R.L.; Wu, D.T. In vitro papillomavirus capsid assembly analyzed by light scattering. *Virology* **2004**, *325*, 320–327. [CrossRef]
74. Bajaj, S.; Banerjee, M. In vitro assembly of polymorphic virus-like particles from the capsid protein of a nodavirus. *Virology* **2016**, *496*, 106–115. [CrossRef]
75. Pierson, E.E.; Keifer, D.Z.; Kukreja, A.A.; Wang, J.C.-Y.; Zlotnick, A.; Jarrold, M.F. Charge Detection Mass Spectrometry Identifies Preferred Non-Icosahedral Polymorphs in the Self-Assembly of Woodchuck Hepatitis Virus Capsids. *J. Mol. Biol.* **2016**, *428*, 292–300. [CrossRef]
76. Patterson, A.; Young, K.; Biever, M.P.; Klein, S.M.; Huang, S.-Y.; DePhillips, P.A.; Jacobson, S.C.; Jarrold, M.F.; Zlotnick, A. Heterogeneity of HPV16 virus-like particles indicates a complex assembly energy surface. *Virology* **2024**, *600*, 110211. [CrossRef]
77. Zlotnick, A.; Aldrich, R.; Johnson, J.M.; Ceres, P.; Young, M.J. Mechanism of Capsid Assembly for an Icosahedral Plant Virus. *Virology* **2000**, *277*, 450–456. [CrossRef] [PubMed]
78. Elrad, O.M.; Hagan, M.F. Mechanisms of Size Control and Polymorphism in Viral Capsid Assembly. *Nano Lett.* **2008**, *8*, 3850–3857. [CrossRef]
79. Nguyen, H.D.; Brooks, C.L. Generalized Structural Polymorphism in Self-Assembled Viral Particles. *Nano Lett.* **2008**, *8*, 4574–4581. [CrossRef]
80. Nguyen, H.D.; Reddy, V.S.; Brooks, C.L., III. Invariant Polymorphism in Virus Capsid Assembly. *J. Am. Chem. Soc.* **2009**, *131*, 2606–2614. [CrossRef]
81. Aznar, M.; Reguera, D. Physical Ingredients Controlling Stability and Structural Selection of Empty Viral Capsids. *J. Phys. Chem. B* **2016**, *120*, 6147–6159. [CrossRef] [PubMed]
82. Grime, J.M.A.; Dama, J.F.; Ganser-Pornillos, B.K.; Woodward, C.L.; Jensen, G.J.; Yeager, M.J.; Voth, G.A. Coarse-grained simulation reveals key features of HIV-1 capsid self-assembly. *Nat. Commun.* **2016**, *7*, 11568. [CrossRef] [PubMed]
83. Waltmann, C.; Asor, R.; Raviv, U.; Olvera de la Cruz, M. Assembly and Stability of Simian Virus 40 Polymorphs. *ACS Nano* **2020**, *14*, 4430–4443. [CrossRef]
84. Spiriti, J.; Conway, J.F.; Zuckerman, D.M. Should Virus Capsids Assemble Perfectly? Theory and Observation of Defects. *Biophys. J.* **2020**, *119*, 1781–1790. [CrossRef]
85. Mohajerani, F.; Tyukodi, B.; Schlicksup, C.J.; Hadden-Perilla, J.A.; Zlotnick, A.; Hagan, M.F. Multiscale Modeling of Hepatitis B Virus Capsid Assembly and Its Dimorphism. *ACS Nano* **2022**, *16*, 13845–13859. [CrossRef]
86. Seitz, I.; Saarinen, S.; Kumpula, E.-P.; McNeale, D.; Anaya-Plaza, E.; Lampinen, V.; Hytönen, V.P.; Sainsbury, F.; Cornelissen, J.J.L.M.; Linko, V.; et al. DNA-origami-directed virus capsid polymorphism. *Nat. Nanotechnol.* **2023**, *18*, 1205–1212. [CrossRef]
87. Zandi, R.; van der Shoot, P.; Reguera, D.; Kegel, W.; Reiss, H. Classical nucleation theory of virus capsids. *Biophys. J.* **2006**, *90*, 1929–1948. [CrossRef]
88. Luque, A.; Reguera, D. Theoretical Studies on Assembly, Physical Stability, and Dynamics of Viruses. *Subcell. Biochem.* **2024**, *105*, 693–741. [CrossRef] [PubMed]

89. Law-Hine, D.; Zeghal, M.; Bressanelli, S.; Constantin, D.; Tresset, G. Identification of a major intermediate along the self-assembly pathway of an icosahedral viral capsid by using an analytical model of a spherical patch. *Soft Matter* **2016**, *12*, 6728–6736. [CrossRef]
90. Timmermans, S.B.P.E.; Ramezani, A.; Montalvo, T.; Nguyen, M.; van der Schoot, P.; van Hest, J.C.M.; Zandi, R. The Dynamics of Viruslike Capsid Assembly and Disassembly. *J. Am. Chem. Soc.* **2022**, *144*, 12608–12612. [CrossRef] [PubMed]
91. Endres, D.; Zlotnick, A. Model-based analysis of assembly kinetics for virus capsids or other spherical polymers. *Biophys. J.* **2002**, *83*, 1217–1230. [CrossRef] [PubMed]
92. Khaykelson, D.; Asor, R.; Zhao, Z.; Schlicksup, C.J.; Zlotnick, A.; Raviv, U. Guanidine Hydrochloride-Induced Hepatitis B Virus Capsid Disassembly Hysteresis. *Biochemistry* **2024**, *63*, 1543–1552. [CrossRef]
93. Singh, S.; Zlotnick, A. Observed Hysteresis of Virus Capsid Disassembly Is Implicit in Kinetic Models of Assembly. *J. Biol. Chem.* **2003**, *278*, 18249–18255. [CrossRef]
94. Starr, C.A.; Barnes, L.F.; Jarrold, M.F.; Zlotnick, A. Hysteresis in Hepatitis B Virus (HBV) Requires Assembly of Near-Perfect Capsids. *Biochemistry* **2022**, *61*, 505–513. [CrossRef]
95. Herrmann, E.; Neumann, A.U.; Schmidt, J.M.; Zeuzem, S. Hepatitis C Virus Kinetics. *Antivir. Ther.* **2000**, *5*, 85–90. [CrossRef]
96. Chen, C.; Kao, C.C.; Dragnea, B. Self-Assembly of Brome Mosaic Virus Capsids: Insights from Shorter Time-Scale Experiments. *J. Phys. Chem. A* **2008**, *112*, 9405–9412. [CrossRef]
97. Perlmutter, J.D.; Hagan, M.F. Mechanisms of Virus Assembly. *Annu. Rev. Phys. Chem.* **2015**, *66*, 217–239. [CrossRef] [PubMed]
98. Peña, J. Virología Física: Teoría de Autoensamblaje de Cápsides Virales. Master's Thesis, Universidad Autónoma Metropolitana-Iztapalapa, CDMX, México, 2024. Available online: https://ixtlan.izt.uam.mx/leo/wp-content/uploads/2024/10/tesis_maestria_jason_pena.pdf (accessed on 15 November 2024).
99. Mendoza, C.; Reguera, D. Shape selection and mis-assembly in viral capsid formation by elastic frustration. *eLife* **2020**, *9*, e52525. [CrossRef]
100. Hagan, M.; Grason, G.M. Equilibrium mechanisms of self-limiting assembly. *Rev. Mod. Phys.* **2021**, *93*, 025008. [CrossRef]
101. Peña, J.; Dagdug, L.; Reguera, D. *On the Influence of Bending Energy on the Assembly of Spherical Viral Capsids*; Universidad Autónoma Metropolitana-Iztapalapa: Mexico City, Mexico; Universitat de Barcelona: Barcelona, Spain, 2025; status (manuscript in preparation; to be submitted).
102. Seung, H.S.; Nelson, D.R. Defects in flexible membranes with crystalline order. *Phys. Rev. A* **1988**, *38*, 1005–1018. [CrossRef] [PubMed]
103. Lidmar, J.; Mirny, L.; Nelson, D.R. Virus shapes and buckling transitions in spherical shells. *Phys. Rev. E* **2003**, *68*, 051910. [CrossRef]
104. Meng, G.; Paulose, J.; Nelson, D.R.; Manoharan, V.N. Elastic Instability of a Crystal Growing on a Curved Surface. *Science* **2014**, *343*, 634–637. [CrossRef] [PubMed]
105. Helfrich, W. Elastic Properties of Lipid Bilayers: Theory and Possible Experiments. *Z. Naturforsch. C J. Biosci.* **1973**, *28*, 693–703. [CrossRef]
106. Nguyen, T.T.; Bruinsma, R.F.; Gelbart, W.M. Elasticity theory and shape transitions of viral shells. *Phys. Rev. E* **2005**, *72*, 051923. [CrossRef]
107. Klug, W.S.; Bruinsma, R.F.; Michel, J.-P.; Knobler, C.M.; Ivanovska, I.L.; Schmidt, C.F.; Wuite, G.J.L. Failure of Viral Shells. *Phys. Rev. Lett.* **2006**, *97*, 228101. [CrossRef]
108. Nguyen, T.T.; Bruinsma, R.F.; Gelbart, W.M. Continuum Theory of Retroviral Capsids. *Phys. Rev. Lett.* **2006**, *96*, 078102. [CrossRef]
109. Zhao, Y.; Ge, Z.; Fang, J. Elastic modulus of viral nanotubes. *Phys. Rev. E* **2008**, *78*, 031914. [CrossRef] [PubMed]
110. Buenemann, M.; Lenz, P. Elastic properties and mechanical stability of chiral and filled viral capsids. *Phys. Rev. E* **2008**, *78*, 051924. [CrossRef] [PubMed]
111. Roos, W.H.; Gibbons, M.M.; Arkhipov, A.; Uetrecht, C.; Watts, N.R.; Wingfield, P.T.; Steven, A.C.; Heck, A.J.R.; Schulten, K.; Klug, W.S.; et al. Squeezing Protein Shells: How Continuum Elastic Models, Molecular Dynamics Simulations, and Experiments Coalesce at the Nanoscale. *Biophys. J.* **2010**, *99*, 1175–1181. [CrossRef] [PubMed]
112. Grason, G.M. Perspective: Geometrically frustrated assemblies. *J. Chem. Phys.* **2016**, *145*, 110901. [CrossRef]
113. Castelnovo, M. Viral self-assembly pathway and mechanical stress relaxation. *Phys. Rev. E* **2017**, *95*, 052405. [CrossRef]
114. Mendoza, C.I.; Reguera, D. *Assembly of Tubes in the Stretching-Dominated Limit*; Universidad Nacional Autónoma de México (UNAM): Mexico City, Mexico; Universitat de Barcelona: Barcelona, Spain, 2025; status (manuscript in preparation; to be submitted).
115. Luque, A. Structure, Mechanical Properties, and Self-Assembly of Virus Capsids. Ph.D. Thesis, Departament de Física Fonamental, Universitat de Barcelona, Barcelona, Spain, 2011.
116. Reguera, D.; Rubí, J.M.; Vilar, J.M.G. The Mesoscopic Dynamics of Thermodynamic Systems. *J. Phys. Chem. B* **2005**, *109*, 21502–21515. [CrossRef]
117. De Groot, S.R.; Mazur, P. *Non-Equilibrium Thermodynamics*; Dover Publications: Mineola, NY, USA, 2011.

118. Landau, L.D.; Lifshitz, E.M. *Statistical Physics, Parts 1 and 2*; Pergamon Press: New York, NY, USA, 1980.
119. Kjelstrup, S.; Bedeaux, D.; Johannessen, E.; Gross, J.; Wilhelmsen, O. *Non-Equilibrium Thermodynamics for Engineering Applications*; World Scientific Publishing Company: Singapore, 2024.
120. Reguera, D.; Rubí, J.M. Kinetic equations for diffusion in the presence of entropic barriers. *Phys. Rev. E* **2001**, *64*, 061106. [CrossRef]
121. Gomila, G.; Pérez-Madrid, A.; Rubí, J.M. Non-equilibrium thermodynamics of thermionic emission processes in abrupt semiconductor junctions, including the effects of surface states. *Phys. A* **1996**, *233*, 208–220. [CrossRef]
122. Pagonabarraga, I.; Pérez-Madrid, A.; Rubí, J.M. Fluctuating hydrodynamics approach to chemical reactions. *Phys. A* **1997**, *237*, 205–219. [CrossRef]
123. Reguera, D.; Rubí, J.M. Homogeneous nucleation in inhomogeneous media. I. Nucleation in a temperature gradient. *J. Chem. Phys.* **2003**, *119*, 9877–9887. [CrossRef]
124. Kjelstrup, S.; Rubí, J.M.; Pagonabarraga, I.; Bedeaux, D. Mesoscopic non-equilibrium thermodynamic analysis of molecular motors. *Phys. Chem. Chem. Phys.* **2013**, *15*, 19405–19414. [CrossRef] [PubMed]
125. Zandi, R.; Reguera, D.; Rudnick, J.; Gelbart, W.M. What drives the translocation of stiff chains? *Proc. Natl. Acad. Sci. USA* **2003**, *100*, 8649–8653. [CrossRef] [PubMed]
126. Kloeden, P.E.; Platen, E. *Numerical Solution of Stochastic Differential Equations*; Springer: Berlin/Heidelberg, Germany, 2011.
127. Vehkamäki, H. *Classical Nucleation Theory in Multicomponent Systems*; Springer: Berlin/Heidelberg, Germany, 2006.
128. Kalikmanov, V.I. *Nucleation Theory*; Springer: Dordrecht, The Netherlands, 2013.

entropy

Article

A New Analytical Formulation for the Electrophoretic Mobility of a Colloidal Sphere

Angela Casarella [1], Simon Gourdin-Bertin [2,†] and Claire Chassagne [3,*]

[1] Department of Architecture and Civil Engineering, Chalmers University of Technology, 412 96 Göteborg, Sweden; angela.casarella@chalmers.se
[2] Independent Researcher, F-75012 Paris, France; simon.n.gourdin@gmail.com
[3] Section of Environmental Fluid Mechanics, Department of Hydraulic Engineering, Delft University of Technology, 2628 CN Delft, The Netherlands
* Correspondence: c.chassagne@tudelft.nl
† Formerly Affiliated to: Laboratoire PHENIX, CNRS, Sorbonne Universités, UPMC Univ Paris 06, Case 51, 4 Place Jussieu, F-75005 Paris, France.

Abstract: A new analytical equation for the electrophoretic mobility of a colloidal sphere, homogeneously charged, is derived. This equation reduces to the well-known Henry's formulation for low surface potentials. For high surface potentials, the equation is compared to the full numerical result. It is found that the equation performs well up to surface potentials of 50 mV. For larger surface potentials, the equation performs well for $\kappa a > 10$, where κ is the inverse of Debye' s length and a the radius of the particle. Differences between analytical and numerical solutions for $\kappa a < 10$ are studied. The case of a particle with a constant surface charge is discussed. In that case, a very simple equation relates the surface charge of the particle to the electrophoretic mobility for $\kappa a > 10$.

Keywords: electrophoresis; colloid; zeta potential; Hückel; Smoluchowski

1. Introduction

The motion of a charged colloidal particle in an electrolyte under the influence of an applied electric field is used routinely in many fields of science to study the changes in interfacial properties of colloids upon changes in solvent (changes in pH or changes in ionic strength, for example). In order to interpret the measured data, that is, the electrophoretic velocity defined as $\mu = U/E_0$ whereby U is the velocity of a colloidal particle subjected to an electric field E_0, one makes use of theories of which the simplest ones date from the work of Smoluchowski and Hückel [1]. A full numerical description of the electrophoretic mobility was proposed in 1978 by O'Brien and White [2]. Numerous analytical approximations have also been developed for different ranges of applicability: high and low ionic strength, low surface charge, and particles coated by polyelectrolytes, for example [1]. Recently, there was regained interest in the physical aspects of electrophoresis. Jayaraman et al. pointed out the unusual fluid dynamics around a charged colloidal particle and provided an overview of different theories [3]. One of the authors of the article even translated the original PhD thesis of Overbeek from Dutch to English to help the modern reader bring this original work into context (see web link in [3]).

In the present article, we would like to make the connection between these theories and the approach taken in the work of Chassagne and Bedeaux [4] to study the polarization of a double layer around a charged colloidal particle. In particular, we would like to make the link between electrophoretic mobility and dipole coefficient β, where β is related to the dipole moment $\mathbf{P}$ (see Equation (41)) generated from the application of the electric field

 https://doi.org/10.3390/e27040336

$\mathbf{E}_0$. The main finding of the present article is Equation (75), where μ is shown to be a linear function of β. This equation is an improved version of a similar equation presented in [4]. In the present article, we provide a mathematical derivation for Equation (75) (which was not conducted in [4]) and discuss the validity of the three main hypotheses formulated to obtain the equation. We hereby extend the analysis conducted by Jayaraman et al. [3] and describe the behavior of the electric potential and electrochemical potentials for specific cases, such as low and high ionic strengths and low and high surface electric charges. The particular case of a particle with a constant surface charge is discussed, which is a case that is often encountered in practice.

2. Definitions

In this section, we define the required variables and set up the relations that will be solved in the following sections. A charged spherical and dielectric particle of radius a, homogeneously charged, is subjected to an oscillating electric field $\mathbf{E}_0 = E_0 \exp(i\omega t)\mathbf{e}_z$ of radial frequency ω (rad/s). In this article, we focus on the case $\omega = 0$. More general equations can be found in [4]. The vector $\mathbf{e}_z$ represents a unity vector along the z-axis, where (x, y, z) designs the usual Cartesian coordinates and $(\mathbf{e}_r, \mathbf{e}_\theta, \mathbf{e}_\varphi)$ are the unit vectors in spherical coordinates. We place ourselves in the frame of reference of the particle, and the origin of the coordinate system is the center of the considered particle, whereby the polar axis ($\theta = 0$) is set parallel to $\mathbf{E}_0$ and hence

$$\mathbf{E}_0 = E_0\mathbf{e}_z = E_0 \cos\theta \mathbf{e}_r - E_0 \sin\theta \mathbf{e}_\theta \tag{1}$$

The particle is immersed in an electrolyte solution composed of one type of positively charged ions and one type of negatively charged ions.

The main assumptions used in the derivation are as follows.

1. The Reynolds number is small so that the inertial terms in the Navier–Stokes equations can be ignored.
2. The fluid is incompressible.
3. The applied electric field is weak compared to the local electric field around the particle so that higher-order terms in E_0 can be neglected.

In the present article, we assume that

$$\Psi_{eq}(a) = \zeta \tag{2}$$

where Ψ_{eq} is the equilibrium potential in the absence of applied electric field. The zeta potential ζ is defined as the electric potential at the surface of shear. By assuming Equation (2) to be valid, we imply that the surface of shear that will be created when the particle is in motion is located on the surface of the particle. We will also assume that there is no Stern layer. The Stern layer is defined as a very small region of space at the interface between the particle and the electrolyte where solvent and ionic properties could deviate from their bulk values. This assumption implies that the equations that will be given in the next sections are valid between $r = a$ (the particle's surface) and infinity. The general relations for the electric potential Ψ and the ionic densities $n_{i,tot}$ $(i = +, -)$ are given by

$$\begin{aligned} \Psi(r,\theta) &= \Psi_{eq}(r) + \delta\Psi(r,\theta) \\ n_{+,tot}(r,\theta) &= n_{+,eq}(r) + \delta n_+(r,\theta) \\ n_{-,tot}(r,\theta) &= n_{-,eq}(r) + \delta n_-(r,\theta) \end{aligned} \tag{3}$$

where $n_{i,eq}$ represents the ionic density of positively/negatively charged ions (number of ions/m^3) of valence z_i and stoichiometric coefficient ν_i in absence of applied electric field. Because of electroneutrality,

$$\sum_i \nu_i z_i = 0 \tag{4}$$

The functions $\delta\Psi$, δn_+, and δn_- represent the contributions due to the presence of the applied electric field.

Due to axial symmetry, one can define

$$\begin{aligned} \delta\Psi(r,\theta) &= (\psi(r) - r)\mathbf{E}_0 \cdot \mathbf{e}_r \\ \delta n_+(r,\theta) &= n_+(r)\mathbf{E}_0 \cdot \mathbf{e}_r \\ \delta n_-(r,\theta) &= n_-(r)\mathbf{E}_0 \cdot \mathbf{e}_r \end{aligned} \tag{5}$$

It follows that the electric field resulting from the application of E_0 is

$$\begin{aligned} \delta E_r &= -\left(\frac{d\psi}{dr} - 1\right) E_0 \cos(\theta) \\ \delta E_\theta &= \left(\frac{\psi}{r} - 1\right) E_0 \sin(\theta) \end{aligned} \tag{6}$$

2.1. The Double Layer

The so-called double layer surrounding the colloidal particle is composed of (a) a first layer composed of the surface charges grafted onto the colloidal particle and (b) a second layer (the diffuse layer) dominated by an excess of counter-ions (and a depletion of co-ions) that are electrically interacting with the surface charges. As done by many authors, the term "double layer" will be used when "diffuse layer" is implied. Beyond the double layer, in the absence of an applied electric field, the concentrations of counter and co-ions are such that electroneutrality (see Equation (4)), is respected.

The thickness of the diffuse layer is given by the Debye length, which is defined by

$$\kappa^{-1} = \sqrt{\frac{\varepsilon_0 \varepsilon_1 kT}{e^2 n_\infty \sum z_i^2 \nu_i}} \tag{7}$$

with

$$n_\infty = \mathrm{Cs} \times \mathrm{N_A} \tag{8}$$

where Cs is the neutral salt concentration in mM (10^{-3} mol/L), $\mathrm{N_A}$ is Avogadro's number, ε_0 is the permittivity of vacuum, and ε_1 is the relative permittivity of the solvent (which will be water in the present article).

Beyond the double layer (in the bulk), we have

$$n_{i,eq}\left(r \gg a + \kappa^{-1}\right) = \nu_i n_\infty$$

2.2. The Electrochemical Potential

The electrochemical potential $\widetilde{\mu}_i$ is defined by

$$\widetilde{\mu}_i = z_i e \Psi + \mu_i \tag{9}$$

The chemical potential μ_i is defined by

$$\mu_i = \mu_i^0 + kT \ln\left(\frac{n_{i,tot}}{n_0}\right) \tag{10}$$

where μ_i^0 and n_0 are reference values, k is the Boltzmann constant, and T is the temperature. Note that it is the fact that the electrochemical potentials are constant in the absence of applied electric field, i.e.,

$$\widetilde{\mu}_{i,eq}(r) = \widetilde{\mu}_{i,eq}(\infty) \tag{11}$$

that leads to the Boltzmann distribution:

$$n_{i,eq} = \nu_i n_\infty \exp\left(\frac{-z_i e \Psi_{eq}}{kT}\right) \tag{12}$$

The electrochemical potential can also be written

$$\begin{aligned} \widetilde{\mu}_i &= \widetilde{\mu}_{i.eq} + \delta\widetilde{\mu}_i \\ &= z_i e(\Psi_{eq} + \delta\Psi) + \mu_i^0 + kT \ln\left(\frac{n_{i,eq} + \delta n_i}{n_0}\right) \end{aligned} \tag{13}$$

One can show that

$$\delta\widetilde{\mu}_i = z_i e \delta\Psi + kT\frac{\delta n_i}{n_{i,eq}} \tag{14}$$

Because of symmetry, we introduce the variable ϕ_i such that

$$\begin{aligned} \delta\widetilde{\mu}_i(r) &= -z_i e[\phi_i(r) + r]E_0 \cos\theta \\ &= z_i e[\psi(r) - r]E_0 \cos\theta + kT\frac{n_i(r)}{n_{i,eq}(r)}E_0 \cos\theta \end{aligned} \tag{15}$$

This implies that

$$n_i = \frac{-z_i e n_{i,eq}}{kT}(\psi + \phi_i) \tag{16}$$

2.3. The Ionic Flux

The ionic flux $\mathbf{J}_i$ is given by

$$\mathbf{J}_i = n_{i,eq}\mathbf{u} - \frac{D_i n_i}{kT}\nabla\widetilde{\mu}_i \tag{17}$$

where $\mathbf{u}$ is the fluid velocity in the reference frame of the particle, and D_i (m^2/s) is the ionic diffusion coefficient. The ionic fluxes at equilibrium are defined by

$$\mathbf{J}_{i,eq} = \frac{-D_i n_{i,eq}}{kT}\nabla\widetilde{\mu}_{i,eq} = \mathbf{0} \tag{18}$$

The fact that the electrochemical potentials are constant in the absence of an applied electric field is correlated to $\mathbf{J}_{i,eq} = \mathbf{0}$, which leads to the Boltzmann distribution, Equation (12). The fluxes due to the application of the electric field are, to first order, given by

$$\delta\mathbf{J}_i = n_{i,eq}\mathbf{u} - \frac{D_i n_{i,eq}}{kT}\nabla\delta\widetilde{\mu}_i \tag{19}$$

2.4. The Velocity

The condition $\nabla \cdot \mathbf{u} = \mathbf{0}$ (which arises from the fact that the fluid is incompressible) is used to express $\mathbf{u}(r,\theta)$ as function of a new function $h(r)$ such that [1]

$$\mathbf{u} = \frac{-2}{r}hE\cos\theta\mathbf{e}_r + \frac{1}{r}\frac{d}{dr}[rh]E_0 \sin\theta\mathbf{e}_\theta \tag{20}$$

The calculations are conducted in the reference frame of the particle, which implies that

$$\begin{aligned} \mathbf{u}(r=a) &= \mathbf{0} \quad \text{(no slip)} \\ \mathbf{u}(r\to\infty) &= -\mu\mathbf{E}_0 \quad \text{(far away, the fluid moves at minus the velocity of the particle)} \end{aligned}$$

where

$$\mu = \frac{U}{E_0} \tag{21}$$

is called electrophoretic mobility (note that its symbol μ should not be confused with μ_i, which is used for the ionic chemical potential).

2.5. Variable Definitions

For convenience, we use the following dimensionless variables:

$$\begin{array}{lcl} \text{no dim.} & & \text{with dim.} \\ \widehat{\Psi}_{eq} & = & e\Psi_{eq}/(kT) \\ \widehat{\psi} & = & \kappa\psi \\ \widehat{\phi}_i & = & \kappa\phi_i \\ x & = & \kappa r \\ \widehat{h} & = & h(e\eta\kappa)/(\varepsilon_0\varepsilon_1 kT) \\ \widehat{M} & = & M(e\eta)/(\kappa^3\varepsilon_0\varepsilon_1 kT) \\ \widehat{\mu} & = & \mu(e\eta)/(\varepsilon_0\varepsilon_1 kT) \\ \widehat{D_i} & = & D_i(e^2\eta)/\left(\varepsilon_0\varepsilon_1 (kT)^2\right) \end{array}$$

3. The Poisson–Boltzmann Equation

The Poisson–Boltzmann equation is given by

$$\nabla^2\Psi = \frac{-1}{\varepsilon_0\varepsilon_1}\sum e z_i n_{i,tot} \tag{22}$$

For the equilibrium part, using Equation (12) and dimensionless variables, one obtains

$$\begin{aligned} \frac{d^2\widehat{\Psi}_{eq}}{dx^2} + \frac{2}{x}\frac{d\widehat{\Psi}_{eq}}{dx} &= \frac{-1}{\sum z_i^2 \nu_i}\sum z_i \nu_i \exp(z_i \widehat{\Psi}_{eq}) \\ &= \sinh(\widehat{\Psi}_{eq}) \quad \text{for a 1–1 electrolyte} \end{aligned} \tag{23}$$

The Poisson–Boltzmann for the potential arising from the applied electric field is obtained by using Equations (12) and (16)

$$\widehat{L}\widehat{\psi} = \frac{d^2\widehat{\psi}}{dx^2} + \frac{2}{x}\frac{d\widehat{\psi}}{dx} - \frac{2\widehat{\psi}}{x^2} = \frac{1}{\sum z_i^2\nu_i}\sum z_i^2\nu_i \exp(-z_i\widehat{\Psi}_{eq})\left[\widehat{\psi}+\widehat{\phi}_i\right] \tag{24}$$

$$= \cosh(\widehat{\Psi}_{eq})\left[\widehat{\psi}+\widehat{\phi}_i\right] \quad \text{for a 1–1 electrolyte} \tag{25}$$

The operator L (with $\widehat{L} = L/\kappa^2$) is defined on an arbitrary function $g(r)$ by

$$Lg(r) = \frac{1}{r}\frac{d^2}{dr^2}(rg(r)) - \frac{2g(r)}{r^2} \tag{26}$$

3.1. Boundary Conditions at $r = a$

The boundary condition for the equilibrium electric potential $\widehat{\Psi}_{eq}$ is given by

$$\widehat{\Psi}_{eq}(x=\kappa a) = \widehat{\Psi}_0 \tag{27}$$

where $\widehat{\Psi}_0 = \widehat{\zeta}$ is the zeta potential in case there is no Stern layer and that the slip plane is located at the surface of the particle.

The boundary condition for the equilibrium electric potential Ψ_{eq} is also given by, following Gauss,

$$\varepsilon_0\varepsilon_2\left(\frac{d\Psi_{2,eq}}{dr}\right)_{r=a} - \varepsilon_0\varepsilon_1\left(\frac{d\Psi_{eq}}{dr}\right)_{r=a} = \sigma_s \tag{28}$$

where σ_s (C/m^2) is the surface charge density, ε_2 is the relative permittivity of the particle, and $\Psi_{eq,2}$ is the electric potential inside the particle. This potential satisfies the Laplace's equation ($\nabla^2\Psi_{2,eq} = 0$). The solution of the Laplace's equation is given in terms of Legendre polynomials (see p. 350 in [5]), and, because $\Psi_{2,eq}$ cannot be singular in $r = 0$ nor be θ-dependent, it follows that $\Psi_{2,eq}$ should be a constant. From the continuity of electric potentials, i.e.,

$$\Psi_{2,eq}(r = a^-) = \Psi_{eq}(r = a^+) \tag{29}$$

it follows that

$$\widehat{\Psi}_{2,eq}(x) = \widehat{\Psi}_0 \tag{30}$$

The surface charge density σ_s can usually be assessed by titration. For most colloids, one usually has $\varepsilon_2 \ll \varepsilon_1$, implying that Equation (28) reduces to

$$\left(\frac{d\Psi_{eq}}{dr}\right)_{r=a} = \frac{-\sigma_s}{\varepsilon_0\varepsilon_1} \tag{31}$$

For specific suspensions such as charged sulfate latex colloidal spheres at a given pH, the surface charge density is not expected to vary when the electrolyte concentration is varied [6]. Equation (31) can be used to estimate the surface charge density σ_s, while the potential $\widehat{\Psi}_0$ can be estimated from electrophoretic mobility data [6]. Discrepancies between measured data and prediction are associated with the thin layer of fluid at the particle/electrolyte interface (Stern layer). In the Stern layer, both the ionic mobility and the dielectric permittivity differ from their bulk values [7], and, for highly charged particles, the use of Equation (12) can lead to unrealistically high ionic concentrations close to the particle's surface, as a simple estimation can show. A Stern layer parameter can then be introduced in the equations to account for the deviation from ideality. We will not consider Stern layers in the present article.

The boundary conditions for $\delta\Psi$ are given by

$$\begin{aligned} \varepsilon_0\varepsilon_2\left(\frac{d(\delta\Psi_2)}{dr}\right)_{r=a} &= \varepsilon_0\varepsilon_1\left(\frac{d(\delta\Psi)}{dr}\right)_{r=a} \\ \delta\Psi_2(a) &= \delta\Psi(a) \end{aligned} \tag{32}$$

$\delta\Psi_2$ is the potential inside the dielectric sphere due to applied electric field and is solution of Laplace's equation ($\Delta\delta\Psi_2 = 0$), which yields

$$\delta\Psi_2 = -E_2 r\cos\theta \tag{33}$$

Eliminating E_2 in the equations leaves

$$\left(\frac{d(\delta\Psi)}{dr}\right)_{r=a} = \frac{\varepsilon_2}{\varepsilon_1}\frac{\delta\Psi(a)}{a} \tag{34}$$

which yields

$$\left(\frac{d\widehat{\psi}}{dx}\right)_{x=\kappa a} - 1 = \frac{\varepsilon_2}{\varepsilon_1}\frac{\widehat{\psi}(\kappa a) - \kappa a}{\kappa a} \tag{35}$$

The magnitude of the electric field inside the colloidal particle is given by

$$E_2 = \left(1 - \frac{\psi(a)}{a}\right) E_0 \tag{36}$$

3.2. Boundary Conditions Far Away

Far away from the particle, one finds, for any electrolyte,

$$\frac{d^2\widehat{\Psi}_{eq}}{dx^2} + \frac{2}{x}\frac{d\widehat{\Psi}_{eq}}{dx} = \frac{-1}{\sum z_i^2 \nu_i} \sum z_i \nu_i \left(1 - z_i \widehat{\Psi}_{eq}\right) = \widehat{\Psi}_{eq} \quad \text{for} \quad x \gg \kappa a + 1$$

yielding

$$\widehat{\Psi}_{eq}^{+}(x) = \widehat{\Psi}_{ap} \frac{\kappa a}{x} \exp(\kappa a - x) \tag{37}$$

where $\widehat{\Psi}_{ap}$ is an apparent surface potential as it would be the surface potential if the relation Equation (37) would hold until $x = \kappa a$ (i.e., $r = a$). Equation (37) holds for any x for low surface potentials ($\widehat{\Psi}_0 = \widehat{\Psi}_{eq}^{+}(\kappa a)$) as then the approximation $\exp(-z_i\widehat{\Psi}_{eq}) = 1 - z_i\widehat{\Psi}_{eq}$ is valid. It follows that, for low surface potentials $\widehat{\Psi}_0$,

$$\begin{aligned} \widehat{\Psi}_{eq}(x) &= \widehat{\Psi}_0 \frac{\kappa a}{x} \exp(\kappa a - x) \quad \text{for any } x \\ \widehat{\Psi}_{ap} &= \widehat{\Psi}_0 \quad \text{for} \quad \widehat{\Psi}_0 \ll 1 \end{aligned} \tag{38}$$

Assymptotically, for $x \to \infty$, one defines

$$\widehat{\Psi}_{eq}^{\infty} = 0 \tag{39}$$

The assymptotic value of ψ^+ is given by (see Appendix A)

$$\widehat{\psi}^{+} = \frac{(\kappa a)^3 \beta}{x^2} \tag{40}$$

where β is the dipole coefficient, related to the dipole moment

$$\mathbf{P} = 4\pi\varepsilon_0\varepsilon_1 a^3 \beta \mathbf{E}_0 \tag{41}$$

generated from the application of the electric field. The dipolar nature of the system composed of the charged colloidal particle and its double layer is illustrated in Appendix D, where the results of COMSOL calculations are plotted. The assymptotic form of the electric potential is given by

$$\delta\psi^{+}(r) = \left(-r + \frac{a^3\beta}{r^2}\right) E_0 \cos\theta \tag{42}$$

The approximate expressions for β are given in Appendix B. More general relations are given in [8].

4. The Conservation of Mass Law (Nernst–Planck)

The law of conservation of mass, for the DC case considered here (which implies that $\partial n_i/\partial t = 0$), can be expressed as

$$\nabla \cdot \mathbf{J}_i = 0 \tag{43}$$

To first order, the ionic flux becomes

$$\begin{aligned} \mathbf{J}_i &= n_{i,eq}\mathbf{u} - D_i n_{i,eq}\nabla\delta\widetilde{\mu}_i \\ \nabla\delta\widetilde{\mu}_i &= \nabla\left(\frac{z_i e}{kT}\delta\Psi + \frac{\delta n_i}{n_{i,eq}}\right) \end{aligned} \tag{44}$$

One can show that

$$\begin{aligned} &\nabla.\left[n_{i,eq}\nabla\delta\widetilde{\mu}_i\right] = \nabla n_{i,eq}\cdot\nabla\delta\widetilde{\mu}_i + n_{i,eq}\nabla^2\delta\widetilde{\mu}_i \\ = & \\ &z_i e n_{i,eq}\left(\frac{z_i e}{kT}\frac{d\Psi_{eq}}{dr}\left(\frac{\partial\phi_i}{\partial r}+1\right) - \left[\frac{\partial^2\phi_i}{\partial r^2} + \frac{2}{r}\frac{\partial\phi_i}{\partial r} - \frac{2}{r^2}\phi_i\right]\right)E_0\cos\theta \end{aligned} \tag{45}$$

This implies that the law of mass conservation can be written

$$\widehat{L\phi_i} = \frac{\partial^2\widehat{\phi_i}}{\partial x^2} + \frac{2}{x}\frac{\partial\widehat{\phi_i}}{\partial x} - \frac{2}{x^2}\widehat{\phi_i} = z_i\frac{d\Psi_{eq}}{dx}\left[\frac{d\widehat{\phi_i}}{dx} - \frac{1}{z_i\widehat{D_i}}\frac{2\widehat{h}}{x} + 1\right] \tag{46}$$

4.1. Boundary Conditions at $r = a$

There is no ionic flux possible perpendicular to the surface of the particle; hence,

$$\begin{aligned} (\mathbf{J}_i\cdot\mathbf{e}_r)_{r=a} &= 0 \\ (\nabla\widetilde{\mu}_i\cdot\mathbf{e}_r)_{r=a} &= 0 \end{aligned} \tag{47}$$

where we have used the no-slip condition

$$\mathbf{u}(r=a) = \mathbf{0} \tag{48}$$

We obtain

$$\left(\frac{d\widehat{\phi_i}}{dx}\right)_{x=\kappa a} = -1 \tag{49}$$

4.2. Boundary Conditions Far Away

In Appendix A, it was found that, for $x \gg \kappa a + 1$,

$$\widehat{\phi}_+^+ = \widehat{\phi}_-^+ = -\frac{(\kappa a)^3\beta}{x^2} \tag{50}$$

5. Navier–Stokes

The Navier–Stokes equation provides the last fundamental equation required to solve the problem:

$$\eta\nabla^2\mathbf{u} - \nabla P = \sum e z_i n_i\nabla\Psi \tag{51}$$

where P is the pressure. We used the fact that the Reynolds number $\mathrm{Re} = |\rho_m(\mathbf{u}\cdot\nabla)\mathbf{u}|/|\eta\nabla^2\mathbf{u}|$ and also $|\rho_m\partial\mathbf{u}/\partial\mathbf{t}|/|\eta\nabla^2\mathbf{u}|$ are very small compared to one. To first order, the Navier–Stokes reduces to

$$\eta\nabla^2\mathbf{u} - \nabla P = \sum e z_i\left[\delta n_i\nabla\Psi_{eq} + n_{i,eq}\nabla\delta\Psi\right] \tag{52}$$

In order to get rid of the pressure term P, we take the curl of the Navier–Stokes equation:

$$\eta\nabla^2(\nabla\times\mathbf{u}) = \sum e z_i\nabla\times[n_i\nabla\Psi] \tag{53}$$

For a discussion about the pressure and the peculiar form of the velocity $\mathbf{u}$, we refer to [3,9]. Equation (53) can be written

$$\eta\nabla^2(\nabla \times \mathbf{u}) = \sum ez_i\left[\nabla\delta n_i \times \nabla\Psi_{eq} + \nabla n_{i,eq} \times \nabla\delta\Psi\right] \quad (54)$$

One can show that

$$\begin{aligned} & \nabla\delta n_i \times \nabla\Psi_{eq} + \nabla n_{i,eq} \times \nabla\delta\Psi \\ = & \frac{-z_i e n_{i,eq}}{kT}\left(\frac{\phi_i}{r} + 1\right)\frac{d\Psi_{eq}}{dr}\sin\theta E_0 \mathbf{e}_\phi \end{aligned} \quad (55)$$

and that

$$\nabla^2(\nabla \times \mathbf{u}) = \left[\frac{\partial^4 h}{\partial r^4} + \frac{4}{r}\frac{\partial^3 h}{\partial r^3} - \frac{4}{r^2}\frac{\partial^2 h}{\partial r^2}\right]\sin\theta E_0 \mathbf{e}_\phi \quad (56)$$

This implies that the Navier–Stokes reduces to

$$\widehat{L}\widehat{L}\widehat{h} = \left[\frac{\partial^4 \widehat{h}}{\partial x^4} + \frac{4}{x}\frac{\partial^3 \widehat{h}}{\partial x^3} - \frac{4}{x^2}\frac{\partial^2 \widehat{h}}{\partial x^2}\right] = \frac{d\widehat{\Psi}_{eq}}{dx}\frac{-1}{\sum z_i^2 \nu_i}\sum z_i^2 \nu_i \exp(-z_i\widehat{\Psi}_{eq})\left(\frac{\widehat{\phi}_i}{x} + 1\right) \quad (57)$$

Note that, in the case that the colloidal sphere is uncharged but placed in a flow field such that $\mathbf{u}(r = a) = \mathbf{0}$ and $\mathbf{u}(r \to \infty) = -\mathbf{U}$, the hydrodynamics are described by the equation

$$LLh = 0$$

The solution of this equation is

$$h(r) = \frac{U}{2}\left(r - \frac{3}{2}a + \frac{a^3}{2r^2}\right) \quad (58)$$

yielding the equation for a Stokes flow around a sphere:

$$\mathbf{u}(r) = -U\left(1 - \frac{3}{2r}a + \frac{a^3}{2r^3}\right)\cos\theta \mathbf{e}_r + U\left(1 - \frac{3}{4r}a + \frac{a^3}{4r^3}\right)\sin\theta \mathbf{e}_\theta \quad (59)$$

5.1. Boundary Conditions at $r = a$

From the no-slip condition,

$$\mathbf{u}(r = a) = \mathbf{0} \quad (60)$$

we obtain

$$\widehat{h}(\kappa a) = 0 \quad (61)$$

$$\left(\frac{d\widehat{h}}{dr}\right)_{x=\kappa a} = 0 \quad (62)$$

5.2. Boundary Conditions Far Away

Far away from the particle,

$$\mathbf{u}(r \to \infty) = -\mu\mathbf{E}_0 \quad (63)$$

yielding

$$\frac{2}{x}\widehat{h} = \widehat{\mu} \text{ for } x \to \infty \tag{64}$$

$$\frac{1}{x}\frac{d}{dx}\left[x\widehat{h}\right] = \widehat{\mu} \text{ for } x \to \infty \tag{65}$$

6. Analytical Solutions

6.1. Full Solution as Function of Integrals

Equation (57) can be solved analytically using boundary conditions Equations (61), (62), (64) and (65). The solution of this fourth-order linear differential equation can be obtained using the method of variation of parameters and the Wronskian of the general solutions (for the method, see [10], p. 331) to yield

$$\begin{aligned}\widehat{h}(x) &= \int_{\kappa a}^{\infty}\left[-\frac{x^3}{30}-\frac{(\kappa a)^5}{20x^2}+\frac{(\kappa a)^3}{12}+y^2\left(\frac{x}{6}+\frac{(\kappa a)^3}{12x^2}-\frac{(\kappa a)}{4}\right)\right]\widehat{M}(y)dy \\ &+\frac{\widehat{\mu}}{2}\left(x-\frac{3}{2}\kappa a+\frac{(\kappa a)^3}{2x^2}\right)+\int_{\kappa a}^{x}\left(\frac{x^3}{30}-\frac{xy^2}{6}+\frac{y^3}{6}-\frac{y^5}{30x^2}\right)\widehat{M}(y)dy\end{aligned} \tag{66}$$

with

$$\widehat{\mu} = \frac{(\kappa a)^2}{9}\int_{\kappa a}^{\infty}\left(1-3\frac{x^2}{(\kappa a)^2}+2\frac{x^3}{(\kappa a)^3}\right)\widehat{M}(x)dx \tag{67}$$

and

$$\begin{aligned}\widehat{M}(x) &= \frac{d\widehat{\Psi}_{eq}}{dx}\frac{-1}{\sum z_i^2\nu_i}\sum z_i^2\nu_i\exp(-z_i\widehat{\Psi}_{eq})\left(\frac{\widehat{\phi}_i}{x}+1\right) \\ &= \frac{d\widehat{\Psi}_{eq}}{dx}\frac{-1}{2}\sum\exp(-z_i\widehat{\Psi}_{eq})\left(\frac{\widehat{\phi}_i}{x}+1\right) \quad \text{for a 1–1 electrolyte}\end{aligned} \tag{68}$$

The term that multiplies the term $\widehat{\mu}/2$ is the same as the one found for the Stokes flow in Equation (58). One can verify that combining Equations (66) and (67) yields the solutions presented by both Ohshima [11] and Jayaraman et al. [3], which they write (inserting Equation (67) in Equation (66))

$$\begin{aligned}\widehat{h}(x) &= -\left(\frac{x^3}{30}+\frac{(\kappa a)^5}{45x^2}-\frac{(\kappa a)^2 x}{18}\right)\int_{\kappa a}^{\infty}\widehat{M}(y)dy+\int_{\kappa a}^{x}\left(\frac{x^3}{30}-\frac{xy^2}{6}+\frac{y^3}{6}-\frac{y^5}{30x^2}\right)\widehat{M}(y)dy \\ &+\int_{\kappa a}^{\infty}\left(\frac{x}{9\kappa a}-\frac{1}{6}+\frac{(\kappa a)^2}{18x^2}\right)y^3\widehat{M}(y)dy\end{aligned} \tag{69}$$

6.2. Approximated Analytical Solution

As a first approximation, we will use the relation Equation (50) assuming that $\widehat{\phi}_i$ can be approximated by $\widehat{\phi_i^+}$ inside the double layer. It follows that

$$\begin{aligned}\widehat{M}(x) &\sim \left(1-\frac{(\kappa a)^3}{x^3}\beta\right)\frac{1}{\sum z_i^2\nu_i}\frac{d}{dx}\left[\sum z_i\nu_i\exp(-z_i\widehat{\Psi}_{eq})\right] \\ \widehat{M}(x) &\sim \left(1-\frac{(\kappa a)^3}{x^3}\beta\right)\frac{d\widehat{\Psi}_{eq}}{dx}\left[\cosh(\widehat{\Psi}_{eq})\right] \quad \text{for a 1–1 electrolyte}\end{aligned} \tag{70}$$

We will now follow the procedure adopted by Ohshima (see p. 83 in [1]), who made the observation that the function

$$\left(\frac{\widehat{\phi}_i}{x}+1\right)\sim\left(1-\frac{(\kappa a)^3}{x^3}\beta\right) \tag{71}$$

is varying very slowly compared to the other functions and can hence be set outside the integral. We note that this assumption is equivalent to assume that there is a local equilibrium and that the electrochemical potentials (we recall that $\delta\widetilde{\mu}_i=-z_ie[\phi_i(r)+r]$) are slowly varying through the double layer. This assumption was already tested in [4], where it was found that it did not hold for low κa.

Using the assumption, we find

$$\widehat{\mu}=\frac{(\kappa a)^2}{9}\left(1-\frac{(\kappa a)^3}{x_1^3}\beta\right)\int_{\kappa a}^{\infty}\left(1-3\frac{x^2}{(\kappa a)^2}+2\frac{x^3}{(\kappa a)^3}\right)\frac{1}{\sum z_i^2\nu_i}\frac{d}{dx}\left[\sum z_i\nu_i\exp(-z_i\widehat{\Psi}_{eq})\right]dx \tag{72}$$

where x_1 is the position where the function to be integrated is maximum. Ohshima has a less general expression as he studies the case where $\widehat{\Psi}_0$ is small. In Ohshima's case, $\beta=-1/2$, which corresponds to the case $\widehat{\Psi}_0\ll 1$ [4].

Integrating by parts, it is found that

$$\widehat{\mu}=\frac{2}{3}\left(1-\frac{(\kappa a)^3}{x_1^3}\beta\right)\int_{\kappa a}^{\infty}\left(\left(1-\frac{x}{(\kappa a)}\right)\right)\frac{x}{\sum z_i^2\nu_i}\left[\sum z_i\nu_i\exp(-z_i\widehat{\Psi}_{eq})\right]dx \tag{73}$$

We now make the approximation that $\widehat{\Psi}_{eq}(x)$ can be approximated by $\widehat{\Psi}_{eq}^{+}(x)$ in the double layer (which also implies that we assume that $\widehat{\Psi}_{app}\sim\widehat{\Psi}_0$) and use Equation (37) to estimate $\widehat{\Psi}_{eq}$. This yields

$$\widehat{\mu}=\frac{-2\kappa a}{3}\widehat{\Psi}_0\left(1-\frac{(\kappa a)^3}{x_1^3}\beta\right)\left[1-\frac{1}{\kappa a}\int_{\kappa a}^{\infty}x\exp(\kappa a-x)dx\right] \tag{74}$$

Integrating by parts, one obtains

$$\widehat{\mu}=\frac{2}{3}\widehat{\Psi}_0\left(1-\frac{(\kappa a)^3}{x_1^3}\beta\right) \tag{75}$$

We have found, inspired by Ohshima [1] and trial and error, that a good estimation of x_1 for a large range of κa and $\widehat{\Psi}_0$ is given by (see Appendix C for a discussion on x_1)

$$x_1=\kappa a+\frac{2.5}{1+2\exp(-\kappa a)\exp\left(-\widehat{\Psi}_0\right)} \tag{76}$$

In the following section, the comparison between Equation (75) and the numerical solution will be discussed. In particular, the following three hypotheses, formulated above, should be studied:

Hypothesis 1. *$\widehat{\phi}_i$ can be approximated by $\widehat{\phi_i^{+}}$ (given by Equation (50)) inside the double layer.*

Hypothesis 2. *$(\widehat{\phi}_i/x+1)$ varies very slowly compared to the other functions in Equation (70).*

Hypothesis 3. *$\widehat{\Psi}_{eq}$ can be approximated by $\widehat{\Psi}_{eq}^{+}$ (given by Equation (37)) in the double layer.*

It can already be anticipated that these conditions hold for low $\widehat{\Psi}_0$ (for which $\widehat{\Psi}_{app}=\widehat{\Psi}_0$) for all κa as this has already been demonstrated by Ohshima [1]. Note that Ohshima uses

$$x_1 = \kappa a + \frac{2.5}{1 + 2\exp(-\kappa a)} \tag{77}$$

which reduces to Equation (76) for low $\widehat{\Psi}_0$.

7. Comparison Between the Use of Equations (76) and (77)

The numerical results presented in this section (symbols) are obtained using a FORTRAN code, which solves the appropriate set of electrokinetic equations using the method from [12], which is an improvement on the Nordsieck method used previously in [4]. In particular, this numerical method enables studying the range of small κa, which was not possible using the Nordsieck method. Additional numerical calculations were performed using the Finite Element software COMSOL Multiphysics v. 6.3 ([13]). Some of the spatial representations of the analyzed quantities for selected values of κa are given in Appendix D. In all cases, the calculations performed using COMSOL match those obtained using FORTRAN. We emphasize that one of the hypotheses of the article is that we do not consider any Stern layer, and that the slip plane is located on the surface of the particle, implying that $\widehat{\Psi}_0 = \widehat{\zeta}$ (the surface electric potential is the zeta potential).

In [4], it was found that Equation (75) using Equation (77) did not perform well at low applied electric field frequencies (see Figure 4A in [4]) but was a good match for high electric field frequencies (see Figure 4C in [4]). The mobility and surface electric potential are linked by a function that we define as Henry's function f_{Henry}:

$$\widehat{\mu} = f_{\text{Henry}} \widehat{\Psi}_0 \tag{78}$$

The name Henry's function is referring to the function originally derived by Henry [14] for which Ohshima provided a simplified version [1]. Henry's (and Oshima's) derivation holds for low $\widehat{\Psi}_0$, for which f_{Henry} is a function of κa only. Henry's function $f_{\text{Henry}} = \widehat{\mu}/\widehat{\Psi}_0$ is given in Figure 1. In the case of low potential ($\widehat{\Psi}_0 = 0.01$, giving $\Psi_0 = 0.01 \times 25$ mV if we assume that $kT/e = 25$ mV), one can use the expressions for the dipolar coefficient given in Appendix B to verify that, in good approximation, $\beta = -1/2$. Using this value for the dipolar coefficient and the fact that $\widehat{\Psi}_0$ is low enables recovering the expression found by Ohshima for Henry's function, as discussed in the previous section. This function (in cyan) is not to be distinguished from the black curve, which represents the original Henry's function. It can be verified that Henry's function (as well as the condition $\beta = -1/2$) holds for potentials up to $\widehat{\Psi}_0 = 0.5$. For higher $\widehat{\Psi}_0$, Henry's function f_{Henry} becomes a function of both κa and $\widehat{\Psi}_0$, as illustrated in Figure 1.

One can see that using Equation (76) for x_1 enables obtaining a better approximation at low frequencies (here, we use zero frequency) than using Equation (77), which was used in [4]. It was verified that using Equation (76) did not change the quality of the prediction at high frequencies and that the match between analytical and numerical solutions in this case is as good as in Figure 4C in [4].

The agreement between numerical calculation and analytical theory is very good for low $\widehat{\Psi}_0$, even for $\widehat{\Psi}_0 = 2$. For higher $\widehat{\Psi}_0$, the agreement for $\kappa a > 10$ is still good but deviates strongly from the numerical calculations at lower κa. The reason for this deviation will be illustrated in the following subsection. Note that the numerical calculations for very low κa go asymptotically to a value of 0.7 instead of $2/3 = 0.66$, which is the Hückel limit. This deviation was also observed for the COMSOL calculations.

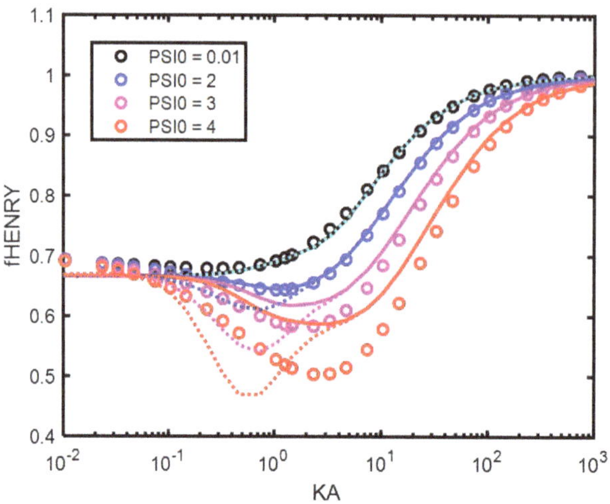

Figure 1. Henry's function f_{Henry} (dimensionless units) as function of κa. The colloidal particle has a radius $a = 1000$ nm and is immersed in a KCl electrolyte. The different surface electric potentials ($\widehat{\Psi}_0 = \widehat{\zeta}$) used are given in the legend. Symbols: numerical results by solving the set of electrokinetic equations. Full curves (except black one): analytical theory using Equations (75) and (76). Black curve: original analytical solution of Henry [14]. Dashed curves: analytical theory using Equations (75) and (77).

8. The Electrophoretic Mobility for Different κa

In order to study the behavior of the electrophoretic mobility as function of ionic strength, we concentrate on the case where $\widehat{\Psi}_0 = 2$, for which (see Figure 1) the agreement between numerical calculation and analytical theory is very good. From Figure 2, upper panel, we can compare the functions $\left(\widehat{\phi_+}/x + 1\right)$ and $\left(\widehat{\phi_-}/x + 1\right)$ with the function $\left(1 - (\kappa a)^3 \beta / x^3\right)$ from which we deduce that Hypothesis 1 is better for higher κa. This holds for all $\widehat{\Psi}_0$ tested ($\widehat{\Psi}_0 = 0.01 - 6$). By comparing the curves $(\widehat{\phi}_i / x + 1)$ with the curve $\left(d\widehat{\Psi}_{eq}/dx\right) \times \cosh(\widehat{\Psi}_{eq})$, one finds that Hypothesis 2 is fulfilled for not too low κa. This holds for all $\widehat{\Psi}_0$ tested ($\widehat{\Psi}_0 = 0.01 - 6$). Hypothesis 3 (the fact that $\widehat{\Psi}_{eq}$ is in good approximation given by $\widehat{\Psi}^+_{eq}$) is fulfilled in all cases, as can be verified by the lower panel of Figure 2. This hypothesis breaks somewhat down for $\widehat{\Psi}_0 > 4$, but, for such high surface potentials and especially at moderate/high κa, the decay of $\widehat{\Psi}_{eq}$ is very fast, leading to the problem of numerical accuracy.

From Figure 1, we observe that, at high surface potential (also at $\widehat{\Psi}_0 = 2$), a minimum appears in f_{Henry}. This minimum is often reported, but its origin is worth discussing. This can be best performed by studying Equation (75). From that equation, one can directly see that the change in curvature is linked to the change in the sign of β: at low κa, we have $\beta > 0$, and β is decreasing when κa is increasing. As $(\kappa a)^3 / x_1^3$ is increasing, the mobility is decreasing, leading to a lowering of $\widehat{\mu}$. One can estimate that, at low κa, $(\kappa a)^3 / x_1^3$ scales as $(\kappa a)^3$. According to the estimation leading to Equation (96), β decreases as $(\kappa a)^{-2}$, which implies that $\widehat{\mu}$ decreases as (κa). At high κa, we have $\beta < 0$, and $|\beta|$ is increasing when κa is increasing, leading to an increase in $\widehat{\mu}$. In Figure 3, both the mobility $\widehat{\mu}$ and the dipolar coefficient β are given as a function of κa. From the points labeled in the figures, one can see that the position where $\beta = 0$ is close to the position where the mobility $\widehat{\mu}$ changes its curvature. Also note how high the dipolar coefficient β becomes at low κa.

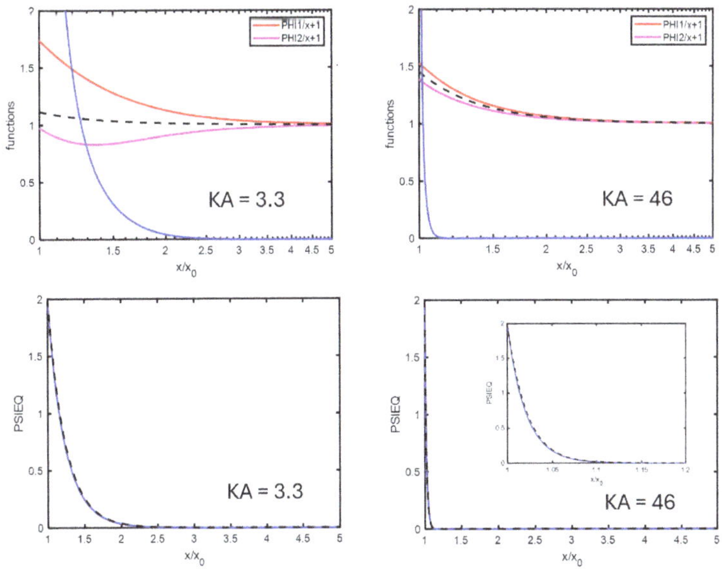

Figure 2. (**Upper panel**): $\left(\widehat{\phi_+}/x+1\right)$ and $\left(\widehat{\phi_-}/x+1\right)$ in red and magenta as function of $x/x_0 = \kappa r/\kappa a = r/a$ for two different κa, as found by numerically solving the set of electrokinetic equations. The blue curve represents the function $\left(d\widehat{\Psi}_{eq}/dx\right) \times \cosh(\widehat{\Psi}_{eq})$, also evaluated numerically. The colloidal particle has a radius $a = 1000$ nm and is immersed in a KCl electrolyte. The black dashed curve represents the function $\left(1-(\kappa a)^3\beta/x^3\right)$, whereby the dipolar coefficient β is the same numerically and analytically (see Figure 3). (**Lower panel**): $\widehat{\Psi}_{eq}(x)$ for different κa as indicated in the figures: blue curves represent the numerical calculations and dashed black curves the analytical approximation Equation (38).

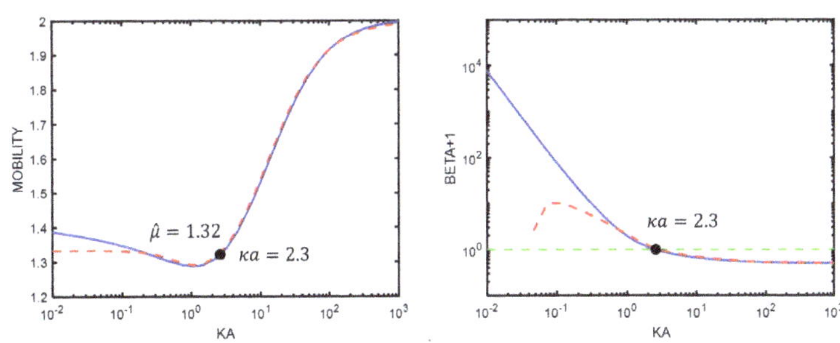

Figure 3. (**Left**): electrophoretic mobility $\widehat{\mu}$ as function of κa. (**Right**): $(\beta+1)$ as function of κa. The blue curves are found numerically. The green dashed line is the function defined by $\beta+1=1$. The red dashed lines represent the curves according to Equations (75) and (A9). The colloidal particle has an electric surface potential $\widehat{\Psi}_0 = 2$, a radius $a = 1000$ nm, and is immersed in a KCl electrolyte. The black dots represent the position where $\beta = 0$.

8.1. The Role of the Double Layer

As mentioned in the previous section, the behavior of the mobility $\widehat{\mu}$ can be directly linked to the dipole coefficient β. Far from the particle and its double layer, the system (particle + double layer) can be seen as an electric dipole, creating a local electric field that influences the particle's velocity (mobility). Because the particle moves with a constant velocity, the sum of all forces exerted on the particle must be zero. This is discussed in detail in [3], where the link with the notations of Overbeek and Wiersema [15,16] is made. We adopt a different approach here. In particular, we do not restrict ourselves to low surface potentials. The colloidal particle is subjected to two forces: one electric ($\mathbf{F}_e$) and one hydrodynamic ($\mathbf{F}_{drag}$). These forces are defined by

$$\begin{aligned} \mathbf{F}_e &\equiv Q_{dyn}\mathbf{E}_0 \\ \mathbf{F}_{drag} &= -6\pi\eta a\mathbf{U} \end{aligned} \tag{79}$$

where $\mathbf{E}_0$ is the applied electric field, which is the electric field far away from the particle as the particle is assumed to be alone in the electrolyte solution. This electric field is different from the electric field found close to the colloidal sphere of charge Q. The dynamic charge Q_{dyn} is defined by the relation given above. The force $\mathbf{F}_{drag}$ is the Stokes drag force and does not account for all the hydrodynamic forces defined when studying the electrokinetic behavior of electrolytes [3]. Any "relaxation force", due to the applied electric field or to the velocity of the particle, is accounted for in the force $\mathbf{F}_e$. By using the balance of forces ($\|\mathbf{F}_e\| = \left\|\mathbf{F}_{drag}\right\|$), one obtains, using Equation (78),

$$Q_{dyn} = 6\pi\eta a\mu = 6\pi a\varepsilon_0\varepsilon_1 f_{\text{Henry}}\Psi_0 \tag{80}$$

The total electric field is defined, for any position around the particle, by

$$\begin{aligned} \widehat{E}_r &= -\left[\frac{d\widehat{\Psi}_{eq}}{dx}\frac{kT\kappa}{eE_0} + \left(\frac{d\widehat{\psi}(x)}{dx} - 1\right)\cos(\theta)\right] \\ \widehat{E}_\theta &= \left(\frac{\widehat{\psi}(x)}{x} - 1\right)\sin(\theta) \end{aligned} \tag{81}$$

The velocities are given by

$$\begin{aligned} \widehat{u}_r &= \frac{-2}{x}\widehat{h} \\ \widehat{u}_\theta &= \frac{1}{x}\frac{d}{dx}\left[x\widehat{h}\right] \end{aligned} \tag{82}$$

8.1.1. Hückel's Approximation ($\kappa a \lll 1$)

When the double layer is thick ($a \ll \kappa^{-1}$), the equilibrium electric potential is given in good approximation by

$$\Psi_{eq}(r) = \Psi_0\frac{a}{r} \tag{83}$$

In this case, the potential decays over distances comparable to the particle size a instead of the double layer thickness κ^{-1}. This potential corresponds to the Coulomb potential around a sphere as if there were no electric double layer. In that case, one has

$$Q_{dyn} = Q \quad \text{for} \quad \kappa a \ll 1 \tag{84}$$

Using Gauss' relation, Equation (31), one finds

$$\Psi_0 = \frac{\sigma_S a}{\varepsilon_0\varepsilon_1} \tag{85}$$

Using the relation between charge and particle surface charge, viz

$$\sigma_s = \frac{Q}{4\pi a^2} \tag{86}$$

one obtains

$$\Psi_0 = \frac{Q}{4\pi a \varepsilon_0 \varepsilon_1 a} \tag{87}$$

From Equation (80), one obtains

$$f_{\text{Henry}} = 2/3 \tag{88}$$

which implies that

$$\begin{aligned} \mu &= \frac{2}{3}\frac{\varepsilon_0 \varepsilon_1}{\eta}\Psi_0 \\ &= \frac{2}{3}\frac{\sigma_s a}{\eta} \end{aligned} \tag{89}$$

For colloidal particles, contrary to ions, it is quite uncommon to have small κa as a simple estimation of the double layer thickness for usual ionic strengths shows. For the purpose of illustration, the following examples are conducted with a 1000 nm colloidal particle at unrealistically low κa.

One case of such small κa is illustrated in Figure 4, where a potential of $\widehat{\Psi}_0 = 0.01$ is used. The curve plotted using Equation (83) is not to be distinguished from the numerical one. In Figure 4, the velocities $\widehat{u}_r$ and $\widehat{u}_\theta$ are plotted. Their values at long distances are $\widehat{u}_r(x \to \infty) = -\widehat{u}_\theta(x \to \infty) = \widehat{\mu} = 0.0068 = 2/3 \times \widehat{\Psi}_0$ (yielding $\widehat{\Psi}_0 = 0.0102$ instead of $\widehat{\Psi}_0 = 0.01$). The same value of $\widehat{\mu} = 0.0068$ was obtained by evaluating the mobility using Equation (67). It can also be seen that, for the whole x range

$$\widehat{\psi}(x) = -\widehat{\phi}_+(x) = -\widehat{\phi}_-(x) = \frac{(\kappa a)^3 \beta}{x^2} \quad \text{with} \quad \beta = -1/2 \tag{90}$$

implying that $\widehat{\psi}^+ = \widehat{\psi}$ and $\widehat{\phi}_i^+ = \widehat{\phi}_i$ (note the small mismatch between analytical and numerical results at low x for $\widehat{\psi}$). The tangential electric field $\widehat{E}_\theta$ is also plotted for $\sin(\theta) = 1$. It can be evaluated that

$$\widehat{E}_\theta(\kappa a) = (\beta - 1)\sin(\theta) = \frac{-3}{2}\sin(\theta) \tag{91}$$

which can be verified from the figure. One can also demonstrate that, for the whole x range

$$\begin{aligned} \delta\widehat{E}_r &= \left(1 + \frac{(\kappa a)^3}{x^3}\right)\cos(\theta) \\ \delta\widehat{E}_\theta &= \left(-1 - \frac{(\kappa a)^3}{2x^3}\right)\sin(\theta) \end{aligned} \tag{92}$$

One also obtains, from Equation (36),

$$E_2 = \frac{-3}{2}E_0 \tag{93}$$

For large surface potentials and/or extremely low κa (for $\kappa a < 1$, see Figure 2), it was observed that the dipolar coefficient would deviate from $\beta = -1/2$ and rapidly increase with decreasing κa, becoming positive and reaching extremely high values. In that case as well, as indicated in Figure 2, Hypothesis 2 does not hold anymore. Despite the very high values of β, as β scales roughly with $(\kappa a)^{-2}$ and the prefactor in Equation (75) with $(\kappa a)^3$,

Equation (89) remains satisfied. A rough estimation to obtain the dependence of β on κ^{-2} can be achieved as follows. The characteristic timescale associated with the double layer is

$$\tau \sim \frac{1}{D\kappa^2} \tag{94}$$

where D is an ionic diffusion coefficient. The velocity associated with the deformation of the double layer (for small κa) is

$$v \sim \frac{Q}{\eta a} E_0 \tag{95}$$

This implies that the dipolar coefficient can be estimated by

$$\beta \sim P \sim Q\frac{v}{\tau} \tag{96}$$

indeed yielding $\beta \sim \kappa^{-2}$.

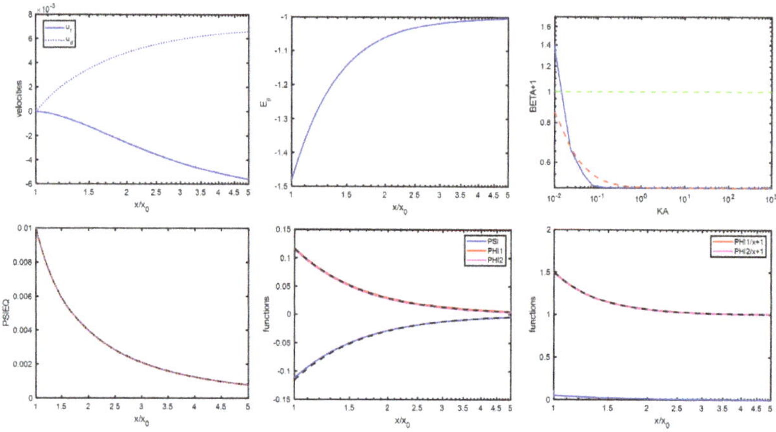

Figure 4. (**Top row**): velocities $\widehat{u}_r$ and $\widehat{u}_\theta$ as function of $x/x_0 = r/a$, electric field $\delta\widehat{E}_\theta(=\widehat{E}_\theta)$, and $(\beta+1)$ as function of κa. The green dashed line is the function defined by $\beta+1=1$. The red dashed lines represent the curves according to Equations (75) and (A9). (**Bottom row**): $\widehat{\Psi}_{eq}$ (red) $\widehat{\phi_+}$ (red), $\widehat{\phi_-}$ (magenta), and $\widehat{\psi}$ (blue) as function of r/a, as found numerically. Dashed and dotted black curves are approximations (see text for details). The numerical functions $\left(\widehat{\phi_+}/x+1\right)$ and $\left(\widehat{\phi_-}/x+1\right)$ in red and magenta as function of $x/x_0 = r/a$ can be observed to decay over the same length as the blue curve, which represents the function $\left(d\widehat{\Psi}_{eq}/dx\right) \times \cosh(\widehat{\Psi}_{eq})$ (also evaluated numerically), which contradicts Hypothesis 2 (Hypotheses 1 and 3 are fulfilled). The colloidal particle has an electric surface potential $\widehat{\Psi}_0 = 0.01$, has a radius $a = 1000$ nm, and is immersed in a KCl electrolyte with $\kappa a = 0.23$.

8.1.2. Smoluchowski' s Approximation ($\kappa a \gg 1$)

When the double layer is thin ($a \gg \kappa^{-1}$), we can approximate that the double layer is not deformed under the influence of the applied electric field. The ionic densities will predominantly vary in the θ direction. When the equilibrium potential is low, it can be given by

$$\Psi_{eq}(r) \simeq \Psi_0 \exp(-\kappa(r-a)) \tag{97}$$

as the characteristic distance over which the electric field is non-zero is $(r-a) \simeq \kappa^{-1}$, which implies that

$$\frac{a}{r} = \frac{a}{a+(r-a)} \simeq \frac{\kappa a}{\kappa a + 1} \simeq 1 \tag{98}$$

Equation (97) is also the distribution found in the case of a planar surface. The main difference with the planar case, as we will see, is that the electrophoretic mobility is mainly influenced by the asymmetric ionic distribution in the θ direction due to the application of an electric field (in the planar case, because of symmetry, there cannot be ionic gradients in the direction parallel to the plane). From Gauss' relation given by Equation (31), it then follows that

$$\Psi_0 = \frac{\sigma_s}{\varepsilon_0\varepsilon_1\kappa} \tag{99}$$

We now do not make the assumption that the potential is low. One can consider the system composed of the charged colloidal particle and its extremely thin double layer as an electroneutral system, for which the electric potential $\delta\Psi$ should obey the Laplace equation and the associated boundary condition (where $a^+ = a + \kappa^{-1}$)

$$\begin{aligned} \nabla^2\delta\Psi(\mathbf{r}) &= 0 \\ (\nabla_r\delta\Psi)_{r=a^+} &= 0 \end{aligned} \tag{100}$$

Solving these two equations yields

$$\begin{aligned} \delta\Psi^+ &= -\left(1 + \frac{1}{2}\frac{a^3}{r^3}\right)\mathbf{E}_0 \cdot \mathbf{r} \\ E_\theta^+ &= \left(\frac{-1}{r}\frac{\partial\delta\Psi}{\partial\theta}\right)_{r=a^+} = \frac{-3}{2}E_0\sin(\theta) \end{aligned} \tag{101}$$

As the pressure does not vary in the tangential (θ) direction, using the fact that $\nabla\Psi = -E_\theta^+\mathbf{e}_\theta$, and using Poisson 's equation in Equation (22), the Navier–Stokes equation, Equation (51), becomes in the tangential direction

$$\eta\frac{\partial^2 u_\theta^+}{\partial r^2} = \varepsilon_0\varepsilon_1\frac{\partial^2\Psi}{\partial r^2}E_\theta^+ \tag{102}$$

Integrating Equation (102) between an arbitrary position r and $r \to \infty$, using the fact that $\partial u_\theta^+/\partial r(r \to \infty) = 0$ and $\partial\Psi/\partial r(r \to \infty) = 0$ yields

$$\eta\frac{\partial u_\theta^+}{\partial r} = \varepsilon_0\varepsilon_1\frac{\partial\Psi}{\partial r}E_\theta^+ \tag{103}$$

Integrating again, this time between $r = a$ and $r = a^+$, making the assumption that $E_\theta = E_\theta^+$ in the double layer (see Figure 5), we find

$$\eta\left[0 - u_\theta^+(a^+)\right] = \varepsilon_0\varepsilon_1[\Psi(a) - 0]E_\theta^+ \tag{104}$$

The tangential fluid velocity $u_\theta^+(a^+)$ can be estimated by realizing that, since there is no net force on the liquid and no pressure gradient is applied, the liquid flow must have a potential nature

$$\mathbf{u}(\mathbf{r}) = -\nabla\Phi(\mathbf{r}) \tag{105}$$

As the fluid is incompressible, $\nabla \cdot \mathbf{u} = \mathbf{0}$ and

$$\nabla^2\Phi(\mathbf{r}) = 0 \tag{106}$$

As the liquid cannot penetrate the colloidal particle,

$$(\nabla_r\Phi)_{r-a} = 0 \tag{107}$$

Combining these equations yields

$$\begin{aligned} \Phi &= -\left(1+\frac{1}{2}\frac{a^3}{r^3}\right)\mathbf{u}_\infty\cdot\mathbf{r} \\ u_\theta^+ &= \left(\frac{-1}{r}\frac{\partial\Phi}{\partial\theta}\right)_{r=a^+} = \frac{-3}{2}u_\infty\sin(\theta) = \frac{3}{2}\mu E_0\sin(\theta) \end{aligned} \tag{108}$$

From Equations (101), (104) and (108), we obtain

$$-\eta\frac{3}{2}\mu E_0\sin(\theta) = \varepsilon_0\varepsilon_1\Psi_0\frac{-3}{2}E_0\sin(\theta) \tag{109}$$

yielding

$$\mu = \frac{\varepsilon_0\varepsilon_1}{\eta}\Psi_0 \tag{110}$$

from which we deduce that

$$f_{\text{Henry}} = 1 \tag{111}$$

A thorough discussion about the Smoluchowski limit is given in [9].

From Equation (80), one obtains

$$Q_{dyn} = 6\pi a\varepsilon_0\varepsilon_1\Psi_0 \tag{112}$$

In order to obtain a relation between Ψ_0 and Q, one uses Equation (A15) found in Appendix B, which yields

$$Q = \frac{4\pi\varepsilon_0\varepsilon_1 kTa}{e}\kappa a\left[2\sinh\left(\frac{\widehat{\Psi}_0}{2}\right) + \frac{4}{\kappa a}\tanh\left(\frac{\widehat{\Psi}_0}{4}\right) - \frac{\widehat{\Psi}_0}{\kappa a}\right] \tag{113}$$

This relation provides a good estimate for $\kappa a \geq 0.5$. In the limit of low $\widehat{\Psi}_0$, one obtains

$$Q = 4\pi\varepsilon_0\varepsilon_1 a\kappa a\Psi_0 \tag{114}$$

from which it is deduced that

$$Q_{dyn} = \frac{3}{2}\frac{Q}{\kappa a} \quad \text{for } \widehat{\Psi}_0 \ll 1 \text{ and } \kappa a \gg 1 \tag{115}$$

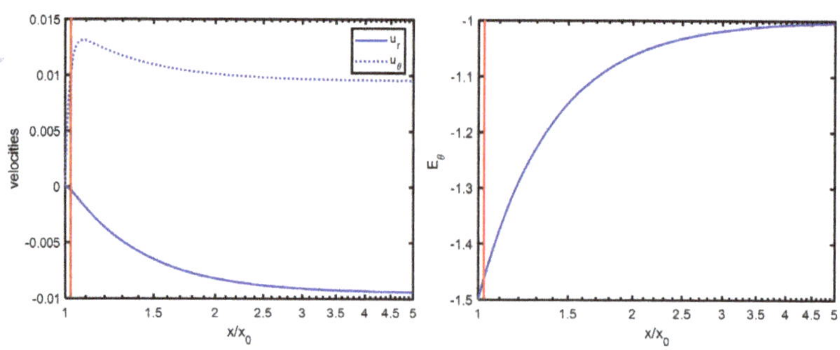

Figure 5. Velocities $\widehat{u}_r$ and $\widehat{u}_\theta$ and electric field $\delta\widehat{E}_\theta(=\widehat{E}_\theta)$ as function of $x/x_0 = r/a$. The red line indicates the position where $x = \kappa a^+ = \kappa a + 1$. The approximation made in the derivation of the Smoluchowski expression, i.e., $E_\theta = E_\theta^+$, seems to be justified. The colloidal particle has an electric surface potential $\widehat{\Psi}_0 = 0.01$, has a radius $a = 1000$ nm, and is immersed in a KCl electrolyte with $\kappa a = 46$.

9. Particle with a Constant Surface Charge

Many articles present theoretical results for the electrophoretic motion of a colloidal particle with a constant surface potential as function of κa. This is done for convenience as a constant surface potential implies a Dirichlet boundary condition (Equation (2)), whereas a constant surface charge implies using a Neumann boundary condition (Equation (31)), which is a bit more complicated to implement. Nonetheless, in practice, colloidal particles tend to have a relatively constant surface charge as a function of ionic strength (for a given pH) and therefore represent an interesting case to study [6,17,18].

In Figure 6, results are shown for a particle of constant surface charge.

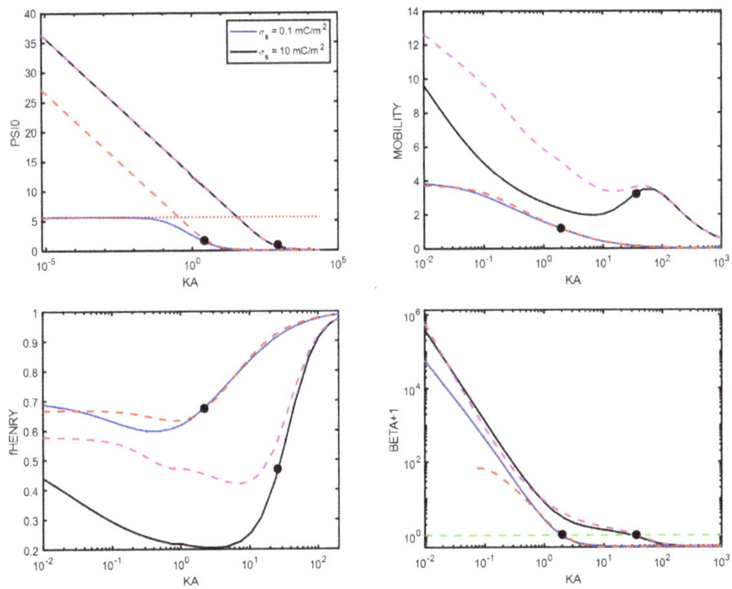

Figure 6. The black and blue curves are found numerically. (**Top row**): surface electric potential $\widehat{\Psi}_0$ and electrophoretic mobility $\widehat{\mu}$ as function of κa. The dotted and dashed lines for $\widehat{\Psi}_0$ represent approximations (see text for details). (**Bottom row**): Henry's function f_{Henry} and $(\beta + 1)$ as function of κa. The green dashed line is the function defined by $\beta + 1 = 1$. The red and magenta dashed curves for $\widehat{\mu}$, f_{Henry}, and β represent the curves according to Equations (75) and (A9). The colloidal particles have a radius $a = 1000$ nm and are immersed in a KCl electrolyte. The black dots represent the position where $\beta = 0$.

One notes that the curve for Henry's function f_{Henry} is very similar to the one given for the constant potential case (see Figure 1) for the same reason that, in the limit of low and high κa, the function reaches Hückel and Smoluchowski limits (which are independent of the fact that one considers a constant surface potential or a constant surface charge).

The approximated dashed functions for high $\widehat{\Psi}_0$ are found using the following relation (see a more accurate formulation, i.e., Equation (A15), in Appendix B)

$$q^* = \frac{e\sigma_s}{\varepsilon_0 \varepsilon_1 kT\kappa} \simeq 2\sinh\left(\frac{\widehat{\Psi}_0}{2}\right) \tag{116}$$

Inverting this equation provides the desired approximation for $\widehat{\Psi}_0$ as function of σ_s and κ. For low surface charge ($\sigma_s = 0.1$ mC/m^2), the surface potential reaches a constant value below $\kappa a = 1$. This can be understood by estimating the potential/charge relation for low $\widehat{\Psi}_0$ by using Equations (31) and (38):

$$\widehat{\Psi}_0 \left(\frac{1}{a} + \kappa \right) = \frac{e\sigma_s}{\varepsilon_0 \varepsilon_1 kT} \tag{117}$$

At low surface charge (implying low surface potential) and low κa (implying no effect of the double layer on the surface potential), the surface potential $\widehat{\Psi}_0$ can be approximated by

$$\widehat{\Psi}_0 \simeq \frac{e\sigma_s a}{\varepsilon_0 \varepsilon_1 kT} \tag{118}$$

For $\sigma_s = 0.1$ mC/m^2, this yields $\widehat{\Psi}_0 \simeq 5.6$, which corresponds to the value found numerically and represented by the red dotted line.

For a particle with a constant surface charge, at low κa, the electrophoretic mobility is given by, using Equations (89) and (117),

$$\mu_{\text{low } \kappa a} = \frac{2}{3} \frac{\sigma_s a}{\eta} \tag{119}$$

This equation is valid for κa values that correspond to physically unrealistic ionic strengths for colloidal particles. In the examples chosen, for the low surface charges of -0.1 mC/m^2 and -0.2 mC/m^2, the relation is valid below a concentration of 10^{-8} mM (!). The limit is not reached for the higher surface charges, not even at 10^{-15} mM. In that case, the curves are superposed below $\kappa a = 1$. At high and increasing κa, the electrophoretic mobility of a particle with constant surface charge is decreasing until reaching zero as its surface electric potential $\widehat{\Psi}_0$ is decreasing rapidly with ionic strength (see Figure 4). For low surface potentials, the decrease is modeled by using Equations (110) and (117)

$$\mu_{\text{high } \kappa a} = \frac{\sigma_s}{\eta \kappa} \tag{120}$$

In Figure 7, it is shown that Equation (120) indeed enables approximating the electrophoretic mobility of colloidal particles for any (constant) surface charge at high κa. Fitting the electrophoretic mobility data for charged colloidal particles with a constant surface charge at high ionic strength with Equation (120) therefore enables obtaining the surface charge without the need for numerical calculations.

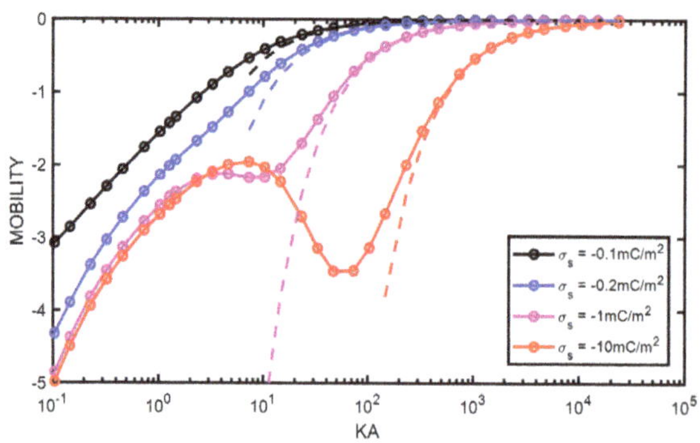

Figure 7. Electrophoretic mobility $\hat{\mu}$ found numerically as function of κa for different values of the (negative) surface charge σ_s as given in the legend. The dashed lines are plotted according to Equation (120). The colloidal particle has a radius $a = 1000$ nm and is immersed in a KCl electrolyte.

10. Conclusions

In the present article, a new analytical equation has been presented for the electrophoretic mobility of a colloidal sphere. It has been shown that the equation predicts the electrophoretic mobility well for the whole range of κa provided that the electric surface potential (where the slip layer is defined) is less than 50 mV ($\widehat{\Psi}_0 = 2$). For higher surface potentials, the analytical prediction deviates from the numerical results below $\kappa a = 10$, and the deviations increase with increasing surface potentials. For experimental conditions, where usually $\kappa a > 10$, the proposed equation is an easy-to-implement alternative to the full numerical solution. From the study of the analytical equation, it was shown that the fact that the electrophoretic mobility decreases with increasing κa for low κa and subsequently increases with increasing κa for high κa is linked to the change in sign of the dipolar coefficient. In the region of intermediate κa, the electrophoretic mobility experiences a change in curvature associated with the presence of a maximum (or minimum, depending on the charge of the particle). As many suspensions consist of colloidal particles with a relatively constant surface charge (for a given pH), it was shown that a simple equation (Equation (120)) enables estimating this surface charge by fitting the electrophoretic mobility as a function of ionic strength at high ionic strength (for $\kappa a > 10$ in Figure 7).

Author Contributions: Conceptualization, A.C.; Methodology, A.C., S.G.-B. and C.C.; Validation, A.C. and C.C.; Formal analysis, S.G.-B. and C.C.; Writing—original draft, C.C.; Writing—review & editing, A.C., S.G.-B. and C.C. All authors have read and agreed to the published version of the manuscript.

Funding: This research received no external funding.

Institutional Review Board Statement: Not applicable.

Data Availability Statement: Data is contained within the article.

Conflicts of Interest: The authors declare no conflicts of interest.

Appendix A. Equivalence with Previous Work

For comparison with the analytical results presented in [4,8], the following variables are introduced:

$$\begin{aligned} n_n &= n_+ + n_- \quad (\text{similarly: } n_{n,eq} = n_{+,eq} + n_{-,eq}) \\ n_c &= n_+ + \left(\frac{z_-}{z_+}\right) n_- \quad (\text{similarly: } n_{c,eq} = n_{+,eq} + \left(\frac{z_-}{z_+}\right) n_{-,eq}) \end{aligned} \tag{A1}$$

We can make the equivalence

$$\begin{aligned} n_{n,eq} &= \nu_+ n_\infty \exp\left(-z_+ \widehat{\Psi}_{eq}\right) + \nu_- n_\infty \exp\left(-z_- \widehat{\Psi}_{eq}\right) \\ &= 2n_\infty \cosh\left(\widehat{\Psi}_{eq}\right) \quad \text{for a 1–1 electrolyte} \\ n_{c,eq} &= \nu_+ n_\infty \left[\exp\left(-z_+ \widehat{\Psi}_{eq}\right) - \exp\left(-z_- \widehat{\Psi}_{eq}\right)\right] \\ &= -2n_\infty \sinh\left(\widehat{\Psi}_{eq}\right) \quad \text{for a 1–1 electrolyte} \end{aligned}$$

Similarly,

$$\begin{aligned} n_n &= \frac{-e}{kT\kappa}\left[z_+ n_{+,eq}\left(\widehat{\psi} + \widehat{\phi_+}\right) + z_- n_{-,eq}\left(\widehat{\psi} + \widehat{\phi_-}\right)\right] \\ n_c &= \frac{-e}{kT\kappa}\left[z_+ n_{+,eq}\left(\widehat{\psi} + \widehat{\phi_+}\right) + \left(\frac{z_-}{z_+}\right) z_- n_{-,eq}\left(\widehat{\psi} + \widehat{\phi_-}\right)\right] \end{aligned} \tag{A2}$$

For a 1–1 electrolyte, these equations reduce to

$$\begin{aligned} n_n &= \frac{\varepsilon_0\varepsilon_1}{e}\kappa\left[\sinh\left(\widehat{\Psi}_{eq}\right)\widehat{\psi} - \frac{1}{2}\left(\exp\left(-\widehat{\Psi}_{eq}\right)\widehat{\phi_+} - \exp\left(\widehat{\Psi}_{eq}\right)\widehat{\phi_-}\right)\right] \\ n_c &= \frac{\varepsilon_0\varepsilon_1}{e}\kappa\left[-\cosh\left(\widehat{\Psi}_{eq}\right)\widehat{\psi} - \frac{1}{2}\left(\exp\left(-\widehat{\Psi}_{eq}\right)\widehat{\phi_+} + \exp\left(\widehat{\Psi}_{eq}\right)\widehat{\phi_-}\right)\right] \end{aligned} \tag{A3}$$

It can be shown that, far from the particle [8],

$$L\psi = \frac{1}{r}\frac{d^2}{dr^2}(r\psi) - \frac{2\psi}{r^2} = \frac{-e}{\varepsilon_0\varepsilon_1}\sum z_i n_i = \frac{-ez_+}{\varepsilon_0\varepsilon_1}n_c$$

with

$$\begin{aligned} n_n^+(r) &= \frac{C_n}{r^2} \\ n_c^+(r) &= C_c\frac{1+\kappa r}{r^2}\exp(-\kappa(r-a)) \simeq 0 \text{ beyond the double layer} \end{aligned} \tag{A4}$$

yielding

$$L\psi^+ = 0 \tag{A5}$$

The solution of this equation is

$$\begin{aligned} \psi^+(r) &= \frac{a^3\beta}{r^2} \\ \widehat{\psi}^+(x) &= \frac{(\kappa a)^3\beta}{x^2} \end{aligned} \tag{A6}$$

where β is the dipolar coefficient. As

$$n_c^+ = \frac{-ez_+\nu_+ n_\infty}{kT\kappa}\left[\left(\widehat{\psi^+} + \widehat{\phi_+^+}\right) - \left(\frac{z_-}{z_+}\right)\left(\widehat{\psi^+} + \widehat{\phi_-^+}\right)\right] \tag{A7}$$

it follows that, using Equation (16) and the fact that $n_i^+ \simeq 0$

$$\phi_+^+(r) = \phi_-^+(r) = \frac{-a^3\beta}{r^2} \tag{A8}$$

Appendix B. An Analytical Expression for β

An analytical expression for β has been derived in [4]. In [8], the link was made between dielectric spectroscopy experiments and dipolar coefficient β. The analytical expression for the dipolar coefficient for $\omega = 0$ for a 1–1 electrolyte is given by

$$\beta = \frac{-K_1 + 2[K_{//} + K_U] + K_\perp}{2K_1 + 2\left[K_{//}(\kappa a/x_0)^3 + K_U(\kappa a/x_1)^3 - K_\perp\right]} \tag{A9}$$

The conductivities K_i are defined by

$$\begin{aligned} K_1 &= \varepsilon_1\varepsilon_0\kappa^2 D_0 \\ K_{//} &= -K_1 I_{n,eq} - \frac{\left[I_{c,eq}^2 - I_{n,eq}^2\right]}{(x_0/\kappa a)^3} \\ K_\perp &= \frac{K_1 I_{n,eq}}{(x_0/\kappa a)^3} \\ K_U &= -K_1 m\widehat{\Psi}_0 I_{c,eq}\left[\frac{1}{4(x_0/\kappa a)^3} - 1\right] \end{aligned} \tag{A10}$$

where

$$\begin{aligned}
x_0 &= 1 + \kappa a + \frac{3}{\kappa a}\exp\left(\frac{-\widehat{\Psi}_0}{2}\right) \\
x_1 &= \kappa a + \frac{2.5}{1 + 2\exp(-\kappa a)\exp\left(-\widehat{\Psi}_0\right)} \\
m &= \frac{2}{3}\frac{\varepsilon_0\varepsilon_1(kT)^2}{\eta D_0 e^2} \\
D_0 &= \frac{z_+ D_+ - z_- D_-}{z_+ - z_-}
\end{aligned} \tag{A11}$$

and

$$\begin{aligned}
I_{c,eq} &= \frac{1}{(\kappa a)^2}\int_{\kappa a}^{x_0} x\sinh(\Psi_{eq})dx \\
I_{n,eq} &= \frac{-1}{(\kappa a)^2}\int_{\kappa a}^{x_0} x\left[\cosh\left(\widehat{\Psi}_{eq}\right) - 1\right]dx
\end{aligned} \tag{A12}$$

Approximated analytical expressions for β, $I_{c,eq}$, and $I_{n,eq}$ can be found in [4,8].

Appendix B.1. Low $\widehat{\Psi}_0$

In the case of low surface potentials, one finds that [8]

$$\begin{aligned}
I_{c,eq} &\sim \frac{\widehat{\Psi}_0}{\kappa a} \\
I_{n,eq} &\sim 0
\end{aligned} \tag{A13}$$

yielding

$$\begin{aligned}
K_{//} &= \left(\widehat{\Psi}_0\right)^2\frac{\kappa a}{x_0^3} \\
K_{\perp} &= 0 \\
K_U &= -K_1 m\left(\widehat{\Psi}_0\right)^2\frac{1}{\kappa a}\left[\frac{1}{4(x_0/\kappa a)^3} - 1\right]
\end{aligned} \tag{A14}$$

Appendix B.2. High $\widehat{\Psi}_0$

For high surface potentials, one finds that [8]

$$\begin{aligned}
I_{c,eq} &\sim \frac{q^*}{\kappa a} = -I_{n,eq} \\
q^* &= 2\sinh\left(\frac{\widehat{\Psi}_0}{2}\right) + \frac{4}{\kappa a}\tanh\left(\frac{\widehat{\Psi}_0}{4}\right) - \frac{\widehat{\Psi}_0}{\kappa a}
\end{aligned} \tag{A15}$$

where q^* is the dimensionless surface charge density, and one finds that

$$\begin{aligned}
K_{//} &= K_1\frac{q^*}{\kappa a} \\
K_{\perp} &= \frac{-K_1}{(x_0/\kappa a)^3}\frac{q^*}{\kappa a} \\
K_U &= -K_1 m\widehat{\Psi}_0\frac{q^*}{\kappa a}\left[\frac{1}{4(x_0/\kappa a)^3} - 1\right]
\end{aligned} \tag{A16}$$

Appendix C. An Estimation of x_1

The length x_1 is defined as the position where the function

$$F(x) = \left(1 - 3\frac{x^2}{(\kappa a)^2} + 2\frac{x^3}{(\kappa a)^3}\right)\frac{1}{\sum z_i^2 \nu_i}\frac{d}{dx}\left[\sum z_i \nu_i \exp(-z_i \widehat{\Psi}_{eq})\right] \quad \text{(A17)}$$

is optimum, see discussion under Equation (72). From the numerical results, the values for x_1 for a large range of κa and $\widehat{\Psi}_0$ were tested. In Figure A1, the values for x_1 for $\widehat{\Psi}_0 = 4$ are presented. Similar figures were obtained for $\widehat{\Psi}_0 = 1, 2, 3, 5$, and 6. The fit given by Equation (76) does not match the numerical result for low κa. For low κa, irrespective of $\widehat{\Psi}_0$, it was found that $x_1 = 1.63$. To better approximate the numerically found x_1, it is necessary to adapt Equation (76) as follows:

$$x_1 = \kappa a + \frac{1.63}{1 + 2\exp(-\kappa a)\exp\left(-\widehat{\Psi}_0/\kappa a\right)} \quad \text{(A18)}$$

Using Equation (A18) (see Figure A1), it is possible to obtain a somewhat better fit at high $\widehat{\Psi}_0$ (= 4 or 6), but it worsens the fit for lower $\widehat{\Psi}_0$ (= 2). Therefore, it was decided to use Equation (76).

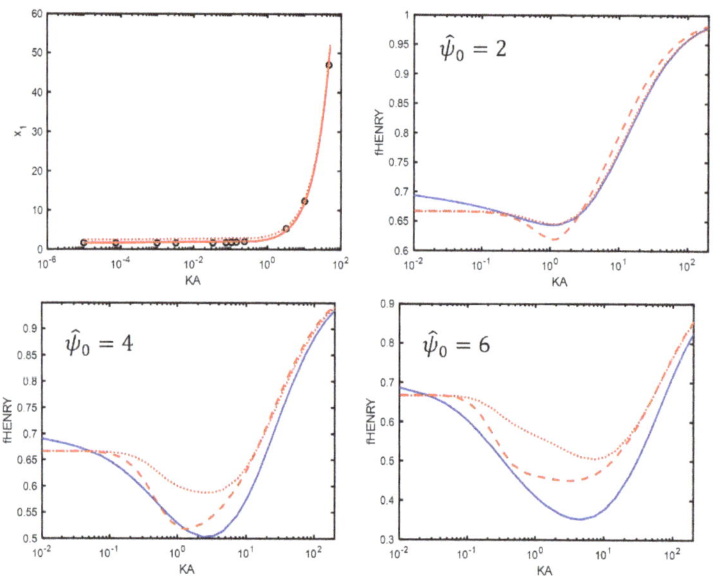

Figure A1. (**Top row, left figure**): length x_1 defined as the position where $F(x)$ defined by Equation (A17) is optimum. The dotted red curve represents Equation (76) and the full red curve represents Equation (A18). (**Other figures**): Henry's functions f_{Henry} for different surface potentials $\widehat{\Psi}_0$, as indicated in each figure. Blue curves: numerical results. Dotted red curve: results using Equations (75) and (76). Dashed red curves: results using Equations (75) and (A18).

Appendix D. Spatial Representation

Some of the spatial representations of the quantities analyzed for selected values of κa are plotted in this section. The numerical calculations were performed using the Finite Element software COMSOL Multiphysics v. 6.3 [13].

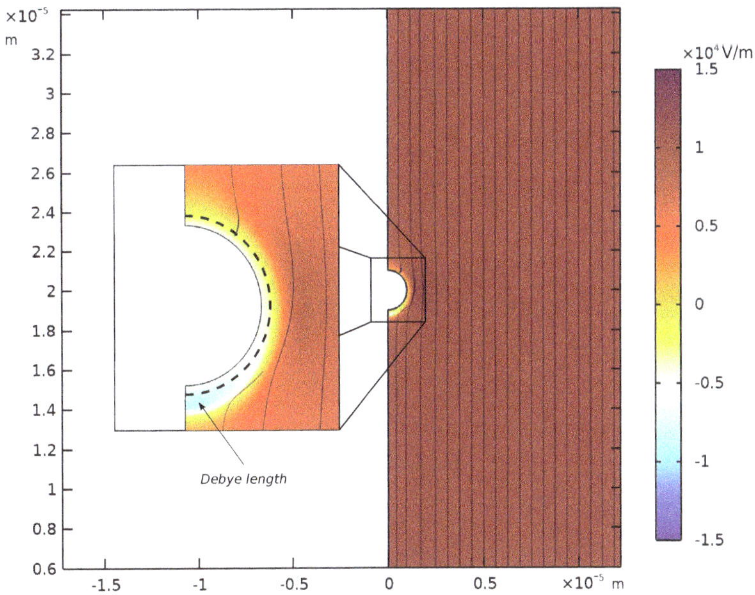

Figure A2. Difference between the magnitude of the electric field resulting from the applied electric field ($\|\delta\mathbf{E}\|$ as described in Equation (6)) and the magnitude of the electric field in the absence of applied electric field ($\|\nabla\Psi_{eq}\|$). The magnitude of the applied electric field is E_0 = 10,000 V/m for a particle radius $a = 1000$ nm, a particle surface charge of 0.1 mC/m^2, and κa = 10.39. The streamlines in the figure represent the total electric field.

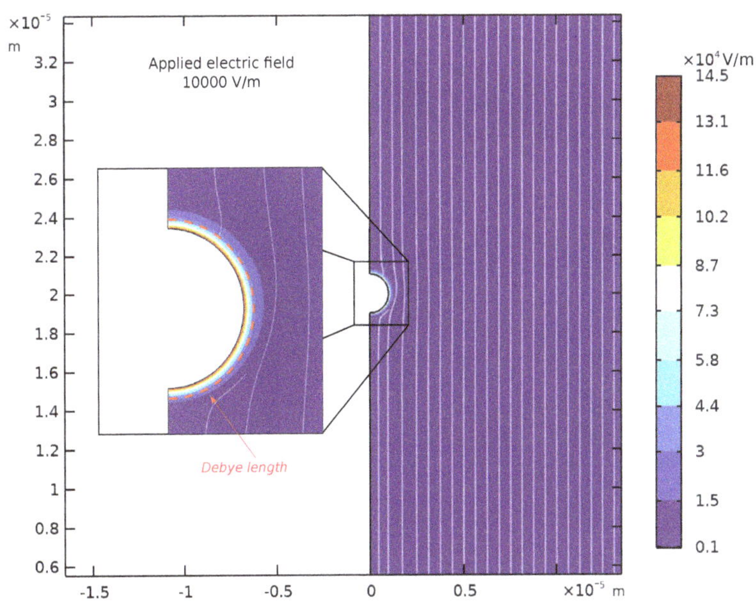

Figure A3. Magnitude and streamlines of the total electric field ($\|\mathbf{E}\|$) resulting from the application of $\mathbf{E}_0 = E_0\mathbf{e}_z$ with E_0 = 10,000 V/m for a particle radius $a = 1000$ nm, a particle surface charge of 0.1 mC/m^2, and κa = 10.39.

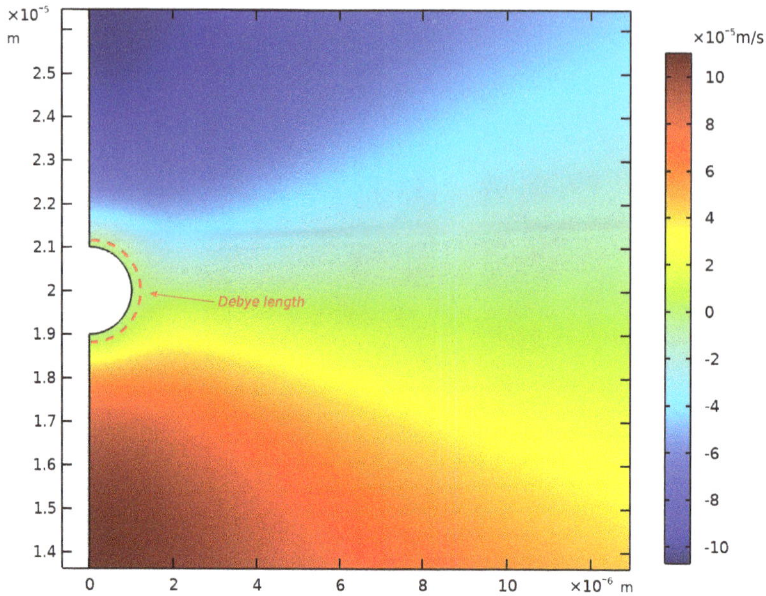

Figure A4. Magnitude of the r-component of the velocity, u_r, resulting from the application of $\mathbf{E}_0 = E_0\mathbf{e}_z$ with E_0 = 10,000 V/m for a particle radius a = 1000 nm, a particle surface charge of 0.1 mC/m^2, and κa = 10.39.

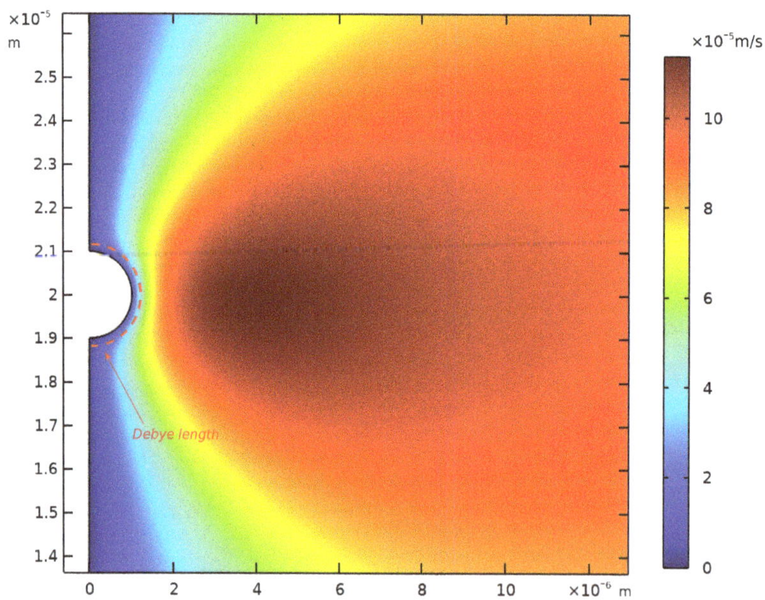

Figure A5. Magnitude of the θ-component of the velocity, u_θ, resulting from the application of $\mathbf{E}_0 = E_0\mathbf{e}_z$ with E_0 = 10,000 V/m for a particle radius a = 1000 nm, a particle surface charge of 0.1 mC/m^2, and κa = 10.39.

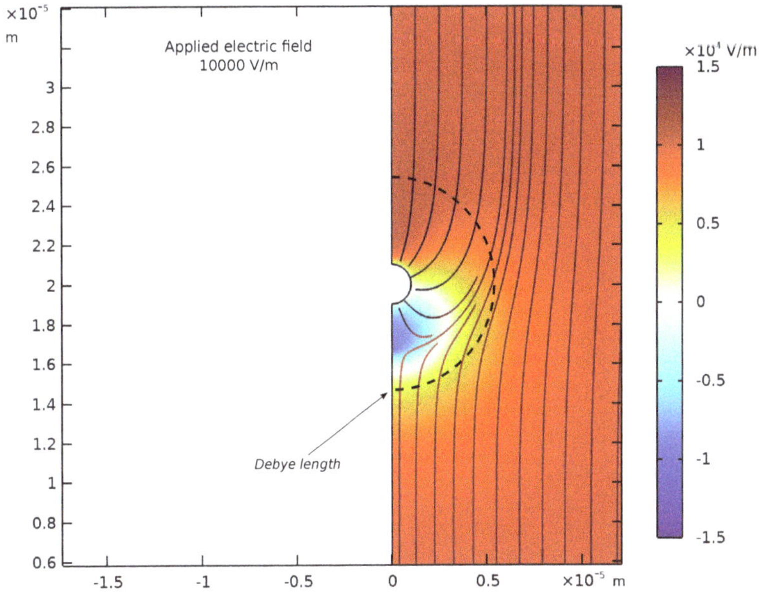

Figure A6. Difference between the magnitude of the electric field resulting from the applied electric field ($\|\delta\mathbf{E}\|$ as described in Equation (6)) and the magnitude of the electric field in the absence of applied electric field ($\|\nabla\Psi_{eq}\|$). The magnitude of the applied electric field is E_0 = 10,000 V/m for a particle radius $a = 1000$ nm, a particle surface charge of 0.1 mC/m^2, and κa = 0.23. The streamlines in the figure represent the total electric field $\mathbf{E}$.

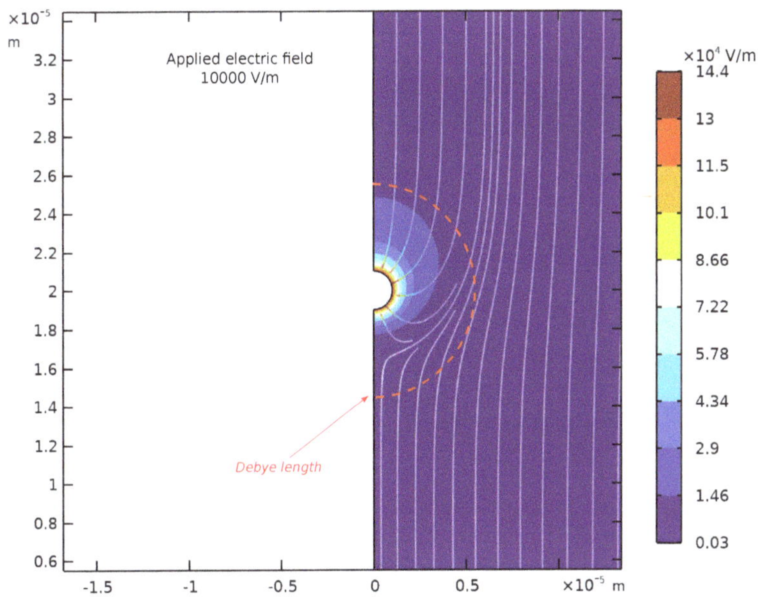

Figure A7. Magnitude and streamlines of the total electric field ($\|\mathbf{E}\|$) resulting from the application of $\mathbf{E}_0 = E_0\mathbf{e}_z$ with E_0 = 10,000 V/m for a particle radius $a = 1000$ nm, a particle surface charge of 0.1 mC/m^2, and κa = 0.23.

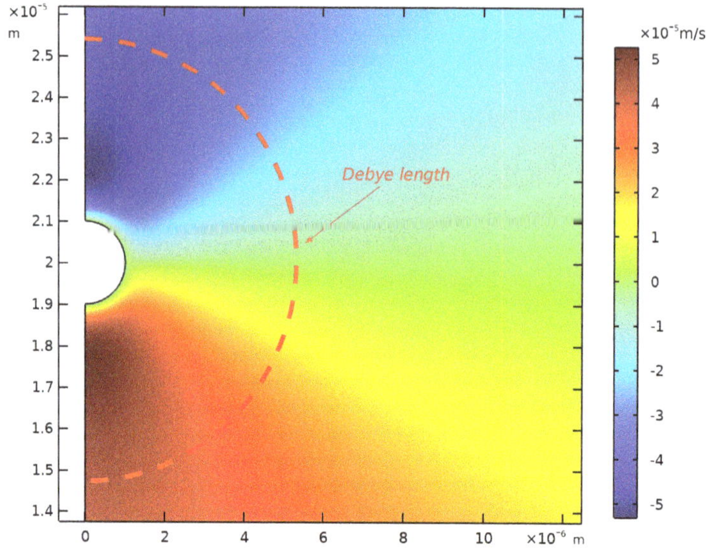

Figure A8. Magnitude of the r-component of the velocity, u_r, resulting from the application of $\mathbf{E}_0 = E_0\mathbf{e}_z$ with E_0 = 10,000 V/m for a particle radius a = 1000 nm, a particle surface charge of 0.1 mC/m^2, and κa = 0.23.

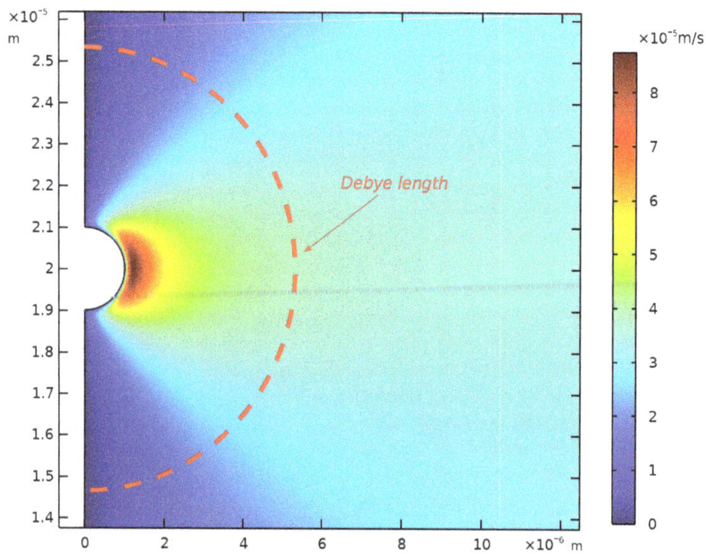

Figure A9. Magnitude of the θ-component of the velocity, u_θ, resulting from the application of $\mathbf{E}_0 = E_0\mathbf{e}_z$ with E_0 = 10,000 V/m for a particle radius a = 1000 nm, a particle surface charge of 0.1 mC/m^2, and κa = 0.23.

References

1. Ohshima, H. *Theory of Colloid and Interfacial Electric Phenomena*; Elsevier: Amsterdam, The Netherlands, 2006; Volume 12.
2. O'Brien, R.W.; White, L.R. Electrophoretic mobility of a spherical colloidal particle. *J. Chem. Soc. Trans. Mol. Chem. Phys.* **1978**, *74*, 1607–1626. [CrossRef]
3. Jayaraman, A.S.; Klaseboer, E.; Chan, D.Y. The unusual fluid dynamics of particle electrophoresis. *J. Colloid Interface Sci.* **2019**, *553*, 845–863. [CrossRef] [PubMed]

4. Chassagne, C.; Bedeaux, D. The dielectric response of a colloidal spheroid. *J. Colloid Interface Sci.* **2008**, *326*, 240–253. [CrossRef] [PubMed]
5. Böttcher, C.J.F.; van Belle, O.C.; Bordewijk, P.; Rip, A.; Yue, D.D. *Theory of Electric Polarization*; American Elsevier Publishing Co., Inc.: New York, NY, USA, 1973.
6. Chassagne, C.; Ibanez, M. Electrophoretic mobility of latex nanospheres in electrolytes: Experimental challenges. *Pure Appl. Chem.* **2012**, *85*, 41–51. [CrossRef]
7. Döpke, M.F.; Hartkamp, R. The importance of specifically adsorbed ions for electrokinetic phenomena: Bridging the gap between experiments and MD simulations. *J. Chem. Phys.* **2021**, *154*, 094701. [CrossRef] [PubMed]
8. Chassagne, C.; Dubois, E.; Jiménez, M.L.; van Der Ploeg, J.M.; Van Turnhout, J. Compensating for electrode polarization in dielectric spectroscopy studies of colloidal suspensions: Theoretical assessment of existing methods. *Front. Chem.* **2016**, *4*, 30. [CrossRef] [PubMed]
9. Klaseboer, E.; Chan, D.Y. On the derivation of the Smoluchowski result of electrophoretic mobility. *J. Colloid Interface Sci.* **2020**, *568*, 176–184. [CrossRef] [PubMed]
10. Jeffrey, A.; Dai, H.H. *Handbook of Mathematical Formulas and Integrals*; Elsevier: Amsterdam, The Netherlands, 2008.
11. Ohshima, H.; Healy, T.W.; White, L.R. Approximate analytic expressions for the electrophoretic mobility of spherical colloidal particles and the conductivity of their dilute suspensions. *J. Chem. Soc. Faraday Trans. 2 Mol. Chem. Phys.* **1983**, *79*, 1613–1628. [CrossRef]
12. Chassagne, C. Dielectric response of a charged prolate spheroid in an electrolyte solution. *Int. J. Thermophys.* **2013**, *34*, 1239–1254. [CrossRef]
13. COMSOL Multiphysics v. 6.3. COMSOL AB, Stockholm, Sweden. Available online: www.comsol.com (accessed on 28 February 2025).
14. Henry, D.C. The cataphoresis of suspended particles. Part I.—The equation of cataphoresis. *Proc. R. Soc. Lond. Ser. A* **1931**, *133*, 106–129.
15. Overbeek, J.T.G.; Wiersema, P.H. *The Interpretation of Electrophoretic Mobilities*; Academic Press: New York, NY, USA, 1967; Volume 2, pp. 1–52.
16. Overbeek, J.T.G.; Loeb, A.L.; Wiersema, P.H. Calculation of the electrophoretic mobility of a spherical colloid particle. *J. Colloid Interface Sci.* **1966**, *22*, 78–99.
17. Kobayashi, M. Electrophoretic mobility of latex spheres in the presence of divalent ions: Experiments and modeling. *Colloid Polym. Sci.* **2008**, *286*, 935–940. [CrossRef]
18. Kaya, A.; Yukselen, Y. Zeta potential of clay minerals and quartz contaminated by heavy metals. *Can. Geotech. J.* **2005**, *42*, 1280–1289. [CrossRef]

entropy

MDPI

Article

Entropy Profiles for Li-Ion Batteries—Effects of Chemistries and Degradation

Julia Wind and Preben J. S. Vie *

Institute for Energy Technology (IFE), NO-2007 Kjeller, Norway; julia.wind@ife.no
* Correspondence: preben.vie@ife.no

Abstract: This paper presents entropy measurements for a large set of commercial Li-ion cells. We present entropy data on full cells with a variety of common Li-ion cell electrode chemistries; graphite, hard carbon, lithium-titanium-oxide (LTO), lithium cobalt-oxide (LCO), nickel manganese cobalt oxides (NMC), nickel cobalt aluminium oxide (NCA), lithium iron-phosphate (LFP), as well as electrodes with mixes of these. All data were collected using an accelerated potentiometric method in steps of approximately 5% State-of-Charge (SoC) across the full SoC window. We observe that the entropy profiles depend on the chemistry of the Li-ion cells, but that they also vary between different commercial cells with the same chemistry. Entropy contributions are quantified with respect to both, their means, positive and negative contributions as well as their SoC variation. In addition, we present how different cyclic ageing temperatures change the entropy profiles for a selected commercial Li-ion cell through ageing. A clear difference in entropy profiles is observed after a capacity loss of 20%. This difference can be attributed to different ageing mechanisms within the Li-ion cells, leading to changes in the balancing of electrodes, as well as changes in the electrode materials.

Keywords: Li-ion battery; entropy; reversible heat; thermal characterization

1. Introduction

Li-ion batteries (LIBs) have emerged as the current major commercial rechargeable battery technology. Since its commercial launch in the 1990s Li-ion batteries have been further developed and steadily improved. Initially, the LIB replaced other secondary batteries as e.g., NiMH in portable electronics (e.g., mobile phones and portable computers), followed by the introduction of LIBs into electric vehicles (EVs) through the Tesla Roadster in 2008, the Mitsubishi i-MiEV and Nissan Leaf in 2010. Since 2010, the LIB has also been used in electric boats, with the first fully electric car ferry starting operation in 2014 [1] and for large-scale electric energy storage [2].

The LIB is known for its very high energy and power density as well as a superior round-cycle efficiency of commonly more than 95%. Currently, there are no other commercial secondary battery technologies with a higher energy density than LIBs. Possibly, the LIB can in future be replaced by the all-solid-state battery (ASSB) based on Li-metal as one electrode with prospects of even higher energy densities. However, this technology is not yet commercially available and will most likely have a much higher initial cost.

Due to its high energy density, as well as consisting of flammable materials (e.g., graphite electrode, polymer separator and electrolyte solvents), the LIB is known to have challenges with its safety properties. Among the most well-known and previously reported safety incidents and issues are the ones reported by the Boeing 787 Dreamliner

(2013) [3] and Samsung Galaxy Note 7 (2016) [4]. These issues were caused by internal faults within the battery cells, as well as issues with the engineering of the overall battery system (Boeing). More recently, reports of fires in "e-scooters" during charging have become a recurring issue [5]. This is quite often linked to the use of low-quality cells in relatively cheap products. There are also several reports of fires in EVs, but the reported number of fires are not higher in EVs compared to fires in normal cars with internal combustion engines. There have also been reported fires in battery-based propulsion systems in ships [6–8] as well as several fires in large stationary energy storage systems [9,10]—all based on LIBs. There are several causes for these fires. Some were caused by a misfitted gasket in the battery liquid cooling system causing a short-circuit of the battery system [6,8], another had a design flaw where saltwater could reverse through the ventilation system and again short-circuit the battery system [7] while others again were caused by internal cell faults or had unknown causes [10].

The major causes of fires in Li-ion-based battery systems can hence be attributed to external factors like e.g., external short-circuits or overheating that may lead to overheating and eventually force the battery into a thermal runaway event. However, fires in LIBs can also arise from internal faults within the battery cell. LIBs can have faults from production that can cause internal short circuits, while battery ageing and abuse may also cause safety-critical events. One well-known factor that can lead to an internal short is lithium plating that can occur during the charging of LIBs at relatively low temperatures and relatively high currents [11–16]. During lithium-plating, the lithium ions are not able to access the graphite sites for intercalation at a high enough rate and will instead be deposited on the surface of the graphite as metallic lithium. The metallic lithium can then grow in a dendritic manner that can potentially penetrate the separator between the anode and cathode and eventually short-circuit the battery from within through a so-called "local short" or "weak short". This could potentially lead to a full fire of the LIB but could also only cause a local short circuit that damages the battery in a smaller localized area. Upon the possible detection or diagnosis of an onset of lithium plating, a major safety risk in LIBs could be mitigated.

During lithium plating, there is an irreversible loss of cyclable lithium within the battery on the graphite anode. This process is commonly known as loss of lithium inventory (LLI) and is one of several known ageing modes in LIBs [17]. Another process leading to LLI is the buildup of the SEI layer (see e.g., [18]) on the anode surface. The two other ageing modes are the loss of active material (LAM), and impedance increase due to reaction kinetics degradation [17]. The detection and identification of rapidly increasing LLI could imply that the LIB is going through a possible hazardous plating of lithium metal on the graphite electrode. The LLI implies a shift in balance between the anode and cathode electrodes and can hence be detected by several different methods. These methods entail e.g., the diagnostic techniques of DVA and ICA (see e.g., [17,19,20]). It is however outside the scope of this paper to elaborate more on these specific techniques. We will in this paper show that changes to the entropy profile through the course of ageing can be used as another technique to reveal LLI, and hence possibly lithium metal plating.

Detailed knowledge of the heat development from a LIB can be crucial for the detection of a sudden safety event, as well as provide a full understanding of the local temperature profile during both charge and discharge. It is well known that the developed heat from a LIB comes from both reversible and irreversible heat sources [21,22]. The irreversible heat is linked to the internal resistance and overpotentials and will always generate heat, while the reversible heat can contribute to both heating and cooling of the battery. The reversible heat stems from the change in entropy in the two electrodes of the LIB and will vary with the Li-ion concentration and thus the battery's state of charge (SoC). Some cooling effects of the entropy can be observed in most LIB temperature profiles. This can be seen, e.g., in Figure 8 in [23] where the large variations in the temperature profiles of both charge and discharge are caused by the variations in the LIB's entropy profile. The lowered temperatures in certain parts of the SoC window illustrate this cooling effect. During discharge, a positive entropy value will contribute to cooling the battery cell, while during charge a negative entropy will do the same. The entropic heat may even balance out or be larger than the irreversible heat contributions caused by, e.g., the internal resistance. The change in entropy is caused by phase transitions in the electrode materials [24–26]. A shift in a cell's temperature profile and hence the entropy profile, can reveal a shift in the electrode balancing of the cell. This in turn could be used as another tool to detect critical LLI in a cell, possibly caused by lithium metal plating [27].

The entropy of the LIB at any SoC can be measured by several methods, e.g., the potentiometric and calorimetric methods. In the potentiometric method, the change in open circuit voltage (OCV) is measured upon a corresponding change in cell temperature [25,28]. The change in entropy is then given by $\Delta S = nF\frac{\partial U_{OCV}}{\partial T}$, where $\frac{\partial U_{OCV}}{\partial T}$ is the derivative of the OCV (U_{OCV}) with respect to temperature (T). The calorimetric method involves measuring the heat from the Li-ion cell during charge and discharge where different models are applied to differentiate between irreversible and reversible heat sources [29–33].

This paper presents the measurements of entropy profiles for a large range of LIBs. A total of 14 different cells were measured using the accelerated potentiometric method [23]. The cells had different anode and cathode chemistries as well as different geometries (i.e., pouch, cylindrical or prismatic) and capacities ranging from 0.4 to 64 Ah. All cells were measured at the beginning of life. In addition, the entropy profile of one specific cell was followed through ageing at different temperatures.

The paper is structured as follows: we first introduce the procedure for the measurement of the entropy profiles (Section 2.1), and present an overview of the tested batteries (Section 2.2). The measured entropy profiles are presented and discussed with respect to electrode chemistries (Section 3.1), and quantified with respect to means, standard deviations as well as positive and negative contributions in Section 3.2. A comparison between entropy profiles at charge and discharge is presented in Section 3.3. Finally, the evolution of an entropy profile for a cell through ageing is presented (Section 3.4) before the results are discussed (Section 4).

2. Experimental/Materials and Methods

2.1. Method for Entropy Measurements

The method applied for obtaining entropy profiles was an accelerated potentiometric method discussed in detail in Hua [23]. The entropy was measured in 5% SoC steps across the full SoC window with an accelerated potentiometric method for both charge and discharge of the cell. The method and apparatus were previously described in full [23]. The LIBs were placed in thick-walled plastic bags before immersion into a high-precision water bath. This enabled good thermal contact and a stable cell temperature. The protocol used for measuring the entropy for both discharge and charge was the following [23]:

1. Charge the cell to the maximum voltage limit for the cell (100% SoC) (CC-CV charge with a 0.05C current cut-off).
2. Allow cell voltage (OCV) to stabilize to a relaxation rate lower than 1.4 mV/h or a maximum relaxation time of 6 h.

3. Cycle the temperature of the water bath from 25 °C to 16 °C, and back to 25 °C via steps of 3 °C increments from 16 °C. At every temperature step, the OCV and temperature were allowed to stabilize before the next temperature step. The temperature was assumed stable when the standard deviation of a five-minute time window was smaller than 0.01 °C. The OCV of the cell was recorded continuously.
4. The SoC was changed in steps of approximately 5% through a constant 0.05C current (CC) discharge or charge. (The cell will in steps of 5% SoC eventually reach the lower voltage limit (0% SoC) and then be charged back to 100% SoC in 5% SoC steps)
5. Loop back to step 2.

The respective maximum and minimum voltage limits for the different cells are presented in Table 1.

For two of the cells (LG JP3 and LG E58C), the measurements were conducted in a slightly different way: instead of the equipment presented in [23], a conventional battery tester (PEC ACT-0550) was used to change SoC and record OCV and temperature. The cell temperature was controlled by a programmable incubator (VWR INCU-Line 150R) and only two different temperatures (25 and 15 °C) were used to measure the change in OCV following the temperature change.

2.2. The Tested Cells

A total of 14 different LIB cells were investigated in this study over the course of approximately 10 years. The set is comprised of cells of a large variety of different chemistries, form factors, energy densities and manufacturers. Cell chemistries include standard graphite, LTO as well as hard carbon on the anode side, combined with cathode chemistries of LCO, different types of NMC, LFP, NCA as well as a blend of NMC532/LCO/LMO. Form factors include pouch, prismatic as well as cylindrical cells, with capacities from very small 0.36 Ah (cylindrical) and up to 64 Ah (pouch). An overview of all cells and relevant details are summarized in Table 1.

Table 1. Overview of the cells measured for full cell entropy and their properties. A separate column gives references to previous literature that has been published by the researchers in the group on the same cell type. The entropy data for the Xalt 31 HE [34] and Melasta cells [35] was published in previous works and is included here for completeness.

Name	Manufacturer & Cell Model	Capacity [Ah]	Voltage Range [V]	Form Factor	Chemistry	Energy Density [Wh/kg]	Comments/ References
Kokam LTO	Kokam SLPB65205130N	11	1.5–2.8	pouch	LMO(?)-LTO	69	Cathode chemistry unconfirmed
Toshiba SciB	Toshiba SciB 20	20	1.5–2.7	prismatic	LMO(?)/-LTO	89	Cathode possibly LMO [36,37]
LG JP3	LGV JP3	64	3.0–4.2	pouch	NMC532-graphite	210	[38,39]
LG E58C	LG E58C	58	3.0–4.2	pouch	NMC622-graphite	242	Cell harvested from a Jaguar I-pace battery
Xalt 31 HE	Xalt 31 HE	31	2.7–4.2	pouch	NMC111-graphite	180	[34,38]
Panasonic UR	Panasonic UR18650W	1.5	2.75–4.2	cylinder	NMC-graphite	120	[40,41]
Enerdel	Enerdel 151105203A 102401	17.5	2.5–4.1	pouch	NMC-hard carbon	147	[42]
Panasonic NCR	Panasonic NCR18650B	3.25	2.5–4.2	cylinder	NCA-graphite	243	[41]
SAFT VL6P	SAFT VL6P	6	2.7–4.0	cylinder	NCA-graphite	74	[40]

Table 1. *Cont.*

Name	Manufacturer & Cell Model	Capacity [Ah]	Voltage Range [V]	Form Factor	Chemistry	Energy Density [Wh/kg]	Comments/ References
SAFT VL30PFe	SAFT VL30PFe	30	2.5–3.8	cylinder	LFP-graphite	91	[40]
Melasta 8C	LPB042126-8C	7	3.0–42	pouch	LCO-graphite	190	[35,39]
Melasta 10C	SLPBB042126-10C	6.55	3.0–42	pouch	LCO-graphite	188	[35,39]
Samsung SDI	ICP103450A	2	2.75–4.2	prismatic	LCO-graphite	-	
GPB ICR10440	Great Power Battery Co Ltd. ICR10440	0.36	3.0–4.2	cylinder	NMC532/ LCO/LMO-graphite	69	[43]

2.3. The Cycle Life Study

In addition to entropy measurements performed at the cells' beginning of life, the entropy profiles of the Xalt 31 HE cell were studied while the cell was aged as part of a cycle life study. The corresponding ageing results for this cell were previously published [38], the entropy profile at 100% SoH (State-of-Health) (100% State-of-Health (SoH) refers to a cell with no loss in cell capacity) was used to exemplify the new methodology presented [23].

In this work, we present ageing results of cells cycled in the full voltage window from 2.7 V to 4.2 V at a constant current of 1C, at three different temperatures of 5, 25 and 45 °C. All charging of the cells was performed with a constant current charge to the maximum voltage and finished with a constant voltage hold until the current was reduced to less than 0.1C. The cells were tested without mechanical compression and were allowed to cool down to at least 1 °C above the set temperature before starting a new charge or discharge during cycling with otherwise a minimum OCV period of 5 min. The cells were all characterised at 25 °C at regular intervals of a maximum of 200 cycles (The term "cycles" refers to total absolute cycles in this algorithm), or earlier when a cell lost 5% capacity during the cycling. The characterisation test included two cycles of testing capacity at both 0.1C and 1C in the full SoC window. The results presented here were part of a larger ageing study where the effects of different temperatures, currents, and voltage limits were scrutinised and have partly been published [38].

2.4. Data Handling

For the direct comparison of the measured entropy profiles and their statistical analysis, the data was treated in the following way before calculating means and standard deviations: entropy data points were resampled to n = 200 equidistant points across the SoC window, interpolated (1d cubic) and smoothed (Gaussian smoothing, sigma = 2). Integrals were calculated using the trapezoidal method. All data handling was conducted in Python, using standard methods as implemented in the scipy (v1.13.0) package.

3. Results and Discussion

An overview of the measured entropy profiles of all the investigated cells at 100% SoH is presented in Figure 1. The respective anode and cathode chemistries for the individual cells are highlighted (for all cell details see Table 1 above). All profiles in the figure were obtained during discharge of the battery. In the following sections, we look at and discuss the similarities and differences of these profiles in more detail.

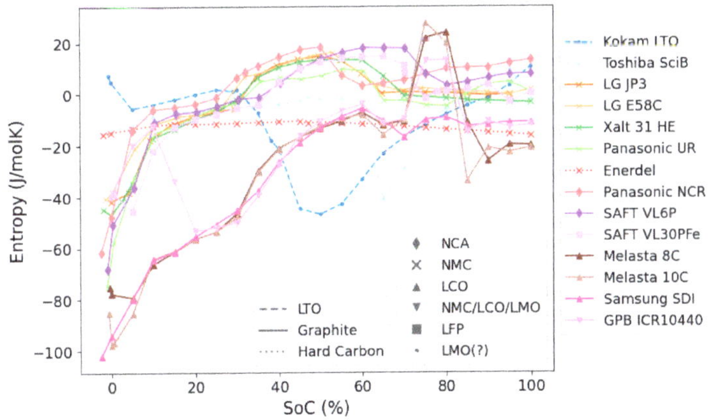

Figure 1. Overview of measured discharge entropy profiles for all cells. Different anode and cathode chemistries are represented by line styles and marker symbols, respectively.

3.1. *Differences in Entropy Profiles by Electrode Chemistry*

Figures 2 and 3 show the entropy profiles grouped by anode and cathode chemistries. Similarities and differences across the two figures allow for the identification and confirmation of the individual electrode chemistries' contribution to the entropy variation of the full cell.

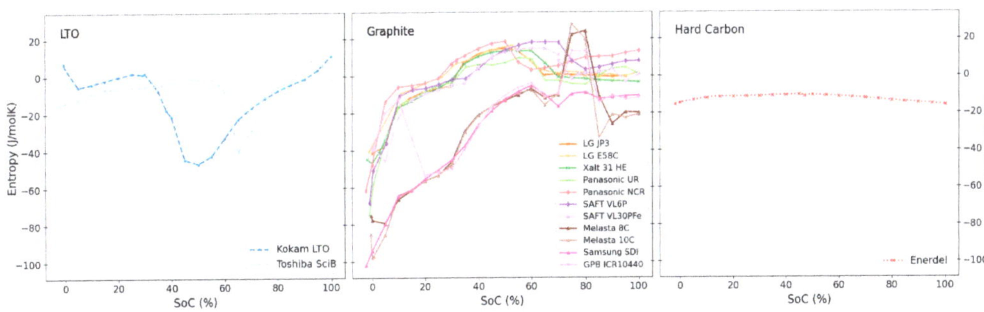

Figure 2. The entropy profiles of all tested full cells sorted by the known anode materials; LTO, graphite or hard carbon.

Hard Carbon. One of the investigated cells (Enerdel) contained a hard-carbon anode, combined with an NMC cathode. For this cell, we observe a profile with almost constant entropy across the entire SoC window (see Figure 2: Hard Carbon). The entropy values across the full SoC window are all within the range of -16.3 to -10.6 J/molK, with a SoC-averaged entropy value of 13 J/molK. This indicates that both electrode active materials, hard carbon and NMC, have little variation in entropy with a change in SoC. For hard carbon specifically, this can be rationalised by the material's almost amorphous structure. The influence of structural disorder on the thermodynamics of lithium intercalation into graphite and disordered carbons was discussed in detail by Reynier et al. [44]. In their work, they point out a clear correlation between the magnitude and variation of the entropy and the state of order in the carbon. In contrast to the characteristic features in the entropy profiles observed for graphite electrodes (see next paragraph), only small entropy values and entropy variations were observed for more disordered carbons. These observations also agree with the findings reported by Takano et al. [25] for an LCO/hard carbon cell:

Takano et al. [25] attributed the main contribution to the cell's entropy profile to the LCO cathode, with negligible contributions from the hard-carbon anode.

Graphite. Cells with graphite anodes exhibit large similarities in their entropy profiles across the range of different investigated cathode chemistries, NMC, NCA, and LFP. The most prominent features in these entropy profiles arise from the different defined intercalation stages of Li into graphite. These intercalation stages are commonly labelled by the number of graphite layers between layers of intercalated lithium and are well established and described in detail in literature (see e.g., [15]). The entropy profiles (see Figure 2: Graphite) show large negative values and a rapid increase at low SoCs (dilute stage 1L), levelling off at around 10% SoC (formation of stage 4). A prominent broad peak is then observed between 40–60% SoC, corresponding to the formation of stage 2 graphite. As all presented profiles arise from full cells, details on the entropy profiles of course also depend on the respective cathode chemistries. Especially NMC (as observed above when paired with hard carbon), LFP and NCA cathode chemistries seem to only contribute very little to the overall entropy profile (see separate discussions on cathode chemistries below). Consequently, the main contribution to these profiles can be attributed to the graphite anodes. The main differences between these profiles are the exact positions of the steep change in entropy at low SoC as well as the position and width of the prominent broad peak (stage 2 formation). These variations are attributed to differences in the balancing between anode and cathode in the respective cells. These profiles are in agreement with earlier literature reports on an NMC/graphite full cell [46], and align with numerous reports on entropy coefficients for graphite half cells (see e.g., summary provided by Gunnarshaug et al. [47] and references therein [24,26,44]). Pairing graphite with LCO-containing cathodes significantly influences the overall entropy profiles [25], leading to overall significantly larger negative entropy values, in addition to the significant LCO peak at around 80% SoC. This will be discussed in more detail in the LCO section below.

LTO. Both LTO-based cells have quite similar distinct entropy profiles that are very different compared to the ones for graphite-based cells. The full LTO profiles show some similarities to the cell with a cathode blend (NMC/LCO/LMO, see Figure 3). The "valley" at around 25% SoC for the cathode blend resembles a similar "valley" at slightly higher SoCs for the LTO cells. The LTO cells' entropy profile is likely largely influenced by the cathode chemistry, hinting towards the cathode chemistry being LMO-based, as indicated by Appiah [36] and Koshtyal et al. [37]. The different SoC windows for the valley can stem from different balancing of the electrodes which is natural due to the substantially different anode chemistries (LTO vs. graphite). Entropy profiles for LTO half-cells are reported to show very low magnitude (around −5 J/molK) and little SoC variation [48,49]. Overall, LTO-based anodes contribute little to a cell's total entropy profile, in agreement with the LTO profiles presented by Viswanathan [26].

LCO. The profiles for LCO-containing cells (see Figure 3: LCO) are characterised by large negative entropy values across the majority of the SoC window. The almost linear increase in entropy up to around 70% SoC (i.e., $x = 0.6$ to $x = 0.8$ in Li_xCO_2) arises from the Li intercalation into the hexagonal LCO phase [44]. This is followed by a peak with positive entropy at around 70–90% SoC. The pronounced peak is commonly attributed to the hexagonal to monoclinic phase transition (around $x = 0.55$ in Li_xCO_2). The monoclinic phase is known to have a significantly higher entropy compared to the hexagonal phase [44]. In our work, all three cells with LCO cathode chemistries show very similar entropy profiles in agreement with literature [26,44], except for the area around 80% SoC in which the two Melasta cells show the pronounced peak that the third cell (Samsung SDI) is lacking. This peak is also the main differentiating feature in the direct comparison of entropy profiles for different types of LCO cells by Viswanathan [26]. However, no further reasoning for this was given. The hexagonal to monoclinic phase transition and consequently the intensity of the peak at 80% SoC was reported to be highly dependent on temperature and the presence of small amounts of different elements (such as e.g., Ni) or excess lithium in the cathode material [25]. This could thus explain the observed difference in this area between the different LCO cells. However, to further pinpoint the causes for the differences between the LCO cells, more detailed experiments, such as e.g., the structural characterisation of the respective active materials would be necessary. In general, LCO cathodes exhibit large overall absolute contributions to the entropy profile (especially compared to other positive cathode chemistries such as NMC, LFP, or NCA) [26,44,50].

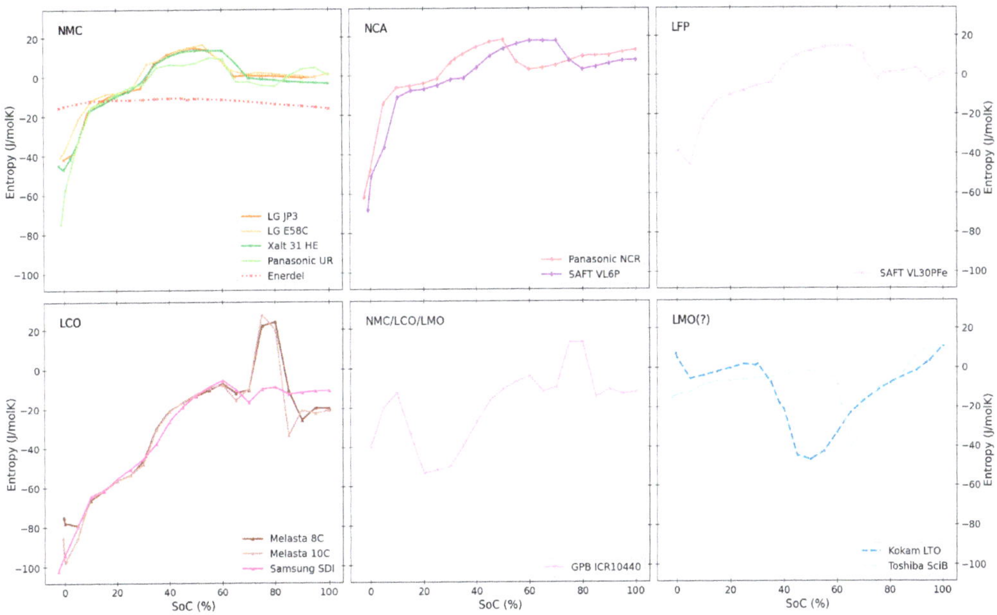

Figure 3. The entropy profiles of all tested full cells sorted by the known cathode materials; NMC, NCA, LFP, LCO, NMC/LCO/LMO (blend) and two cells with likely (but unconfirmed) LMO cathodes.

NMC. As already indicated by the NMC/hard carbon and NMC/graphite profiles, NMC seems to contribute very little to the overall entropy profile of the cell. This agrees with literature reports [26,46,51–53] and indicates that NMC is thermodynamically more stable compared to LCO. As discussed above, the measured NMC/graphite profiles all closely resemble each other and are dominated by the characteristic graphite features. Minor differences can be attributed to changes in electrode balancing, or different transition metal ratios. Sturm et al. [52] hint towards slightly higher entropy values especially at low lithiation degrees for high-Ni NMC (NMC811) as compared to e.g., NMC111. However, these differences are small in relation to the entropy contribution from the graphite counter electrode. Here we observe slightly higher entropy within the lower SoC range for the NMC with the highest Ni content (LG E58C with NMC622), and lower entropy values for the Panasonic cell with unknown Ni:Mn:Co ratios above ~35% SoC. The subtlety of these differences does not allow for any further conclusions from the available full-cell data. A systematic investigation of the half-cell entropies for different Ni:Mn:Co ratios, comparable to the one presented by Schlueter et al. [54] for LMO cathodes, could bring valuable insights to this topic and is suggested for future work.

LFP. The obtained entropy profile for the LFP/graphite cell agrees well with the ones reported in literature [55,56]. The main contribution is, again, attributed to the graphite electrode with very small, almost SoC-independent contributions from the LFP cathode. Entropy profiles for LFP cathodes (half-cell data) were shown to have low magnitudes and only subtle variations with SoC, varying between −6 J/molK at low SoC to around −3 J/molK at higher SoC [56].

NCA. Similarly to NMC and LFP cathode chemistries, NCA is also observed to have only minor contributions to the overall entropy profile upon pairing with graphite, in agreement with literature reports [57,58].

NMC/LMO/LCO (and LMO). The blended-cathode cell resembles selected parts of several entropy profiles of other cells. Visually we observe a distinct similarity with the pure LCO-based cells at higher SoC (cf. peak at 80% SoC). However, at lower SoC, the profile is more similar to the LTO-based cells, which we found indications of having an LMO cathode. LMO cathodes were also reported to show small overall entropy with a characteristic SoC variation dependent on Li:Mn ratios [26,54,59]: Across most Li:Mn ratios, the main features are a positive peak with its maximum around $x = 0.65$ (x in $Li_xMn_2O_4$), followed by a valley around $x = 0.55$, both related to transitions from disordered solid solutions to an ordered phase at $x = 0.5$ [54]. The intensity and exact location of these minima and maxima strongly depend on the exact Li:Mn ratios as well as the balancing of the electrodes within the respective full cells.

The discussed characteristic entropy features by electrode chemistry and references to previous literature reports are summarised in Table 2.

Table 2. Summary of characteristic entropy features by electrode chemistries and comparison to previously reported entropy data for similar chemistries.

Chemistry	Cell Labels (This Work)	Features and Comments	References
LTO	Kokam LTO, Toshiba SciB	Small overall contribution and little SoC variation	Half-cell: [48,49] Full cell: [26]
Graphite	LG JP3, LG E58C, XALT 31 HE, Panasonic UR, Panasonic NCR, SAFT VL6P, SAFT VL30PFe, Melasta 8C, Melasta 10C, Samsung SDI, GPB ICR10440	Characteristic entropy profile following intercalation stages of Li into graphite (see text)	Full cell: [25,46,55,60,61] Half cell: [24,26,44,46,47]
Hard Carbon	Enerdel	Rather featureless, overall low magnitude values with very little variation with SoC	Full cell: [25] Half cell: [44]
NMC	LG JP3, LG E58C, XALT 31 HE, Panasonic UR, Enerdel	Low magnitude entropy, little SoC variation; detailed half-cell entropy measurements necessary to establish differences between different NMC compositions	Full cell: [26,46,51–53,60] Half cell: [46,52]
NCA	Panasonic NCR, SAFT VL6P	Low magnitude entropy, little SoC variation	Full cell: [57,58]
LFP	SAFT VL30PFe	Low magnitude entropy, little SoC variation	Full cell: [55,56,61] Half cell: [56]
LCO	Melasta 8C, Melasta 10C, Samsung SDI	Large negative entropy across almost the entire SoC window, except for pronounced peak at around 70–90% SoC (monoclinic-hexagonal phase transition)	Full cell: [26,44] Half cell: [25]
LMO	GPB ICR10440 (part of cathode mix in), Kokam LTO, Toshiba SciB	Low magnitude entropy with characteristic features: maximum around $x = 0.65$, minimum around $x = 0.55$ (x in $Li_xMn_2O_4$)	Half or full cell: [26,49,54,59]

3.2. *Quantitative Comparison of Entropy Variation*

Figure 4a summarises and directly compares the SoC-averaged entropy values for all the measured cells. The entropy variation across the entire SoC window is quantified by the standard deviation and illustrated by the added error bars in Figure 4a. Note that for increased comparability, the obtained entropy data was resampled to the same number of equidistant points across the SoC window for the calculation of both means and standard deviations, as described in the methodology section (Section 2.4).

LCO-based cells clearly exhibit the largest overall negative entropy values. For cells with graphite anodes that are paired with cathode chemistries with moderate entropy changes, such as NMC, NCA, and LFP, the entropy across the entire SoC window almost averages out to zero. All cell chemistries, except for NMC/hard carbon (Enerdel), show a significant variation of the entropy with SoC, illustrated by the large standard deviation included in Figure 4a. While providing a quick and simple overview, these average values and standard deviations across the entire SoC window only give very limited insights into potential heating and cooling needs specific to the prevalent battery chemistry.

In Figure 4b we present the split of the overall average entropy into its positive and negative contributions. For this, we make use of normalised integrals across the entire SoC. The normalised integral values are directly related to the entropic heat generated (negative) or consumed (positive) during a full discharge from 100% to 0% SoC. If the total integral value is negative, there is a net entropic heat release during discharge, while there is a net entropic heat consumption if the total integral is positive. The split into negative and positive contributions illustrates the additional heat generated and consumed during battery discharge across the full SoC window (see, e.g., temperature profile during discharge in Figure 8 [23]).

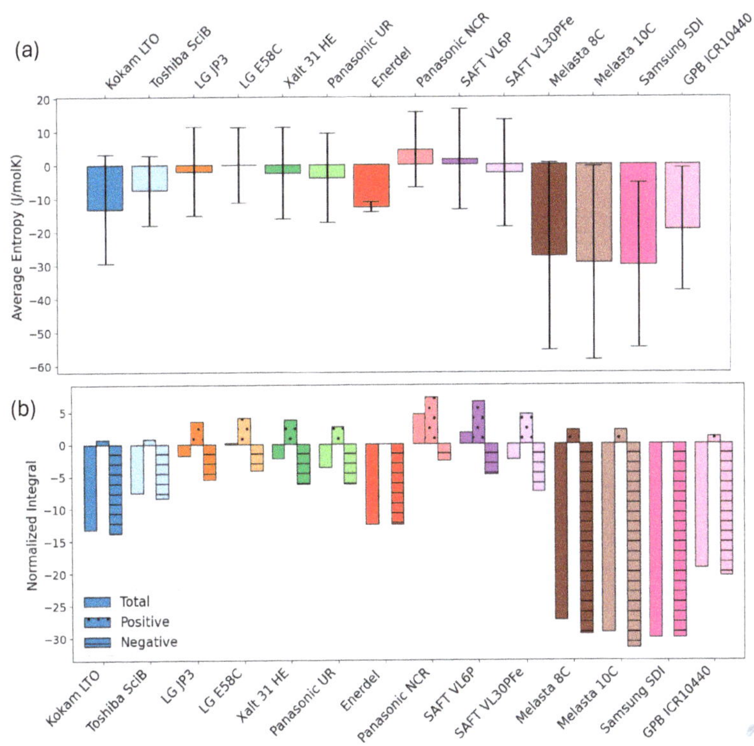

Figure 4. (**a**): Mean entropy values across the entire SoC window including standard deviation during discharge. (**b**): Normalized integrals (J/molK) over the full SoC window for the measured discharge entropy values and their respective positive and negative contributions for all measured cells (cf. Table 3). Colours match the overview plots in Figures 1–3.

For even more detailed insights, Figure 5 illustrates positive and negative contributions to the overall entropy and their SoC dependence. This overview clearly highlights the relevant SoC ranges for entropy-related heating and cooling effects. LCO-based cells show negative entropy contributions across the majority of the SoC window, with positive contributions only located at around 80% SoC. NMC, LFP, and NCA cathodes have very moderate absolute entropy changes compared to other cathode materials such as LCO or LMO, in agreement with [56]. For these materials, graphite constitutes for the main contribution and SoC variation, with significant negative entropy contributions at lower SoCs (<30%), positive contributions between around 30–70% SoC, and rather small entropy contributions at higher SoCs. This leads to an overall small average entropy in the full cells. Note here that the exact SoC values of course depend on the balancing of the electrodes in the respective cells.

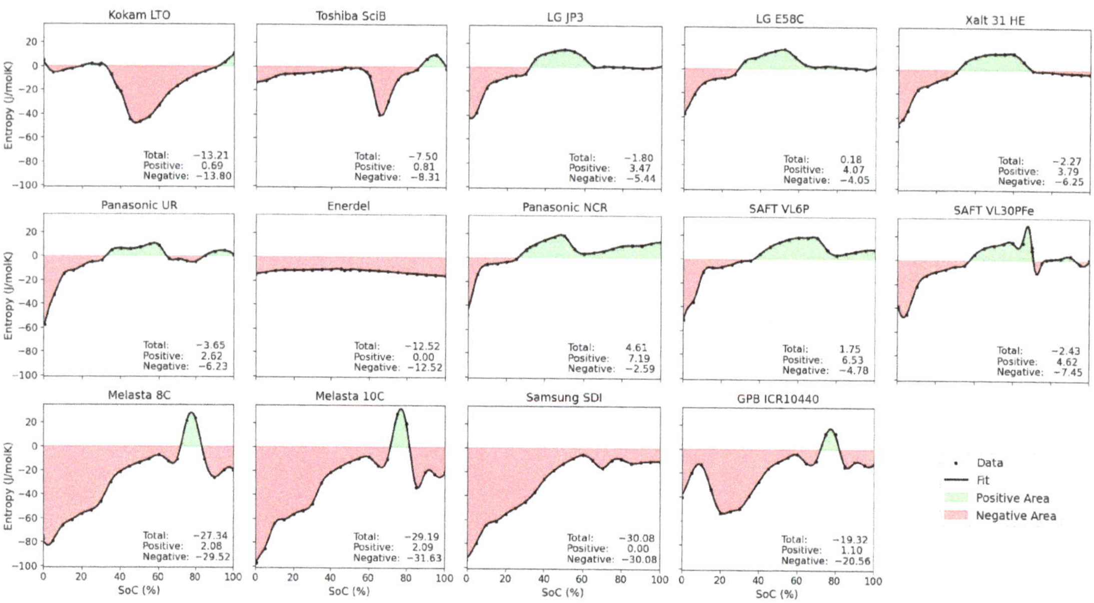

Figure 5. Illustration of the split of positive (green areas, cooling effect upon discharge) and negative (red areas, heating effect upon discharge) contributions to the overall entropy of the cells. The stated values are the obtained normalised integrals across (i) the entire SoC window, (ii) SoC areas with positive and (iii) negative entropy values. All units are J/molK.

The obtained values for normalised integrals, the respective positive and negative contributions as well as means and standard deviations are compiled in Table 3. Note that differences between the sum of positive and negative integrals and the overall integral is due to numerical errors.

Table 3. Tabulated values for normalised integrals, means, and standard deviations (* Integrals refer to normalised integrals and are calculated on resampled, interpolated and smoothed data; ** mean entropy and standard deviation (stdv) calculated from resampled values across SoC 0–100%.) All units are J/molK.

Cell Label	Chemistry	Integral *	Positive Integral *	Negative Integral *	Mean **	Stdv **
Kokam LTO	LMO(?)-LTO	−13.21	0.69	−13.80	−13	16
Toshiba SciB	LMO(?)-LTO	−7.50	0.81	−8.31	−7	11

Table 3. *Cont.*

Cell Label	Chemistry	Integral *	Positive Integral *	Negative Integral *	Mean **	Stdv **
LG JP3	NMC532-graphite	−1.80	3.47	−5.44	−2	13
LG E58C	NMC622-graphite	0.18	4.07	−4.05	0	11
Xalt 31 HE	NMC111-graphite	−2.27	3.79	−6.25	−2	14
Panasonic UR	NMC-graphite	−3.65	2.62	−6.23	−4	13
Enerdel	NMC-hard carbon	−12.52	0	−12.52	−13	2
Panasonic NCR	NCA-graphite	4.61	7.19	−2.59	5	11
SAFT VL6P	NCA-graphite	1.75	6.53	−4.78	2	15
SAFT VL30PFe	LFP-graphite	−2.43	4.62	−7.45	−3	16
Melasta 8C	LCO-graphite	−27.34	2.08	−29.52	−27	28
Melasta 10C	LCO-graphite	−29.19	2.09	−31.63	−29	29
Samsung SDI	LCO-graphite	−30.08	0	−30.08	−30	25
GPB ICR10440	NMC532/ LCO/LMO-graphite	−19.32	1.10	−20.56	−19	19

3.3. Difference in Charge and Discharge Entropy

Both charge and discharge entropies were measured for three selected cells, the two LG cells as well as the XALT cell, all with NMC/graphite chemistry. The obtained profiles are shown in Figure 6. A similar hysteresis effect between charge and discharge entropy was observed for all three cells. The charge entropy profiles are shifted to slightly lower SoCs, with an increasing accumulating hysteresis effect with increasing SoC. Hysteresis typically refers to OCV hysteresis, the remaining difference between the charge and discharge voltage curves of a battery when approaching very small currents or long relaxation times (see e.g., [62–64]). OCV hysteresis also leads to entropy hysteresis, i.e., the difference in entropy variation with SoC based on whether the data was collected during charging or discharge [58]. Hysteresis in entropy data was reported in several literature studies (e.g., [55,61,65]). However, there is still no clear consensus in the literature on the "true" origins of entropy hysteresis. Schmid et al. attribute the hysteresis behaviour to the potentiometric measurement methodology, speculating that a calorimetric approach should result in identical charge and discharge entropies. Similarly, Bedürftig et al. [60,61] were not able to fully clarify the origins of entropy hysteresis, also pointing towards the need for further research for more definite conclusions.

With respect to the magnitudes of the measured entropy profiles, neither charge nor discharge profiles are consistently shifted to lower or higher values with respect to each other. This observation is in perfect agreement with the reported entropy profiles for charge and discharge for the LG JP3 cell presented by Reiter [60], and other literature reports such as e.g., [60,61]. However, it is worth noting, that the absolute values for charge entropy coefficients for all measured cells in this work are almost consistently lower than the discharge entropy coefficients across the entire SoC window. This indicates slightly lower contributions of entropic heat during the charging process.

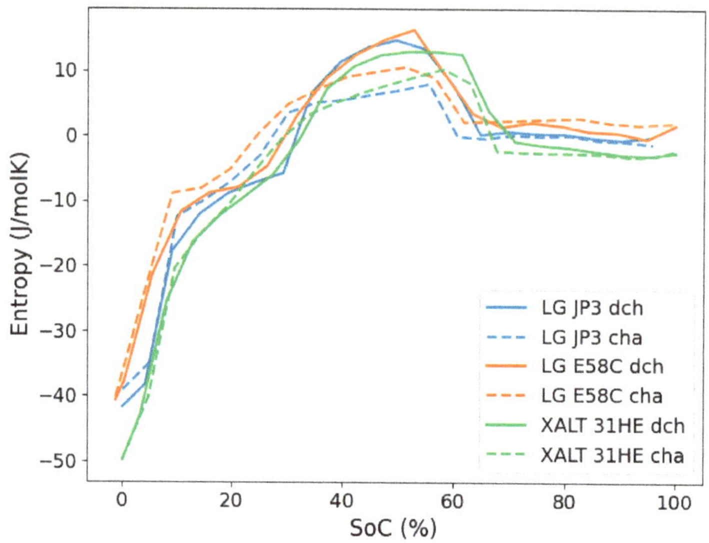

Figure 6. Direct comparison of charge (dashed lines) and corresponding discharge (solid lines) entropies for three selected cells.

3.4. *Ageing & Degradation*

For the XALT cells (NMC/graphite), the evolution of the entropy profiles was followed throughout the lifetime of the cell when cycled at different temperatures, 5, 25 and 45 °C. Entropy profiles at selected SoHs and the corresponding lifetime data are shown in Figure 7. For all conditions, the most pronounced change is the shift of the entropy profiles towards higher SoCs, often accompanied by a horizontal stretch, with the largest shift caused by cycling at 5 °C. While cycling at low and intermediate temperatures largely preserved the main characteristic features and maximum observed entropy values, cyclic ageing at 45 °C led to a significant broadening and flattening of the main features.

As reported in Spitthoff et al. [38], the cells show very poor cycling performance at 5 °C even at 1C. The two cells within this study were cycled at an even higher C-rate of 1.5C and exhibited a very steep decrease in SoH, dropping down to 77% SoH after only 6 equivalent full cycles (EFCs) for cell 1 and 68% SoC after 14 EFCs for cell 2 respectively. The corresponding resistance increase was reported to be very low, likely owing to the low number of undergone cycles. The prevalent degradation mechanism was assumed to be irreversible lithium plating, leading to a significant loss of lithium inventory (LLI), in agreement with the observed shift and stretch of the entropy profile indicative of LLI [57]. The otherwise almost unchanged entropy signature of the cell indicates that the structural integrity of the graphite and thus the prevalent intercalation mechanisms are largely preserved.

At higher temperatures, loss of active material (LAM) on both electrodes was identified as the main cause of degradation. This can be related to structural changes/degradation of the materials and thus explain the significant change in the shape of the entropy profile at low SoH that is most pronounced for the cells cycled at 45 °C/1C and 25 °C/1.5C respectively. Inhomogeneous degradation is likely to further exaggerate this effect.

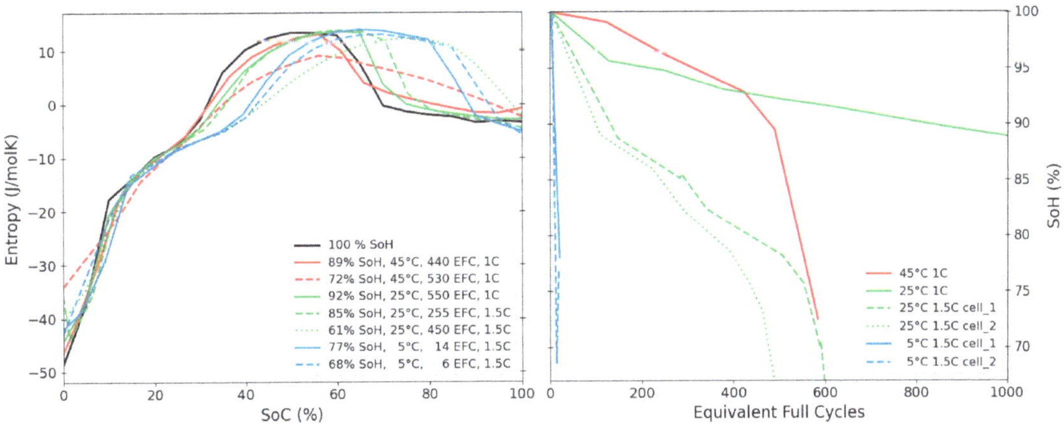

Figure 7. Left: Evolution of entropy changes for the XALT cell upon ageing at different temperatures and C-rates. Right: Corresponding lifetime data.

This example clearly illustrates that ageing under different conditions can have a significant influence on the entropy profiles of a battery cell and can yield insights into the prevalent degradation mechanisms and in combinations with, e.g., ICA and DVA analysis contribute to revealing the underlying ageing modes and mechanisms during LIB ageing [17,19,20]. In this case, the ageing mechanism occurring when cycling at 5 °C can lead to a safety hazard, that is of extreme importance to reveal if that occurred in a real-life application. A detailed discussion of ageing and degradation mechanisms is outside the scope of this work.

4. Conclusions

Within this work, we presented a comprehensive collection of entropy profiles for a total of 14 common commercially available battery cells. The presented data provides valuable insights into the batteries' thermal characteristics with respect to the reversible heat evolution. The results clearly show that reversible heat can significantly contribute to the overall heat generation in a battery during operation. Consequently, it is in general not enough to only consider heating effects related to irreversible heat such as resistance and overpotentials to understand and describe a Li-ion cell's accurate temperature profile during charge or discharge.

A direct comparison based on battery chemistry illustrates significant differences between individual electrode chemistries and their combinations into full cells. By looking into different ways of quantifying the entropic contributions, we clearly illustrate that a simple average entropy value across the entire SoC window is not enough to appropriately characterise the reversible heat evolution. Even average entropy values close to zero do not necessarily imply negligible reversible heat contributions to the total heat generation in the cell. Detailed knowledge of entropy, especially its variation with SoC is crucial for the development of appropriate thermal models and thermal management strategies during battery operation in modules and systems. Significantly different heating and cooling rates might be necessary depending on the selected SoC window. This is particularly relevant for LCO and graphite-based electrodes, where significant heating and cooling effects are expected across large SoC ranges. Monitoring the entropy profile of a cell during ageing at different temperatures showed significant shifts also in the cell's thermal balance, clearly illustrating the need for updated thermal management strategies upon cell ageing.

We hope that the provided data will contribute to the further improvement of both, thermal battery modelling and the development of appropriate thermal management strategies for battery cells, modules and systems.

Supplementary Materials: The following supporting information can be downloaded at: https://www.mdpi.com/article/10.3390/e27040364/s1.

Author Contributions: Conceptualisation, J.W. and P.J.S.V.; experiments and formal analysis, P.J.S.V.; visualisation, J.W.; writing—original draft, review, and editing, J.W. and P.J.S.V. All authors have read and agreed to the published version of the manuscript.

Funding: The authors acknowledge the Research Council of Norway and partner companies of the projects Safety and modelling of aged Li-ion Batteries (BattMarine) (Grant No. 281005), MoreIsLess—design of electrodes for Li-ion batteries with optimised balance of energy and power (Grant No. 324077), 2ND LIFE—The value of second life batteries in the future energy system (Grant No. 320760), Life and Safety for Li-ion batteries in Maritime conditions (Grant No. 228739) and Research and Demonstration of Key Technologies for Reliable and Efficient Application of New Energy Vehicles in China and Norway (Grant No. 304213) for financial support.

Institutional Review Board Statement: Not applicable.

Data Availability Statement: The original data presented in the study are included in the article's Supplementary Material; further inquiries can be directed to the corresponding author/s. This includes: .csv of entropy profiles.

Conflicts of Interest: The authors declare no conflicts of interest. The funders had no role in the design of the study; in the collection, analyses, or interpretation of data; in the writing of the manuscript; or in the decision to publish the results.

References

1. Sæther, S.R.; Moe, E. A green maritime shift: Lessons from the electrification of ferries in Norway. *Energy Res. Soc. Sci.* **2021**, *81*, 102282. [CrossRef]
2. Chen, T.M.; Jin, Y.; Lv, H.Y.; Yang, A.T.; Liu, M.Y.; Chen, B.; Xie, Y.; Chen, Q. Applications of Lithium-Ion Batteries in Grid-Scale Energy Storage Systems. *Trans. Tianjin Univ.* **2020**, *26*, 208–217. [CrossRef]
3. Williard, N.; He, W.; Hendricks, C.; Pecht, M. Lessons Learned from the 787 Dreamliner Issue on Lithium-Ion Battery Reliability. *Energies* **2013**, *6*, 4682–4695. [CrossRef]
4. BBC. Samsung Confirms Battery Faults as Cause of Note 7 Fires. *BBC*, 23 January 2017.
5. Warren, J. E-Bike Ignites Every Two Days, Says Fire Brigade. *BBC News*, 28 June 2023.
6. Stensvold, T. Fergebrann: Brann- og Eksplosjonsårsak Fortsatt en Dåte, in Teknisk Ukeblad. 2019. Available online: https://www.tu.no/artikler/fergebrann-brann-og-eksplosjonsarsak-fortsatt-en-gate/476597 (accessed on 18 March 2025).
7. Norwegian Safety Investigation Authority. Report: Fire on Board 'MS Brim' in the Outer Oslofjord on 11 March 2021. 2022. Available online: https://nsia.no/Marine/Published-reports/2022-08?pid=SHT-Report-ReportFile&attach=1 (accessed on 18 March 2025).
8. Maritime Battery Forum. Fire on Board the Car-Ferry MF Ytterøyningen. 2019. Available online: https://www.maritimebatteryforum.com/reports-documents-1/fire-on-board-the-car-ferry-mf-ytteryningen-2019 (accessed on 18 March 2025).
9. McKinnon, M.B.; DeCrance, S.; Kerber, S. *Four Firefigheters Injured In Lithium-Ion Battery Energy Storage System Explosion—Arizona*; Underwriters Laboratories: Columbia, MD, USA, 2020; p. 66.
10. Na, Y.U.; Jeon, J.W. Unraveling the Characteristics of ESS Fires in South Korea: An In-Depth Analysis of ESS Fire Investigation Outcomes. *Fire* **2023**, *6*, 389. [CrossRef]
11. Smith, A.J.; Fang, Y.; Mikheenkova, A.; Ekstrom, H.; Svens, P.; Ahmed, I.; Lacey, M.J.; Lindbergh, G.; Furo, I.; Lindstrom, R.W. Localized lithium plating under mild cycling conditions in high-energy lithium-ion batteries. *J. Power Source* **2023**, *573*, 233118. [CrossRef]
12. Xie, Y.C.; Wang, S.; Li, R.H.; Ren, D.S.; Yi, M.C.; Xu, C.S.; Han, X.B.; Lu, L.G.; Friess, B.; Offer, G.; et al. Inhomogeneous degradation induced by lithium plating in a large-format lithium-ion battery. *J. Power Source* **2022**, *542*, 231753. [CrossRef]
13. Anseán, D.; Dubarry, M.; Devie, A.; Liaw, B.Y.; García, V.M.; Viera, J.C.; González, M. Operando lithium plating quantification and early detection of a commercial $LiFePO_4$ cell cycled under dynamic driving schedule. *J. Power Source* **2017**, *356*, 36–46. [CrossRef]

14. Burns, J.C.; Stevens, D.A.; Dahn, J.R. In-Situ Detection of Lithium Plating Using High Precision Coulometry. *J. Electrochem. Soc.* **2015**, *162*, A959–A964. [CrossRef]
15. Petzl, M.; Danzer, M.A. Nondestructive detection, characterization, and quantification of lithium plating in commercial lithium-ion batteries. *J. Power Source* **2014**, *254*, 80–87. [CrossRef]
16. Ratnakumar, B.V.; Smart, M.C. Lithium Plating Behavior in Lithium-ion Cells. In *Rechargeable Lithium-Ion Batteries*; Winter, M., Doughty, D.H., Zaghib, K., Abraham, K.M., Ogumi, Z., Dudney, N.J., Eds.; Electrochemical Society Inc.: Pennington, NJ, USA, 2010; Volume 25, pp. 241–252.
17. Dubarry, M.; Ansean, D. Best practices for incremental capacity analysis. *Front. Energy Res.* **2022**, *10*, 1023555. [CrossRef]
18. Lu, M.; Cheng, H.; Yang, Y. A comparison of solid electrolyte interphase (SEI) on the artificial graphite anode of the aged and cycled commercial lithium ion cells. *Electrochim. Acta* **2008**, *53*, 3539–3546. [CrossRef]
19. Dubarry, M.; Svoboda, V.; Hwu, R.; Liaw, B.Y. Incremental capacity analysis and close-to-equilibrium OCV measurements to quantify capacity fade in commercial rechargeable lithium batteries. *Electrochem. Solid State Lett.* **2006**, *9*, A454–A457. [CrossRef]
20. Lewerenz, M.; Marongiu, A.; Warnecke, A.; Sauer, D.U. Differential voltage analysis as a tool for analyzing inhomogeneous aging: A case study for $LiFePO_4$ | Graphite cylindrical cells. *J. Power Source* **2017**, *368*, 57–67. [CrossRef]
21. Bernardi, D.; Pawlikowski, E.; Newman, J. A General Energy-Balance for Battery Systems. *J. Electrochem. Soc.* **1985**, *132*, 5–12. [CrossRef]
22. Thomas, K.E.; Newman, J. Thermal modeling of porous insertion electrodes. *J. Electrochem. Soc.* **2003**, *150*, A176–A192. [CrossRef]
23. Hua, J.S.; Vie, P.J.S.; Wind, J. Methods for faster estimation of the entropy profile of a lithium-ion battery: A comparison of accelerated potentiometry and the estimation of entropy through thermal signatures. *Electrochim. Acta* **2025**, *509*, 145289. [CrossRef]
24. Reynier, Y.; Yazami, R.; Fultz, B. The entropy and enthalpy of lithium intercalation into graphite. *J. Power Source* **2003**, *119*, 850–855. [CrossRef]
25. Takano, K.; Saito, Y.; Kanari, K.; Nozaki, K.; Kato, K.; Negishi, A.; Kato, T. Entropy change in lithium ion cells on charge and discharge. *J. Appl. Electrochem.* **2002**, *32*, 251–258. [CrossRef]
26. Viswanathan, V.V.; Choi, D.; Wang, D.H.; Xu, W.; Towne, S.; Williford, R.E.; Zhang, J.G.; Liu, J.; Yang, Z.G. Effect of entropy change of lithium intercalation in cathodes and anodes on Li-ion battery thermal management. *J. Power Source* **2010**, *195*, 3720–3729. [CrossRef]
27. Prosser, R.; Offer, G.; Patel, Y. Lithium-Ion Diagnostics: The First Quantitative In-Operando Technique for Diagnosing Lithium Ion Battery Degradation Modes under Load with Realistic Thermal Boundary Conditions. *J. Electrochem. Soc.* **2021**, *168*, 030532. [CrossRef]
28. Thomas, K.E.; Bogatu, C.; Newman, J. Measurement of the entropy of reaction as a function of state of charge in doped and undoped lithium manganese oxide. *J. Electrochem. Soc.* **2001**, *148*, A570–A575. [CrossRef]
29. Geng, Z.Y.; Groot, J.; Thiringer, T. A Time- and Cost-Effective Method for Entropic Coefficient Determination of a Large Commercial Battery Cell. *IEEE Trans. Transp. Electrif.* **2020**, *6*, 257–266. [CrossRef]
30. Hu, Y.; Choe, S.Y. Simultaneous and continuous characterization of reversible and irreversible heat of lithium-ion battery using wavelet transform technique. *Electrochim. Acta* **2021**, *375*, 137973. [CrossRef]
31. Eddahech, A.; Briat, O.; Vinassa, J.M. Thermal characterization of a high-power lithium-ion battery: Potentiometric and calorimetric measurement of entropy changes. *Energy* **2013**, *61*, 432–439. [CrossRef]
32. Murashko, K.A.; Mityakov, A.V.; Mityakov, V.Y.; Sapozhnikov, S.Z.; Jokiniemi, J.; Pyrhonen, J. Determination of the entropy change profile of a cylindrical lithium-ion battery by heat flux measurements. *J. Power Source* **2016**, *330*, 61–69. [CrossRef]
33. Damay, N.; Forgez, C.; Bichat, M.P.; Friedrich, G. A method for the fast estimation of a battery entropy-variation high-resolution curve—Application on a commercial LiFePO4/graphite cell. *J. Power Source* **2016**, *332*, 149–153. [CrossRef]
34. Hua, J.; Vie, P.J.S.; Wind, J. Extending an Empirical Degradation Model for Li-Ion Batteries into Second-Life Applications. In Proceedings of the 245th ECS Meeting, San Fransicso, CA, USA, 26–30 May 2024.
35. Spitthoff, L.; Wahl, M.S.; Lamb, J.J.; Shearing, P.R.; Vie, P.J.S.; Burheim, O.S. On the Relations between Lithium-Ion Battery Reaction Entropy, Surface Temperatures and Degradation. *Batteries* **2023**, *9*, 249. [CrossRef]
36. Appiah, W.A.; Park, J.; Byun, S.; Ryou, M.-H.; Lee, Y.M. A Mathematical Model for Cyclic Aging of Spinel $LiMn_2O_4$/Graphite Lithium-Ion Cells. *J. Electrochem. Soc.* **2016**, *163*, A2757. [CrossRef]
37. Koshtyal, Y.; Rumyantsev, A. *Lithium-Ion Cells: Materials and Applications*; MDPI: Basel, Switzerland, 2024. [CrossRef]
38. Spitthoff, L.; Vie, P.J.S.; Wahl, M.S.; Wind, J.; Burheim, O.S. Incremental capacity analysis (dQ/dV) as a tool for analysing the effect of ambient temperature and mechanical clamping on degradation. *J. Electroanal. Chem.* **2023**, *944*, 117627. [CrossRef]
39. Spitthoff, L.; Wahl, M.S.; Vie, P.J.S.; Burheim, O.S. Thermal transport in lithium-ion batteries: The effect of degradation. *J. Power Source* **2023**, *577*, 233149. [CrossRef]
40. Lian, T.; Vie, P.J.S.; Gilljam, M.; Forseth, S. (Invited) Changes in Thermal Stability of Cyclic Aged Commercial Lithium-Ion Cells. *ECS Trans.* **2019**, *89*, 73–81. [CrossRef]
41. Lian, T. Sikkerhetsaspekter i Litiumionbatterier. Master Thesis, Universitetet i Oslo, Oslo, Norway, 2017.
42. Richter, F.; Vie, P.J.S.; Kjelstrup, S.H.; Burheim, O.S. Measurements of ageing and thermal conductivity in a secondary NMC-hard carbon Li-ion battery and the impact on internal temperature profiles. *Electrochim. Acta* **2017**, *250*, 228–237. [CrossRef]

43. Nazer, N.; Yartys, V.; Azib, T.; Latroche, M.; Cuevas, F.; Forseth, S.; Vie, P.; Denys, R.; Sørby, M.; Hauback, B. In operando neutron diffraction study of a commercial graphite/(Ni, Mn, Co) oxide-based multi-component lithium ion battery. *J. Power Source* **2016**, *326*, 93–103.
44. Reynier, Y.F.; Yazami, R.; Fultz, B. Thermodynamics of lithium intercalation into graphites and disordered carbons. *J. Electrochem. Soc.* **2004**, *151*, A422–A426. [CrossRef]
45. Asenbauer, J.; Eisenmann, T.; Kuenzel, M.; Kazzazi, A.; Chen, Z.; Bresser, D. The success story of graphite as a lithium-ion anode material—Fundamentals, remaining challenges, and recent developments including silicon (oxide) composites. *Sustain. Energy Fuels* **2020**, *4*, 5387–5416. [CrossRef]
46. Yun, F.L.; Jin, W.R.; Tang, L.; Li, W.C.; Pang, J.; Lu, S.G. Analysis of Capacity Fade from Entropic Heat Coefficient of Li[NixCoyMnz]O-2/Graphite Lithium Ion Battery. *J. Electrochem. Soc.* **2016**, *163*, A639–A643. [CrossRef]
47. Gunnarshaug, A.F.; Vie, P.J.S.; Kjelstrup, S. Review-Reversible Heat Effects in Cells Relevant for Lithium-ion Batteries. *J. Electrochem. Soc.* **2021**, *168*, 050522. [CrossRef]
48. Jalkanen, K.; Vuorilehto, K. Entropy change characteristics of $LiMn_{0.67}Fe_{0.33}PO_4$ and $Li_4Ti_5O_{12}$ electrode materials. *J. Power Source* **2015**, *273*, 351–359. [CrossRef]
49. Lu, W.; Belharouak, I.; Liu, J.; Amine, K. Electrochemical and thermal investigation of $Li_4/3Ti_5/3O_4$ spinel. *J. Electrochem. Soc.* **2007**, *154*, A114–A118. [CrossRef]
50. Hudak, N.S.; Davis, L.E.; Nagasubramanian, G. Cycling-Induced Changes in the Entropy Profiles of Lithium Cobalt Oxide Electrodes. *J. Electrochem. Soc.* **2015**, *162*, A315–A321. [CrossRef]
51. Williford, R.E.; Viswanathan, V.V.; Zhang, J.G. Effects of entropy changes in anodes and cathodes on the thermal behavior of lithium ion batteries. *J. Power Source* **2009**, *189*, 101–107. [CrossRef]
52. Sturm, J.; Rheinfeld, A.; Zilberman, I.; Spingler, F.B.; Kosch, S.; Frie, F.; Jossen, A. Modeling and simulation of inhomogeneities in a 18650 nickel-rich, silicon-graphite lithium-ion cell during fast charging. *J. Power Source* **2019**, *412*, 204–223. [CrossRef]
53. Marcicki, J.; Yang, X.G. Model-Based Estimation of Reversible Heat Generation in Lithium-Ion Cells. *J. Electrochem. Soc.* **2014**, *161*, A1794–A1800. [CrossRef]
54. Schlueter, S.; Genieser, R.; Richards, D.; Hoster, H.E.; Mercer, M.P. Quantifying structure dependent responses in Li-ion cells with excess Li spinel cathodes: Matching voltage and entropy profiles through mean field models. *Phys. Chem. Chem. Phys.* **2018**, *20*, 21417–21429. [CrossRef]
55. Schmidt, J.P.; Weber, A.; Ivers-Tiffee, E. A novel and precise measuring method for the entropy of lithium-ion cells: Delta S via electrothermal impedance spectroscopy. *Electrochim. Acta* **2014**, *137*, 311–319. [CrossRef]
56. Jalkanen, K.; Aho, T.; Vuorilehto, K. Entropy change effects on the thermal behavior of a $LiFePO_4$/graphite lithium-ion cell at different states of charge. *J. Power Source* **2013**, *243*, 354–360. [CrossRef]
57. Wojtala, M.E.; Zulke, A.A.; Burrell, R.; Nagarathinam, M.; Li, G.C.; Hoster, H.E.; Howey, D.A.; Mercer, M.P. Entropy Profiling for the Diagnosis of NCA/Gr-SiOx Li-Ion Battery Health. *J. Electrochem. Soc.* **2022**, *169*, 100527. [CrossRef]
58. Osswald, P.J.; del Rosario, M.; Garche, J.; Jossen, A.; Hoster, H.E. Fast and Accurate Measurement of Entropy Profiles of Commercial Lithium-Ion Cells. *Electrochim. Acta* **2015**, *177*, 270–276. [CrossRef]
59. Yazami, R.; Reynier, Y.; Fultz, B. Entropymetry of Lithium Intercalation in Spinel Manganese Oxide: Effect of Lithium Stoichiometry. *ECS Trans.* **2006**, *1*, 87. [CrossRef]
60. Reiter, A.; Schütte, M.F.; Rosenmüller, C.; Stroebl, F.; Lehner, S.; Sauer, D.U.; Bohlen, O. Exploring the Effects of Aging, Temperature and Hysteresis on the Entropy Variation of Lithium-Ion Batteries. 2025. Available online: https://ssrn.com/abstract=5142673 (accessed on 18 March 2025).
61. Bedurftig, B.; Oldenburger, M.; Hufner, T.; Richter, E.; Braatz, R.D.; Gruhle, A.; Findeisen, R. Measuring the reversible heat of lithium-ion cells via current pulses for modeling of temperature dynamics. *J. Power Source* **2021**, *506*, 230110. [CrossRef]
62. Dreyer, W.; Jamnik, J.; Guhlke, C.; Huth, R.; Moskon, J.; Gaberscek, M. The thermodynamic origin of hysteresis in insertion batteries. *Nat. Mater.* **2010**, *9*, 448–453. [CrossRef] [PubMed]
63. Petzl, M.; Danzer, M.A. Advancements in OCV Measurement and Analysis for Lithium-Ion Batteries. *IEEE Trans. Energy Convers.* **2013**, *28*, 675–681. [CrossRef]
64. Wind, J.; Vie, P.J.S. Revisiting Pulse-Based OCV Incremental Capacity Analysis for Diagnostics of Li-Ion Batteries. *Batteries* **2024**, *10*, 277. [CrossRef]
65. Tardy, E.; Thivel, P.X.; Druart, F.; Kuntz, P.; Devaux, D.; Bultel, Y. Internal temperature distribution in lithium-ion battery cell and module based on a 3D electrothermal model: An investigation of real geometry, entropy change and thermal process. *J. Energy Storage* **2023**, *64*, 107090. [CrossRef]

entropy

MDPI

Article

Local Equilibrium Approximation in Non-Equilibrium Thermodynamics of Diffusion

Kim R. Kristiansen † **and Bjørn Hafskjold** *

PoreLab, Department of Chemistry, Norwegian University of Science and Technology (NTNU), N-7491 Trondheim, Norway; kim.kristiansen@ntnu.no

* Correspondence: bjorn.hafskjold@ntnu.no

† Current address: Ocean GeoLoop AS, N-7652 Verdal, Norway.

Abstract: Local equilibrium approximation (LEA) is a central assumption in many applications of non-equilibrium thermodynamics involving the transport of energy, mass, and momentum. However, assessing the validity of the LEA remains challenging due to the limited development of tools for characterizing non-equilibrium states compared to equilibrium states. To address this, we have developed a theory based on kinetic theory, which provides a nonlinear extension of the telegrapher's equation commonly discussed in non-equilibrium frameworks that extend beyond LEA. A key result of this theory is a steady-state diffusion equation that accounts for the constraint imposed by available thermal energy on the diffusion flux. The theory is suitable for analysis of steady-state composition profiles and can be used to quantify the deviation from the local equilibrium. To validate the theory and test LEA, we performed molecular dynamics simulations on a two-component system where the two components had identical physical properties. The results show that deviation from the local equilibrium can be systematically quantified, and for the diffusion process we have studied here, we have confirmed that LEA remains accurate even under extreme concentration gradients in gas mixtures.

Keywords: local equilibrium approximation; non-equilibrium entropy; diffusion; kinetic theory; molecular dynamics; telegrapher equation

1. Introduction

The concept of a local equilibrium in non-equilibrium thermodynamics comes up in analyses of transport processes such as heat and mass transport. The term "local equilibrium" is not well defined [1]. Moreover, depending on whether a perturbation of an equilibrium state is temporal or spacial, the concept of local equilibrium in the perturbed state may have different meanings. Our use of the term is related to so-called linear non-equilibrium thermodynamics, where one assumes a local equilibrium, meaning that the Gibbs–Duhem equation is assumed valid with local values of the variables. The assumption is closely related to assuming that the local non-equilibrium entropy can be approximated by its equilibrium value in the entropy balance, which is the basis for finding the entropy production and coupled flux–force relations [2]. Whereas entropy is well understood and modeled for thermodynamic systems at equilibrium, the situation for systems out of equilibrium is less clear [3]. Usually, the assumption is introduced as an approximation and justified with successful consistency tests as "circumstantial evidence". This is not satisfactory, and the extremely large forces and fluxes used in computer simulations that are necessary for acceptable signal-to-noise ratios in the results raise a need for a quantitative examination of local equilibrium approximation (LEA).

An alternative approach is to *assume the opposite, viz. a local non-equilibrium*, derive the corresponding theory whenever possible, and determine the deviation from the local equilibrium. A powerful criterion for this deviation is the difference in equilibrium and non-equilibrium entropy. Several theories for non-equilibrium entropy have been proposed for heat transport and diffusion, see, e.g., [4,5], and references therein. Cimmelli et al. [4] have given a very thorough discussion of four different theories and compared their pros and cons. They discuss the theories' relation to kinetic theory, valid for low-density gases and conclude that in general, continuum theories that are compatible with kinetic theory are less universal, but more predictable than other theories.

In the present work, we pursue developing a theory for transport processes that do not depend on LEA. The theory is based on the Boltzmann equation and developed for diffusion in a binary gas mixture. It is then used to analyze results from molecular dynamics (MD) simulations of binary diffusion in a Lennard–Jones/spline (LJs) system [6]. We use an MD method introduced by Holian for diffusion simulations, which will be described in detail in Section 3. (Information about this method was communicated by B. L. Holian to W. G. Hoover in 1973, see Hoover and Ashurst [7] and Allen and Tildesley, p.370 [8]. A very similar method has been used by Dong and Cao to compute self-diffusion coefficients [9]).

The two components are physically identical but they are identified as two types, "1" and "2" or "red" and "blue". Therefore, the system's properties are those of a one-component system, which simplifies the analysis and enables comparison with self-diffusion data computed with other methods. This is one of the simplest cases one can imagine for a study of a transport process. Our intention is to force the diffusion process as much as the system allows in an attempt to provoke a deviation from the local equilibrium. The results are compared with what we obtain from Fick's law and with data from equilibrium simulations (mean square displacement).

According to Darken's second equation [10], the mutual diffusion coefficient D in a binary mixture relates to the self-diffusion coefficients D_1 and D_2, as

$$D = (x_1 D_2 + x_2 D_1)\Gamma \,. \tag{1}$$

where x_1 is the mole fraction of component 1 ($x_2 = 1 - x_1$) and $\Gamma = \frac{1}{k_B T}\left(\frac{\partial \mu_1}{\partial \ln x_1}\right)_{P,T}$ is the thermodynamic factor. In our ideal case, $\Gamma = 1$ and $D_1 = D_2$, so that the diffusion coefficient we compute by generating a diffusive flux in the two-component system is simply the self-diffusion coefficient. This enables a comparison between the binary diffusion coefficient, generated with extreme fluxes and forces, and self-diffusion coefficients generated at global equilibrium.

2. Theory

2.1. Fick's Law

In most common applications, diffusion is usually expressed in the form of Fick's law, which states that the diffusion flux $\mathbf{J}_i$of component i is proportional to the gradient in the concentration n_i of the corresponding component in the mixture

$$\mathbf{J}_i = -D\nabla n_i \tag{2}$$

where D is the diffusion coefficient (here, we use the flux of particles and the number density n_i instead of the usual molar flux and molar concentration). Fick's law for gas mixtures can be obtained ab initio from kinetic theory in the short mean free path limit, meaning that it holds for descriptions on length scales much longer than the mean free path. The divergence of Equation (2) at constant temperature and mass density, combined with the mass conservation law in a fluid at rest,

$$\partial_t n_i + \nabla \cdot \mathbf{J}_i = 0\,, \tag{3}$$

yields the well-known diffusion equation, under the additional approximation that D is independent of n_i,

$$\partial_t n_i = D\nabla^2 n_i\,. \tag{4}$$

Fick's law can also be obtained using the non-equilibrium thermodynamics methodology. Considering a binary mixture at isothermal conditions, the entropy differential is

$$\mathrm{d}s = -\frac{\mu_1}{T}\mathrm{d}x_1 - \frac{\mu_2}{T}\mathrm{d}x_2 = -\frac{\mu}{T}\mathrm{d}x_1 \tag{5}$$

where s is the entropy per particle, T the temperature, and μ_i and x_i are the chemical potential and mole fraction of component i, respectively. We used $\mathrm{d}x_1 = -\mathrm{d}x_2$ and introduced $\mu = \mu_1 - \mu_2$. In order to keep the analysis relatively simple, we will in the following neglect the Dufour effect in the special case we consider here; in other words, we will neglect the heat flux accompanying diffusion. Using the mass conservation law, we obtain the entropy balance

$$n\partial_t s = \frac{\mu}{T}\nabla \cdot \mathbf{J}_1 = \nabla \cdot \frac{\mu \mathbf{J}_1}{T} - \mathbf{J}_1 \cdot \nabla\frac{\mu}{T} = -\nabla \cdot \mathbf{J}_s + \sigma\,, \tag{6}$$

where n is the total number density. We have here identified the entropy flux $\mathbf{J}_s = -\mu\mathbf{J}_1/T$ and the entropy production rate $\sigma = \mathbf{J}_1 \cdot (-\nabla\mu/T)$. The latter is expressed on the standard form as a product of the flux $\mathbf{J}_1$ and its conjugate driving force $-\nabla\mu/T$. Assuming that this driving force is proportional to $\mathbf{J}_1$ by some proportionality factor $r > 0$ guarantees the positivity of the entropy-production rate, and gives the following relation at isothermal conditions:

$$r\mathbf{J}_1 = -\nabla\frac{\mu}{T} \implies \mathbf{J}_1 = -\frac{1}{rT}\frac{\partial\mu}{\partial n_1}\nabla n_1 = -D\nabla n_1 \implies r = \frac{\partial\mu/\partial n_1}{TD}\,. \tag{7}$$

The physical meaning of r is a resistivity relating the force ($\nabla(\mu/T)$) to the flux ($\mathbf{J}$). We arrive at Fick's law by the general requirement that the flux is a linear combination of the thermodynamic driving forces in the system, where in this case $-\nabla\mu/T$ is the sole driving force. The identification of D from r clarifies the connection.

2.2. Generalized Diffusion Equation

While Fick's law is adequate for describing diffusion processes when the system is everywhere in the local equilibrium, significant departures from the local equilibrium expose its unphysical features. For example, according to Fick's law, a local change in the composition or flux anywhere in the system will *immediately* affect the evolution of the physical state everywhere else in the system. That is, Fick's law propagates information across the spatial domain at an unphysical infinite speed, due to a chain of instantaneous molecular collisions arising as an artifact of neglecting the finite length of flight paths between collisions [11]. We seek a more general evolution equation of diffusion from the Boltzmann equation, which is not constrained by assumptions of a local equilibrium, and accounts fully for the discrete nature of molecular collisions, propagating information at speeds consistent with molecular velocities and collision dynamics. (Strictly speaking, this rectifies only one out of two mechanisms that violate causality. The resulting classical distribution of molecular velocities remains unbounded in momentum space, and so arbitrarily large velocities still have finite though vanishingly small probability amplitudes. A relativistic theory would be required to fully resolve the causality problem, but we stick to the classical formulation here.)

The momentum balance for the general case (several components with different properties, flowing system) was given by Snell et al. [12]. According to their work, the general equation of motion for component i in an isothermal multicomponent mixture, as obtained from the Boltzmann equation, is

$$\partial_t(\rho_i \mathbf{u}_i) = \boldsymbol{\nabla} \cdot [\mathbf{A}_i - \mathbf{B}_i - \rho_i(\mathbf{u}_i \otimes \mathbf{u} + \mathbf{u} \otimes \mathbf{u}_i - \mathbf{u} \otimes \mathbf{u})] + x_i \sum_k \xi_{ik} x_k (\mathbf{u}_k - \mathbf{u}_i) - n_i \boldsymbol{\nabla} \mu_i \quad (8)$$

where ρ_i and $\mathbf{u}_i$ are the mass density and the mean velocity of component i, respectively, $\mathbf{u}$ is the local barycentric velocity of the mixture, and ξ_{ik} is a friction coefficient describing the exchange of momentum between components i and k. (The wotk by Snell et al. is based on the methodology used in the earlier works of Bearman and Kirkwood, where effective forces are introduced to create a pseudo-equilibrium state that emulates a true non-equilibrium state. Since the local state variables are, by construction, independent of the imposed forces, those state variables are assumed equal to the state variables in the local equilibrium)

The tensor $\mathbf{A}_i$ denotes viscous contributions,

$$\mathbf{A}_i = x_i \sum_k \left[\left(\zeta_{ik} - \frac{2}{3}\eta_{ik} \right) \mathbf{I} \boldsymbol{\nabla} \cdot (x_k \mathbf{u}_k) + \eta_{ik} \left(\boldsymbol{\nabla}(x_k \mathbf{u}_k) + (\boldsymbol{\nabla}(x_k \mathbf{u}_k))^{\mathrm{T}} \right) \right], \quad (9)$$

where ζ_{ik} and η_{ik} are coefficients related to the bulk and shear viscosity, respectively. We refer the interested reader to [12] for their definitions. The tensor $\mathbf{B}_i$ represents the traceless part of the diffusion momentum outer product (indicated by the symbol $\overline{\otimes}$):

$$\mathbf{B}_i = \rho_i \overline{(\mathbf{u}_i - \mathbf{u}) \otimes (\mathbf{u}_i - \mathbf{u})} = \rho_i \left[(\mathbf{u}_i - \mathbf{u}) \otimes (\mathbf{u}_i - \mathbf{u}) - \frac{1}{3} \mathbf{I} \mathrm{Tr}[(\mathbf{u}_i - \mathbf{u}) \otimes (\mathbf{u}_i - \mathbf{u})] \right]. \quad (10)$$

2.2.1. Application to the Present System

When the two components are physically identical (which will be considered from here on), the viscosity coefficients ζ_{ik} and η_{ik} in Equation (9) are independent of the pair ik. Moreover, when the fluid is stagnant, all contributions to $\mathbf{A}_i$ that are proportional to $\boldsymbol{\nabla}\mathbf{u}$ vanish. Equation (8) can then be reduced to

$$\partial_t(\rho_i \mathbf{u}_i) + \boldsymbol{\nabla} \cdot (\rho_i \overline{\mathbf{u}_i \otimes \mathbf{u}_i}) = x_i \sum_k \xi_{ik} x_k (\mathbf{u}_k - \mathbf{u}_i) - n_i \boldsymbol{\nabla} \mu_i \,. \quad (11)$$

We identify the fluxes $\mathbf{J}_i = n_i \mathbf{u}_i$. The difference between the equations of motion for the two components is

$$\begin{aligned} \partial_t(m_1 \mathbf{J}_1 - m_2 \mathbf{J}_2) + \boldsymbol{\nabla} \cdot \left(\frac{m_1 \overline{\mathbf{J}_1 \otimes \mathbf{J}_1}}{n_1} - \frac{m_2 \overline{\mathbf{J}_2 \otimes \mathbf{J}_2}}{n_2} \right) \\ = 2 x_1 x_2 \xi \left(\frac{\mathbf{J}_2}{n_2} - \frac{\mathbf{J}_1}{n_1} \right) - n_1 \boldsymbol{\nabla} \mu_1 + n_2 \boldsymbol{\nabla} \mu_2 \,. \end{aligned} \quad (12)$$

where m_i is the molecular mass of component i and the friction coefficients are symmetric, $\xi_{ik} = \xi_{ki} = \xi$. We now use that $\mathbf{J}_2 = -\mathbf{J}_1$ and $\boldsymbol{\nabla}\mu_1 = \Gamma \frac{k_B T}{x_1} \boldsymbol{\nabla} x_1$. Equation (12) can then be reduced to

$$\overline{m} \partial_t \mathbf{J}_1 + \frac{\overline{m}}{n} \boldsymbol{\nabla} \cdot \left[\overline{\mathbf{J}_1 \otimes \mathbf{J}_1} \left(\frac{c_1}{x_1} - \frac{1 - c_1}{1 - x_1} \right) \right] = -\frac{\xi \mathbf{J}_1}{n} - \Gamma n k_B T \boldsymbol{\nabla} x_1 \quad (13)$$

where $\overline{m} = (m_1 + m_2)/2$ is the average molecular mass, and $c_1 = m_1/(m_1 + m_2)$. In the system we consider here, $m_1 = m_2 = m$, so that $c = 1/2$, which is used in the following. We see that under conditions where the terms on the left-hand side of Equation (13) are small, that is when temporal and spatial variations in the flux are small, we reproduce Fick's law. This allows us to identify the friction coefficient ξ in terms of the diffusion coefficient:

$$\mathbf{J}_1 = -\frac{\Gamma n^2 k_B T}{\xi} \boldsymbol{\nabla} x_1 = -nD \boldsymbol{\nabla} x_1 \implies \xi = \frac{\Gamma n k_B T}{D} . \tag{14}$$

Using the product rule, we can split the term involving the divergence of the traceless kinetic energy tensor:

$$\begin{aligned} &\boldsymbol{\nabla} \cdot \left[\overline{\mathbf{J}_1 \otimes \mathbf{J}_1} \left(\frac{1}{x_1} - \frac{1}{1 - x_1} \right) \right] \\ &= \left(\frac{1}{x_1} - \frac{1}{1 - x_1} \right) \boldsymbol{\nabla} \cdot \overline{\mathbf{J}_1 \otimes \mathbf{J}_1} - \left(\frac{1}{x_1^2} + \frac{1}{(1 - x_1)^2} \right) \overline{\mathbf{J}_1 \otimes \mathbf{J}_1} \cdot \boldsymbol{\nabla} x_1 . \end{aligned} \tag{15}$$

Inserting Equation (15) into Equation (13), we find

$$\begin{aligned} &\tau \left[\partial_t \mathbf{J}_1 + \frac{1}{2n} \left(\frac{1}{x_1} - \frac{1}{1 - x_1} \right) \boldsymbol{\nabla} \cdot \overline{\mathbf{J}_1 \otimes \mathbf{J}_1} \right] + \mathbf{J}_1 \\ &= -D \left[1 - \frac{m}{2\Gamma n^2 k_B T} \left(\frac{1}{x_1^2} + \frac{1}{(1 - x_1)^2} \right) \overline{\mathbf{J}_1 \otimes \mathbf{J}_1} \cdot \right] \boldsymbol{\nabla} n_1 \end{aligned} \tag{16}$$

where we have identified the characteristic time $\tau = mn/\xi = mD/(\Gamma k_B T)$. We note that when the terms involving the divergence of the traceless kinetic energy tensor can be neglected, this reduces to

$$\tau \partial_t \mathbf{J}_1 + \mathbf{J}_1 = -D \boldsymbol{\nabla} n_1 , \tag{17}$$

the divergence of which leads to the well-known telegrapher's equation for n_1 when combined with Equation (3) for a stagnant fluid and gradients in D and τ are neglected. Equation (17) is a damped wave equation describing waves propagating with a characteristic speed $\sqrt{D/\tau}$ and attenuated over a characteristic time τ. Equation (2) corresponds to Equation (17) in the limit $\tau \to 0$, which corresponds to a diverging wave speed that leads to the unphysically instantaneous propagation of information across the domain. The telegrapher's equation takes into account the inertia of molecular motion that resists rapid changes to the diffusion flux, and its connection to non-equilibrium thermodynamics is thoroughly discussed in the work by Jou, Casas-Vázquez and Lebon [13]. There, the authors arrive at equations of the telegrapher type by means of the thirteen moment approximation, which is on the same order as Equation (16) with respect to the order of moments retained in the approximation to the full Boltzmann equation. Their work leads us to expect that the relationship between τ and D at this level of approximation should be corrected for higher order correlations in order to match real data, justifying an a priori unknown correction factor α such that the true relaxation time is $\alpha\tau$. The examples explicitly calculated by Jou et al. by continued fraction expansions yield constant α-values [13].

For the sake of simplicity, we will in the following restrict our attention to self-diffusion along a single dimension in three-dimensional space. Introducing also the correction factor α, Equation (16) is replaced by the one-dimensional equation for either component

$$\alpha\tau \left[\partial_t J_i + \frac{1}{3n} \left(\frac{1}{x_i} - \frac{1}{1 - x_i} \right) \nabla J_i^2 \right] + J_i = -D \left[1 - \frac{\alpha m J_i^2}{3\Gamma n^2 k_B T} \left(\frac{1}{x_i^2} + \frac{1}{(1 - x_i)^2} \right) \right] \nabla n_i . \tag{18}$$

Rearranging Equation (18), we find that it can be written in the form

$$r_i J_i = -\left[1 - \frac{\alpha m J_i^2}{3\Gamma n^2 k_B T}\left(\frac{1}{x_i^2} + \frac{1}{(1-x_i)^2}\right)\right]\frac{\nabla \mu_{i,\text{eq}}}{T} - \frac{\alpha m}{n x_i T}\left[\partial_t J_i + \frac{1}{3n}\left(\frac{1}{x_i} - \frac{1}{1-x_i}\right)\nabla J_i^2\right], \tag{19}$$

with the identified resistivity $r_i = (\partial\mu_{i,\text{eq}}/\partial n_i)/(TD)$, cf. Equation (7). The substitution of ∇n_i in Equation (18) with $\nabla \mu_i$ in Equation (19) was made by using $rD\nabla n_i = \frac{1}{TD}\frac{\partial \mu_{i,\text{eq}}}{\partial n_i} D\nabla n_i = \frac{1}{T}\nabla \mu_i$. In Equation (19), we specified that the chemical potential $\mu_{i,\text{eq}}$ is the equilibrium value, i.e., that obtained from the equation of state to distinguish it from the non-equilibrium value to be introduced below.

We can now identify the right-hand side of Equation (19) as a generalized driving force proportional to the flux, $X_i = r_i J_i$. In order to express the entropy production in terms of the flux–force pairs, we must identify a non-equilibrium entropy differential ds that reproduces the derived equation of motion. This can be achieved by considering the diffusion fluxes as independent non-equilibrium state variables in accordance with the extended irreversible thermodynamics framework [13].

$$\text{d}s = -\frac{\mu_1}{T}\text{d}x_1 - \frac{\mu_2}{T}\text{d}x_2 + B_1 \text{d}J_1 + B_2 \text{d}J_2 \tag{20}$$

where μ_i is now a generalized non-equilibrium chemical potential of component i, and the B_i are *a priori* unknown coefficients. In order to have correspondence with the equilibrium entropy, we must have that $\lim_{J_i \to 0} \mu_i = \mu_{i,\text{eq}}$. Noting that the corresponding entropy balance equation takes the form

$$\partial_t s + \nabla J_s = \sigma\,, \tag{21}$$

the connection with the equation of motion is made by assuming that the entropy production rate is a bilinear sum $\sigma = J_1 X_1 + J_2 X_2$ with fluxes J_i and forces X_i. That is, we have

$$\begin{aligned}\sigma = -\sum_{i=1}^{2}\Bigg\{ & J_i\left[1 - \frac{\alpha m J_i^2}{3\Gamma n^2 k_B T}\left(\frac{1}{x_i^2} + \frac{1}{(1-x_i)^2}\right)\right]\nabla\frac{\mu_{i,\text{eq}}}{T} \\ & + J_i \frac{\alpha m}{n x_i T}\left[\partial_t J_i + \frac{1}{3n}\left(\frac{1}{x_i} - \frac{1}{1-x_i}\right)\nabla J_i^2\right]\Bigg\}.\end{aligned} \tag{22}$$

We can now determine the coefficients B_i by comparing this form of the entropy production rate to that which can be derived from the proposed entropy differential. We note that the contribution from the time derivative of J_i to σ is

$$\sigma \propto -\frac{\alpha m J_i}{n x_i T}\partial_t J_i\,, \tag{23}$$

which allows us to deduce that

$$n\partial_t s \propto -\frac{\alpha m J_i}{n x_i T}\partial_t J_i \implies \text{d}s \propto -\frac{\alpha m J_i}{n^2 x_i T}\text{d}J_i \implies B_i = -\frac{\alpha m J_i}{n^2 x_i T}\,. \tag{24}$$

Now, we use that $x_1 + x_2 = 1$ to simplify Equation (20) to

$$\text{d}s = -\frac{\mu}{T}\text{d}x_1 + B_1\text{d}J_1 + B_2\text{d}J_2 \tag{25}$$

where $\mu = \mu_1 - \mu_2$, equivalent to the definition in Equation (5). We can then obtain the J_i-dependence of μ through symmetry of mixed derivatives

$$-\frac{\partial \mu/T}{\partial J_i} = \frac{\partial^2 s}{\partial J_i \partial x_1} = \frac{\partial^2 s}{\partial x_1 \partial J_i} = \frac{\partial B_i}{\partial x_1} = -\frac{\alpha}{n^2 T}\frac{\partial}{\partial x_1}\frac{m J_i}{x_i} = \begin{cases} \frac{\alpha m J_1}{n^2 x_1^2 T} & i = 1 \\ -\frac{\alpha m J_2}{n^2 x_2^2 T} & i = 2 \end{cases} \tag{26}$$

Combined with the correspondence requirement to s_{eq} and that $J_1 + J_2 = 0$, we find

$$-\frac{\mu}{T} = -\frac{\mu_{\text{eq}}}{T} + \int_0^{J_1} \mathrm{d}J' \frac{\alpha m J'}{n^2 T}\left(\frac{1}{x_1^2} + \frac{1}{(1-x_1)^2}\right) = -\frac{\mu_{\text{eq}}}{T} + \frac{\alpha m J_1^2}{2n^2 T}\left(\frac{1}{x_1^2} + \frac{1}{(1-x_1)^2}\right) \quad (27)$$

We identify the second term on the right-hand side as the kinetic energy associated with species interdiffusion, multiplied by a factor α/T. This implies that in the non-equilibrium situation, the equilibrium chemical potential is replaced by a generalized chemical potential where a term proportional to the kinetic energy of diffusion is subtracted. This agrees with the assessment by de Groot and Mazur, who arrived at a similar generalized driving force by considering the necessity of subtracting the kinetic energy of diffusion from the typically defined internal energy to obtain the *truly* internal energy [14]. We may quantify the departure from the local equilibrium using the deviation of the local chemical potential from its equilibrium value:

$$\mu_{\text{eq}} - \mu = \frac{\alpha m J_1^2}{2n^2}\left(\frac{1}{x_1^2} + \frac{1}{(1-x_1)^2}\right) = \alpha\left(\frac{m u_1^2}{2} + \frac{m u_2^2}{2}\right), \quad (28)$$

which reproduces the de Groot and Mazur result in the special case $\alpha = 1$. We restrict our attention to the steady-state equation of motion, where $\partial_t J_1 = 0$ and $\nabla J_1 \propto \partial_t x_1 = 0$. This now reads

$$J_1 = -D\left[1 - \frac{\mu_{\text{eq}} - \mu}{\Gamma E_k}\right]\nabla n_1 \quad (29)$$

where $E_k = 3k_B T/2$ is the average molecular translational kinetic energy. Fick's law with constant D predicts a constant composition gradient in steady state, according to Equation (4). According to Equation (29), we expect to see a curved composition profile when the local deviation in chemical potential from the local equilibrium value is significant relative to ΓE_k. The mechanism at play here is that of the diffusion process approaching its speed limit dictated by the local distribution of molecular velocities. Whereas a Fickian model would violate this speed limit by allowing the diffusion flux to increase beyond the rate of transport permissible by the ballistic motion of molecules at a given temperature, this generalized formulation properly enforces said speed limit in a continuous and physically consistent manner. Investigating the curvature of the steady-state composition profile resulting from a large imposed diffusion flux allows us to properly quantify the deviation from the local equilibrium.

2.2.2. The Non-Equilibrium Entropy

We note that for fixed x_i, we can use the result for B_i, Equation (24), in Equation (20) and integrate to give

$$s - s_{\text{eq}} = -\sum_i \frac{\alpha\, m J_i^2}{2n^2 x_i T} = -\frac{\alpha m J_1^2}{2n^2 T}\left(\frac{1}{x_1} + \frac{1}{1-x_1}\right) = -\frac{\alpha m J_1^2}{2n^2 x_1(1-x_1)T}, \quad (30)$$

which shows that the non-equilibrium entropy s is always smaller than the equilibrium value, which is consistent with the second law. This form of the non-equilibrium entropy is also in agreement with that derived by Jou et al. [13], which means that the nonlinear contribution to the flux, not included in the theory of Jou et al., does not directly influence the value of the local entropy. The absolute deviation in the entropy from the local equilibrium value has a minimum at $x_1 = 0.5$.

3. MD Simulations

Non-equilibrium molecular dynamics (NEMD) simulations were carried out to provide data for use with the theory. The layout of the MD system is shown in Figure 1.

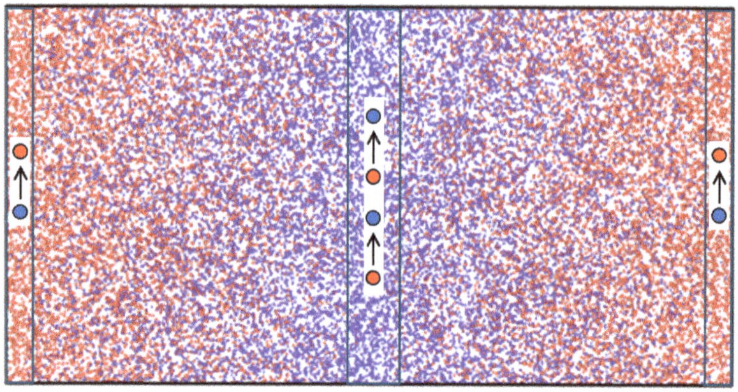

Figure 1. Layout of MD box used in the simulations. The particle types are swapped as indicated in the end and central regions of the MD box (the swapping layers).

The system consisted of $N = 32,768$ Lennard–Jones/spline (LJs) particles [6] in a rectangular cuboid with aspect ratio $L_x : L_y : L_z = 2 : 1 : 1$. The LJs potential is a smoothly truncated Lennard–Jones potential with cut-off distance $r_c \approx 1.7\sigma$. The MD box had periodic boundary conditions in all three directions. The box was divided into 64 layers of equal thickness in the x-direction for recording local data of temperature, density, composition, etc. The particles were of two types, 1 ("red") and 2 ("blue"), with exactly the same mass and potential parameters [8].

Concentration gradients and diffusive mass fluxes were generated with an in-house NEMD code using the MEX algorithm [1]. The MEX algorithm works so that a randomly picked particle of type 2 is changed to type 1 in one of the control volumes at the box ends (see Figure 1) while a particle of type 1 is changed to type 2 in the central volume at the same time. The overall concentration was thus kept constant. The identity swapping was tried at regular intervals, which varied between every time step and every 100th time step. The swapping acts as source and sink terms at the boundaries (ends and center of the MD box). At steady state, the mass flux in the bulk fluid is related to the swapping frequency, ν = number of swaps/time, $J_i = \nu_i/2A$, where A is the cross-sectional area of the box. By changing the swapping frequency, the diffusion flux was controlled from zero (equilibrium) to maximum (saturated control volumes, i.e., $x_1 = 1$ at the box ends and $x_1 = 0$ in the box center). If there were no candidate particles to swap in either control volume, the swapping was skipped and the actual swapping frequency was reduced accordingly. The reported mass fluxes in Table 1 were computed from $J_1 = \langle n_1 u_1 \rangle$ for each layer in the fluid and then averaged over the layers in the bulk. The logged swapping frequency was used for consistency check. Typical swapping frequencies varied from 9 to 22 (in LJ units).The mass current is half the swapping frequency (due to the fact that particles move both left and right from the swapping layers), and the mass flux is the current divided by the cross-sectional area. For the results reported here, the time between each swap was on average between 0.02 and 0.06, which is three orders of magnitude shorter than the characteristic time $\tau = mD/(\Gamma k_B T) \approx 10$. We can therefore expect that the non-continuous particle swapping at the boundaries leads to a mass flux that is smooth in time (which was also observed).

The simulations started from an FCC lattice configuration with $N_1 = N_2 = N/2$. To prevent heating of the system due to the entropy production, the system temperature was controlled by thermostating each of the 64 layers to the same set temperature and constant density was assured by rapid equilibration of the pressure in the fluid. The overall density was controlled by the total number of particles in the fixed box volume. The thermostat was a simple velocity scaling and shifting subject to set the local momentum to zero [1]. The scaling factor was the same in x-, y-, and z-directions, but the shift was direction specific. The thermostat was activated every 20 time steps.

One may argue that enforcing a constant temperature through the system has an impact on properties derived from temperature gradients such as Soret and Dufour effects. This should be considered for systems where the components do not have identical properties, but for the present system, such coupled effects vanish.

Six cases were generated with the density and temperature as listed in Table 1. All values of physical properties reported here are in Lennard–Jones units. The mean free path, λ, was estimated from elementary kinetic theory as $\lambda = 1/(\pi\sqrt{2}n)$ and compared with the half box length, $L_x/2$. Case 1 is a very dilute (ideal) gas which is not well suited for MD simulations due to the fact that the particles are mostly in free flight; the mean free path is of the same order as $L_x/2$ (the distance between the swapping layers) in this case. Nevertheless, this low-density case was included here because of the kinetic theory basis for the analysis. Case 6 is close to the LJs system's dew point (at $n = 0.026$ for $T = 0.7$). When the mean free path is much larger than the range of the potential, particles may overlap at the end of a time step. To avoid strong accelerations if this happens, the particles were given a reflecting "shield" at $r_{ik} = 0.8$.

Table 1. Thermodynamic states and imposed mass flux used in this work. All values are in Lennard–Jones units. The values of $|\nabla x_1|_{\max}$ are the maximum difference in x_1, 1.0, divided by $L_x/2$. The actual values obtained in the simulations are given in Table 2.

Case	n	T	$L_x/2$	$\lvert\nabla x_1\rvert_{\max} \times 10^3$	λ	$J_1 \times 10^4$
1	0.001	0.7	254	3.9	225	1.780 ± 0.005
2	0.005	0.7	148	6.8	45.0	4.88 ± 0.02
3	0.01	0.7	118	8.5	22.5	6.64 ± 0.01
4	0.01	1.0	118	8.5	22.5	9.136 ± 0.008
5	0.01	2.0	118	8.5	22.5	16.09 ± 0.01
6	0.02	0.7	94	10.6	11	10.17 ± 0.03

Table 2. Simulation results for the six cases. The diffusion coefficient D_{Fick} and D were determined from Equations (2) and (29), respectively. The values for ∇x_1 represent the central part of the mole fraction profiles in Figure 2, used to determine D_{Fick}. The values of D_{MSD} were determined in independent equilibrium simulations. For the α-values, please see the text.

Case	$\nabla x_1 \times 10^3$	D_{Fick}	D	D_{MSD}	α
1	-2.19 ± 0.03	81 ± 1	114 ± 2	109.7 ± 0.5	2.1
2	-5.16 ± 0.05	18.9 ± 0.1	20.8 ± 0.2	21.73 ± 0.05	2.2
3	-7.28 ± 0.03	9.1 ± 0.1	9.63 ± 0.05	10.73 ± 0.03	2.6
4	-7.10 ± 0.05	12.87 ± 0.04	13.8 ± 0.1	14.77 ± 0.05	2.4
5	-6.66 ± 0.04	24.20 ± 0.02	26.4 ± 0.1	27.5 ± 0.1	2.1
6	-9.95 ± 0.03	5.11 ± 0.03	5.27 ± 0.02	5.19 ± 0.01	2.2

A total of 4,000,000 time steps of length $\delta t = 0.002$ in Lennard–Jones units were used to produce data for analysis. The final 2,000,000 time steps were used to represent steady state with $J = 0$.

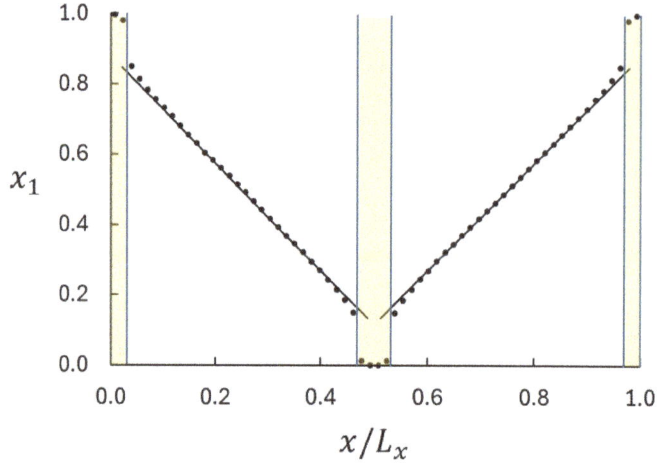

Figure 2. Simulation results for mole fraction profile for Case 2. The straight lines are fitted to the central part of the profiles ($0.15 < x/L_x < 0.35$ and $0.65 < x/L_x < 0.85$) and used to estimate the Fick's law diffusion coefficient from Equation (2). The positions of the swapping layers are marked yellow. The jumps in x_1 near the ends and in the center of the MD box occur between the swapping layers and the bulk. Note the slight curvatures of the profile in the bulk near the boundaries. The curvature signifies deviation from the local equilibrium and is the basis for analysis with Equation (29). The profiles for the other cases are qualitatively the same as shown here.

4. Results and Discussion

We shall now analyze the binary diffusion data using the theory developed in Section 2.2. Central in this analysis is the curvature of the mole fraction profile and the deviation from the local equilibrium. A typical mole fraction profile is shown in Figure 2. The following are three important features: (i) the profile appears to be linear in the central part of the bulk regions, (ii) the profile is slightly curved near the ends of the bulk regions, and (iii) there are significant jumps in x_1 between the swapping volumes and the bulk. We shall focus on the profile in the bulk near the boundaries where the profile is slightly curved. Equation (29) shows that for J_1 and D to be constant in the bulk region, the fraction $(\mu_{\mathrm{eq}} \quad \mu)/\Gamma E_k$ must follow from

$$\frac{\mu_{\mathrm{eq}} - \mu}{\Gamma E_k} = 1 + \frac{constant}{\nabla x_1} \tag{31}$$

where $constant = J_1/(nD) > 0$ and $\nabla x_1 < 0$. In other words, there is a direct relation between the nonlinearity in the mole fraction profile and $\mu_{\mathrm{eq}} - \mu$. Hence, we use the curvature in the mole fraction profile to analyze the deviation from the local equilibrium.

The jumps in x_1 are consequences of the large imposed mass flux and the resolution of the data acquisition in the x-direction. Similar jumps are seen for temperature profiles in simulations of heat transport [15,16]. A weaker imposed mass flux would bring the system closer to equilibrium and show smaller jumps, but the objective here is to bring the system as far from equilibrium as possible. Clearly, the maximum difference in composition, and therefore the maximum mass flux, is obtained when the swapping layers are saturated with one of the components.

A central question in this work is to what extent LEA is valid. Equation (28) gives a direct measure of the deviation from the equilibrium chemical potential. We see that $\mu_{\mathrm{eq}} - \mu$ is positive. From Equation (29), we see that Fick's law will always underestimate the value of D. The difference can be quantified with the second term in the square bracket in

Equation (29). If the fraction $(\mu_{\text{eq}} - \mu)/\Gamma E_k$ is small compared to 1, the local equilibrium is a good approximation, reducing Equation (29) to Fick's law. Otherwise, the approximation is poor. Exactly what we mean by an acceptable approximation is a matter of choice, but the important point is that whatever choice is made, it can be quantified and communicated. Equation (28) shows that the deviation from equilibrium depends on x_1 with the minimum value at $x_1 = 0.5$. The minimum value of $(\mu_{\text{eq}} - \mu)/\Gamma E_k$ is then $8\alpha m J^2/(3n^2kT)$ with $\Gamma = 1$.

Lack of local equilibrium will depend on density, temperature, and the mass flux. Figure 3 shows the deviations from the local equilibrium for four densities used in this study at $T = 0.7$ and maximum imposed mass flux (saturated control volumes). The right endpoint of each graph represents the minimum value of $|\nabla x_1|$, found in the central part of the bulk region where the system is closest to equilibrium. The maximum value of $|\nabla x_1|$ is the left endpoint (near the swapping layers) where the deviation from equilibrium is largest. LEA appears to be very poor for the lowest density, at best approximately 25% (in the central part of the bulk) and worse towards the swapping layers. In this region, the diffusive velocity of the rare component is large, $u_i = J_i/n_i$, and the right-hand side of Equation (28) becomes large. Furthermore, for $n = 0.001$, molecular collisions are rare and mass transport is to a large extent ballistic. The situation is much better for higher densities. For $n = 0.02$, the deviation from equilibrium is approximately 2 % in the central part of the bulk, rising to about 30 % near the swapping layers.

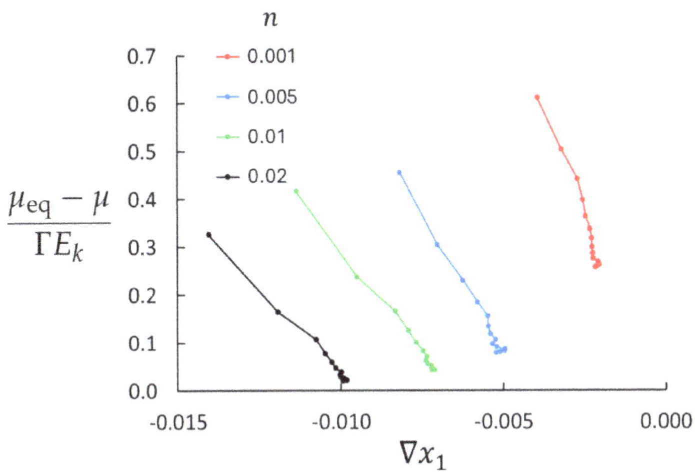

Figure 3. Simulation results for the deviation from the local equilibrium according to Equations (28) and (29) for four different densities (Case 1, 2, 3, and 6). All cases have $T = 0.7$.

Figure 4 shows the deviations from the local equilibrium for three temperatures ($T = 0.7, 1.0$, and 2.0), $n = 0.01$, and maximum mass flux. LEA is less than 4% off in the central part of the bulk for the highest temperature and better for the lower temperatures.

LEA will clearly be better for reduced mass flux. According to Equations (28) and (29), the $(\mu_{\text{eq}} - \mu)/\Gamma E_k$ is proportional to J_1^2. We simulated Case 4 for six additional imposed fluxes. The results are shown in Figure 5. In addition to the reduced value of $(\mu_{\text{eq}} - \mu)/\Gamma E_k$ with reduced flux, it is worth noting that the range spanned by ∇x_1 is also reduced with reduced flux, meaning that the mole fraction profile is less curved.

The parameter α in Equation (28), which is implicit in Equation (29), was determined as follows. In principle, D may be a function of x_1 and ∇x_1. However, the system is an ideal mixture of particles with identical physical properties at uniform density and temperature. The diffusion coefficient in Equation (29) must therefore be independent of x_1 and ∇x_1. The value of α was adjusted to satisfy this condition. An example of the relation between α and D is shown in Figure 6. A value of α that is too large (small) gives too much (little) emphasis on the curvature in ∇x_1, especially near the ends of the mole fraction range (cf. Figure 2). By fitting a linear function to D vs. ∇x_1, α was determined as the value giving zero slope (=2.4 for Case 4 shown in Figure 6). We note that the value of D for the largest values of ∇x_1, where the x_1-profile is linear, is not very sensitive to the value of α. This procedure gave almost the same value for α for the six cases studied here (Table 2).

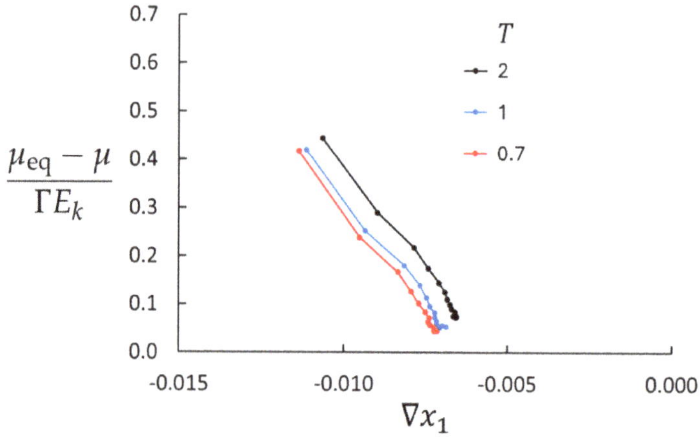

Figure 4. Simulation results for the deviation from the local equilibrium according to Equations (28) and (29) for three temperatures (Case 3, 4, and 5). All cases have $n = 0.01$.

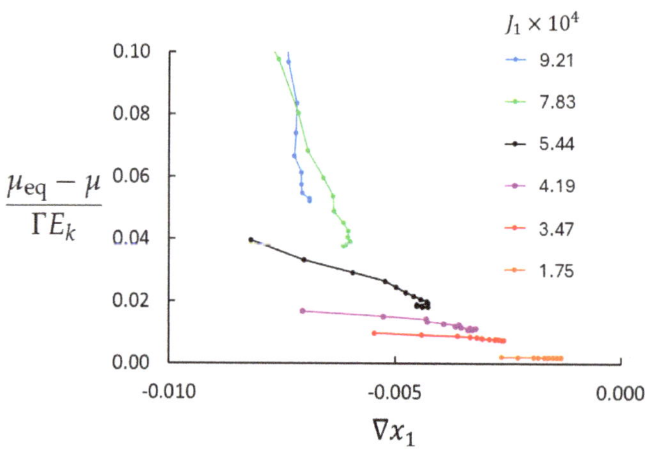

Figure 5. Deviation from the local equilibrium according to Equations (28) and (29) for six cases with different imposed mass fluxes. All results are for $n = 0.01$ and $T = 0.7$ (Case 4). Note that the graphs for the two highest mass fluxes are incomplete; top values are larger than the axis maximum value.

We note that the condition on D applies to the special case of identical components, in a more general case one must expect that D depends on the composition.

Numerical values for the diffusion coefficient obtained from Equation (29) for the six cases are shown in Table 2. The coefficient obtained from Fick's law is consistently smaller than D and the agreement between D_{Fick} and D is better for higher densities with fixed temperature and for higher temperature with fixed density. Both trends can be understood in terms of the system's ability to dissipate the effect of the perturbation.

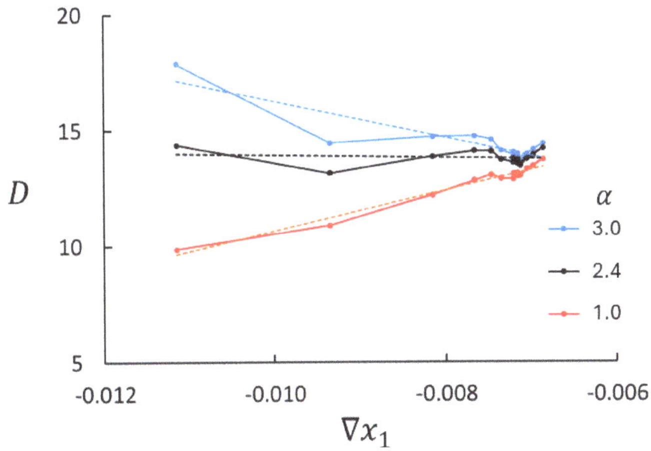

Figure 6. The diffusion coefficient as function of mole fraction for three different values of α, $n = 0.01$ and $T = 0.7$ (Case 4). The dashed lines are fitted linear functions to the MD data.

As mentioned in the Introduction, the main purpose of this work is to develop a tool for analysis of LEA in non-equilibrium thermodynamics. We considered a very special case, mutual diffusion of an ideal system in gas phase, and used kinetic theory as the basis. As it stands, the method has limited value for higher densities, which requires an improved version of the kinetic theory. Another limitation is that we approximated Γ, μ_i, and s_{eq} with ideal-gas values. These properties must be determined from a more realistic equation of state, which is available for the LJs model [17].

The theory was developed for binary diffusion at low density in general and simplified for a system with identical components. Application to binary systems with different components is therefore possible. For instance, analysis of mutual diffusion in a binary mixture with two different components will be possible with the same tool, although it will require more information to evaluate the terms $\mathbf{A}_i$ and $\mathbf{B}_i$ in Equations (9) and (10). The conclusions with respect to LEA may or may not be the same as we draw in this work, but the point is that the theory presented here can be used for the analysis. The results presented in Table 2 suggest that increased density will improve LEA due to more frequent molecular collisions.

Another interesting application is heat transport, where the same considerations of the combination of ballistic and diffusive transport mechanisms apply.

Equation (30) gives us a tool to compute the difference between the non-equilibrium and equilibrium entropy. We found for $(s - s_{\text{eq}})/s_{\text{eq}}$:

$$\frac{s - s_{\text{eq}}}{s_{\text{eq}}} = -\frac{\alpha m}{2\, s_{\text{ST}}\; n^2 x_1(1 - x_1)T} J_1^2 \tag{32}$$

where we used the Sackur–Tetrode ideal gas entropy, s_{ST}, as representing s_{eq}. Using Case 2 as an example and assuming that the LJs model represents argon with the potential parameters, $\epsilon/k_{\text{B}} = 120$ K and $\sigma = 0.34$ nm, we find that the molar equilibrium entropy is 125 J mol^{-1} K^{-1} and the difference between the molar non-equilibrium and equilibrium entropies is 0.5 J mol^{-1} K^{-1} for $x_1 = 0.5$. The relative deviation is $(s - s_{\text{eq}})/s_{\text{eq}} = 0.004$ for $x_1 = 0.5$ and 0.01 for $x_1 = 0.9$ (the extreme value in the bulk, *cf* Figure 2). These results are additional demonstrations of the excellence of LEA in this diffusion case.

The importance of the kinetic energy of diffusion [14] can be illustrated by the ratio between the kinetic energy of diffusion and the thermal energy,

$$\text{ratio} = \frac{m}{3k_{\text{B}}T}\left(\frac{J_1}{n_1}\right)^2\left(\frac{1}{x_1^2}+\frac{1}{(1-x_1)^2}\right). \tag{33}$$

Again using Case 2 as an example, we found that the ratio is 0.04 and 0.46 for $x_1 = 0.5$ and $x_1 = 0.9$, respectively. This result shows that even if the kinetic energy of diffusion is large compared with the thermal energy, the deviation from the local equilibrium is small and LEA is good.

A more detailed analysis of the velocities should include the velocity distributions. A local Maxwellian distribution would indicate a local equilibrium. An analysis of the velocity part of the Boltzmann H-function can be used to quantify the deviation from equilibrium [1]. The velocity distributions can also be used for a discussion of the local temperature and the role of the kinetic energy of diffusion in computing the local temperature, which we have not considered here. Based on the thermodynamic results we have, which show that the system is close to equilibrium, we expect that the velocity distribution for each component will be Gaussian in the y- and z-directions, a shifted Gaussian in the x-direction, and Gaussians with zero mean for the combined distributions of the two components in all three directions. The variance of the distributions will give us the local kinetic temperature.

5. Conclusions

Using a theoretical framework based on the Boltzmann equation, we quantitatively assessed local equilibrium approximation (LEA) commonly used in non-equilibrium thermodynamics. Our analysis, which applies to both equilibrium and non-equilibrium conditions, was developed for an ideal diffusion process in a binary system composed of two physically identical components. This approach yields a nonlinear generalization of the telegrapher's equation. In particular, the steady-state equation we derive, Equation (29), provides a novel method for determining a corrected relaxation time from steady-state measurements, which are generally easier to analyze than transient phenomena. We then employ the relaxation time correction factor, α, to quantify the deviation from equilibrium by comparing the equilibrium entropy with its non-equilibrium counterpart, Equation (30). This framework offers a quantitative measure of the deviation from the local equilibrium and enhances the assessment of LEA in non-equilibrium thermodynamics.

Our results indicate that LEA holds well for most conditions except at the lowest density examined (0.001 in Lennard–Jones units). At low density, a large mean free path and a high diffusion kinetic energy relative to the system's thermal energy lead to significant deviations from the local equilibrium. In contrast, for higher gas-phase densities, LEA remains a robust approximation for the system considered. Notably, even under conditions of large fluxes and forces where, for argon, the mass fluxes and concentration gradients reach approximately 10^4 mol m^{-1} s^{-1} and 10^{10} mol m^{-4}, respectively (Case 2), the estimated diffusion coefficient for argon at about 1 bar and 85 K is on the order of 10^{-6} m^2 s^{-1}, which agrees well with experimental data [18].

Author Contributions: K.R.K. developed the theory. B.H. did the computer simulations. Both authors contributed equally to the analysis, discussion, and writing the paper. All authors have read and agreed to the published version of the manuscript.

Funding: This work was partly supported by the Research Council of Norway through its Centers of Excellence funding scheme, project number 262644, PoreLab.

Institutional Review Board Statement: Not applicable.

Data Availability Statement: Data used in this study are available on request to the corresponding author.

Acknowledgments: The computer simulations were performed with resources provided by Department of Chemistry at The Norwegian University of Science and Technology-NTNU. We are grateful to Johannes Salomonsen Løken for providing a LAMMPS input script for computing the MSD diffusion coefficient. Most of all, we thank Signe Kjelstrup for wonderful collaboration and inspiration over the years.

Conflicts of Interest: Author Kim R. Kristiansen is currently employed by the company Ocean GeoLoop AS. The remaining authors declare that the research was conducted in the absence of any commercial or financial relationships that could be construed as a potential conflict of interest.

References

1. Hafskjold, B.; Ratkje, S.K. Criteria for local equilibrium in a system with transport of heat and mass. *J. Stat. Phys.* **1995**, *78*, 463–494. [CrossRef]
2. Kjelstrup, S.; Bedaux, D.; Johannessen, E.; Gross, J. *Non-Equilibrium Thermodynamics for Engineers*; World Scientific Publishing: Singapore , 2017; p. 156.
3. Lieb, E.H.; Yngvason, J. The entropy concept for non-equilibrium states. *Proc. R. Soc. A Math. Phys. Eng. Sci.* **2013**, *469*, 20130408. [CrossRef] [PubMed]
4. Cimmelli, V.A.; Jou, D.; Ruggeri, T.; Ván, P. Entropy principle and recent results in non-equilibrium theories. *Entropy* **2014**, *16*, 1756–1807. [CrossRef]
5. Lattanzio, C.; Mascia, C.; Plaza, R.G.; Simeoni, C. Kinetic schemes for assessing stability of traveling fronts for the Allen–Cahn equation with relaxation. *Appl. Numer. Math.* **2019**, *141*, 234–247. [CrossRef]
6. Holian, B.L.; Evans, D.J. Shear viscosities away from the melting line: A comparison of equilibrium and nonequilibrium molecular dynamics. *J. Chem. Phys.* **1983**, *78*, 5147–5150. [CrossRef]
7. Hoover, W.G.; Ashurst, W.T. Nonequilibrium molecular dynamics. In *Theoretical Chemistry: Advances and Perspectives*; Eyring, H., Henderson, D., Eds.; Academic Press: New York, NY, USA, 1975; Volume 1, pp. 1–51.
8. Allen, M.P.; Tildesley, D.J. *Computer Simulation of Liquids*; Oxford University Press: Oxford, UK, 2017.
9. Dong, R.Y.; Cao, B.Y. Application of the uniform source-and-sink scheme to molecular dynamics calculation of the self-diffusion coefficient of fluids. *Int. J. Numer. Methods Eng.* **2012**, *92*, 229–237.
10. Darken, L.S. Diffusion, mobility and their interrelation through free energy in binary metallic systems. *Trans. AIME* **1948**, *175*, 184–194.
11. Aranovich, G.; Donohue, M. Limitations and generalizations of the classical phenomenological model for diffusion in fluids. *Mol. Phys.* **2007**, *105*, 1085–1093. [CrossRef]
12. Snell, F.M.; Aranow, R.; Spangler, R.A. Statistical-Mechanical Derivation of the Partial Molecular Stress Tensors in Isothermal Multicomponent Systems. *J. Chem. Phys.* **1967**, *47*, 4959–4971.
13. Jou, D.; Casas-Vázquez, J.; Lebon, G. *Extended Irreversible Thermodynamics*; Springer: Berlin/Heidelberg, Germany, 2010.
14. de Groot, S.R.; Mazur, P. *Non-Equilibrium Thermodynamics*; Dover Publications: Garden City, NY, USA, 1984.
15. Hafskjold, B.; Ikeshoji, T.; Ratkje, S.K. On the molecular mechanism of thermal diffusion in liquids. *Mol. Phys.* **1993**, *80*, 1389–1412. [CrossRef]
16. Wang, S.C.; Liang, X.G.; Xu, X.H. Thermal conductivity of silicon nanowire by nonequilibrium molecular dynamics simulations. In Proceedings of the International Conference on Micro/Nanoscale Heat Transfer, Tainan, Taiwan, 6–9 June 2008; Volume 42924, pp. 1155–1161.
17. Van Westen, T.; Hammer, M.; Hafskjold, B.; Aasen, A.; Gross, J.; Wilhelmsen, Ø. Perturbation theories for fluids with short-ranged attractive forces: A case study of the Lennard–Jones spline fluid. *J. Chem. Phys.* **2022**, *156*, 104504. [CrossRef] [PubMed]
18. Hutchinson, F. Self-diffusion in argon. *J. Chem. Phys.* **1949**, *17*, 1081–1086. [CrossRef]

entropy

MDPI

Article

Energetic Variational Modeling of Active Nematics: Coupling the Toner–Tu Model with ATP Hydrolysis

Yiwei Wang

Department of Mathematics, University of California, Riverside, CA 92507, USA; yiweiw@ucr.edu

Abstract: We present a thermodynamically consistent energetic variational model for active nematics driven by ATP hydrolysis. Extending the classical Toner–Tu framework, we introduce a chemo-mechanical coupling mechanism in which the self-advection and polarization dynamics are modulated by the ATP hydrolysis rate. The model is derived using an energetic variational approach that integrates both chemical free energy and mechanical energy into a unified energy dissipation law. The reaction rate equation explicitly incorporates mechanical feedback, revealing how active transport and alignment interactions influence chemical fluxes and vice versa. This formulation not only preserves consistency with non-equilibrium thermodynamics but also provides a transparent pathway for modeling energy transduction in active systems. We also present numerical simulations demonstrating the positive energy transduction under a specific choice of model parameters. The new modeling framework offers new insights into energy transduction and regulation mechanisms in biologically related active systems.

Keywords: active nematics; nonequilibrium thermodynamics; energetic variational approach; Toner-Tu model; ATP hydrolysis

1. Introduction

Active matter refers to a class of materials composed of active units that convert chemical energy into mechanical work [1–4]. Examples of active matter include living and non-living systems, such as collections of molecular motor, cells, bacteria, synthetic microswimmers, and self-propelled colloids [5–8]. In recent years, active matter has gained significant attention due to its potential for various biomedical applications, such as drug delivery and soft robotics [9–11].

One of the fundamental goals in studying of active matter is to obtain a quantitative, mechanistic understanding of various chemo-mechanical processes in different spatial-temporal scales in these systems [1]. For example, (1) how mechanical properties like viscoelasticity and stiffness are affected by biological, chemical, and physical cues such as cell-cell contacts, biochemical signaling, temperature, and voltage [12,13] and (2) how to quantify the energy transduction, entropy production, and heat generation [1,14–19]. Building computational models for active matter is an indispensable way to achieve this goal. Despite the development of many mathematical models for active materials [3,20,21], most existing models are developed by adding an active stress or force to existing passive material models. Detailed biochemical interactions, such as ATP hydrolysis, are not included explicitly in these models. As a consequence, these models may provide limited insight into the origins of activity and the process of energy transduction.

Based on fundamental principles of non-equilibrium thermodynamics [22,23], energetic variational approach (EnVarA) [24–26], together with other variational principles such

https://doi.org/10.3390/e27080801

as Onsager's variational principle [27,28] and the general equation for the non-equilibrium reversible-irreversible coupling (GENERIC) [29,30], offer a promising way to develop better models for active matter. These variational principles have been successfully applied to develop thermodynamically consistent models for passive soft matter, including systems with chemo-mechanical coupling [31] and non-isothermal effects [32]. The main idea of these approaches is to derive a mathematical model for a complex system based on variations of a free energy and a dissipation functional, which is related to the rate of entropy production [26]. Through the choice of free energy and the dissipation, the coupling and competition of various multiscale, multiphysics, chemo-mechanical processes can be included, enabling the quantification of energy consumption, the rate of entropy production, and heat generation in the developed model.

However, extending these variational principles to active matter is challenging as they were originally developed for closed and dissipative systems. The functionality of active matter requires continuously injection of energy through chemical energy, mechanical force, electric fields, or light [1,16,33] to maintain the out-of-equilibrium states, they need to be viewed as either open systems or a subsystem for a significant large close system [34]. Although substantial efforts have been made in this direction recently [16,35–41], further development of modeling and computational tools is still needed to gain deeper insights on active matter.

The purpose of this paper is to explore the possibility of constructing a thermodynamically consistent model for active matter using the energetic variational approach. The key idea is to explicitly incorporate the chemical reactions that power the system, i.e., ATP hydrolysis, into the model. Although the resulting chemo-mechanical system is dissipative, i.e., the total free energy decreases over time, the mechanical free energy may increase due to energy transduction from chemical to mechanical forms. Existing active matter models can be recovered as special cases by prescribing a constant ATP hydrolysis rate, which represents continuous energy input. To illustrate this new modeling framework, we focus on a variant of the Toner–Tu model [42], which describes the large-scale dynamics of self-propelled particles via coarse-grained velocity and density fields. The developed approach can be used to revise more complex active soft matter systems. This variational framework provides a new perspective for modeling active matter in a thermodynamically consistent manner and enables a quantitative description of energy transduction.

The remainder of the paper is organized as follows. In Section 2, we briefly review the classical Toner–Tu model and its variants, drawing connections to models for nematic liquid crystals. We also introduce the EnVarA framework and discuss why Toner–Tu-type models cannot be directly derived from this framework. In Section 3, we develop a new thermodynamically consistent chemo-mechanical model using EnVarA. Numerical simulation results are presented in Section 4. Finally, concluding remarks are provided in Section 5.

2. Preliminaries

2.1. The Toner–Tu Model and Its Variants

In this section, we briefly review the Toner–Tu model and its variants. The Toner–Tu model can be viewed as a continuum-level description inspired by the Vicsek model [43], which captures the collective behavior of self-propelled particles with alignment interactions. In the Vicsek model, each particle i at position $\boldsymbol{x}_i(t)$ moves with a constant speed v_0 in the direction of its heading $\boldsymbol{\theta}_i(t)$. The direction of moving is updated at each time step according to

$$\boldsymbol{\theta}_i(t+1) = \langle \boldsymbol{\theta}_i(t) \rangle_R + \boldsymbol{\xi}_i(t) \tag{1}$$

where $\langle \boldsymbol{\theta}_i(t) \rangle_R$ is the average orientation vector of particles located in a circle of radius R surrounding particle, and $\boldsymbol{\xi}_i$ is a random vector with orientation obtained from a uniform distribution $[-\sigma\pi, \sigma\pi]$ with σ being the strength of the noise. The new position of particle i is updated by

$$\boldsymbol{x}_i(t+1) = \boldsymbol{x}_i(t) + v_0 \hat{\boldsymbol{\theta}}_i \tag{2}$$

where $\hat{\boldsymbol{\theta}}_i$ is the unit vector in the direction of $\boldsymbol{\theta}_i$. Mathematically, the Vicsek model can be written as [44]

$$\begin{cases} \mathrm{d}X_t^i = \mathrm{d}V_t^i \mathrm{d}t \\ \mathrm{d}V_t^i = -P(V_t^i)\left(\frac{1}{N}K(X_t^i - X_t^j)(V_t^i - V_t^j)\right)\mathrm{d}t + \sqrt{2}P(V_t^i) \circ \mathrm{d}W_t \end{cases} \tag{3}$$

where $P(\boldsymbol{v})$ is the projection operator on the tangent space at $\frac{\boldsymbol{v}}{|\boldsymbol{v}|}$ to the unit sphere, given by

$$P(\boldsymbol{v}) = I - \frac{\boldsymbol{v} \otimes \boldsymbol{v}}{|\boldsymbol{v}|^2} \tag{4}$$

and K is the interaction kernel. In the original Vicsek model,

$$K(x) = \begin{cases} 1, & |x| \le R \\ 0, & |x| > R. \end{cases} \tag{5}$$

The term $\sqrt{2}P(V_t^i) \circ \mathrm{d}W_t$ introduces angular noise into the dynamics. Here, $\mathrm{d}W_t$ is a standard Brownian motion in velocity space, and the Stratonovich integral (denoted by $\circ \mathrm{d}W_t$) preserves the geometric structure. The activity in the Vicsek model is encoded in the self-propulsion at constant speed v_0, which leads to rich collective behavior, such as flocking and phase separation, not seen in equilibrium systems [43,45].

Although the Toner–Tu model is not derived as a formal hydrodynamic limit of the Vicsek model, it is built by leveraging its key symmetries and conservation laws [42]. The original Toner–Tu model, proposed in [42], is formulated as

$$\begin{cases} \partial_t n + \nabla \cdot (n\boldsymbol{v}) = 0 \\ \partial_t \boldsymbol{v} + \lambda_1 (\boldsymbol{v} \cdot \nabla)\boldsymbol{v} = (\alpha - \beta|\boldsymbol{v}|^2)\boldsymbol{v} - \nabla P(n) + D_L \nabla(\nabla \cdot \boldsymbol{v}) + D_1 \Delta \boldsymbol{v} + D_2 (\boldsymbol{v} \cdot \nabla)^2 \boldsymbol{v} + \boldsymbol{f} \end{cases} \tag{6}$$

where $n(\boldsymbol{x}, t)$ is the number density of active particles, and $\boldsymbol{v}$ is the velocity or the polarization field, and $\boldsymbol{f}$ is the Gaussian random noise. The phenomenological parameters β, D_1, D_2 and D_L are all positive, while $\alpha < 0$ is the disordered phase and $\alpha > 0$ in the ordered state. In the ordered phase, the α and β terms simply make the local $\boldsymbol{v}$ have a nonzero magnitude $\sqrt{\alpha/\beta}$, corresponding to the constant speed assumption in the original Vicsek model. The pressure $P(n)$ is assumed to be a function of density $n(\boldsymbol{x}, t)$, given by

$$P(n) = \sum_{\alpha=1}^{\infty} \sigma_\alpha (n - n_0)^\alpha \tag{7}$$

where n_0 is the mean of the local number density and σ_α are coefficients in the pressure expansion.

In the special case where $\lambda_1 = 1$ and the Gaussian noise $\boldsymbol{f}$ is absent, the second equation of (6) can be formally interpreted as a momentum equation in hydrodynamic models [42], and one can derive the following energy dissipation relation:

$$\begin{aligned} \frac{\mathrm{d}}{\mathrm{d}t} \int \frac{1}{2}|\boldsymbol{v}|^2 \mathrm{d}x = -\int \Big[&(\beta|\boldsymbol{v}|^2 - \alpha)|\boldsymbol{v}|^2 + D_L |\nabla \cdot \boldsymbol{v}|^2 + D_1 |\nabla \boldsymbol{v}|^2 \\ &- D_2 \Big((\boldsymbol{v} \cdot \nabla)^2 \boldsymbol{v} \Big) \cdot \boldsymbol{v} - P(n)(\nabla \cdot \boldsymbol{v}) \Big] \mathrm{d}x. \end{aligned} \tag{8}$$

Clearly, the right-hand side of this energy law does not exhibit Galilean invariance or satisfy the frame indifference. Moreover, the total energy not always decay with respect to time. These features highlight the non-equilibrium and active nature of the model, distinguishing it from conventional hydrodynamic systems.

Different variants of the Toner–Tu model have been proposed [46,47]. For instance, [46] proposed a model of the form

$$\begin{cases} \partial_t n + v_0 \nabla \cdot (n\boldsymbol{p}) = D\Delta n. \\ \partial_t \boldsymbol{p} + \lambda \boldsymbol{p} \cdot \nabla \boldsymbol{p} = \alpha \boldsymbol{p} - \beta |\boldsymbol{p}|^2 \boldsymbol{p} + K_1 \nabla(\nabla \cdot \boldsymbol{p}) + K_2 \Delta \boldsymbol{p} - \xi \nabla c + \boldsymbol{f}, \end{cases} \tag{9}$$

where $n(\boldsymbol{x}, t)$ is the local concentration, and $\boldsymbol{p}$ is the polarization field. The model is similar to Toner–Tu model with $P(n) = \xi(n - n_0)$. The main discrepancy lies in the absence of $(\boldsymbol{v} \cdot \nabla)^2$ in the second equation and the inclusion of a diffusion term in the number density equation. The model is referred as Parabolic–Parabolic Toner–Tu (PPTT) model in [47].

Although the structure of PPTT model resembles that of the original Toner–Tu model, it provides a different physical interpretation. In the Toner–Tu model, $\boldsymbol{v}$ represents the velocity field of self-propelled particles and simultaneously encodes their polarization direction. In contrast, the PPTT model views $\boldsymbol{p}$ as a polarization field. If we define $\boldsymbol{u} = \lambda \boldsymbol{p}$ as the background velocity, we can interpret $\partial_t \boldsymbol{p} + (\boldsymbol{u} \cdot \nabla)\boldsymbol{p}$ as material derivative of the director field $\boldsymbol{p}$ [48], as the right-hand side is a relaxation dynamics of $\boldsymbol{p}$ (without the noise term), which minimizing a free energy

$$\mathcal{F}[p] = \int -\frac{\alpha}{2}|\boldsymbol{p}|^2 + \frac{\beta}{4}|\boldsymbol{p}|^4 + K_1(\nabla \cdot \boldsymbol{p})^2 + K_2|\nabla \boldsymbol{p}|^2 + \xi \nabla n \cdot \boldsymbol{p} \tag{10}$$

The free energy (10) can be viewed as a special case of the Ericksen-type free energy, commonly used to describe nematic liquid crystals with variable degree of ordering [48,49], with an additional coupling term between the director field and the number density. Because of this connection, the PPTT type models can also be interpreted as continuum models for active nematics, even though they are often referred to as models for active polar fluids [21] due to the use of a vector order parameter that, in principle, lacks head-to-tail symmetry.

Without loss of generality, in the following, we will assume $\alpha = \beta > 0$ and $K_1 = 0$. We also consider the case without the noise term. Under these assumptions, the PPTT model (9) reduces to the following simplified form:

$$\begin{cases} \partial_t n + \nabla \cdot (v_0 \boldsymbol{p} n) = \eta \Delta n \\ \partial_t \boldsymbol{p} + \lambda(\boldsymbol{p} \cdot \nabla)\boldsymbol{p} = \gamma(\Delta \boldsymbol{p} - \frac{1}{\epsilon^2}(|\boldsymbol{p}|^2 - 1)\boldsymbol{p}) - \zeta \nabla c \end{cases} \tag{11}$$

In the case that $v_0 = \lambda = \zeta = 0$, the system (11) can be viewed as a passive nematic system, which satisfies the energy dissipation law

$$\frac{\mathrm{d}}{\mathrm{d}t}\mathcal{F}(n, \boldsymbol{p}) = -\int \eta n |\nabla \ln n|^2 + \frac{1}{\gamma}|\partial_t \boldsymbol{p}|^2 \mathrm{d}\boldsymbol{x}, \tag{12}$$

where the free energy is given by

$$\mathcal{F}(n, \boldsymbol{p}) = \int n(\ln n - 1) + \frac{1}{2}|\nabla \boldsymbol{p}|^2 + \frac{1}{4\epsilon^2}(|\boldsymbol{p}|^2 - 1)^2 \mathrm{d}\boldsymbol{x} \tag{13}$$

The energy dissipation law (12) can be viewed as an L^2-gradient flow with respect to the orientational order parameter $\boldsymbol{p}$ and a Wasserstein-type gradient flow or diffusion with respect to the number density n [25]. There is no coupling between number density $n(\boldsymbol{x}, t)$

and the orientational order parameter $\boldsymbol{p}$. If one further assume that n is constant, then the system reduces to a phenomenological model for passive nematic systems associated with modified Oseen–Frank free energy under the assumption of one constant approximation, as the $\frac{1}{4\epsilon^2}(|\boldsymbol{p}|^2 - 1)^2$ term can be viewed as a penalty term for the unit length constraint [48].

Remark 1. *Similarly, "dry" active nematics with a tensor order parameter defined as* $\mathbf{Q}(\boldsymbol{x}, t)$ *can be written as [21,50,51]*

$$\begin{aligned} &\partial_t n + \nabla \cdot (\zeta \nabla \cdot Q + J_{\text{passive}}) = 0, \\ &\frac{1}{\gamma}\partial_t Q = -\frac{\delta \mathcal{F}[n, Q]}{\delta Q} + f, \end{aligned} \tag{14}$$

Here, $\mathbf{Q}$ *is a symmetric, traceless matrix, and the free energy functional* $\mathcal{F}[n, \mathbf{Q}]$ *is given by*

$$\mathcal{F}[n, Q] = \int \left[n \ln n + \left(\frac{A}{2}|Q|^2 + \frac{B}{3}\mathrm{tr}(Q^3) + \frac{C}{4}|Q|^4 \right) + \frac{L}{2}|\nabla Q|^2 \right] \mathrm{d}\boldsymbol{x}, \tag{15}$$

which consists of the classical Landau–de Gennes free energy [52] and an entropic contribution from the number density. The term $\zeta \nabla \cdot Q$ *represents the active flux in the density evolution. An advantage of using the tensor order parameter* $\mathbf{Q}$ *is that it naturally preserves the head-to-tail symmetry of rod-like molecules. The mathematical structure of* (14) *is similar to the simplified Toner–Tu model* (11)*. The primary difference lies in the absence of a self-advection term in the evolution equation of* $\mathbf{Q}$*, which plays a key role in the original Toner–Tu dynamics.*

Despite their success in capturing key phenomenology of active polar fluids or active nematics, existing Toner–Tu-type models suffer from several important limitations. One major drawback is the absence of a well-defined energy dissipation law. These models are typically constructed phenomenologically, guided by symmetry considerations and coarse-graining arguments, rather than derived by non-equilibrium thermodynamics-based variational principles. As a result, they generally lack a clear connection to an underlying free energy functional. This lack of thermodynamic consistency makes it difficult to interpret or control the balance between energy input (due to activity) and dissipation (due to friction, viscosity, or diffusion). Consequently, while these models provide valuable insight into large-scale collective dynamics, they remain limited in their ability to describe energy transduction.

In many biological and synthetic active systems, activity originates from the conversion of chemical energy (e.g., ATP hydrolysis, fuel consumption) [19,53], which naturally suggests the inclusion of reaction kinetics. However, without a consistent framework, it becomes unclear how to incorporate these reactions systematically due to the two-way coupling between the chemical and mechanical parts. To overcome the limitations of phenomenological modeling approaches, we aim to derive models of active nematics that incorporate chemo-mechanical coupling within the energetic variational approach (EnVarA) framework, which has proven to be a powerful tool for constructing thermodynamically consistent models of complex systems.

2.2. Energetic Variational Approach for Chemo-Mechanical Systems

In this subsection, we briefly introduce EnVarA, a general framework that offers a unified and thermodynamically consistent methodology for modeling complex systems.

The framework of EnVarA is rooted in the non-equilibrium thermodynamics [22,23], especially the seminal works of Rayleigh [54] and Onsager [55,56]. The key idea is to describe a complex system by an energy dissipation law. The energy dissipation law, together with the kinematics of the employed variables, describe all the physics and assumptions in the system. The dynamics and the constitutive equation can be derived through variational procedures. In more detail, an energy dissipation law (for an isothermal closed system) can be written as

$$\frac{\mathrm{d}}{\mathrm{d}t}E^{\text{total}} = -\triangle(t), \tag{16}$$

where E^{total} is the total energy, including both the kinetic energy $\mathcal{K}$ and the Helmholtz free energy $\mathcal{F}$, and $\triangle(t) \geq 0$ is the rate of the energy dissipation which is equal to the rate of entropy production in this case.

The existence of energy dissipation law (16) follows directly from the first and second laws of thermodynamics [57]. The first law expresses conservation of energy as

$$\frac{\mathrm{d}}{\mathrm{d}t}(\mathcal{K}+\mathcal{U}) = \delta W + \delta Q, \tag{17}$$

where $\mathcal{K}$ and $\mathcal{U}$ denote the kinetic and internal energy, respectively, and δW, δQ represent the rate of mechanical work and heat exchange. These quantities are path-dependent and are not exact differentials. To quantify the heat exchange, one needs to consider the second law of thermodynamics, which states

$$T\mathrm{d}S = \delta Q + \triangle, \tag{18}$$

where T is the absolute temperature, S is the entropy, and $\triangle \geq 0$ denotes the rate of entropy production. For an isothermal and mechanically isolated system (i.e., $\delta W = 0$ and T constant), subtracting (18) from (17) yields the energy dissipation law (16), as the free energy $\mathcal{F} = \mathcal{U} - TS$. For non-isothermal systems, the situation becomes much more complicated. We refer interested readers to [26,32] for detailed treatments.

Starting with an energy dissipation law (16), EnVarA derives the dynamics of the systems through the least action principle (LAP) and the maximum dissipation principle (MDP). The LAP, which states the equation of motion for a Hamiltonian system can be derived from the variation in the action functional $\mathcal{A} = \int_0^T \mathcal{K} - \mathcal{F}\mathrm{d}t$ with respect to the flow map $x(\mathbf{X},t)$, gives a unique procedure to derive the conservative force for the system. For a fixed $\mathbf{X}$, the flow map $x(\mathbf{X},t)$ can be viewed as the trajectory of a particle initially at position $\mathbf{X}$. For fixed t, the flow map $x(\mathbf{X},t)$ defines a diffeomorphism from a reference domain Ω_0 to the current domain Ω_t. The MDP, variation in the dissipation potential $\mathcal{D}$, which equals to $\frac{1}{2}\triangle$ in the linear response regime, with respect to x_t, i.e., the velocity, gives the dissipation force for the system. In turn, the force balance condition leads to the evolution equation of the system

$$\frac{\delta\mathcal{D}}{\delta x_t} = \frac{\delta\mathcal{A}}{\delta x}. \tag{19}$$

The variational procedure guarantees the resulting evolution equation is thermodynamically consistent, which is crucial for establishing the well-posedness of the resulting system.

As an example, we consider how to model a generalized diffusion using EnVarA. A generalized diffusion is concerned with a conserved quantity $\rho(x,t)$ satisfying the kinematics, i.e., the conservation of mass,

$$\partial_t\rho + \nabla\cdot(\rho\boldsymbol{u}) = 0, \tag{20}$$

where $\boldsymbol{u}$ is the average velocity. The energy dissipation law of a generalized diffusion can be written as

$$\frac{\mathrm{d}}{\mathrm{d}t}\mathcal{F}[\rho] = -\int_\Omega \eta(\rho)|\boldsymbol{u}|^2\,\mathrm{d}\boldsymbol{x}, \quad \mathcal{F}[\rho] = \int_\Omega \omega(\rho)\,\mathrm{d}\boldsymbol{x}, \tag{21}$$

where $\omega(\rho)$ is the free energy density, and $\eta(\rho)$ is the friction coefficient. We emphasize that throughout this paper, we always consider a non-dimensionalized setting.

Due to the kinematics (20), the free energy can be reformulated as a functional of $\boldsymbol{x}(\mathbf{X},t)$ in Lagrangian coordinates, i.e.,

$$\mathcal{F}[x] = \int_{\Omega_0} \omega\left(\frac{\rho_0(\mathbf{X})}{\det \mathsf{F}}\right) \det \mathsf{F}\, \mathrm{d}\mathbf{X}. \tag{22}$$

where $\rho_0(\mathbf{X})$ is the initial density at the reference domain Ω_0 and $\mathsf{F} = \nabla_{\mathbf{X}}\boldsymbol{x}(\mathbf{X},t)$ is the deformation tensor. A direct computation shows that

$$\begin{aligned}\delta\mathcal{A} &= -\delta\int_0^T\int_{\Omega_0} \omega(\rho_0(X)/\det F)\det F\,\mathrm{d}\mathbf{X}\mathrm{d}t \\ &= -\int_0^T\int_{\Omega_0}\left(-\frac{\partial\omega}{\partial\rho}\left(\frac{\rho_0(X)}{\det F}\right)\cdot\frac{\rho_0(X)}{\det F}+\omega\left(\frac{\rho_0(X)}{\det F}\right)\right)\times(F^{-\mathrm{T}}:\nabla_{\mathbf{X}}\delta x)\det F\;\mathrm{d}\mathbf{X}\mathrm{d}t,\end{aligned}$$

where $\delta\boldsymbol{x}(\mathbf{X},t)$ is the test function satisfying $\delta\boldsymbol{x}\cdot\mathbf{n} = 0$ with $\mathbf{n}$ is the outer normal of Ω in Eulerian coordinates (here we will not distinguish $\tilde{\delta x}(\boldsymbol{x}(\mathbf{X},t),t) = \delta\boldsymbol{x}(\mathbf{X},t)$ and $\delta\boldsymbol{x}(\mathbf{X},t)$ without ambiguity). Moving on to Eulerian coordinates, we have

$$\delta\mathcal{A} = -\int_0^T\int_\Omega(-\frac{\partial\omega}{\partial\rho}\rho+\omega)\nabla\cdot(\delta x)\mathrm{d}\boldsymbol{x} = \int_0^T\int_\Omega -\nabla(\frac{\partial\omega}{\partial\rho}\rho-\omega)\cdot\delta\boldsymbol{x}\mathrm{d}\boldsymbol{x}\mathrm{d}t, \tag{23}$$

which indicates that

$$\frac{\delta\mathcal{A}}{\delta\boldsymbol{x}} = -\nabla(\frac{\partial\omega}{\partial\rho}\rho-\omega) = -\rho\nabla\mu,$$

where $\mu = \frac{\delta\mathcal{F}}{\delta\rho}$ is the chemical potential. For the dissipation part, since $\mathcal{D} = \frac{1}{2}\int\eta(\rho)|\boldsymbol{u}|^2\mathrm{d}\boldsymbol{x}$ it is easy to compute that $\frac{\delta\mathcal{D}}{\delta\boldsymbol{u}} = \eta(\rho)\boldsymbol{u}$. As a consequence, we have the force balance equation

$$\eta(\rho)\boldsymbol{u} = -\rho\,\nabla\mu. \tag{24}$$

Combining the force balance Equation (24) with the kinematics (20), one obtain a generalized diffusion equation

$$\partial_t\rho = \nabla\cdot\left(\frac{\rho^2}{\eta(\rho)}\nabla\mu\right). \tag{25}$$

Remark 2. *For the L^2-gradient flow*

$$\frac{\mathrm{d}}{\mathrm{d}t}\mathcal{F}[\varphi]\mathrm{d}\boldsymbol{x} = -\int_\Omega\frac{1}{\gamma}|\partial_t\varphi|^2\mathrm{d}\boldsymbol{x}\,, \tag{26}$$

one may apply the EnVarA by viewed φ as generalized coordinates [27], which leads to

$$\frac{1}{\gamma}\partial_t\varphi = -\frac{\delta\mathcal{F}}{\delta\varphi} \tag{27}$$

Alternatively, one can impose the kinematics $\partial_t\varphi + \boldsymbol{u}\cdot\nabla\varphi = 0$ [58] and rewrite the energy dissipation law as

$$\frac{\mathrm{d}}{\mathrm{d}t}\mathcal{F}[\varphi]\mathrm{d}\boldsymbol{x} = -\int_\Omega\frac{1}{\gamma}|\boldsymbol{u}\cdot\nabla\varphi|^2\mathrm{d}\boldsymbol{x}\,, \tag{28}$$

By performing the EnVarA with respect to the flow map $\boldsymbol{x}(\mathbf{X},t)$ and the velocity $\boldsymbol{u}(\boldsymbol{x},t)$, one will end up with, at least formally,

$$\frac{1}{\gamma}(\nabla\varphi\cdot\boldsymbol{u})\nabla\varphi = \frac{\delta\mathcal{F}}{\delta\varphi}\nabla\varphi\,, \tag{29}$$

which is equivalent to (27) when $\nabla\varphi \neq 0$.

Classical EnVarA approach is developed for mechanical systems, and the variation are taken with respect to the flow map $x(\mathbf{X}, t)$ and its time derivative, i.e., the velocity $u(x,t)$. In [31], we formulate a reaction kinetics with detailed balance in an energetic variational form by using the reaction trajectory $\boldsymbol{R}$ [59], which is an analogy to the flow map in mechanical system. Roughly speaking, the reaction trajectory $\boldsymbol{R}$ accounts for the number of forward reaction has occurred by time t. The reaction rate $\boldsymbol{r}$ is defined as $\partial_t \boldsymbol{R}$, which is the reaction velocity [60]. The reaction trajectory is known as extent of reaction in physical chemistry and was introduced by de Donder in 1920s [61].

For a general reversible chemical reaction system containing N species $\{X_1, X_2, \ldots X_N\}$ and M reactions can be represented by

$$\alpha_1{}^l \mathrm{X}_1 + \alpha_2{}^l \mathrm{X}_2 + \ldots \alpha_N{}^l \mathrm{X}_N \underset{\mathrm{k}_l^-}{\overset{\mathrm{k}_l^+}{\rightleftharpoons}} \beta_1{}^l \mathrm{X}_1 + \beta_2{}^l \mathrm{X}_2 + \ldots \beta_N{}^l \mathrm{X}_N, \quad l = 1, \ldots, M.$$

The relation between species concentration $\boldsymbol{c}(t) \in \mathbb{R}_+^N$ and the reaction trajectory $\boldsymbol{R}(t) \in \mathbb{R}^M$ is given by $\boldsymbol{c} = \boldsymbol{c}_0 + \gamma \boldsymbol{R}(t)$, where $\boldsymbol{c}_0$ is the initial concentration, and $\gamma \in \mathbb{R}^{N \times M}$ is the stoichiometric matrix with $\gamma_{il} = \beta_i^l - \alpha_i^l$. Within the reaction trajectory, we can describe the chemical kinetics by energy dissipation law in terms of $\boldsymbol{R}$ and $\boldsymbol{R}_t$:

$$\frac{\mathrm{d}}{\mathrm{d}t}\mathcal{F}_{\text{chem}}[\boldsymbol{R}] = -\triangle_{\text{chem}}[\boldsymbol{R}, \boldsymbol{R}_t], \tag{30}$$

where $\triangle_{\text{chem}}[\boldsymbol{R}, \boldsymbol{R}_t]$ is the rate of energy dissipation due to the chemical reaction procedure. Since the linear response assumption may not be valid for chemical reactions [22,62], it is often assumed that $\mathcal{D}_{\text{chem}}[\boldsymbol{R}, \boldsymbol{R}_t] = (\boldsymbol{\Gamma}(\boldsymbol{R}, \boldsymbol{R}_t), \boldsymbol{R}_t) = \sum_{l=1}^{M} \Gamma_l(\boldsymbol{R}, \boldsymbol{R}_t)\partial_t R_l \geq 0$, which leads to the reaction rate can be derived as [31,63]:

$$\Gamma_l(\boldsymbol{R}, \boldsymbol{R}_t) = -\frac{\delta \mathcal{F}_{\text{chem}}}{\delta R_l}, \tag{31}$$

In this formulation, the choice of the free energy determines the chemical equilibrium, while the choice of the dissipation functional $\mathcal{D}_{\text{chem}}[\boldsymbol{R}, \partial_t \boldsymbol{R}]$ determines the reaction rate.

As an example, we consider a simple chemical reaction:

$$\alpha_1 X_1 + \alpha_2 X_2 \underset{\mathrm{k}^-}{\overset{\mathrm{k}^+}{\rightleftharpoons}} \beta_3 X_3, \tag{32}$$

where α_i and β_i are the stoichiometric coefficients. Let c_i denote the concentration of species X_i. According to the law of mass action (LMA), the reaction rate is given by

$$r = k^+ c_1^{\alpha_1} c_2^{\alpha_2} - k^- c_3^{\beta_3}. \tag{33}$$

This system satisfies the detailed balance condition, meaning there exists an equilibrium state $(c_1^\infty, c_2^\infty, c_3^\infty)$ with all $c_i^\infty > 0$ such that

$$k^+ (c_1^\infty)^{\alpha_1} (c_2^\infty)^{\alpha_2} = k^- (c_3^\infty)^{\beta_3}. \tag{34}$$

A corresponding free energy functional can be defined as [64]

$$\mathcal{F}_{\text{chem}}[\boldsymbol{c}] = \int_\Omega \sum_{i=1}^{3} c_i \left(\ln\left(\frac{c_i}{c_i^\infty} \right) - 1 \right) \mathrm{d}\boldsymbol{x} = \int_\Omega \sum_{i=1}^{3} (c_i(\ln c_i - 1) + c_i \sigma_i) \, \mathrm{d}\boldsymbol{x}, \tag{35}$$

where $\sigma_i = -\ln c_i^\infty$ can be interpreted as an internal energy contribution for species X_i, reflecting its chemical activity. Clearly, the free energy determines

$$\frac{k^+}{k^-} = \frac{(c_3^\infty)^{\beta_3}}{(c_1^\infty)^{\alpha_1}(c_2^\infty)^{\alpha_2}}. \tag{36}$$

We introduce the stoichiometric vector $\gamma = (-\alpha_1, -\alpha_2, \beta_3)$. Then, $c_i(t) = c_i(0) + \gamma_i R(t)$ with $R(t)$ being the reaction trajectory. We can view the free energy (35) as the functional of $R(t)$. To determine the dynamics, we impose the dissipation as

$$\triangle_{\text{chem}} = \Gamma(R, R_t)R_t\,, \tag{37}$$

where $\Gamma(R, R_t)$ is chosen such that $\Gamma(R, R_t)R_t \geq 0$. Then, we have

$$\Gamma(R, R_t) = -\frac{\delta \mathcal{F}_{\text{chem}}}{\delta R} = -\sum_{i=1}^{3} \gamma_i \mu_i, \quad \mu_i = \frac{\delta \mathcal{F}_{\text{chem}}}{\delta c_i} = \ln c_i + \sigma_i\,. \tag{38}$$

Here, $-\gamma_i\mu_i$ is known as the chemical affinity, which is the driven force in chemical reaction [60]. One can interpret (38) as the force balance equation for the chemical part of the system. To derive the classical law of mass action (33), one can choose

$$\Gamma(R, R_t) = \ln\left(\frac{R_t}{k^- c_3^{\beta_3}} + 1\right), \tag{39}$$

which leads to $R_t = k^- c_3^{\beta_3}(\exp(-\gamma_i\mu_i) - 1) = k^+ c_1^{\alpha_1} c_2^{\alpha_2} - k^- c_3^{\beta_3}$ by using (36). Another choice of $\Gamma(R, R_t)$ may lead to different reaction kinetics [31].

The energetic variational formulation of chemical reactions opens a new door to model a general chemo-mechanical system in a unified way. For example, we can add the diffusion, a mechanical process, to (32) by assuming c_i satisfies the kinematics:

$$\partial_t c_i + \nabla \cdot (c_i \boldsymbol{u}_i) = \gamma_i R_t, \quad i = 1, 2, \ldots 3, \tag{40}$$

where $\boldsymbol{u}_i(\boldsymbol{x}, t)$ is the average velocity of each species due to its own diffusion, and $R(\boldsymbol{x}, t)$ is the reaction trajectory at each space location $\boldsymbol{x}$. Here, R_t is the partial derivative of $R(\boldsymbol{x}, t)$ with respect to t. Then, the reaction–diffusion equation can be modeled through the energy dissipation law [31]:

$$\frac{\mathrm{d}}{\mathrm{d}t}\mathcal{F}_{\text{chem}}[\boldsymbol{c}] = -\int_\Omega \left[\sum_{i=1}^{3} \eta_i(c_i)|\boldsymbol{u}_i|^2 + \Gamma(R, R_t)R_t\right] \mathrm{d}\boldsymbol{x}. \tag{41}$$

We can employ EnVarA to obtain equations for the reaction and diffusion part, respectively, i.e., to obtain the "force balance equation" of the chemical and mechanical subsystems, which leads to

$$\begin{cases} \eta_i(c_i)\boldsymbol{u}_i = -c_i \nabla \mu_i, \quad i = 1, 2, \ldots 3, \\ \Gamma(R, \partial_t R) = -\sum_{i=1}^{3} \gamma_i \mu_i \end{cases} \tag{42}$$

By taking $\eta_i(c_i) = \frac{1}{D_i} c_i$, we have a reaction–diffusion system

$$\partial_t c_i = D_i \Delta c_i + \gamma_i R_t\,. \tag{43}$$

More mechanical effects can be included by modifying the energy and the dissipation.

2.3. Energy Dissipation Analysis on a Simplified Toner–Tu Model

A natural question is whether the Toner–Tu type models can be derived using the energetic variational approach. To this end, we first perform an energy dissipation analysis on a Toner–Tu model with the following form

$$\begin{cases} \partial_t n + \nabla \cdot (v_0 \boldsymbol{p} n) = \eta \nabla \cdot (n \nabla \mu_n), & \mu_n = \frac{\delta \mathcal{F}}{\delta n} \\ \partial_t \boldsymbol{p} + \lambda (\boldsymbol{p} \cdot \nabla) \boldsymbol{p} = -\gamma \mu_{\boldsymbol{p}}, & \mu_{\boldsymbol{p}} = \frac{\delta \mathcal{F}}{\delta \boldsymbol{p}} \end{cases} \tag{44}$$

where the free energy $\mathcal{F}[n, \boldsymbol{p}]$ is given by

$$\mathcal{F}[n, \boldsymbol{p}] = \int n \ln n + \frac{1}{2} |\nabla \boldsymbol{p}|^2 + \frac{1}{4\epsilon^2} (|\boldsymbol{p}|^2 - 1)^2 + \xi \nabla c \cdot \boldsymbol{p} \mathrm{d}\boldsymbol{x}. \tag{45}$$

This model (44) corresponds to the simplified Toner–Tu model (11), with an additional term $-\eta \nabla \cdot (n \nabla (\xi \nabla \cdot \boldsymbol{p}))$ in the number density equation to ensure variational consistency with the free energy (45). Since $\nabla \cdot \boldsymbol{p}$ is often small due to the choice of the free energy, it is reasonable to neglect this term in practice. By a direct calculation, we have the following energy dissipation relation

$$\begin{aligned} \frac{\mathrm{d}}{\mathrm{d}t} \mathcal{F}[n, \boldsymbol{p}] &= \int (\mu_n \partial_t n + \mu_{\boldsymbol{p}} \cdot \partial_t \boldsymbol{p}) \mathrm{d}\boldsymbol{x} \\ &= \int \mu_n [-\nabla \cdot (v_0 n \boldsymbol{p}) + \eta \nabla \cdot (n \nabla \mu_n)] + \mu_{\boldsymbol{p}} \cdot [-\lambda (\boldsymbol{p} \cdot \nabla) \boldsymbol{p} - \gamma \mu_{\boldsymbol{p}}] \mathrm{d}\boldsymbol{x} \\ &= \int -\eta n |\nabla \mu_n|^2 - \gamma |\mu_{\boldsymbol{p}}|^2 + v_0 n \boldsymbol{p} \cdot \nabla \mu_n - \lambda \mu_{\boldsymbol{p}} \cdot ((\boldsymbol{p} \cdot \nabla) \boldsymbol{p}) \mathrm{d}\boldsymbol{x}. \end{aligned} \tag{46}$$

Clearly, the system is no longer strictly dissipative if $v_0 \neq 0$ and $\lambda \neq 0$, due to the appearance of two active terms arising from the advection of the density and the convective derivative in the polarization equation. The energy dissipation relation (46) can be written as

$$\begin{aligned} \frac{\mathrm{d}}{\mathrm{d}t} \mathcal{F}(n, \boldsymbol{p}) = &- \int \left(n |\boldsymbol{u} - v_0 \boldsymbol{p}|^2 + \frac{1}{\gamma} |\partial_t \boldsymbol{p} + \lambda (\boldsymbol{p} \cdot \nabla) \boldsymbol{p}|^2 \right) \mathrm{d}\boldsymbol{x} \\ &+ \int v_0 n \boldsymbol{p} \cdot \nabla \mu_n - \lambda \mu_{\boldsymbol{p}} \cdot ((\boldsymbol{p} \cdot \nabla) \boldsymbol{p}) \mathrm{d}\boldsymbol{x}. \end{aligned} \tag{47}$$

where $\boldsymbol{u}$ is the transport velocity for the number density such that the number density satisfies the kinematics $\partial_t n + \nabla \cdot (n \boldsymbol{u}) = 0$. However, one cannot apply the energetic variational approach to (47) to derive the underlying equation.

Remark 3. *Formally, one may start with the energy dissipation law*

$$\frac{\mathrm{d}}{\mathrm{d}t} \mathcal{F}(n, \boldsymbol{p}) = - \int \left(n |\boldsymbol{u} - v_0 \boldsymbol{p}|^2 + \frac{1}{\gamma} |\partial_t \boldsymbol{p} + \lambda (\boldsymbol{p} \cdot \nabla) \boldsymbol{p}|^2 \right) \mathrm{d}\boldsymbol{x}, \tag{48}$$

and apply the energetic variational approach to derive the force balance equations

$$\begin{aligned} & n(\boldsymbol{u} - v_0 \boldsymbol{p}) = -n \nabla \mu_n, \\ & \frac{1}{\gamma} \partial_t \boldsymbol{p} + \lambda (\boldsymbol{p} \cdot \nabla) \boldsymbol{p} = -\mu_{\boldsymbol{p}}. \end{aligned} \tag{49}$$

However, as shown in the earlier computation (46), the system (49) does not yield the original energy dissipation law (48). The discrepancy arises because $\boldsymbol{p}$ *is not a true velocity field. Consequently,* $\boldsymbol{u} - v_0 \boldsymbol{p}$ *cannot be interpreted as a relative drag velocity, and the term* $\partial_t \boldsymbol{p} + \lambda (\boldsymbol{p} \cdot \nabla) \boldsymbol{p}$ *does not represent material transport as in classical liquid crystal models [48]. In other words, the presence of active transport terms destroys the variational structure, which is preserved only in the case* $v_0 = \lambda = 0$.

To address this issue, one possible strategy is to introduce an auxiliary velocity field $\mathbf{V}$ *representing the macroscopic flow and consider the modified energy dissipation law*

$$\frac{\mathrm{d}}{\mathrm{d}t}\mathcal{F}(n, \boldsymbol{p}) = -\int \left(n|\boldsymbol{u} - \kappa \boldsymbol{V}|^2 + \frac{1}{\gamma}|\partial_t \boldsymbol{p} + (\boldsymbol{V} \cdot \nabla)\boldsymbol{p}|^2 \right) \mathrm{d}\boldsymbol{x}, \tag{50}$$

together with the constraint $\boldsymbol{V} = \lambda \boldsymbol{p}$. *This constraint can be relaxed, for example, by enforcing* $\boldsymbol{V} \times \boldsymbol{p}$ *to remain small. Nevertheless, it remains challenging to incorporate this condition within a thermodynamically consistent variational framework and to derive a well-posed system involving both* $\boldsymbol{V}$ *and* $\boldsymbol{p}$. *We leave a detailed investigation of this direction for future work.*

This limitation motivates the development of alternative strategies for uncovering the variational structure of Toner–Tu-type models by explicitly incorporate the chemical reaction that powers active transport into the model.

3. Toner–Tu Model with ATP Hydrolysis

In the previous section, we discussed the challenges in formulating a variational structure for Toner–Tu type models, primarily due to the presence of active transport terms. Without explicitly specifying the energy source driving the activity, the system behaves as an open system lacking a well-defined energy dissipation law and thus cannot be derived by the energetic variational framework.

In this section, we show that by explicitly incorporating the chemical reaction for the origin of the activity into the model, we can have a thermodynamically consistent dissipative model. The simplified Toner–Tu model (44) can be obtained by setting the reaction rate as a constant. The chemo-mechanical coupling not only quantifies the energetic cost of activity but also ensures that the system becomes passive in the absence of ATP consumption.

3.1. Model Derivation

To explain the idea, we assume that the primary energy source for the activity comes from the ATP hydrolysis and consider the following form of the reaction

$$\mathrm{ATP} \rightleftharpoons \mathrm{ADP} + \mathrm{P}. \tag{51}$$

In the current study, we neglect electrodynamic effects of ATP hydrolysis, even though ATP, ADP, and phosphate are all charged species and the reaction involves water and proton release. We assume the free energy change for the hydrolysis of each ATP molecule is given by

$$\triangle \mu = \mu_{\mathrm{ATP}} - \mu_{\mathrm{ADP}} - \mu_{\mathrm{P}}, \tag{52}$$

which indeed in the chemical affinity of ATP hydrolysis [41]. Here, μ_{ATP}, μ_{ADP}, and μ_{P} are the chemical potentials of ATP, ADP, and P, respectively, which are associated with a free energy

$$\mathcal{F}_{\mathrm{chem}}(\boldsymbol{c}) = \int \sum_{i=1}^{3} c_i(\ln c_i - 1) + c_i \sigma_i \mathrm{d}\boldsymbol{x}. \tag{53}$$

where c_1, c_2, and c_3 are the concentrations of ATP, ADP and P, respectively, and σ_i represent the internal energy associated with each chemical species, which contribute to the activity of each species. Following the general framework of modeling reaction–diffusion equation, we assume c_i satisfies the kinematics

$$\partial_t c_i + \nabla \cdot (c_i \boldsymbol{u}_i) = \gamma_i R_t, \quad i = 1, 2, 3, \tag{54}$$

where $\boldsymbol{u}_i$ is the transport velocity of c_i, $R(x,t)$ is the reaction trajectory of the ATP hydrolysis (51). The stoichiometric vector $(\gamma_1, \gamma_2, \gamma_3) = (-1, 1, 1)$ in this case and $\triangle\mu = -\sum_{i=1}^{3} \gamma_i \mu_i$.

The key idea of coupling the ATP hydrolysis with the Toner–Tu model (44) is to assume that the coefficients in the active transport, i.e., μ_0 and λ in (44), depend on the reaction rate of ATP hydrolysis, i.e., R_t. Both coefficients go to zero when the ATP hydrolysis stops, i.e., the reaction rate is zero. We can derive the overall model from the energy dissipation law

$$\begin{aligned}\frac{\mathrm{d}}{\mathrm{d}t}(\mathcal{F}(n,\boldsymbol{p}) + \xi \mathcal{F}_{\text{chem}}(\boldsymbol{c})) = -\int \Bigg(\frac{n}{\eta}|\boldsymbol{u} - v_0 R_t \boldsymbol{p}|^2 + \frac{1}{\gamma}|\partial_t \boldsymbol{p} + \lambda R_t (\boldsymbol{p}\cdot\nabla)\boldsymbol{p}|^2 \\ + \xi \left(\sum_{i=1}^{3} \frac{c_i}{\eta_i}|\boldsymbol{u}_i|^2 + \Gamma(R, R_t) R_t \right) \Bigg) \mathrm{d}x\end{aligned} \tag{55}$$

where $\mathcal{F}(n,\boldsymbol{p})$ is the mechanical free energy defined in (45), ξ represents the ratio between the chemical free energy and the mechanical free energy. In the current study, we take $\xi = 1$ without loss of generality.

By performing the energetic variational approach to each component following the steps introduced in the previous section, we obtain the following "force" balance equations for both the mechanical and chemical parts:

$$\begin{aligned}
& n(\boldsymbol{u} - v_0 R_t \boldsymbol{p}) = -\eta n \nabla \mu_n, \quad \mu_n = \frac{\delta \mathcal{F}}{\delta n} \\
& \partial_t \boldsymbol{p} + \lambda R_t (\boldsymbol{p}\cdot\nabla)\boldsymbol{p} = -\gamma \boldsymbol{\mu}_p, \quad \boldsymbol{\mu}_p = \frac{\delta \mathcal{F}}{\delta \boldsymbol{p}} \\
& c_i \boldsymbol{u}_i = -\eta_i \nabla \mu_i, \quad \mu_i = \frac{\delta \mathcal{F}_{\text{chem}}}{\delta c_i} \\
& \Gamma(R, R_t) - \frac{n}{\eta}(\boldsymbol{u} - v_0 R_t \boldsymbol{p}) \cdot (v_0 \boldsymbol{p}) + \frac{1}{\gamma}(\partial_t \boldsymbol{p} + \lambda R_t (\boldsymbol{p}\cdot\nabla)\boldsymbol{p}) \cdot (\lambda (\boldsymbol{p}\cdot\nabla)\boldsymbol{p}) = -\sum_{i=1}^{3} \gamma_i \mu_i
\end{aligned} \tag{56}$$

The reaction rate equation can be written as

$$\Gamma(R, R_t) = -\sum_{i=1}^{3} \gamma_i \mu_i - n \nabla \mu_n \cdot (v_0 \boldsymbol{p}) + \lambda \boldsymbol{\mu}_p \cdot (\boldsymbol{p}\cdot\nabla)\boldsymbol{p}\,, \tag{57}$$

which can be reduced to the reaction rate Equation (38) if $\lambda = v_0 = 0$. The Equation (57) reveals the chemo-mechanical coupling, in which the reaction rate is modulated by the mechanical process. One can choose

$$\Gamma(R, R_t) = \ln\left(\frac{R_t}{k^- c_2 c_3} + 1 \right),$$

which recovers the law of mass action in the absence of mechanical coupling (i.e., $\lambda = v_0 = 0$). The corresponding reaction rate equation is given by

$$\begin{aligned}R_t &= k^- c_2 c_3 \exp\left(-\textstyle\sum_{i=1}^{3} \gamma_i \mu_i - n \nabla \mu_n \cdot (v_0 \boldsymbol{p}) + \lambda \boldsymbol{\mu}_p \cdot (\boldsymbol{p}\cdot\nabla)\boldsymbol{p} \right) - k^- c_2 c_3. \\ &= k^+ \exp(\lambda \boldsymbol{\mu}_p \cdot (\boldsymbol{p}\cdot\nabla)\boldsymbol{p} - n \nabla \mu_n \cdot (v_0 \boldsymbol{p})) c_1 - k^- c_2 c_3,\end{aligned} \tag{58}$$

where $k^+ = k^- \exp(\sigma_1 - \sigma_2 - \sigma_3)$. The reaction rate is modulated by the mechanical process. In the case that close to equilibrium, i.e., R_t is small, we can expand the reaction rate Equation (57) to the first order of R_t, which leads to a linear response assumption [55]

$$\Gamma(R, R_t) = \Lambda(\boldsymbol{c}) R_t\,,$$

where $\Lambda(\boldsymbol{c}) > 0$ is a positive function of the chemical species concentration $\boldsymbol{c} = (c_1, c_2, c_3)$. In principle, the exact form of $\Gamma(R, R_t)$ can not be determined through mathematics alone; a data-driven approach is often required to infer it from experimental data.

The final dynamics are governed by the following system:

$$\begin{cases} \partial_t n + \nabla \cdot (v_0 R_t \boldsymbol{p} n) = \eta \nabla \cdot (n \nabla \mu_n), \quad \mu_n = \frac{\delta \mathcal{F}}{\delta n} \\ \partial_t \boldsymbol{p} + \lambda R_t (\boldsymbol{p} \cdot \nabla) \boldsymbol{p} = -\gamma \mu_p, \mu_p = \frac{\delta \mathcal{F}}{\delta \boldsymbol{p}}, \\ \partial_t c_i = \eta_i \nabla \cdot (c_i \nabla \mu_u) + \gamma_i R_t \\ \Gamma(R, R_t) = \triangle \mu - v_0 n (\boldsymbol{u} \cdot \boldsymbol{p}) + \lambda \mu_{\boldsymbol{p}} \cdot (\boldsymbol{p} \cdot \nabla) \boldsymbol{p}, \end{cases} \tag{59}$$

where $\mathcal{F}(n, \boldsymbol{p})$ is the mechanical free energy defined in (45) and $\triangle \mu = \mu_{\text{ATP}} - \mu_{\text{ADP}} - \mu_{\text{P}}$ is the chemical affinity. The model is reduced to the simplified Toner–Tu model (44) if R_t is a constant. It is worth highlighting that the final system (59) exhibits a two-way chemo-mechanical coupling. The active transport in the mechanical subsystem is modulated by the chemical reaction rate and the reaction rate itself is influenced by the mechanical interactions. A key advantage of the energetic variational approach is that it ensures this bidirectional coupling remains thermodynamically consistent by construction.

Remark 4. *Unlike the issue that we mentioned in remark 2.2, we can prove that the final Equation (59) satisfies the energy dissipation law (55), since* $\boldsymbol{p}R_t$ *becomes a certain "velocity", induced by the chemical reaction. Indeed, by a direct calculation, we have*

$$\begin{aligned} \frac{\mathrm{d}}{\mathrm{d}t}(\mathcal{F}[n, \boldsymbol{p}] + \mathcal{F}_{\text{chem}}[\boldsymbol{c}]) &= \int \mu_n \partial_t n + \boldsymbol{\mu}_p \cdot \partial_t \boldsymbol{p} + \sum_{i=1}^{3} \mu_i \partial_t c_i \,\mathrm{d}x \\ &= \int \mu_n (\eta \nabla \cdot (n \nabla \mu_n) - \nabla \cdot (v_0 R_t \boldsymbol{p} n)) + \boldsymbol{\mu}_p \cdot (-\gamma \mu_p - \lambda R_t (\boldsymbol{p} \cdot \nabla) \boldsymbol{p}) \\ &\quad + \sum_{i=1}^{3} \mu_i (\eta_i \nabla \cdot (c_i \nabla \mu_u) + \gamma_i R_t) \mathrm{d}x \\ &= \int -\eta n |\nabla \mu_n|^2 - \gamma |\mu_p|^2 - \eta_i c_i |\nabla \mu_i|^2 + R_t (\sum_{i=1}^{3} \gamma_i \mu_i + v_0 n \nabla \mu_n \cdot \boldsymbol{p} - \lambda \boldsymbol{\mu}_{\boldsymbol{p}} \cdot (\boldsymbol{p} \cdot \nabla \boldsymbol{p})) \\ &= \int -\eta n |\nabla \mu_n|^2 - \gamma |\mu_p|^2 - \eta_i c_i |\nabla \mu_i|^2 - \Lambda |R_t|^2 \mathrm{d}x \leq 0 \end{aligned} \tag{60}$$

Remark 5. *It is straightforward to apply the same framework to revise the "dry" active nematic models with* $\boldsymbol{Q}$*-tensor as order parameter [50] using the energy dissipation law*

$$\frac{\mathrm{d}}{\mathrm{d}t}\mathcal{F}[\boldsymbol{Q}, n] + \mathcal{F}_{\text{chem}}[\boldsymbol{c}] = -\int c|\boldsymbol{u} - \zeta R_t \nabla \cdot \boldsymbol{Q}|^2 + \frac{1}{\gamma}|\partial_t \boldsymbol{Q}|^2 + \sum_{i=1}^{n} \frac{c_i}{\eta_i}|\boldsymbol{u}_i|^2 + \Gamma(R, R_t) R_t \mathrm{d}\boldsymbol{x}\,,$$

where n *is the number density of the liquid crystal molecules,* $\boldsymbol{Q}$ *is the tensor order parameter, and* $\mathcal{F}[\boldsymbol{Q}, n]$ *is the free energy for the mechanical part, and* $\mathcal{F}_{\text{chem}}[\boldsymbol{c}]$ *is the free energy associated with the ATP hydrolysis procedure, defined in (53). We will study this model in future work.*

3.2. Energy Transduction and Efficiency

A key feature of the proposed model is its ability to quantify the energy transduction, the free energy released by ATP hydrolysis into the mechanical activity. This is achieved through the coupling of the reaction coordinate R_t with the active transport coefficient in the number density and the polarization equations.

The total free energy of the system consists of two parts: the mechanical free energy $\mathcal{F}(n, \boldsymbol{p})$, and the chemical free energy $\mathcal{F}_{\text{chem}}(\boldsymbol{c})$. The previous analysis shows that the term

$$R_t(\nu_0 n \boldsymbol{p} \cdot \nabla \mu_n - \lambda \boldsymbol{\mu}_p \cdot ((\boldsymbol{p} \cdot \nabla)\boldsymbol{p})) \tag{61}$$

may contribute to the increase in the mechanical energy $\mathcal{F}(n, \boldsymbol{p})$. Hence, the energy transduction happened when

$$\nu_0 n \boldsymbol{p} \cdot \nabla \mu_n - \lambda \boldsymbol{\mu}_p \cdot ((\boldsymbol{p} \cdot \nabla)\boldsymbol{p}) > 0, \quad R_t > 0 \tag{62}$$

In the meantime, the energy released by the chemical reaction is given by $-R_t(\sum_{i=1}^{3} \gamma_i \mu_i) = R_t \Delta\mu > 0$. Hence, following the definition of efficiency [19,53], we can define the local chemo-mechanical transduction efficiency in our model as

$$\eta_{\text{loc}}(\boldsymbol{x}, t) = \frac{\nu_0 n \boldsymbol{p} \cdot \nabla \mu_n - \lambda \boldsymbol{\mu}_p \cdot ((\boldsymbol{p} \cdot \nabla)\boldsymbol{p})}{\Delta\mu}, \tag{63}$$

where $\Delta\mu$ is the chemical affinity associated with ATP hydrolysis. This dimensionless quantity measures the proportion of chemical energy that is effectively transduced into mechanical contributions at each spatial location. For a global efficiency measure, one may define

$$\eta_{\text{tot}}(t) = \frac{\int_\Omega R_t(\nu_0 n \boldsymbol{p} \cdot \nabla \mu_n - \lambda \boldsymbol{\mu}_p \cdot ((\boldsymbol{p} \cdot \nabla)\boldsymbol{p}))\mathrm{d}\boldsymbol{x}}{\int_\Omega R_t \Delta\mu \,\mathrm{d}\boldsymbol{x}}, \tag{64}$$

which quantifies the net transduction efficiency over the entire domain.

It is important to note that even when the transduction efficiency is positive, the mechanical energy $\mathcal{F}(n, \boldsymbol{p})$ may not increase in time, due to dissipation in both the n and $\boldsymbol{p}$ dynamics. The increasing of free energy only occurs when the chemical driving overcomes the dissipative losses in the system.

Remark 6. *Without proper choice of model parameters, one may have* $R_t < 0$ *or* $\nu_0 n \boldsymbol{p} \cdot \nabla \mu_n - \lambda \boldsymbol{\mu}_p \cdot ((\boldsymbol{p} \cdot \nabla)\boldsymbol{p}) < 0$*, which may lead to an increase in the chemical energy* $\mathcal{F}_{\text{chem}}$*, indicating that mechanical energy is being converted back into chemical energy. Although this scenario is non-physical in the current setting, it highlights the importance of parameter selection to ensure that the model reflects the correct direction of energy transduction.*

4. Numerics

In this section, we provide some numerical simulation result to new Toner–Tu model with ATP hydrolysis. For simplicity, we consider the linear response case for $\Gamma(R, R_t)$ and assume that $\Lambda(\boldsymbol{c})$ is a constant, denoted by $\Lambda > 0$. This assumption is valid when the chemical reaction is close to equilibrium, and the reaction rate is small. Then, the final system can be written as

$$\begin{cases} \partial_t n + \nabla \cdot (v_0 R_t \boldsymbol{p} n) = \eta \nabla \cdot (n \nabla \mu_n), \quad \mu_n = \ln n - \xi(\nabla \cdot \boldsymbol{p}) \\ \partial_t \boldsymbol{p} + \lambda R_t (\boldsymbol{p} \cdot \nabla)\boldsymbol{p} = -\gamma \mu_{\boldsymbol{p}}, \mu_{\boldsymbol{p}} = -\Delta \boldsymbol{p} + \frac{1}{\epsilon^2}(|\boldsymbol{p}|^2 - 1)\boldsymbol{p} + \xi \nabla c, \\ \partial_t c_i = \eta_i \nabla \cdot (c_i \nabla \mu_u) + \gamma_i R_t \\ \Lambda R_t = \triangle \mu - v_0 n (\nabla \mu_n \cdot \boldsymbol{p}) + \lambda \mu_{\boldsymbol{p}} \cdot (\boldsymbol{p} \cdot \nabla)\boldsymbol{p}, \\ \triangle \mu = \ln\left(\frac{c_1}{c_2 c_3}\right) + \sigma_1 - \sigma_2 - \sigma_3 \end{cases} \tag{65}$$

where σ_1, σ_2 and σ_3 are constant, related to the activity of ATP, ADP and P.

We consider the computational domain $\Omega = [0,1]^2$ with periodic boundary conditions applied to all variables. We use the standard finite difference method with explicit Euler to solve the model. The grid size is $\Delta x = 1/50$, and the temporal stepsize is $\Delta t = 10^{-4}$. We will develop a structure-preserving numerical discretization for the variational model in feature work. The initial condition is given by

$$\boldsymbol{p}(x,y,0) = (\cos(2\pi x), \sin(2\pi y)), \quad n(x,y,0) = n_0 + n_1(\sin(2\pi x) - \cos(2\pi y)), \tag{66}$$

where $n_0 = 1$ and $n_1 = 0.3$. The initial concentrations of the chemical species are set as $c_1(x,y) = 1$, and $c_2(x,y) = c_3(x,y) = 0.01$. Model parameters are chosen as follows: $v_0 = 0.1$, $\lambda = 0.01$, $\Lambda = 10$, $\gamma = 1$, $\epsilon = 0.1$, $\sigma = (1, 0.1, 0.1)$, and $\eta = \eta_i = 0.1$ for all diffusion coefficients. We emphasize that these parameters are selected for illustrative and numerical purposes only; they are not based on physical measurements or calibration to experimental data. The model parameters and the initial condition were chosen primarily to ensure positive chemo-mechanical efficiency and numerical stability. From numerical side, certain parameters, particularly diffusion coefficients and coupling strengths, must be sufficiently small to avoid numerical instabilities. This constraint is purely numerical and not intrinsic to the model. From the modeling perspective, we observed that to maintain the positivity of both the efficiency and the reaction rate R_t, both λ and v_0 must be small.

Figure 1 shows the simulation result, visualized by the number density $n(\boldsymbol{x}, t)$, the concentration of ATP $c_1(\boldsymbol{x}, t)$, and the polarization field $\boldsymbol{p}(\boldsymbol{x}, t)$ (colored by $|\boldsymbol{p}|$). Figure 2 shows the evolution of $\mathcal{F}(n, \boldsymbol{p})$, $\mathcal{F}_{\text{chem}}(\boldsymbol{c})$, and the efficiency η_{tot} over time. While $\mathcal{F}(n, \boldsymbol{p})$ generally decreases due to dissipation in both n and $\boldsymbol{p}$, the efficiency plot demonstrates effective energy transduction from chemical to mechanical forms during the transition.

While the current simulations are proof-of-concept and do not exhibit striking emergent phenomena in active nematics, they confirm the model's thermodynamic consistency and demonstrate positive chemo-mechanical energy transduction. The simplicity of the vector-based polarization field may limit the richness of the observed patterns, and future extensions using Q-tensor formulations or biologically calibrated parameters may reveal more complex defect dynamics.

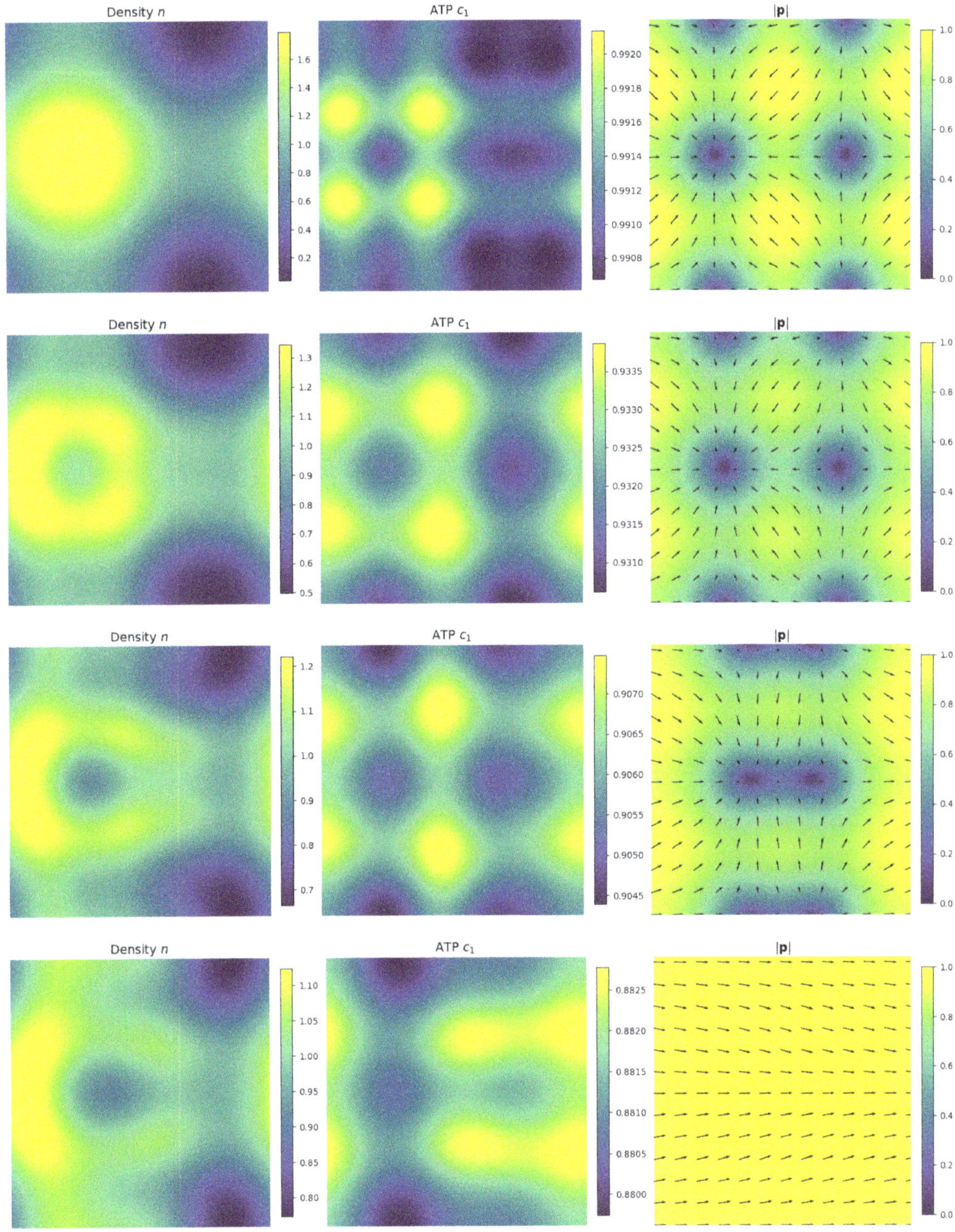

Figure 1. Snapshots of the number density $n(\boldsymbol{x}, t)$, ATP concentration $c_1(\boldsymbol{x}, t)$, and polarization field $\boldsymbol{p}(\boldsymbol{x}, t)$ (colored by $|\boldsymbol{p}|$) at $t = 0.01, 0.1, 0.15$, and 0.2 (from top to bottom).

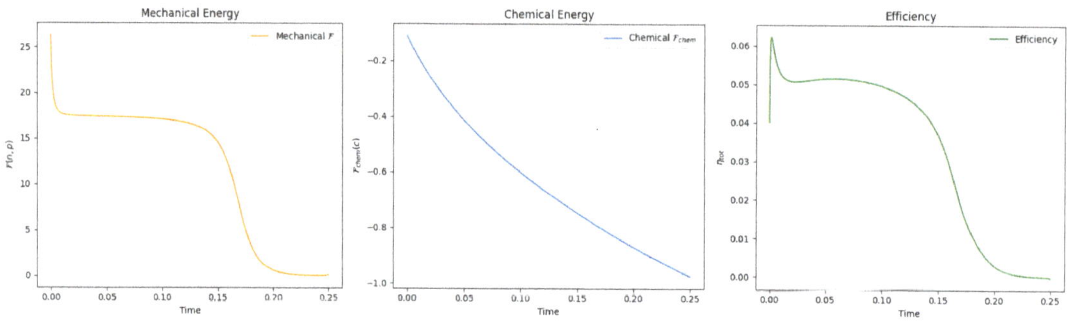

Figure 2. Evolution of $\mathcal{F}(n, \boldsymbol{p})$, $\mathcal{F}_{\text{chem}}(\boldsymbol{c})$ and efficiency with respect to time.

5. Conclusions

In this work, we developed a thermodynamically consistent extension of the simplified Toner–Tu model by explicitly incorporating ATP hydrolysis as the energetic source of activity. Our model is constructed using an energetic variational approach within a non-equilibrium thermodynamic framework, ensuring that the total free energy, comprising both mechanical and chemical components, monotonically decreases over time, consistent with the first and second laws of thermodynamics. The formulation captures a two-way coupling between chemical and mechanical processes: the reaction rate of ATP hydrolysis is influenced by mechanical alignment, while active transport is, in turn, modulated by the local reaction rate. Although the full system is dissipative, the free energy of the mechanical subsystem may increase, and the mechanical dynamics can exhibit out-of-equilibrium behavior due to the transduction of chemical energy into mechanical work. In the absence of ATP consumption, the model naturally reduces to a passive system, providing a clear energetic interpretation of activity.

We also analyzed the transduction of chemical energy into mechanical work, identifying the conditions under which ATP hydrolysis fuels transport and alignment. A simple expression for energy conversion efficiency was derived. We also present numerical simulations demonstrating the positive energy transduction in the active system. In future work, we plan to conduct a more detailed mathematical analysis of solution behavior, perform comprehensive numerical investigations across broader parameter regimes, and develop structure-preserving discretization schemes that better reflect the underlying variational principles.

This work provides a variational and physically grounded framework for studying active matter powered by biochemical reactions. The chemo-mechanical coupling we introduced can be extended to more complex settings, including hydrodynamic interactions, anisotropic geometries, and micro-macro coupling. It also opens new doors for investigating energy transduction and efficiency in biologically inspired active matter systems.

Funding: Y.W. is partially support by NSF DMS-2410740.

Institutional Review Board Statement: Not applicable.

Data Availability Statement: The data presented in this study are available on request from the corresponding author

Acknowledgments: Y.W. would like to thank Chun Liu for his insightful guidance and many stimulating discussions.

Conflicts of Interest: The authors declare no conflicts of interest.

References

1. Das, M.; Schmidt, C.F.; Murrell, M. Introduction to active matter. *Soft Matter* **2020**, *16*, 7185–7190. [CrossRef]
2. Doostmohammadi, A.; Ladoux, B. Physics of liquid crystals in cell biology. *Trends Cell Biol.* **2022**, *32*, 140–150. [CrossRef]
3. Marchetti, M.C.; Joanny, J.F.; Ramaswamy, S.; Liverpool, T.B.; Prost, J.; Rao, M.; Simha, R.A. Hydrodynamics of soft active matter. *Rev. Mod. Phys.* **2013**, *85*, 1143. [CrossRef]
4. Needleman, D.; Dogic, Z. Active matter at the interface between materials science and cell biology. *Nat. Rev. Mater.* **2017**, *2*, 1–14. [CrossRef]
5. Aranson, I. Bacterial active matter. *Rep. Prog. Phys.* **2022**, *85*, 076601. [CrossRef] [PubMed]
6. Saintillan, D.; Shelley, M.J. Instabilities, pattern formation, and mixing in active suspensions. *Phys. Fluids* **2008**, *20*, 123304. [CrossRef]
7. Sanchez, T.; Chen, D.T.; DeCamp, S.J.; Heymann, M.; Dogic, Z. Spontaneous motion in hierarchically assembled active matter. *Nature* **2012**, *491*, 431–434. [CrossRef]
8. Wu, Y.; Jiang, Y.; Kaiser, A.D.; Alber, M. Self-organization in bacterial swarming: Lessons from myxobacteria. *Phys. Biol.* **2011**, *8*, 055003. [CrossRef]
9. Cianchetti, M.; Laschi, C.; Menciassi, A.; Dario, P. Biomedical applications of soft robotics. *Nat. Rev. Mater.* **2018**, *3*, 143–153. [CrossRef]
10. Özkale, B.; Sakar, M.S.; Mooney, D.J. Active biomaterials for mechanobiology. *Biomaterials* **2021**, *267*, 120497. [CrossRef]
11. Shah, A.; Malik, M.S.; Khan, G.S.; Nosheen, E.; Iftikhar, F.J.; Khan, F.A.; Shukla, S.S.; Akhter, M.S.; Kraatz, H.B.; Aminabhavi, T.M. Stimuli-responsive peptide-based biomaterials as drug delivery systems. *Chem. Eng. J.* **2018**, *353*, 559–583. [CrossRef]
12. Alber, M.; Buganza Tepole, A.; Cannon, W.R.; De, S.; Dura-Bernal, S.; Garikipati, K.; Karniadakis, G.; Lytton, W.W.; Perdikaris, P.; Petzold, L.; et al. Integrating machine learning and multiscale modeling—Perspectives, challenges, and opportunities in the biological, biomedical, and behavioral sciences. *NPJ Digit. Med.* **2019**, *2*, 115. [CrossRef]
13. Liu, A.P.; Appel, E.A.; Ashby, P.D.; Baker, B.M.; Franco, E.; Gu, L.; Haynes, K.; Joshi, N.S.; Kloxin, A.M.; Kouwer, P.H.; et al. The living interface between synthetic biology and biomaterial design. *Nat. Mater.* **2022**, *21*, 390–397. [CrossRef]
14. Kjelstrup, S.; Lervik, A. The energy conversion in active transport of ions. *Proc. Natl. Acad. Sci. USA* **2021**, *118*, e2116586118. [CrossRef]
15. Blumenfeld, L. The physical aspects of energy transduction in biological systems. *Q. Rev. Biophys.* **1978**, *11*, 251–308. [CrossRef] [PubMed]
16. O'Byrne, J.; Kafri, Y.; Tailleur, J.; van Wijland, F. Time irreversibility in active matter, from micro to macro. *Nat. Rev. Phys.* **2022**, *4*, 167–183. [CrossRef]
17. Pearce, D.J.; Martínez-Prat, B.; Ignés-Mullol, J.; Sagués, F. Topological defects lead to energy transfer in active nematics. *arXiv* **2024**, arXiv:2411.18214. [CrossRef]
18. Wachtel, A.; Rao, R.; Esposito, M. Free-energy transduction in chemical reaction networks: From enzymes to metabolism. *J. Chem. Phys.* **2022**, *157*, 024109. [CrossRef]
19. Yang, X.; Heinemann, M.; Howard, J.; Huber, G.; Iyer-Biswas, S.; Le Treut, G.; Lynch, M.; Montooth, K.L.; Needleman, D.J.; Pigolotti, S.; et al. Physical bioenergetics: Energy fluxes, budgets, and constraints in cells. *Proc. Natl. Acad. Sci. USA* **2021**, *118*, e2026786118. [CrossRef]
20. Saintillan, D.; Shelley, M.J. Theory of active suspensions. In *Complex Fluids in Biological Systems: Experiment, Theory, and Computation*; Springer: Berlin/Heidelberg, Germany, 2014; pp. 319–355.
21. Shaebani, M.R.; Wysocki, A.; Winkler, R.G.; Gompper, G.; Rieger, H. Computational models for active matter. *Nat. Rev. Phys.* **2020**, *2*, 181–199. [CrossRef]
22. De Groot, S.R.; Mazur, P. *Non-Equilibrium Thermodynamics*; Courier Corporation: Chelmsford, MA, USA, 2013.
23. Kjelstrup, S.; Bedeaux, D. *Non-Equilibrium Thermodynamics of Heterogeneous Systems*; World Scientific: Singapore, 2008.
24. Liu, C. An introduction of elastic complex fluids: An energetic variational approach. In *Multi-Scale Phenomena in Complex Fluids: Modeling, Analysis and Numerical Simulation*; World Scientific: Singapore, 2009; pp. 286–337.
25. Giga, M.H.; Kirshtein, A.; Liu, C. Variational modeling and complex fluids. In *Handbook of Mathematical Analysis in Mechanics of Viscous Fluids*; Springer: Berlin/Heidelberg, Germany, 2017; pp. 1–41.
26. Wang, Y.; Liu, C. Some recent advances in energetic variational approaches. *Entropy* **2022**, *24*, 721. [CrossRef]
27. Doi, M. Onsager's variational principle in soft matter. *J. Phys. Condens. Matter* **2011**, *23*, 284118. [CrossRef]
28. Wang, Q. Generalized onsager principle and it applications. In *Frontiers and Progress of Current Soft Matter Research*; Springer: Singapore, 2021; pp. 101–132.
29. Grmela, M.; Öttinger, H.C. Dynamics and thermodynamics of complex fluids. I. Development of a general formalism. *Phys. Rev. E* **1997**, *56*, 6620. [CrossRef]
30. Öttinger, H.C.; Grmela, M. Dynamics and thermodynamics of complex fluids. II. Illustrations of a general formalism. *Phys. Rev. E* **1997**, *56*, 6633. [CrossRef]

31. Wang, Y.; Liu, C.; Liu, P.; Eisenberg, B. Field theory of reaction-diffusion: Law of mass action with an energetic variational approach. *Phys. Rev. E* **2020**, *102*, 062147. [CrossRef]
32. Liu, C.; Sulzbach, J.E. The Brinkman-Fourier system with ideal gas equilibrium. *Discret. Contin. Dyn. Syst.* **2021**, *42*, 425–462. [CrossRef]
33. Morrow, S.M.; Colomer, I.; Fletcher, S.P. A chemically fuelled self-replicator. *Nat. Commun.* **2019**, *10*, 1011. [CrossRef]
34. Ge, H.; Qian, H. Dissipation, generalized free energy, and a self-consistent nonequilibrium thermodynamics of chemically driven open subsystems. *Phys. Rev. E* **2013**, *87*, 062125. [CrossRef]
35. Ackermann, J.; Ben Amar, M. Onsager's variational principle in proliferating biological tissues, in the presence of activity and anisotropy. *Eur. Phys. J. Plus* **2023**, *138*, 1103. [CrossRef]
36. Klamser, J.U.; Kapfer, S.C.; Krauth, W. Thermodynamic phases in two-dimensional active matter. *Nat. Commun.* **2018**, *9*, 5045. [CrossRef]
37. Mirza, W.; Torres-Sánchez, A.; Vilanova, G.; Arroyo, M. Variational formulation of active nematic fluids: Theory and simulation. *New J. Phys.* **2025**, *27*, 043025. [CrossRef]
38. Gaspard, P. The non-equilibrium thermodynamics of active suspensions. *arXiv* **2025**, arXiv:2505.21009. [CrossRef]
39. Markovich, T.; Fodor, É.; Tjhung, E.; Cates, M.E. Thermodynamics of active field theories: Energetic cost of coupling to reservoirs. *Phys. Rev. X* **2021**, *11*, 021057. [CrossRef]
40. Nardini, C.; Fodor, É.; Tjhung, E.; Van Wijland, F.; Tailleur, J.; Cates, M.E. Entropy production in field theories without time-reversal symmetry: Quantifying the non-equilibrium character of active matter. *Phys. Rev. X* **2017**, *7*, 021007. [CrossRef]
41. Wang, H.; Qian, T.; Xu, X. Onsager's variational principle in active soft matter. *Soft Matter* **2021**, *17*, 3634–3653. [CrossRef]
42. Toner, J.; Tu, Y. Long-range order in a two-dimensional dynamical XY model: How birds fly together. *Phys. Rev. Lett.* **1995**, *75*, 4326. [CrossRef]
43. Vicsek, T.; Czirók, A.; Ben-Jacob, E.; Cohen, I.; Shochet, O. Novel type of phase transition in a system of self-driven particles. *Phys. Rev. Lett.* **1995**, *75*, 1226. [CrossRef]
44. Bolley, F.; Cañizo, J.A.; Carrillo, J.A. Mean-field limit for the stochastic Vicsek model. *Appl. Math. Lett.* **2012**, *25*, 339–343. [CrossRef]
45. Ginelli, F. The physics of the Vicsek model. *Eur. Phys. J. Spec. Top.* **2016**, *225*, 2099–2117. [CrossRef]
46. Gowrishankar, K.; Ghosh, S.; Saha, S.; Mayor, S.; Rao, M. Active remodeling of cortical actin regulates spatiotemporal organization of cell surface molecules. *Cell* **2012**, *149*, 1353–1367. [CrossRef]
47. Choi, Y.P.; Kang, K.; Lee, W. Global existence and asymptotic stability for the Toner-Tu model of flocking. *arXiv* **2024**, arXiv:2403.09114. [CrossRef]
48. Lin, F.H.; Liu, C. Nonparabolic dissipative systems modeling the flow of liquid crystals. *Commun. Pure Appl. Math.* **1995**, *48*, 501–537. [CrossRef]
49. Ericksen, J.L. Liquid crystals with variable degree of orientation. *Arch. Ration. Mech. Anal.* **1991**, *113*, 97–120. [CrossRef]
50. Narayan, V.; Ramaswamy, S.; Menon, N. Long-lived giant number fluctuations in a swarming granular nematic. *Science* **2007**, *317*, 105–108. [CrossRef] [PubMed]
51. Ramaswamy, S.; Simha, R.A.; Toner, J. Active nematics on a substrate: Giantnumber fluctuations and long-time tails. *Europhys. Lett.* **2003**, *62*, 196. [CrossRef]
52. De Gennes, P.G.; Prost, J. *The Physics of Liquid Crystals*; Number 83; Oxford University Press: Oxford, UK, 1993.
53. Parmeggiani, A.; Jülicher, F.; Ajdari, A.; Prost, J. Energy transduction of isothermal ratchets: Generic aspects and specific examples close to and far from equilibrium. *Phys. Rev. E* **1999**, *60*, 2127. [CrossRef]
54. Rayleigh, L. Some General Theorems Relating to Vibrations. *Proc. Lond. Math. Soc.* **1873**, *4*, 357–368.
55. Onsager, L. Reciprocal relations in irreversible processes. I. *Phys. Rev.* **1931**, *37*, 405. [CrossRef]
56. Onsager, L. Reciprocal relations in irreversible processes. II. *Phys. Rev.* **1931**, *38*, 2265. [CrossRef]
57. Ericksen, J.L. *Introduction to the Thermodynamics of Solids*; Chapman and Hall: New York, NY, USA, 1992.
58. Liu, C.; Wang, Y. A variational Lagrangian scheme for a phase-field model: A discrete energetic variational approach. *SIAM J. Sci. Comput.* **2020**, *42*, B1541–B1569. [CrossRef]
59. Oster, G.F.; Perelson, A.S. Chemical reaction dynamics. *Arch. Ration. Mech. Anal.* **1974**, *55*, 230–274. [CrossRef]
60. Kondepudi, D.; Prigogine, I. *Modern Thermodynamics: From Heat Engines to Dissipative Structures*; John Wiley & Sons: Hoboken, NJ, USA, 2014.
61. De Donder, T. *Leçons de Thermodynamique et de Chimie Physique*; Gauthier-Villars et cie: Paris, France, 1920.
62. Beris, A.N.; Edwards, B.J. *Thermodynamics of Flowing Systems: With Internal Microstructure*; Number 36; Oxford University Press: Oxford, UK, 1994.

63. Liu, C.; Wang, C.; Wang, Y. A structure-preserving, operator splitting scheme for reaction-diffusion equations with detailed balance. *J. Comput. Phys.* **2021**, *436*, 110253. [CrossRef]
64. Mielke, A. A gradient structure for reaction–diffusion systems and for energy-drift-diffusion systems. *Nonlinearity* **2011**, *24*, 1329. [CrossRef]

MDPI AG
Grosspeteranlage 5
4052 Basel
Switzerland
Tel.: +41 61 683 77 34

Entropy Editorial Office
E-mail: entropy@mdpi.com
www.mdpi.com/journal/entropy

www.ingramcontent.com/pod-product-compliance
Lightning Source LLC
Chambersburg PA
CBHW061606220926
51850CB00005B/434
9783725853953